Nanocomposite Materials: Characterization and Applications

Nanocomposite Materials: Characterization and Applications

Edited by
Rich Falcon

WILLFORD PRESS
www.willfordpress.com

Published by Willford Press,
118-35 Queens Blvd., Suite 400,
Forest Hills, NY 11375, USA

ISBN: 978-1-68285-383-2

Cataloging-in-Publication Data

Nanocomposite materials : characterization and applications / edited by Rich Falcon.
p. cm.
Includes bibliographical references and index.
ISBN 978-1-68285-383-2
1. Nanocomposites (Materials). 2. Nanostructured materials. 3. Composite materials. I. Falcon, Rich.
TA418.9.N35 N35 2017
620.5--dc23

For information on all Willford Press publications
visit our website at www.willfordpress.com

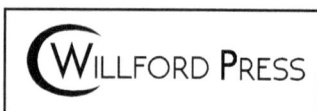

WILLFORD PRESS

Printed in the United States of America.

Contents

Chapter 20   **Nanocomposite toughness, strength and stiffness: role of filler geometry** 188
Israel Greenfeld and H. Daniel Wagner

Chapter 21   **Evaluation of fracture behavior of polyethylene/CaCO$_3$ nanocomposite using essential work of fracture (EWF) approach**............................203
Meymanat S. Mohsenzadeh, Mohammad Mazinani and Seyed Mojtaba Zebarjad

Chapter 22   **Bioactive nanocomposites for dental application obtained by reactive suspension method**..212
Oussama Boumezgane, Federica Bondioli, Sergio Bortolini, Alfredo Natali, Aldo R. Boccaccini, Elena Boccardi and Massimo Messori

Permissions

List of Contributors

Index

Preface

This book provides comprehensive insights into the characterization and applications of nanocomposite materials. Nanocomposite materials are preferred in several industries because of their superior properties in comparison to conventional materials. The properties of nanocomposite materials are based on their individual parents, morphology and interfacial characteristics. Applications of nanocomposites can be found in various industries like food packaging, fuel tanks, flammability reduction, etc. Comprehensive insights have been provided in this book that would help the reader in understanding the rapidly developing field of nanocomposites. Students, researchers, experts and all associated with nanocomposite materials will benefit alike from the book.

This book unites the global concepts and researches in an organized manner for a comprehensive understanding of the subject. It is a ripe text for all researchers, students, scientists or anyone else who is interested in acquiring a better knowledge of this dynamic field.

I extend my sincere thanks to the contributors for such eloquent research chapters. Finally, I thank my family for being a source of support and help.

Editor

A study on the fatigue performance of a glass fiber-epoxy polymer nanocomposite under random loads

N. Jagannathan, A. R. Anilchandra and C. M. Manjunatha

Fatigue and Structural Integrity Group, Structural Technologies Division, CSIR-National Aerospace Laboratories, Bangalore 560017, India

Abstract A glass-fiber reinforced plastic (GFRP) nanocomposite containing 10 wt-% silica nano particles in the epoxy matrix was fatigue tested under a standard helicopter random load sequence, Helix-32. Fatigue life was determined at various reference stresses. The stiffness variation and the matrix crack density in the test specimen were monitored at regular intervals during the fatigue test. The random load fatigue life of the GFRP nanocomposite was about four times higher than that of its neat counterpart over the entire range of reference stress levels investigated. The suppressed matrix cracking and reduced crack/delamination growth rate in nanocomposite were responsible for fatigue life enhancement. Further, the random load fatigue life was predicted by empirical method using constant fatigue life diagrams. Three different damage accumulation models, namely, Palmgren–Miner (PM), Broutman–Sahu (BS) and Hashin–Rotem (HR), were used. All the three models predicted similar results, and a good correlation was observed between experimental and predicted fatigue life.

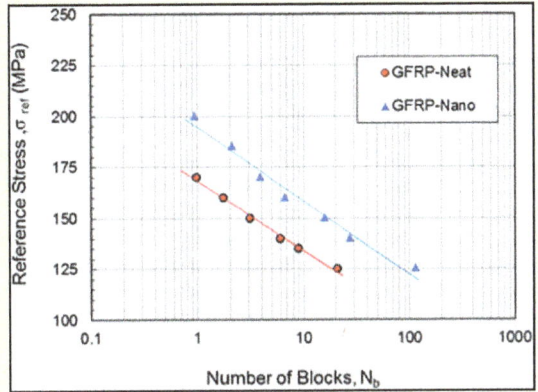

Keywords Nanocomposite, Glass fiber, Random loads, Fatigue, Matrix cracks

Introduction

Continuous fiber reinforced polymer (FRP) composites are fast replacing structural metallic alloys in several applications because of their high specific strength, stiffness, and tailorability. The volume of polymer composites used in airframe construction is continuously increasing in the last decade. Recent airframes are built with up to about 50% of composites. The polymer composites are used in various other structures as well such as wind turbine blades, ship hull, etc. The structural composites invariably experience fatigue loads in service. The durability of the structure thus depends on the fatigue properties of these composites.

It has been well established that when composites are subjected to cyclic fatigue loads, several types of microstructural damages initiate and grow leading to fatigue failure. Matrix cracks, disbond between fiber and matrix,

delamination between the layers are some of the most common damages observed in composites. Understanding the behavior of these damages under externally applied loads would assist in defining the safe life of composite structures. Any improvements in fatigue life by controlling the damages to delay their initiation and/or reduce their growth rates would be advantageous in terms of enhancing the safe life of the structure.

One of the possible ways to improve the fatigue life of polymer composites further is to incorporate second phase fillers in the epoxy. Recent studies have shown that composites, which contain nano fillers exhibit improved mechanical, electrical and thermal properties.[1–3] The presence of various types of nano-sized particulate, fibrous and layered fillers in the epoxies and FRPs has been observed to enhance the fatigue properties.

Tremendous improvements in fatigue life have been observed in bulk epoxies modified with nano fillers. The presence of 10 wt-% silica nanoparticles in a thermosetting epoxy polymer has been shown to improve

*Corresponding author, email manjucm@nal.res.in

the fatigue life by three to four times.[4,5] The energy dissipating mechanisms of nano particle debonding – void formation and plastic deformation of voids – have been observed to enhance the fatigue life in the epoxy. Addition of small amounts of carbon nanotubes (CNT) to an epoxy increases its fatigue life by about 10–15 times.[6,7] Similarly, carbon nanofiber (CNF) has been observed to enhance fatigue life[8] by over 350%. Layered fillers such as clays have been shown to increase the fatigue life of epoxies significantly.[9] Some investigators have experimented use of multiple nano fillers to produce hybrid epoxy composites with tremendous improvements in fatigue behavior. For e.g. Shokrieh et al.[10] added graphene nanosheet and CNF to an epoxy resin and observed a remarkable improvement in flexural fatigue life of epoxy resin, by about 37 times.

The fatigue life of FRPs employing the nano-modified epoxy matrix also show enhanced fatigue life when compared to their counterpart with neat epoxy matrix. Addition of 10 wt-% silica nanoparticles improves the fatigue life of a GFRP composite by three to four times.[4] Presence of low amounts of silica nanoparticles and multi-wall carbon nanotubes (MWCNT) increase the fatigue life of a GFRP by several orders of magnitude.[11] Use of 1.5–3.0% SiC nanoparticles in epoxy matrix of a CFC has been reported to enhance flexural fatigue life significantly.[12] Various mechanisms such as suppressed matrix cracking, reduced crack growth rates, delayed initiation of delamination etc. have been suggested to enhance fatigue life in these FRP nanocomposites.

Incorporation of well-dispersed CNT in the epoxy matrix has been shown to improve the fatigue life of CFRP by over an order of magnitude.[13] Carbon nanotubes also increase fatigue performance of GFRP significantly.[14] Modification of epoxy with CNF also shows enormous improvements in fatigue life of FRPs.[8,15] Carbon fiber reinforced polymer (CFRP) with hybrid-modified epoxy matrix containing MWCNT and graphene has been shown to enhance fatigue life through enormous plastic deformation of the matrix during cyclic loads. Presence of layered nano fillers such as nanoclay in the epoxy matrix of FRPs has been observed to enhance fatigue life of CFC composites. For e.g., Zhou et al.[16] observed significant improvements in fatigue life of CFC with 2 wt-% nanoclay. Khan et al.[17] showed that nanoclay suppresses and delays delamination damage growth and eventual failure by improving the fiber/matrix interfacial bond and through the formation of nanoclay-induced dimples in CFC.

The structural fatigue loads in service are mostly random in nature. Most of the fatigue studies conducted on FRP nanocomposite have been limited to constant amplitude fatigue. Authors have recently shown that GFRP nanocomposite containing 10 wt-% silica nano particles in the epoxy matrix exhibit improved constant amplitude fatigue life.[4] It may be noted that the fatigue life improvement observed under constant amplitude fatigue loads may never be similar to improvements under random loads owing to presence of load sequence effects.[18–20] Hence, the main aim of this investigation was to study the fatigue performance of GFRP nanocomposite under a standard helicopter random load sequence. An attempt was also made to predict the random load fatigue life through empirical method and compare with experimental observations.

Experimental

Materials

Two types of glass fiber reinforced polymer (GFRP) composites were considered in this investigation, namely (i) GFRP composite with neat LY556 epoxy-matrix (GFRP-neat) and (ii) GFRP composite with 10 wt-% silica nanoparticle-modified LY556 epoxy-matrix (GFRP-nano). The details of the fabrication of these GFRP composite laminates are shown in Ref.[4] However, the main details are briefly mentioned here.

The silica nanoparticles were obtained as a master batch of 40 wt-% in LY556 epoxy resin. The calculated quantities of master batch resin-silica particle mix and pure LY556 epoxy resin was mixed with curing agent methylhexahydrophthalic acid anhydride to produce resin mixture containing 10 wt-% silica nanoparticles. The silica particles of about 20 nm in diameter were uniformly distributed throughout the epoxy.[4] The resin mix was then infused in to a E-glass fibre fabric under vacuum to produce QI Lay-up [(+45/45/0/90)$_s$]$_2$ GFRP composite laminates. The laminate was cured at 100°C for 2 h and post-cured at 150°C for 10 h. The GFRP laminates fabricated were about 2.6 mm thick with fibre volume fraction of about 57%. The static tensile and compressive properties of the composite laminates[21] are shown in Table 1.

Fatigue testing

Fatigue tests on GFRP composites were conducted under a standard helicopter rotor random load sequence,[22,23] Helix-32 shown in Fig. 1 This particular random load sequence was considered because the GFRP composites are used in helicopter rotor blades. One block of the Helix-32 load sequence consists of 291 725 load reversals at 31 different stress levels, which is equivalent of 140 h of usage. The stress level shown in Fig. 1 is a normalised value and the actual stress sequence for experiments was obtained by multiplying all the peak/trough points with a constant reference stress, σ_{ref}. Random load fatigue tests were performed on GFRP composites with different reference stress levels ranging from 125 to 200 MPa. The random load block with specific reference stress was repeatedly applied to the test specimens until failure and the fatigue life, expressed as the number of blocks to failure, was determined.

The fatigue test specimens of length 150 mm and a constant rectangular cross-section of 12 mm × 2.6 mm were cut and prepared from the laminates. The specimens were bonded with end-tabs so that the gage length of the specimen was about 15 mm. All the fatigue tests were conducted in a computer-controlled 50 kN servo-hydraulic test machine. Tests were performed using linear ramp waveform with an average frequency of 5 Hz. When the specimen failed in-between any block, the fraction of the block completed was determined as the ratio of the number of reversals applied until then to the total number of reversals in the block.

The stiffness of the specimen during fatigue testing was determined from the load-displacement data obtained at regular intervals during the fatigue test. Also, the specimen was dismounted at regular intervals to obtain matrix crack pattern in the specimen as explained in Manjunatha et al.[4] Both stiffness and matrix crack evaluation were carried out until complete failure of the specimen.

Table 1   Mechanical properties of the GFRP composites[21]

Type of test	Mechanical property	Material		% change
		GFRP-neat	GFRP-nano	
Tension	σ_{UTS} (MPa)	365 ± 13	382 ± 12	+4.65
	E_T (GPa)	17.5 ± 0.1	18.8 ± 1.7	+7.42
Compression	σ_{UCS} (MPa)	355 ± 47	361 ± 28	+1.69
	E_C (GPa)	21.3 ± 0.4	22.6 ± 0.4	+6.10

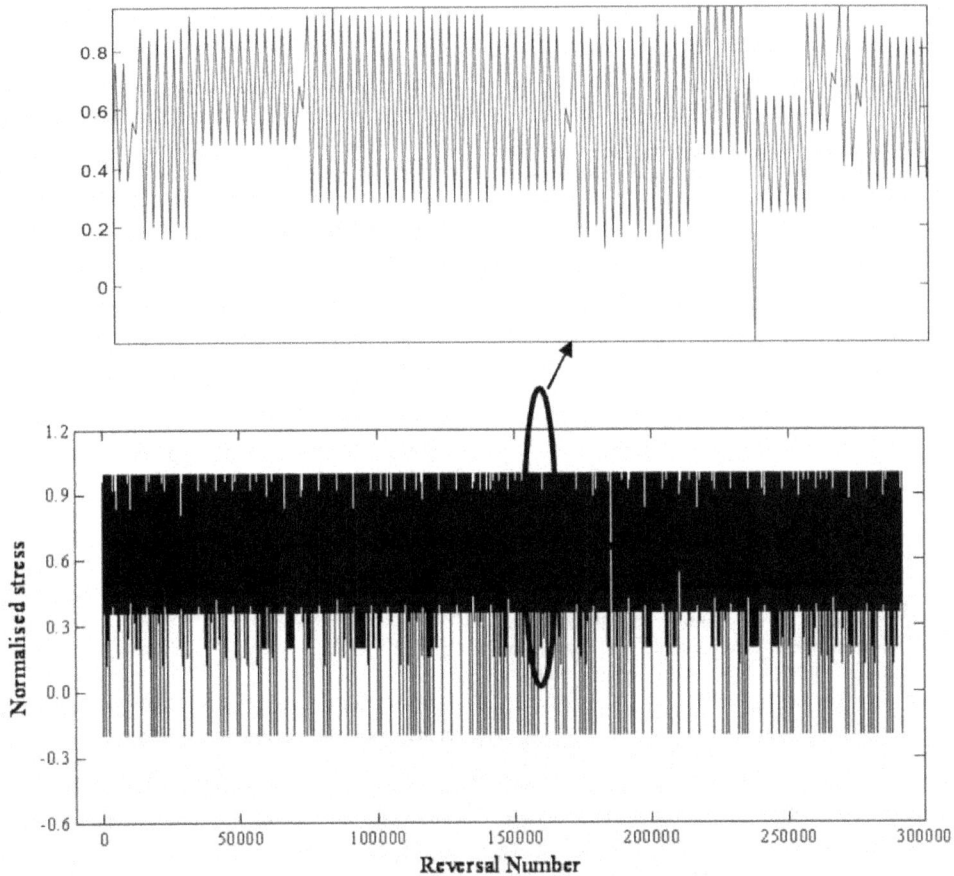

Figure 1   The standard Helix-32 random load sequence[23]

Results and discussion

Experimental observations

The experimentally determined fatigue life of GFRP composites under Helix-32 random load sequence is shown in Fig. 2. As expected, the fatigue life of both GFRP-neat and GFRP-nano composites increased with decreasing reference stress, a similar trend observed by many others in GFRP composites subjected to random loads.[21,24,25] However, for any given reference stress, the GFRP-nano composite shows higher fatigue life compared to its neat counterpart. The fatigue life enhancement observed is about four times over the entire range of reference stress levels investigated.

The variation of the normalized stiffness of the specimen as a function of applied load blocks is shown in Fig. 3. These stiffness reduction trends are similar to those generally

observed in FRP composites.[26-28] The stiffness reduction curves exhibit distinctly different stages. However, the stiffness degradation rate is much more severe, particularly in Stages I and II, in GFRP-neat composite compared to that of nanocomposite.

Typical photographs of the matrix cracks observed on the surface of the composites after application of one complete load block are shown for both GFRP composites in Fig. 4. Similar matrix cracking in GFRP composite under fatigue loads has been reported earlier.[29] For the same number of fatigue load cycles i.e., one complete Helix-32 load block, the GFRP-neat composite shows much more severe matrix cracking compared to that shown by GFRP-nano composite.

The matrix crack density was quantified as the number of cracks per unit length by several measurements on such photographs taken at regular intervals during the fatigue

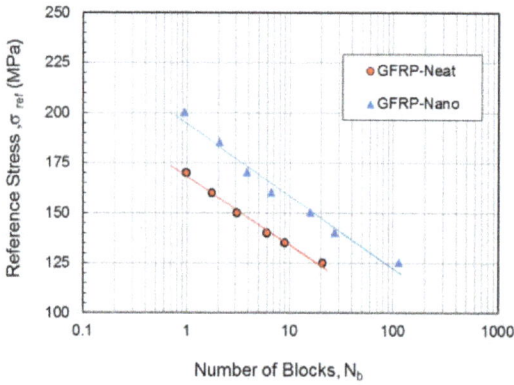

Figure 2 Experimental fatigue life of the glass-fiber reinforced plastic (GFRP) composites determined under Helix-32 random load sequence

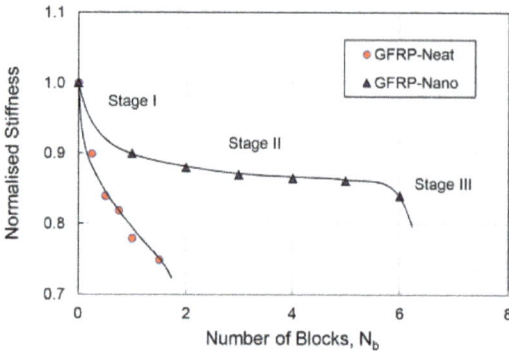

Figure 3 The variation of the normalized stiffness with applied random load blocks determined for the glass-fiber reinforced plastic (GFRP) composites under Helix-32 random load sequence, $\sigma_{ref} = 160\,MPa$

test. The variation of matrix crack density with applied load blocks determined for a test with $\sigma_{ref} = 160\,MPa$ is shown in Fig. 5. In both the GFRP composites, the crack density increased with applied load blocks, similar to observation made by others.[26,29] Under constant amplitude fatigue loads,

the crack density has been shown to increase and reach a saturation.[4,26,29] This saturation level, which is termed as characteristic damage state (CDS), depends on the stress level. However, crack density does not show indication of saturation under spectrum loads (Fig. 5). This may be owing to varying stress level of individual load cycles in the spectrum. Nevertheless, the results do clearly show that for any given applied load block, the matrix crack density is much lower in GFRP-nano composite compared to GFRP-neat composite. Thus, suppression of matrix cracking was clearly visible in GFRP-nano composite under the Helix-32 random load sequence.

The progressive fatigue damage under cyclic loads in FRP composites has been studied in detail.[26,27] It has been generally observed during fatigue cycling that matrix cracks initiate and grow, leading to initiation of disbonds and delaminations, and the subsequent growth of these cracks and delaminations leads to final failure. The stiffness loss in Stage I and II results primarily from matrix cracking. [26,27,29] The nanocomposite appears to suppress the formation of matrix cracks (Fig. 4) thereby showing reduced stiffness loss compared to neat composite (Fig. 3). Silica nano particle-modified epoxy has been shown to exhibit reduced fatigue crack growth rate by more than an order of magnitude. [30,31] The slow growth of matrix cracks in nano-modified epoxy matrix delays initiation of disbonds and delaminations. Furthermore, reduced rate of crack and delamination growth delays the final failure. Thus, several micro-mechanisms, such as (i) suppressed matrix cracking, (ii) reduced crack growth rates in the matrix, (iii) delayed initiation of disbonds and delamination, and (iv) lowered growth rate of delamination, appear to result in enhanced fatigue life of GFRP nanocomposites.

Random load fatigue life prediction

An effort was made to predict the fatigue life of GFRP composites and compare with experimental results. The present available fatigue life prediction models can be classified into three major categories,[19,32] namely empirical, phenomenological and physics based damage models. Empirical models rely on experimental data (stress levels, stress ratio

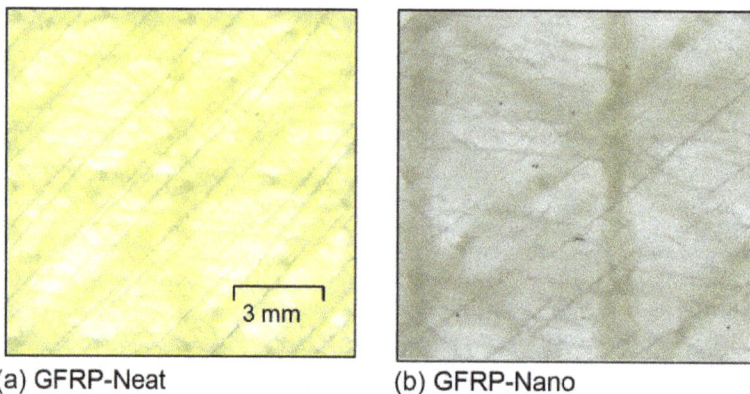

(a) GFRP-Neat   (b) GFRP-Nano

Figure 4 Typical photographs showing matrix cracks in the glass-fiber reinforced plastic (GFRP) composites subjected to Helix-32 random load sequence after application of one complete load block, $\sigma_{ref} = 160\,MPa$

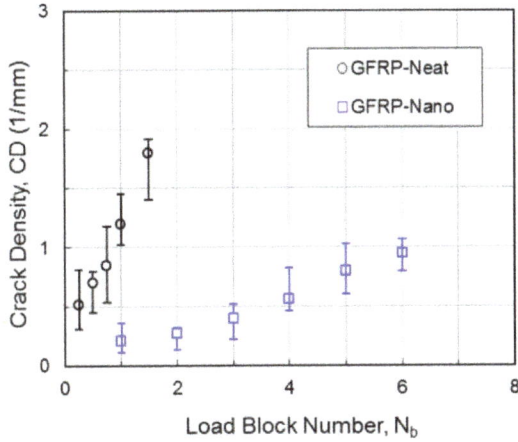

Figure 5 Variation of matrix crack density with applied random load blocks in glass-fiber reinforced plastic (GFRP) composites, σ_{ref} = 160 MPa

or frequency) without considering the inherent damage mechanisms. Phenomenological models use experimentally measurable phenomena like residual stiffness or strength as a damage matrix. In physics-based damage models, one or more appropriately chosen damage variables are introduced to account for deterioration of composite properties. The macroscopic mechanical property degradation is correlated to underlying damage mechanisms through sound physical modelling. Although these are models that can be used as simple design tools, still no robust model is available to accurately predict the response under fatigue loads.

The general approach to the prediction of the fatigue life by empirical models is shown schematically in Fig. 6. The fatigue life prediction was carried out following this procedure and the details of each of the steps followed are shown below:

Cycle counting

The individual load cycles in the random load sequence was counted using rain flow counting technique as per ASTM standard specifications.[33]

Estimation of N_f for each of the counted cycle

The number of cycles to failure, N_f, for each of the rain flow counted load cycle was obtained from the constant life diagram (CLD) of the material. The CLDs for these materials[21] are shown in Fig. 7. However, this procedure requires invariably an interpolation technique to estimate N_f. Although there are many interpolation techniques available, Vassilopoulos et al. have shown that piece-wise linear interpolation technique is more appropriate in prediction of N_f from CLDs.[34] The detailed procedure of a piece-wise interpolation technique may be found elsewhere.[19] In the present investigation, a program was written to interpolate using piece-wise interpolation technique and N_f was obtained for each counted cycle from CLDs.

Estimation of damage fraction for each cycle

The damage fraction for each of the rain flow counted load cycles was calculated as the ratio of applied cycle to N_f for that load amplitude i.e., $D_i = n_i/N_{f,i}$

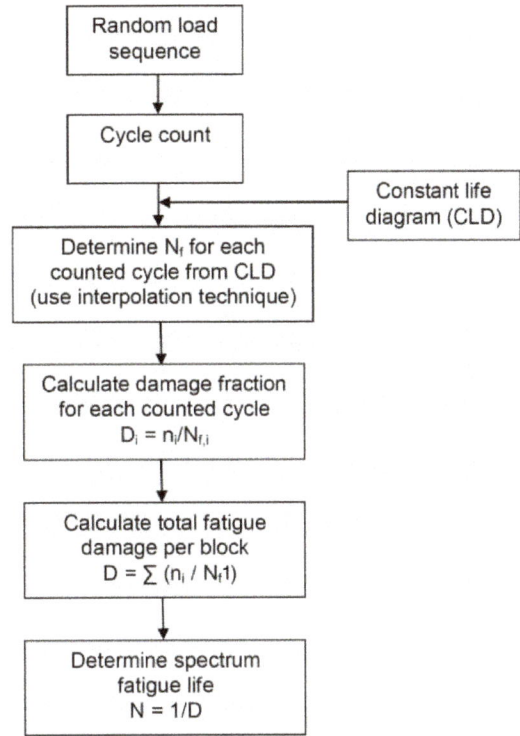

Figure 6 A flow chart showing the procedure used for prediction of random load fatigue life

Estimation of total damage per block

The total fatigue damage for the entire load block was estimated by summation of the damage fraction of all the individual load cycles in the entire block. There are various damage accumulation models used for fatigue life estimation of polymer composites.[19,32] In the present study, three different damage accumulation models were employed.

(a) Palmgren–Miner (PM) rule[35] is a linear damage accumulation law and is given by

$$D = \sum n_i/N_{f,i} \tag{1}$$

Where, D = total fatigue damage per block, $n_i/N_{f,i}$ = damage fraction of ith load cycle.

(b) Hashin–Rotem (HR) model[36] is a non-linear damage accumulation law which is given by

$$D_i = D^{\frac{\log \sigma_{p,i}}{\log \sigma_{p,i-1}}} + \frac{n_i}{N_{f,1}}; D_1 = n_1/N_f1 \tag{2}$$

Where, $\sigma_{p,i}$ = maximum stress in the ith load step, $n_i/N_{f,i}$ = damage fraction of ith load cycles.

(c) The Broutman–Sahu (BS) law[37] is a residual strength based model and is given by

$$S_r = S_u - \sum_i (S_u - \sigma_{p,i}) \frac{n_i}{N_f1} \tag{3}$$

Where, S_r = residual strength, S_u = ultimate strength $\sigma_{p,i}$ = maximum stress in the ith load step, $n_i/N_{f,i}$ = damage fraction of ith load cycles. The specimen is assumed to fail when

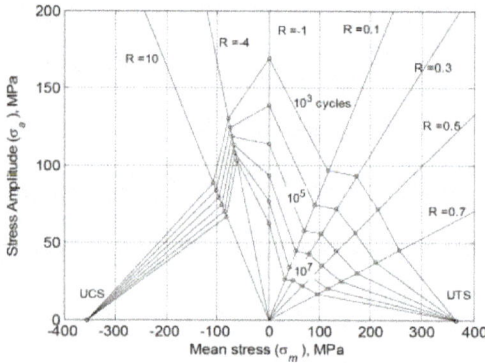

(a) GFRP-neat composite

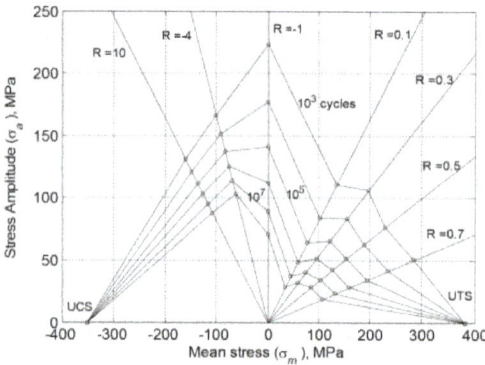

(b) GFRP-nano composite

Figure 7 Constant life diagrams (CLDs) of glass-fiber reinforced plastic (GFRP) composites[21]

the residual strength is lower than the maximum stress of the counted cycle.

Estimation of fatigue life

The material is assumed to fail when the total damage fraction reaches 1.0 and, hence, the fatigue life under the random load-sequence is equal to the reciprocal of the estimated total accumulated damage per load block.

The predicted fatigue life under Helix-32 load sequence using the above procedure and empirical models along with experimental results are shown in Fig. 8. The damage evolution under all three empirical models for a specific test of $\sigma_{ref} = 160$ MPa was determined for both GFRP composites. The damage evolution curves for GFRP-nano composite are shown in Fig. 9. As expected, PM model shows linear damage accumulation behaviour. The HR model shows a random non-linear accumulation of damage with load blocks. The power value of damage D in each cycle (according to equation (2)) varies randomly depending on the load sequence. The linear reduction in residual strength was observed in BS Model similar to PM model. The final random load fatigue life estimated from each of these models was almost similar although with minor difference as shown in Table 2. Similar variations on life predictions by empirical models have been reported by Post et al.[19]

It may be observed that the fatigue life predicted by these models is in good agreement with the experimental results for

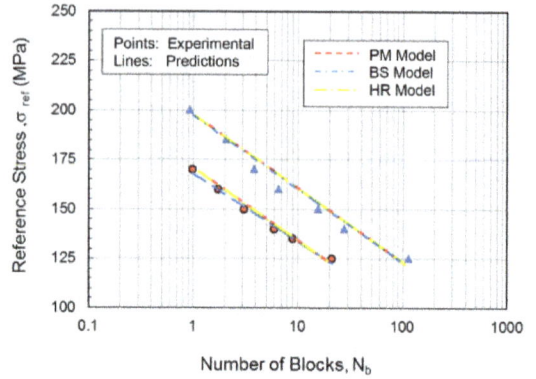

Figure 8 Comparison of experimental and predicted fatigue life of glass-fiber reinforced plastic (GFRP) composites under Helix-32 random load sequence

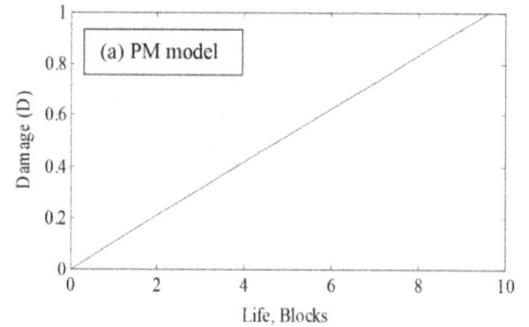

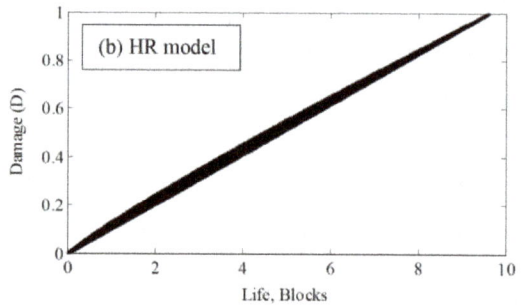

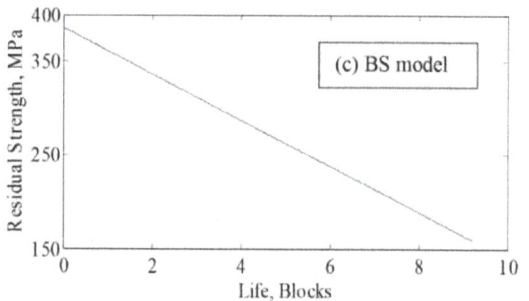

Figure 9 The predicted damage evolution by empirical models under Helix-32 random load sequence with $\sigma_{ref} = 160$ MPa for glass-fiber reinforced plastic (GFRP)-nano composite

Table 2 Predicted fatigue life under Helix-32 random loads with σ_{ref} = 160 MPa

Empirical model	Predicted fatigue life, N_b	
	GFRP-neat	GFRP-nano
Palmgren-Miner (PM)	1.99	9.60
Hashin-Rotem (HR)	1.99	9.61
Broutman-Sahu (BS)	1.90	9.20

both GFRP-neat and GFRP-nano composites (Fig. 8). All the three different models appear to predict similar fatigue life in both composites. It is interesting to note that the fatigue life enhancement in GFRP nanocomposite with 10 wt-% silica nano particles in the epoxy matrix under a wind turbine random load sequence is also about four times,[21] similar to the present investigation. Furthermore, use of different types of empirical models does not appear to show any variation in the fatigue life enhancement factor. These observations suggest that there may not be any load sequence effect under Helix-32 in this material. However, further systematic studies are required to study the insensitivity of the fatigue life to load sequence in this material.

Conclusions

Based on the results obtained in this study, following conclusions may be drawn:

1. Under Helix-32 random load sequence, the fatigue life of the GFRP-nano composite containing 10 wt-% silica nanoparticles in the epoxy matrix is about four times higher than its neat counterpart. Micro-mechanisms of suppressed matrix cracking and reduced matrix crack/delamination growth rate owing to addition of the silica nanoparticles are responsible for improved fatigue life in nanocomposite.

2. The fatigue life under Helix-32 random load sequence was predicted by empirical method using three different damage accumulation rules i.e., PM, BS, and HR theories. The predicted fatigue life is in good agreement with experimental observations for both GFRP-neat and GFRP-nano composites. All three damage accumulation laws, i.e. linear, non-linear, and strength-based rules predict similar results indicating probably the absence of load sequence effects under Helix-32 loads in this material.

Conflicts of interest

The authors have no conflicts of interest to declare.

Acknowledgments

The authors from wish to thank Mr Shyam Chetty, Director and Dr. Satish Chandra, Head, Structural Technologies Division, CSIR-NAL, Bangalore, India for their support during this work. The material was fabricated at Dept. of Mech. Engg., Imperial College, London, UK. The authors wish to thank Professor A. J. Kinloch and Dr. A. C. Taylor for their support and encouragement. The technical support staff members of the Materials Evaluation Lab, FSIG, STTD, CSIR-NAL are thanked for their assistance in the experimental work.

References

1. F. Hussain, M. Hojjati, M. Okamoto and R. Gorga: *J. Compos. Mater.*, 2006, **40**, (17), 1511–1575.
2. E. Thostenson, C. Li and T. Chou: *Compos. Sci. Tech.*, 2005, **65**, 491–516.
3. R. Sengupta, M. Bhattacharya, S. Bandyopadhyay and A. K. Bhowmick: *Progr Poly Sci*, 2011, **36**, (5), 638–670.
4. C. M. Manjunatha, A. C. Taylor, A. J. Kinloch and S. Sprenger: *Compos. Sci. Tech.*, 2010, **70**, (1), 193–199.
5. C. M. Manjunatha, A. C. Taylor, A. J. Kinloch and S. Sprenger: *J. Mater. Sci.*, 2009, **44**, (16), 4487–4490.
6. N. Yu, Z. Zhang and S. He: *Mater. Sci. Engg. A.*, 2008, **494**, (1–2), 380–384.
7. M. Loos, J. Yang, D. Feke and I. Manas-Zloczower: *Poly. Engg. Sci.*, 2012, **52**, 1882–1887.
8. D. R. Bortz, C. Merino and I. Martin-Gullon: *Compos. Sci. Tech.*, 2011, **71**, (1), 31–38.
9. Y. Zhou, V. Rangari, H. Mahfuz, S. Jeelani and P. Mallick: *Mater. Sci. Engg. A.*, 2005, **402**, 109–117.
10. M. Shokrieh, M. Esmkhani, A. Haghighatkhah and Z. Zhao: *Mater. Des.*, 2014, **62**, 401–408.
11. L. Boger, J. Sumfleth, H. Hedemann and K. Schulte: *Compos. A Appl. Sci. Manufact.*, 2010, **41**, (10), 1419–1424.
12. N. Chisholm, H. Mahfuz, V. Rangari, A. Ashfaq and S. Jeelani: *Compos. Struct.*, 2005, **67**, 115–124.
13. J. Fenner and I. Daniel: *Compos. A Appl. Sci. Manufact.*, 2014, **65**, 47–56.
14. C. Grimmer and C. Dharan: *J. Mater. Sci.*, 2008, **43**, 4487–4492.
15. F. Zhou, F. Pervin, S. Jeelani and P. Mallick: *J. Mater. Proc. Tech.*, 2008, **198**, 445–453.
16. Y. Zhou, M. Hosur, S. Jeelani and P. Mallick: *J. Mater. Sci.*, 2012, **47**, (12), 5002–5012.
17. S. Khan, A. Munir, R. Hussain and J. Kim: *Compos. Sci. Tech.*, 2010, **70**, (14), 2077–2085.
18. E. Gamstedt and B. Sjogren: *Int. J. Fat.*, 2002, **24**, 437–446.
19. N. L. Post, S. Case and J. J. Lesko: *Int. J. Fat.*, 2008, **30**, 2064–2086.
20. W. Van Paepegem and J. Degrieck: *Mech. Adv. Mater. Struct.*, 2002, **9**, 19–35.
21. C. M. Manjunatha, R. Bojja and N. Jagannathan: *Mater. Perfor. Charact.*, 2014, **3**, (1), 327–341.
22. P. Heuler and H. Klatschke: *Int. J. Fat.*, 2005, **27**, 974–990.
23. P. Edwards and J. Darts: 'HELIX and FELIX- final definition of two standardised fatigue loading sequence for helicopter rotors', NLR Technical Report 84043U, 1984.
24. T. Philippidis and A. Vassilopoulos: *Compos. A Appl. Sci. Manuf.*, 2004, **35**, 657–666.
25. C. M. Manjunatha, R. Bojja, N. Jagannathan, A. J. Kinloch and A. C. Taylor: *Int. J. Fat.*, 2013, **54**, 25–31.
26. S. Case and K. Reifsnider: 'Fatigue of composite materials', in 'Comprehensive structural integrity', Vol. 4, cyclic loading and fatigue', 1st edn, (ed. I. Milne *et al.*), 2003, Elsevier, Amsterdam.
27. R. Talreja: *Proc. Roy. Soc. Lond. A*, 1989, **378**, 559–567.
28. J. Tate and A. Kelkar: *Compos. B Engg.*, 2008, **39**, (3), 548–555.
29. A. Gagel, D. Lange and K. Schulte: *Compos. A*, 2006, **37**, 222–228.
30. H.-Y. Liu, G.-T. Wang and Y.-W. Mai: *Compos. Sci. Tech.*, 2012, **72**, (13), 1530–1538.
31. G.-T. Wang, H.-Y. Liu, N. Saintier and Y.-W. Mai: *Engg. Fail. Analy.*, 2009, **16**, 2635–2645.
32. J. Degrieck and W. V. Paepegem: *Appl. Mech. Rev.*, 2001, **54**, (4), 279–300.
33. 'Standard practices for cycle counting in fatigue analysis', E1049, in 'Annual book of ASTM standards', Vol. **15.03**; 2013, West Conshohocken, PA, ASTM.
34. A. Vassilopoulos, B. Manshadi and T. Keller: *Int. J. Fat.*, 2010, **32**, 659–669.
35. M. Miner: *J. Appl. Mech.*, 1945, **12A**, 159–164.
36. Z. Hashin and A. Rotem: *J. Mater. Sci. Engg.*, 1978, **34**, 147–160.
37. L. Broutman and S. Sahu: 'New theory to predict cumulative fatigue damage', in 'Fiberglass reinforced plastics, composite materials: testing and design', 170–188; 1972, West Conshohocken, PA, ASTM International.

2

Evaluation of a new processing method for improved nanocomposite dispersions

B. M. Cromer, E. B. Coughlin and A. J. Lesser*

Department of Polymer Science and Engineering, University of Massachusetts Amherst, 120 Governors Drive, Amherst, MA 01003, USA

Abstract Herein, a new process referred to as melt-mastication (MM) is used for the first time and evaluated for dispersing nanoparticles in polymers. Compared to a conventional melt processing (CMP) technique, MM produces higher mixing torque and therefore shear in the melt during processing, resulting in the fragmentation of micrometer-scale agglomerates of exfoliated graphene nanoplatelets (xGnP) in polypropylene. The efficacy of MM compared to CMP is evaluated using quantitative stereological techniques. Stereology reveals a correlation between the steady state process torque and the spatial size distribution of agglomerates. Finally, a mechanism for agglomerate fragmentation is proposed and discussed with respect to the results.

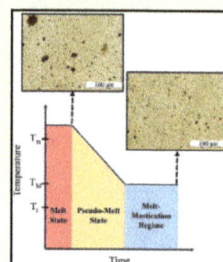

Keywords Nanocomposites, Particle dispersion, Particle reinforcement, Stereological characterization

Abbreviations CMP: conventional melt processing, HTGPC: high-temperature gel permeation chromatography, iPP: isotactic polypropylene, MM: melt-mastication, PNC: polymer nanocomposite, TM: mastication temperature, xGnP: exfoliated graphene nanoplatelets.

Introduction

Over the past two decades, significant interest in polymer nanocomposites (PNCs) has been driven by the potential for significant thermal and physical property improvements.[1–3]. Unlike traditional filled-polymer composites, PNCs derive property improvements from a relatively small concentration of a well-dispersed, high-surface area discrete phase.[4] Several PNCs have been reported using various discrete phase materials, including fumed silica,[5–7] titanium dioxide,[8–10] layered mineral silicates,[2,11–13] carbon black, and graphite-derived particles, such as graphite oxide,[14] thermally reduced graphite oxide,[15–17] and expanded graphite.[14,18]

There are many potential applications for polyolefin PNCs, including automotive materials, packaging materials, electronic applications, optical applications, and applications where accelerated polymer crystallization kinetics are desired.[6,8,14,15,19–31] However, it is difficult to prepare well-dispersed PNCs because of the strong tendency for nanoparticles to agglomerate, especially during conventional melt processing (CMP). The unfavorable interaction energy between the nanoparticle and polymer–matrix provides a driving force for nanoparticle agglomeration during melt processing.[16,22,32] Several research groups have proposed chemical and/or physical techniques to improve the

nanoparticle dispersion in PNCs. Chemical techniques are intended to optimize the polymer–matrix interaction energy. This can be accomplished through modifying the nanoparticle surface,[16] modifying the polymer resin, or through addition of a compatibilizer.[6,26] *In situ* polymerization has shown success for limited systems; however, scale-up to commercial volumes is not trivial.[29,33] Physical techniques utilize non-conventional processing conditions and equipments to promote fragmentation of nanoparticle agglomerates, thereby improving dispersion into polyolefin resins. Examples of physical techniques include solid-state shear pulverization,[30,31] solid-state melt extrusion,[34] and solid-state ball milling,[35] which generated well-dispersed graphite-polyolefin PNCs compared to similar PNCs prepared by CMP conditions. Additionally, others improved polyolefin PNC dispersions by increasing the mixing shear, through changing various process conditions.[12,13,26,36]

Methods to characterize the PNC dispersion state are also critical to understanding the structure–process–property relationships of these materials. Current characterization techniques either directly or indirectly characterize the dispersion state. Direct methods, such as microscopy, absolutely portray the dispersion state in a localized region of the composite, but do not represent the dispersion state in the entire material. In contrast, indirect methods characterize an intrinsic material property related to dispersion, such as crystallization kinetics,[23,30,31] thermal and electrical

---

*Corresponding author, email ajl@polysci.umass.edu

conductivity,[22,35] or rheology.[37] Indirect methods are useful for comparing the relative dispersion state between similar samples, but do not directly describe the size and spacing of nanoparticle agglomerates. Accordingly, Johnson and Salt-ykov developed a stereological analysis method to combine the benefits of direct and indirect characterization tech-niques. When applied to PNCs, quantitative stereology pro-vides an opportunity to calculate the 3D spatial distribution of nanoparticles using information extracted from multiple 2D microscopy images.

Herein, we evaluate a new method referred to as melt-mastication (MM) for the first time. Melt-mastication was used to achieve improved nanocomposite dispersions by fractur-ing larger nanoparticle clusters during the mixing process. With MM, an isotactic polypropylene-exfoliated graphene nanoplatelet (iPP-xGnP) system is subjected to high shear and compressive forces through a low-temperature proces-sing method. The dispersion efficiency of MM is evaluated with a quantitative stereological technique and compared to the dispersion efficiency of a CMP technique. Finally, we introduce a new model to describe the mechanics of agglomerate fragmentation and dispersion. This model in-corporates a scale effect not captured in conventional models to date.[38–40] The results from this model are discussed in comparison to the results obtained from the present study.

Experimental methods

Materials

Exfoliated Graphene Nanoplatelets (xGnP-c-750, 750 $m^2 g^{-1}$) were purchased from XGSciences, Inc (Lansing, MI, USA). and used as received. Isotactic polypropylene (iPP, PP9999SS) was provided by ExxonMobil ($T_m = 165°C$, $T_c = 111°C$ by differential scanning calorimetry). Irganox 1010 and Irgafos 168 were purchased from Ciba and used as received.

Processing of polyolefin nanocomposites via melt-mastication

Melt-mastication is achieved by compounding a polyolefin nanocomposite melt over a three-step temperature process. In the first step, the polyolefin and nanoparticle are melt compounded at a temperature above the melting tem-perature (T_m) of the polyolefin. The nanocomposite melt is then cooled under continuous mixing to the mastication temperature (T_M), followed by isothermal mixing at T_M, where $T_c < T_M < T_m$. The generalized temperature profile for MM is illustrated in Fig. 1. Processing at T_M requires a significant increase in mixing torque, which subjects the nanocomposite melt to intense shear and compressive forces, producing enhanced break-up of larger nanoparticle agglomerates.

In this study, MM was evaluated for an iPP-xGnP (2 wt-%) nanocomposite system. First, iPP, xGnP, and 0.05 wt-% process stabilizers are compounded at 200°C stock temperature for 5 min at 70 rev min^{-1} in a 50 mL batch mixer (Brabender ATR Plasticorder®). Then, the nanocomposite melt is cooled at 3°C min^{-1} under continuous mixing until it reaches T_M, fol-lowed by isothermal mixing at T_M. The mastication step was performed at either 154, 165, or 180°C at 70 rev min^{-1} mixing rate. The total process time for all samples was 25 min.

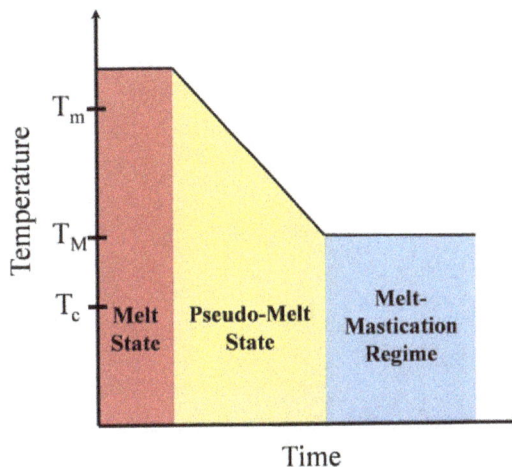

Figure 1   Three-step temperature profile for melt-mastication

For reference, five control samples of identical compo-sition were prepared by CMP, or melt processing at 200°C at various mixing rates. Isotactic polypropylene, xGnP, and 0.05 wt-% process stabilizers were combined and processed at 200°C for 25 min at 70, 100, 130, 160, or 190 rev min^{-1} in the same 50 mL batch mixer (Brabender ATR Plasticorder®). The mixing torque, stock temperature, and mixing rate were recorded for all samples. Degradation of iPP during MM and CMP was mitigated through the addition of antioxidant stabilizers, demonstrated in Appendix 1.

Characterization

The molecular weight of iPP was characterized with high-temperature gel permeation chromatography (HTGPC) on a Polymer Labs PL-220 GPC. Samples were dissolved and analyzed in 1,2,4-trichlorobenzene at 145°C against poly-styrene standards. The absolute molecular weights were calculated via the Mark–Houwink equation.

Transmission optical microscopy (OM) was conducted on an Olympus optical microscope with DP71 digital camera. Thin sample sections were prepared at room temperature by a glass knife microtome technique on a Reichert-Jung FC4 ultramicrotome. Samples were analyzed at room tempera-ture in transmission mode.

TEM was conducted on a JEOL JEM-2000FX transmission electron microscope with LaB$_6$ electron source, at accel-erating voltage 200 kV. Thin (∼ 40 nm) sample sections were prepared at room temperature using a Leica CryoUl-tramicrotome and Microstar diamond knife, then imaged on 400 mesh copper grids.

Quantitative stereology was performed on OM images in order to determine the 3D size distribution of xGnP agglomerates. Images were analyzed using ImageJ image processing software (National Institutes of Health, Bethesda, MD, USA).[41] Images similar to Fig. 2A were converted to binary via a consistent threshold function, and then the section area of each xGnP agglomerate was calculated. Sample section areas <1 μm^2 were not considered because of the resolution limits of the optical microscope. Each image contained ∼800–1200 distinguishable sections, and each

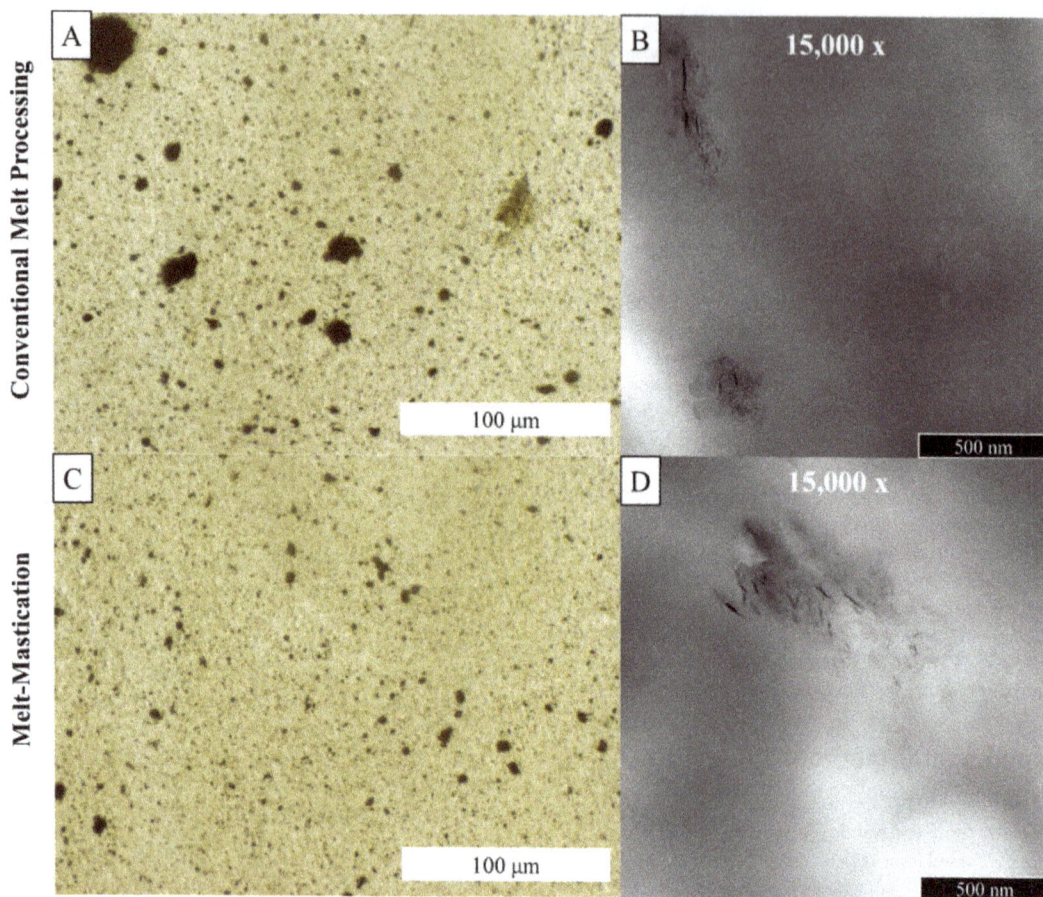

Figure 2 Optical microscopy (OM) and TEM images of isotactic polypropylene (iPP)-exfoliated graphene nanoplatelets (xGnP) nanocomposites. *a* OM, 200°C_70 rev min^{-1}, *b* TEM, 200°C_70 rev min^{-1} *c* OM, 154°C_70 rev min^{-1} *d* TEM, 154°C_70 rev min^{-1}

sample was analyzed by five images for a total of ~4000–6000 sections per sample. Stereological analysis was performed on the sections data according to a procedure developed by Johnson and Saltykov.[42–46]

Results and discussion

Qualitative comparison

Herein, we evaluate a new method referred to as MM to achieve improved dispersion in a 2 wt-% iPP-xGnP nanocomposite compared to an analogous nanocomposite prepared by CMP. Optical microscopy in transmission mode was used to qualitatively compare the dispersions of a sample prepared by MM (154°C_70 rev min^{-1}) and a sample produced by CMP (200°C_70 rev min^{-1}), shown in Fig. 2. Both samples were processed for 25 min at 70 rev min^{-1}. Optical microscopy shows that each sample comprises discrete, black xGnP nanoparticle agglomerates within a continuous, transparent iPP phase. The agglomerate sections range from ~0.5 to 15 μm in radius. Qualitatively, sample 154°C_70 rev min^{-1} appears to have improved dispersion compared to sample 200°C_70 rev min^{-1}. In particular, the population of large agglomerates [radius, $r > 5$ μm]

apparent in sample 200°C_70 rev min^{-1} is absent in sample 154°C_70 rev min^{-1}. The effect of MM on smaller ($r < 0.5$ μm) nanoparticle agglomerates is not clear from this technique because of the resolution limitations of the instrument. Similarly, transmission electron microscopy was used to evaluate the dispersion state of the smaller ($r < 0.5$ μm) population fraction of nanoparticle agglomerates. At 15,000 × magnification, agglomerates appear to be the assemblies of primary platelet particles. The contrast of each primary platelet particle is affected by its orientation relative to the beam direction; platelets oriented parallel to the beam direction show the most contrast. The boundaries of each agglomerate are difficult to define because of the random orientation of primary particles; therefore, the agglomerate sizes cannot be accurately determined. From limited qualitative analysis, there is no discernable distinction between samples 154°C_70 rev min^{-1} and 200°C_70 rev min^{-1} across several TEM images.

Quantitative stereology

Quantitative stereology is used to calculate the 3-D spatial size distribution of xGnP agglomerates and derive a statistical distinction between the distributions of samples

154°C_70 rev min^{-1} and 200°C_70 rev min^{-1}. Johnson and Saltykov demonstrated a stereological method to calculate the 3-D spatial size distribution of discrete spheres trapped in a continuous medium from measurements on random test planes.[45] In the present study, quantitative stereology is applied to 2-D transmission OM images in order to derive the 3D size distribution of discrete xGnP agglomerates within a continuous iPP phase.

The 2-D section size distribution is derived from OM images using image processing software. Samples 154°C_70 rev min^{-1} and 200°C_70 rev min^{-1} were each sectioned and imaged at five different regions in order to better represent the dispersion state in the entire material. Each image was analyzed with ImageJ software in order to derive the 2-D section size distribution, and the resulting distributions were combined and plotted in Fig. 3.[41] The data are plotted on a histogram with logarithmically spaced bins so as to provide finer subdivisions in the small radii range, where the majority of sections exist. The distribution of agglomerate section sizes is expected to follow a lognormal distribution, which appears as a Gaussian distribution for logarithmically spaced bin sizes.[46] However, the distribution of sections below $r = 0.5\,\mu m$ is truncated, because of the resolution limitations of the optical microscope. Despite this limitation, quantitative stereology still demonstrates significant differences between the agglomerate section distributions of samples 154°C_70 rev min^{-1} and 200°C_70 rev min^{-1}. In particular, the distribution of sample 154°C_70 rev min^{-1} contains the highest number of sections for $r < 1.7\,\mu m$, whereas sample 200°C_70 rev min^{-1} contains the most sections for $r > 1.7\,\mu m$. That is, the low temperature processing condition apparently produces smaller agglomerates of nanoparticles when compared to the same system processed at higher temperature and similar shear rate.

In order to establish a more quantitative evaluation of actual particle size, the Johnson and Saltykov analysis method was applied to the 2-D distributions in order to calculate the 3-D spatial distribution of agglomerates. These resulting distributions are plotted in Fig. 4. The 3-D spatial distributions are expected to follow Gaussian statistics; however, the distribution is again truncated below $r = 0.5\,\mu m$. A Gaussian curve was also fitted to each sample distribution for regimes that were measured. The Gaussian curves corresponding to this fit are also shown in Fig. 4.

Note that the differences in the distributions of samples 200°C_70 rev min^{-1} and 154°C_70 rev min^{-1} are reflected in the standard deviations of the fitted curves (0.39 and 0.31 μm, respectively). The contrast between these two samples is further illustrated by the differential distribution, shown in the inset of Fig. 4. Within each bin size, positive values represent agglomerate populations greater than sample 200°C_70 rev min^{-1}, whereas negative values represent agglomerate populations greater than sample 154°C_70 rev min^{-1}. It follows that Fig. 4 shows which agglomerate size populations were eliminated, and which were created during MM, relative to sample 200°C_70 rev min^{-1}. Considering that all samples had identical xGnP loading, it is reasonable to assume that the nanoparticle agglomerate volume between samples is identical. Accordingly, the elimination of few, large ($r < 1.7\,\mu m$) agglomerates from sample 200°C_70 rev min^{-1} should create many small ($r < 1.7\,\mu m$) agglomerates in sample 154°C_70 rev min^{-1}. The effect of MM becomes clearer when the data are presented as volume fraction of xGnP agglomerates within each bin size (Fig. 5). The larger ($r < 2\,\mu m$) agglomerate volume fractions present in sample 200°C_70 rev min^{-1} shifts to a smaller ($r > 2\,\mu m$) agglomerate fractions. These results indicate that the process conditions of MM promote fragmentation of larger nanoparticle clusters into smaller ($r < 1.7\,\mu m$) clusters.

Evaluation of processing parameters

The processing parameters were varied across six additional samples in order to ascertain the effect of process conditions on the spatial size distribution of xGnP agglomerates. Each

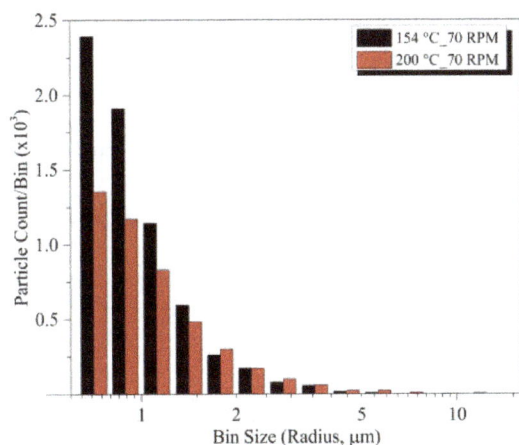

Figure 3   Two-dimensional (2-D) section size distribution of exfoliated graphene nanoplatelets (xGnP) agglomerates from optical microscopy

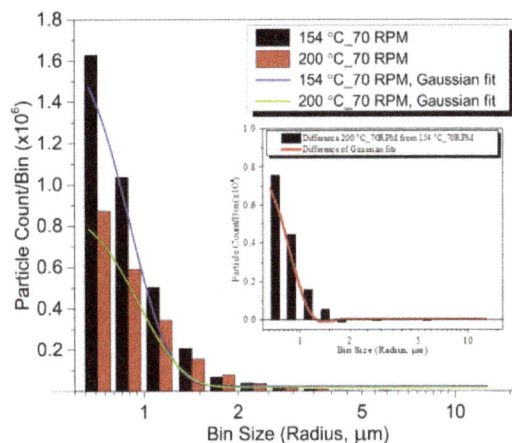

Figure 4   Spatial size distribution of exfoliated graphene nanoplatelets (xGnP) agglomerates per cubic millimeters from Johnson and Saltykov analysis. °C Black: Sample 154°C_70 rev min^{-1}. Red: Sample 200°C_70 rev min^{-1}. Each distribution is fitted to a Gaussian function. Inset: differential particle size distribution

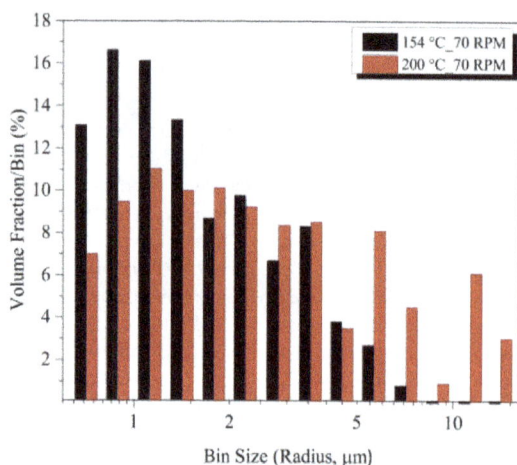

Figure 5 Volume fraction size distribution of exfoliated graphene nanoplatelet (xGnP) agglomerates from Johnson and Saltykov analysis. Black: sample 154°C_70 rev min^{-1}. Red: sample 200°C_70 rev min^{-1}

sample was processed by either MM or CMP, followed by characterization by OM and analysis with the same stereological treatment described in the previous section. The process conditions investigated were T_M and mixing rate. For samples prepared by MM, T_M was varied from 154 to 200°C, and the mixing rate was maintained at 70 rev min^{-1} for all samples. A second sample series was prepared using CMP, where the temperature was maintained at 200°C and the mixing rate was varied from 70 to 190 rev min^{-1}. All samples were processed for 25 min. Table 1 summarizes the experimental parameters of each sample, along with the steady state mixing torque measured during processing. Longer process times did not improve the dispersion quality, and resulted in iPP degradation. Mixing rates >190 rev min^{-1}, and process temperatures below 154°C also caused excessive iPP degradation.

Careful inspection of the results shows that the largest agglomerates are most affected by the steady state mixing torque. Both the agglomerate dispersity and maximum agglomerate size decrease with increasing mixing torque. Figure 6A demonstrates an inverse correlation between distribution standard deviation and the steady state mixing torque. The deviation increased with increasing T_M, and decreased slightly with increasing mixing rate. Similarly,

Fig. 6B shows the maximum agglomerate size decreases with increasing steady state mixing torque, and increasing mixing rate. Sample 154°C/70 rev min^{-1} showed the lowest standard deviation and lowest maximum agglomerate size, with the highest steady state mixing torque.

Even though the stress field of the mixing process is very complex, the measured torque is an integral result of the stresses generated during mixing. Given the fact that the boundary conditions do not change during each process condition (e.g. different shear rates or temperatures), the stress fields and their magnitudes generated for each condition are proportional to the measured torque. Consequently, changes in the agglomerate size dispersion resulting from changing process conditions can be primarily attributed to changes in the magnitude of the applied stress field either through changes in the shear rate (RPMs), or changes in the temperature (viscosity of the matrix).

Model for agglomerate fragmentation

Current models for the dispersive mixing of solid agglomerates during polymer processing do not capture the scale effect observed in this study.[39,40,47] That is, models based on stress analysis, fracture mechanics, or fluid mechanics do not predict the preferential fragmentation of large agglomerates with increasing mixing torque, seen in Fig. 6. Instead, current models based on stress analysis predict that agglomerate fragmentation is independent of agglomerate size. These models assume fragmentation occurs when the hydrodynamic stress imparted by the matrix overcomes the cohesive strength of the agglomerate. Because both hydrodynamic stress and cohesive strength depend on the size of the agglomerate, stress analysis models do not predict a size dependence. Other models based on fracture mechanics and probability theory do predict a scale effect by considering that larger bodies have a greater probability of containing a critical flaw size compared to smaller bodies.[38,48,49] Models based on fracture mechanics are useful for describing fragmentation of brittle systems; however, they are not applicable to the present study because they require prior knowledge of the flaw size distributions. Additional models based on fluid mechanics have been developed to describe the fragmentation of immiscible fluids in definable flow fields.[50–54] Models based on fluid mechanics incorporate the capillary number of the droplet, as well as the viscosity ratio between the droplet phase and the continuous matrix phase. These models predict that

Table 1 Processing parameters for isotactic polypropylene (iPP)-exfoliated graphene nanoplatelet (xGnP) nanocomposites prepared by melt-mastication (MM) or conventional melt processing (CMP)

Sample name	T_M (°C)	Mixing rate (RPM)	Steady state mixing torque (Nm)
MM			
154°C_70 rev min^{-1}	154	70	13.2
165°C_70 rev min^{-1}	165	70	6.5
180°C_70 rev min^{-1}	180	70	3.5
CMP			
200°C_70 rev min^{-1}	200	70	2.6
200°C_100 rev min^{-1}	200	100	2.8
200°C_130 rev min^{-1}	200	130	3.1
200°C_160 rev min^{-1}	200	160	3.5
200°C_190 rev min^{-1}	200	190	3.5

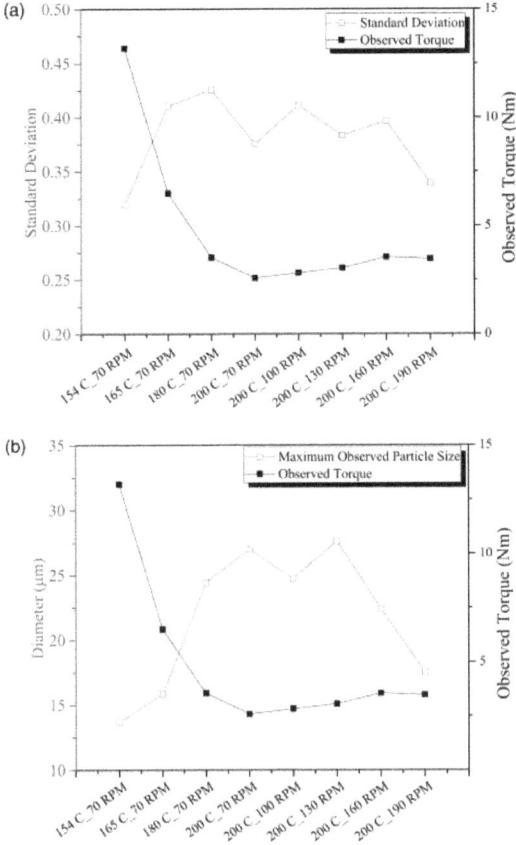

(a)

(b)

Figure 6 Particle size distribution characteristics and observed torque for each sample. *a* Gaussian standard deviation. *b* Maximum particle size

droplet size is a function of shear rate, but only when the viscosity ratio between the droplet and matrix phases is $\leq 1.$[55] In the present study, we anticipate that the stiffness of the nanoparticle agglomerates is much greater than the viscosity of the matrix, therefore we consider these models inappropriate for our system.

Herein, we present a model for agglomerate fragmentation in order to capture the scale effect observed in our results. For the case of polymer nanocomposites, the concentration of reinforcement necessary to achieve property improvements is usually very low (<6 vol.-%). Consequently, fragmentation of agglomerates caused by agglomerate-agglomerate interactions such as collisions is not expected to be a major contributor and will not be considered here. The primary mechanism for agglomerate fragmentation is expected to result from the stress field imposed on the agglomerates arising from the viscous flow of the polymer–matrix.

In the present model, we propose that agglomerate fragmentation will occur at a critical condition where the strain energy stored in the agglomerate meets or exceeds the cohesive energy necessary to produce agglomerate fragmentation. This condition is shown in equation (1)

$$U \geq \Gamma S' \tag{1}$$

where U is the strain energy, Γ is the surface energy, and S' is the surface area of the deformed agglomerate. The strain energy is an integral sum of the strain energy density u, where u is a contraction of the stress σ_{ij} and strain ε_{ij} acting on the agglomerate (i.e., $u = \sigma_{ij}\varepsilon_{ij}/2$) over the agglomerate volume V. If we consider that only distortional energy is imposed on the agglomerate, and the particle does not undergo any volume change, the stress and strain tensors can be written in terms of the octahedral shear stress and strain, respectively. This is shown in equation (2)

$$U = \int_V u\,dv = \frac{1}{2}\int_V \tau_{oct}\gamma_{oct}\,dv = \frac{G}{2}\int_V \gamma_{oct}^2\,dv \tag{2}$$

where τ_{oct} and γ_{oct} are the octahedral shear stress and strain imparted by the matrix onto the agglomerate, and G is the shear modulus of the agglomerate. Finally, an expression for the total strain energy in the agglomerate may be obtained by integrating the mean octahedral strain over a spherical shape of radius R.

$$U = \left(\frac{G}{2}\gamma_{oct}^2\right)\left(\frac{4}{3}\pi R^3\right) = \frac{2\pi G\gamma_{oct}^2 R^3}{3} \tag{3}$$

Authors now consider the condition where the strain energy imposed on the agglomerate will deform the agglomerate, thereby increasing its interfacial surface area from S to S' as shown in Fig. 7. Once S' reaches a critical value sufficient to create two or more smaller particles, the agglomerate is considered unstable and fragmentation occurs. In accordance with the criterion in equation (1), agglomerate fragmentation is favourable when the total strain energy reaches or exceeds a critical surface energy. The lowest energy condition for this to occur would be fragmentation of one agglomerate into two smaller agglomerates (Fig. 7). Assuming the conservation of agglomerate volume, the critical radii R' of the two smaller agglomerates can be related to R of the original agglomerate by

$$\frac{4}{3}\pi R^3 = 2\left(\frac{4}{3}\pi R'^3\right)$$

$$R' = \frac{R}{2^{1/3}} \tag{4}$$

Similarly, the critical surface area can be described in terms of R

$$S = \frac{8\pi R^2}{2^{2/3}} = 2^{7/3}\pi R^2 \tag{5}$$

Rewriting equation (1) using equations (3) and (5) yields an expression for the critical radius size

$$U = \Gamma S'$$

$$\frac{2\pi G\gamma_{oct}^2 R^3}{3} = \Gamma\left(\frac{8\pi R^2}{\sqrt[3]{4}}\right)$$

$$R = \frac{2^{4/3}3\Gamma}{G\gamma_{oct}^2} \tag{6}$$

Applying the definition of shear stress, $\tau_{oct} = \theta\gamma = G\gamma_{oct}$

$$R = \frac{2^{4/3}\Gamma G}{\left(\eta_\gamma^\bullet\right)^2} \tag{7}$$

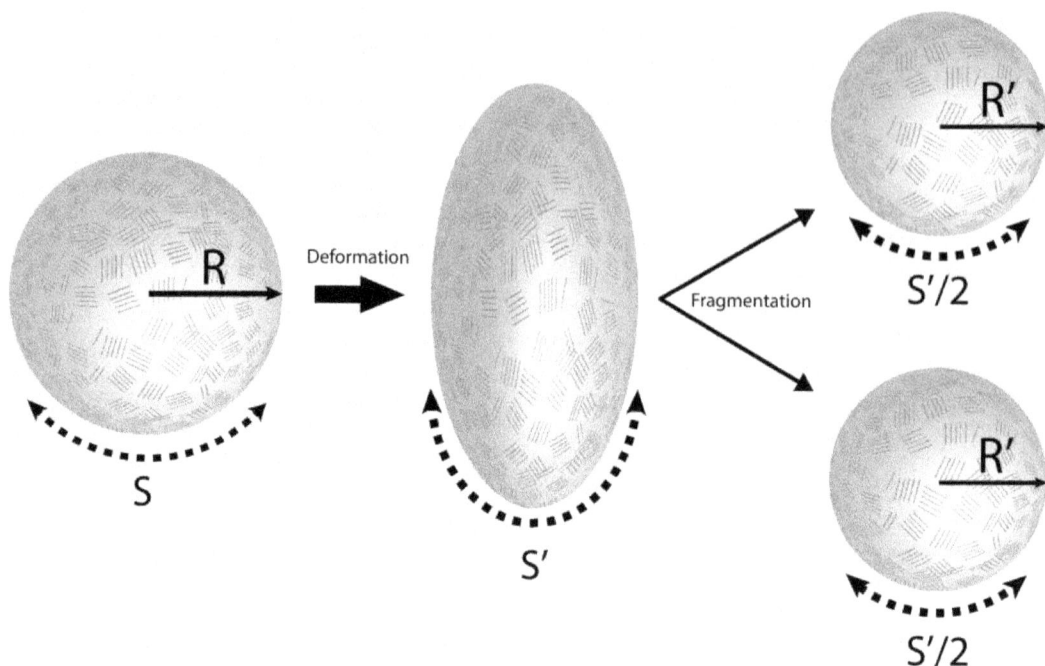

Figure 7   Schematic for fragmentation of exfoliated graphene nanoplatelet (xGnP) nanoplatelet agglomerates

Accordingly, this model predicts that the agglomerate size is inversely proportional to the square of shear stress. It follows that processing conditions that increase shear stress, such as MM, will promote agglomerate fragmentation and generate smaller agglomerate sizes. Figure 6 shows that both the maximum agglomerate size and dispersity decrease with increasing mixing torque; which is proportional to shear stress. Owing to the complex nature of the stress field of the mixing process as well as the resolution limitations of the optical microscope, it is not possible to predict or measure the complete agglomerate size distributions in the current study. However, the available results in Fig. 6 as well as the well-recognized size dependent phenomena of agglomerate fragmentation in particle grinding and comminution studies support the present model.[38,48,49]

Conclusions

Presented is an evaluation of a new polyolefin nano-composite processing strategy called MM to achieve improved nanoparticle dispersions, as compared to CMP. Melt-mastication was accomplished with a three-step temperature program, and the mixing rate and mastication temperature (T_M) were systematically varied to ascertain the conditions that enhance nanoparticle agglomerate fragmentation. The spatial size distribution of xGnP nanoparticle agglomerates was calculated from 2D OM images using a quantitative stereological treatment. The results show that the dispersity and size of the agglomerate distribution decreased with decreasing T_M and increasing mixing rate. A mechanism for agglomerate fragmentation was proposed and discussed with respect to the results. The dispersion quality and mixing torque were most improved when samples were treated with MM.

Conflicts of interest

The authors have no conflicts of interest to disclose.

Acknowledgements

Financial support was provided by the Department of Defense (DoD) through the National Defense Science and Engineering Graduate Research Fellowship (NDSEG) Program to B. M. Cromer. Funding was also provided by the Center for UMass/Industry Research on Polymers (CUMIRP) Cluster M: Mechanics of Polymers and Composites and Cluster F: Fire Safe Polymers and Polymer Composites. Authors also thank ExxonMobil for kindly providing materials.

References

1. Y. Kojima, A. Usuki, M. Kawasumi, A. Okada, Y. Fukushima, T. Kurauchi and O. Kamigaito: *J. Mater. Res.*, 2011, **8**, 1185–1189.
2. A. Usuki, Y. Kojima, M. Kawasumi, A. Okada, Y. Fukushima, T. Kurauchi and O. Kamigaito: *J. Mater. Res.*, 1993, **8**, 1179–1184.
3. J. Jancar, J. F. Douglas, F. W. Starr, S. K. Kumar, P. Cassagnau, A. J. Lesser, S. S. Sternstein and M. J. Buehler: *Polymer*, 2010, **51**, 3321–3343.
4. R. A. Vaia and H. D. Wagner: *Mater. Today*, 2004, **7**, 32–37.
5. P. Cassagnau: *Polymer*, 2008, **49**, 2183–2196.
6. V. Vladimirov, C. Betchev, A. Vassiliou, G. Papageorgiou and D. Bikiaris: *Compos. Sci. Technol.*, 2006, **66**, 2935–2944.
7. S. Chung, W. Hahm, S. Im and S. Oh: *Macromol. Res.*, 2002, **10**, 221–229.
8. S. Matteucci, V. A. Kusuma, S. Swinnea and B. D. Freeman: *Polymer*, 2008, **49**, 757–773.
9. B. Wetzel, F. Haupert, K. Friedrich, M. Q. Zhang and M. Z. Rong: *Polym. Eng. Sci.*, 2002, **42**, 1919–1927.
10. R. J. Nussbaumer, W. R. Caseri, P. Smith and T. Tervoort: *Macromol. Mater. Eng.*, 2003, **288**, 44–49.
11. J. M. Pochan: *Macromolecules*, 2007, **40**, 290–296.
12. A. Vermogen, K. Masenelli-Varlot, R. Séguéla, J. Duchet-Rumeau, S. Boucard and P. Prele: *Macromolecules*, 2005, **38**, 9661–9669.
13. W. Lertwimolnun and B. Vergnes: *Polym. Eng. Sci.*, 2006, **46**, 314–323.

14. G. Zheng, J. Wu, W. Wang and C. Pan: *Carbon*, 2004, **42**, 2839–2847.
15. H. Kim and C. W. Macosko: *Macromolecules*, 2008, **41**, 3317–3327.
16. P. Steurer, R. Wissert, R. Thomann and R. Mülhaupt: *Macromol. Rapid Commun.*, 2009, **30**, 316–327.
17. H. Kim, S. Kobayashi, M. A. AbdurRahim, M. J. Zhang, A. Khusainova, M. A. Hillmyer, A. A. Abdala and C. W. Macosko: *Polymer*, 2011, **52**, 1837–1846.
18. D. Cho, S. Lee, G. Yang, H. Fukushima and L. T. Drzal: *Macromol. Mater. Eng.*, 2005, **290**, 179–187.
19. R. Eller: 'Current states and future potential for polyolefins, TPOs, and TPEs in the global automotive market'; Int. Polyolefins Conf., Houston, TX, USA, February 2012, Society of Plastics Engineers.
20. H. -D. Huang, P. -G. Ren, J. Chen, W. -Q. Zhang, X. Ji and Z. -M. Li: *J. Memb. Sci.*, 2012, **409-410**, 156–163.
21. H. Kim, A. A. Abdala and C. W. Macosko: *Macromolecules*, 2010, **43**, 6515–6530.
22. X. Jiang and L. T. Drzal: *Compos. A Appl. Sci. Manuf.*, 2011, **42**, 1840–1849.
23. K. Kalaitzidou, H. Fukushima, P. Askeland and L. T. Drzal: *J. Mater. Sci.*, 2007, **43**, 2895–2907.
24. J. J. George, S. Bhadra and A. K. Bhowmick: *Polym. Compos.*, 2009, **31**, 218–225.
25. S. C. Tjong, G. D. Liang and S. P. Bao: *Scripta Mater.*, 2007, **57**, 461–464.
26. M. Modesti, A. Lorenzetti, D. Bon and S. Besco: *Polym. Degrad. Stab.*, 2006, **91**, 672–680.
27. J. Liang, Y. Wang, Y. Huang, Y. Ma, Z. Liu, J. Cai, C. Zhang, H. Gao and Y. Chen: *Carbon*, 2009, **47**, 922–925.
28. M. B. Bryning, M. F. Islam, J. M. Kikkawa and A. G. Yodh: *Adv. Mater.*, 2005, **17**, 1186–1191.
29. H. Kim, H. Thomas Hahn, L. M. Viculis, S. Gilje and R. B. Kaner: *Carbon*, 2007, **45**, 1578–1582.
30. J. M. Torkelson and K. Wakabayashi: 'Polymer-graphite nano-composites via solid-state shear pulverization', US Patent 7906053 B1, published 15 March 2011.
31. K. Wakabayashi, P. J. Brunner, J. Masuda, S. A. Hewlett and J. M. Torkelson: *Polymer*, 2010, **51**, 5525–5531.
32. W. R. Caseri: *Mater. Sci. Technol.*, 2006, **22**, 807–817.
33. D. Bikiaris: *Materials*, 2010, **3**, 2884–2946.
34. K. Wakabayashi, M. B. Stephen and M. D. Boches: 'Process for producing exfoliated and/or dispersed polymer composites and/or nanocomposites via solid-state/melt extrusion (ssme)', US patent 0113135 A1, published 9 May 2013.
35. X. Jiang and L. T. Drzal: *J. Appl. Polym. Sci.*, 2012, **124**, 525–535.
36. K. Wang, S. Liang, R. Du, Q. Zhang and Q. Fu: *Polymer*, 2004, **45**, 7953–7960.
37. C. Bartholome, E. Beyou, E. Bourgeat-Lami, P. Cassagnau, P. Chaumont, L. David and N. Zydowicz: *Polymer*, 2005, **46**, 9965–9973.
38. M. Härtelt, H. Riesch-Oppermann and O. Kraft: *Microsyst. Technol.*, 2011, **17**, 325–335.
39. I. Manas-Zloczower, A. Nir and Z. Tadmor: *Rubber Chem Technol.*, 1982, **55**, 1250–1285.
40. H. Rumpf: 'The strength of granules and agglomerates', in 'Agglomeration', (ed. W. A. Knepper), 379–413; 1962, New York, NY, Wiley-Interscience.
41. C. A. Schneider, W. S. Rasband and K. W. Eliceiri: *Nat. Methods*, 2012, **9**, 671–675.
42. H. A. Schwartz: *Met. Alloy.*, 1934, **5**, 139.
43. R. T. DeHoff and F. N. Rhines: 'Quantitative microscopy'; 1968, New York, NY, McGraw-Hill Book Company.
44. S. A. Saltykov: 'The determination of the size distribution of particles in an opaque material from a measurement of the size distribution of their sections', in 'Proc. Second Int. Cong. Stereol', 163–173; 1967, Berlin, Germany, Springer Berlin Heidelberg.
45. S. A. Saltykov: 'Stereometric metallography', 2nd edn; 1958, Moscow, Metallurgizdat.
46. E. E. Underwood: 'Quantitative stereology'; 1970, Reading, MA, Addison-Wesley Publishing Company.
47. A. Nir and A. Acrivos: *J. Fluid Mech.*, 2006, **59**, 209.
48. G. R. McDowell: *Soils Found.*, 2002, **42**, 139–145.
49. G. R. McDowell and J. P. de Bono: *Géotechnique Lett.*, 2013, **3**, 166–172.
50. G. I. Taylor: 'Proc. R. Soc. London', 501–523; 1934.
51. U. Sundararaj, C. W. Macosko, R. J. Rolando and H. T. Chan: *Polym. Eng. Sci.*, 1992, **32**, 1814–1823.
52. E. Van Hemelrijck, P. Van Puyvelde, S. Velankar, C. W. Macosko and P. Moldenaers: *J. Rheol*, 2004, **48**, 143.
53. L. Levitt and C. W. Macosko: *Macromolecules*, 1999, **32**, 6270–6277.
54. C. W. Macosko: *Macromol. Symp.*, 2000, **149**, 171–184.
55. F. Rumscheidt and S. Mason: *J. Colloid Sci.*, 1961, **16**, 238–261.
56. A. Addeo: 'Polypropylene handbook', 2nd edn; 2005, Munich, Carl Hanser Verlag.

Appendix 1: Degradation Studies

Degradation of iPP during MM and CMP was mitigated through the addition of antioxidant process stabilizers. Polypropylene degradation is known to occur through a complex process, usually involving oxidation of a tertiary carbons followed by β-scission.[56] Oxygen is critical to the initiation and propagation of radicals in iPP, and the con-figuration of the batch mixer exposes iPP to air during MM. Potential iPP degradation due to MM was evaluated with HTGPC, and the results are summarized in Table A1. Without antioxidant process stabilizers, the molecular weight of iPP treated with MM decreases relative to the virgin material, suggesting thermo-oxidative degradation occurred. For-tunately, iPP treated with process stabilizers shows the same molecular weight as the virgin material (Table A1). All samples are treated with antioxidants, in addition to the antioxidants present in the commercial resin.

Table A1   Degradation study of Melt-Masticated isotactic polypropylene

Samples	Mw (kg mol⁻¹)	Dispersity (Å)
Virgin iPP	134.6	3.5
Melt-Masticated iPP	52.5	2.5
Melt-Masticated iPP + Stabilizers[a]	133.1	3.4

[a] 0.025 wt-% Irganox1010 + 0.025 wt-% Irgafos 168

3

Nanocomposite hydrogels based on embedded PLGA nanoparticles in gelatin

Irmina Samba[1], Rebeca Hernandez[1], Nicoletta Rescignano*[1], Carmen Mijangos[1] and Josè Maria Kenny[1,2]

[1]Instituto de Ciencia y Tecnología de Polimeros, ICTP-CSIC, Juan de la Cierva 3, 28006 Madrid, Spain
[2]Materials Engineering Center, UdR INSTM, University of Perugia, Strada di Pentima 4 05100, Terni, Italy

Abstract Novel nanocomposites based on gelatin polymer networks containing poly(lactic-co-glycolide) (PLGA) nanoparticles (PLGA-NPs) were prepared and morphological, chemical and rheological properties were investigated. The successful incorporation of PLGA nanoparticles into the gelatin gels was confirmed by field emission scanning electron microscope (FESEM) and infrared spectroscopy (FT-IR). FESEM microscopy showed also a more homogeneous pore structure of the nanocomposite

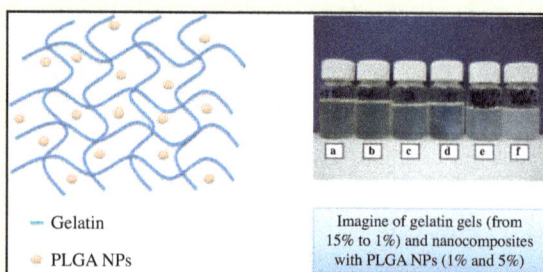

— Gelatin

— PLGA NPs

Imagine of gelatin gels (from 15% to 1%) and nanocomposites with PLGA NPs (1% and 5%)

with respect to the primary gel. The introduction of PLGA nanospheres at a 1% w/v with respect to gelatin weight does not influence the elastic modulus of the pristine gelatin but an increase in the amount of PLGA spheres determine a negative effect on the elastic modulus. The prepared PLGA-NPs gelatin gels with biocompatible and biodegradable properties are very interesting from both applied and fundamental perspectives for a future application in biomedical and food fields for the control release of active molecules.

Keywords Poly (DL-Lactide-co-Glycolide), Biodegradable nanoparticles, Gelatin, Gel and nanocomposite

Introduction

Gelatin is a natural polymer derived from collagen and is commonly used for pharmaceutical and medical applications because of its biodegradability[1] and biocompatibility in physiological environments.[2] The isoelectric point of gelatin can be modified during the fabrication process to yield either a negatively charged acidic gelatin or a positively charged basic gelatin at physiological pH. This property theoretically allows electrostatic interactions between a charged biomolecule and gelatin of the opposite charge, forming polyion complexes.[2] Various forms of gelatin carrier matrices were reported for controlled release applications while characterization studies have shown that gelatin carriers are able to sorb charged biomolecules such as proteins and plasmid DNA.[3,4] It has been shown that the crosslinking density of gelatin hydrogels affects their degradation rate in vivo leading to a similar profile for the biomolecule release rate from gelatin carriers, suggesting that complexed gelatin/biomolecule fragments are released by enzymatic degradation of the carrier.[5]

The most commonly used biodegradable synthetic polymers for particles formation in biomedical field are saturated poly(α-hydroxy esters), including poly(lactic acid) (PLA) and poly(glycolic acid) (PGA), as well as poly(lactic-co-glycolide) (PLGA) copolymers.[6,7] The chemical properties of these polymers allow hydrolytic degradation through de-esterification. Once degraded, the monomeric components of each polymer are removed by natural pathways. PGA is converted to metabolites or eliminated by other mechanisms, and PLA can be cleared through the tricarboxylic acid cycle. Owing to these properties PLA and PGA have been used in biomedical products and devices, which have been approved by the US Food and Drug Administration.[8] Generally, the co-polymer PLGA is preferred for the fabrication of bone substitute constructs, compared with its constituent homopolymers, as it offers superior control of the degradation properties by varying the ratio between its monomers. PLGA, for instance, has a wide range of degradation rates, governed by the composition of chains, both hydrophobic and hydrophilic balance and crystallinity.[8]

The use of nanocomposites based on a hydrogel matrix and biopolymeric nanoparticles is expanding significantly in recent years for control release applications.

---
*Correspondence author, email nicoletta@ictp.csic.es

Rescignano et al.[9] described a nanocomposite hydrogel prepared by the incorporation of biocompatible and biodegradable PLLA nanoparticles into semi-interpenetrating polymer hydrogels of a natural polymer such as alginate and thermosensitive poly(N-isopropylacrylamide). This kind of systems has been used to stem different aspects like the double control of release or the vehiculation of nanodevices towards the target of interest. In fact, the addition of polymer nanoparticles into polymer gels also allows the local delivery of hydrophobic drugs when these are loaded into nanoparticles.[10] These nanoparticles have several advantages over the conventional bulk hydrogels for controlling the release of high molecular weight biomolecules. The drug would diffuse out first through the nanoparticles and from them to the hydrogel network; then, they will be further released from the bulk hydrogel network. Therefore, compared with the bulk hydrogel or pure nanoparticles, in the case of a nanocomposite hydrogel, two barriers can control the drug release. The aim of the research reported here is the preparation and rheological characterization of natural gelatin gels containing PLGA nanoparticles. The viscoelastic properties of gelatin gels at different concentrations will be compared with the corresponding gelatin gels containing PLGA nanoparticles.

Experimental part

Materials

Poly(lactic-co-glycolic)acid with a 50/50 ratio (PLA/PGA) was obtained from Absorbables Polymers/Lactel (Durect Corporation, UK) (Mw 91600–120 000 g mol^{-1}). Polyvinyl alcohol (Mw 31 000–50 000 g mol^{-1}, 87–89% hydrolyzed) was used as surfactant and chloroform (CHCl$_3$) from Aldrich was used as solvent. The gelatin used was derived from porcin skin, type A, and was purchased from Sigma-Aldrich. Millipore water (resistivity 18.2 mΩ cm) was used for the preparation of PLGA nanospheres while distilled water (resistivity 15.0 mΩ cm) was used for the preparation of the gelatin gel.

Preparation of PLGA nanoparticles

PLGA nanoparticles were prepared by a double emulsion water/oil/water method with subsequent solvent evaporation. Around 0.125 g of PLGA were dissolved in 5 mL of chloroform for 2 h under magnetic agitation at room temperature. The solution was emulsified with 2.5 mL of Millipore water using a sonicator (SONICS Vibra Cell) at 30% amplitude for 15 min. A PVA aqueous solution (2% w/v) was prepared by dissolving about 0.8 g in 40 mL of Millipore water. This solution was added to the first emulsion and mixed for 15 min with sonication for the formation of the second emulsion. For the solvent evaporation, the second emulsion was transferred in 200 mL of 0.2% w/v PVA aqueous solution in Millipore water and was magnetically stirred over night at room temperature. The nanospheres were collected by centrifugation at 3500 rev min^{-1} for 30 min and then washed four times with Millipore water and finally lyophilized to get a powder.[6]

Preparation of gelatin gels with embedded PLGA nanoparticles

The gelatin was dissolved at various concentrations (5, 10 and 15% (w/v)) in distilled water, by stirring the system for 30 min at 50°C. Then, gels were prepared on Petri dishes (50 mm diameter) and maintained at 4°C for 18 h prior to analysis. For the preparation of nanocomposite hydrogels, aqueous suspensions of PLGA nanoparticles (1, 2.5 and 5% w/w PLGA with respect to gelatin weight) were dispersed into the gelatin aqueous solution (1% w/v) under constant stirring at 50°C during 30 min. Gels were prepared on Petri dishes (50 mm diameter) and maintained at 4°C for 18 h prior to analysis. Nanocomposite hydrogels were denoted as PLGA1-gelatin and PLGA5-gelatin where 1 and 5 denotes the concentration (% w/w) of PLGA with respect to the gelatin weight.

Attenuated total reflectance Fourier transform infrared spectroscopy (ATR-FTIR)

Experiments were made on lyophilized samples with a Perkin Elmer Spectrum One ATR-FTIR spectrometer over a diamond crystal. The wavenumber sweep was 4000–650 cm^{-1} with a resolution of 4 cm^{-1}.

Morphological characterization

The morphology of the polymeric nanocomposite and nanoparticles was analyzed by field emission scanning electron microscopy (FESEM Supra 25, Zeiss, Germany). Samples were deposited onto FTO (fluorine-doped tin oxide) substrates using a drop casting method, allowing them to dry at room temperature for 24 h and gold coated by an Agar automatic sputter coating.

Rheological characterization

Rheological properties of gels were carried out in a Rheometer AR-1000 using parallel plates of 20 mm. All measurements were carried out at $T = 10$°C. The linear viscoelasticity region was determined by means of oscillatory torque sweeps carried out between 0.1 and 1000 µN m^{-1} at a frequency of 1 Hz frequency sweeps between 10 and 0.1 Hz at a fixed torque in the linear viscoelastic region.

Results and discussion

Morphology of nanocomposite gels

Figure 1 a–e shows the FESEM images corresponding to cross-sections performed through the thickness of the gelatin 1% w/v and the nanocomposite hydrogel, PLGA5-gelatin. Figure 1f reports an image of the synthesized PLGA NPs. As previously reported,[6] PLGA nanoparticles show a spherical shape. No aggregation phenomena after drying were observed; the average diameter was around 130 nm as previously demonstrated by Rescignano et al.[6] A cross-linked morphology can be observed in the images of pure gel and nanocomposites; however, the gelatin 1% w/v is characterized by a more homogeneous and dense pore structure than PLGA5-gelatin. This might suggest that the presence of PLGA nanoparticles interferes somehow with the formation of the gelatin hydrogel as it will be ascertained through rheological measurements. Figure 1d shows the image of the

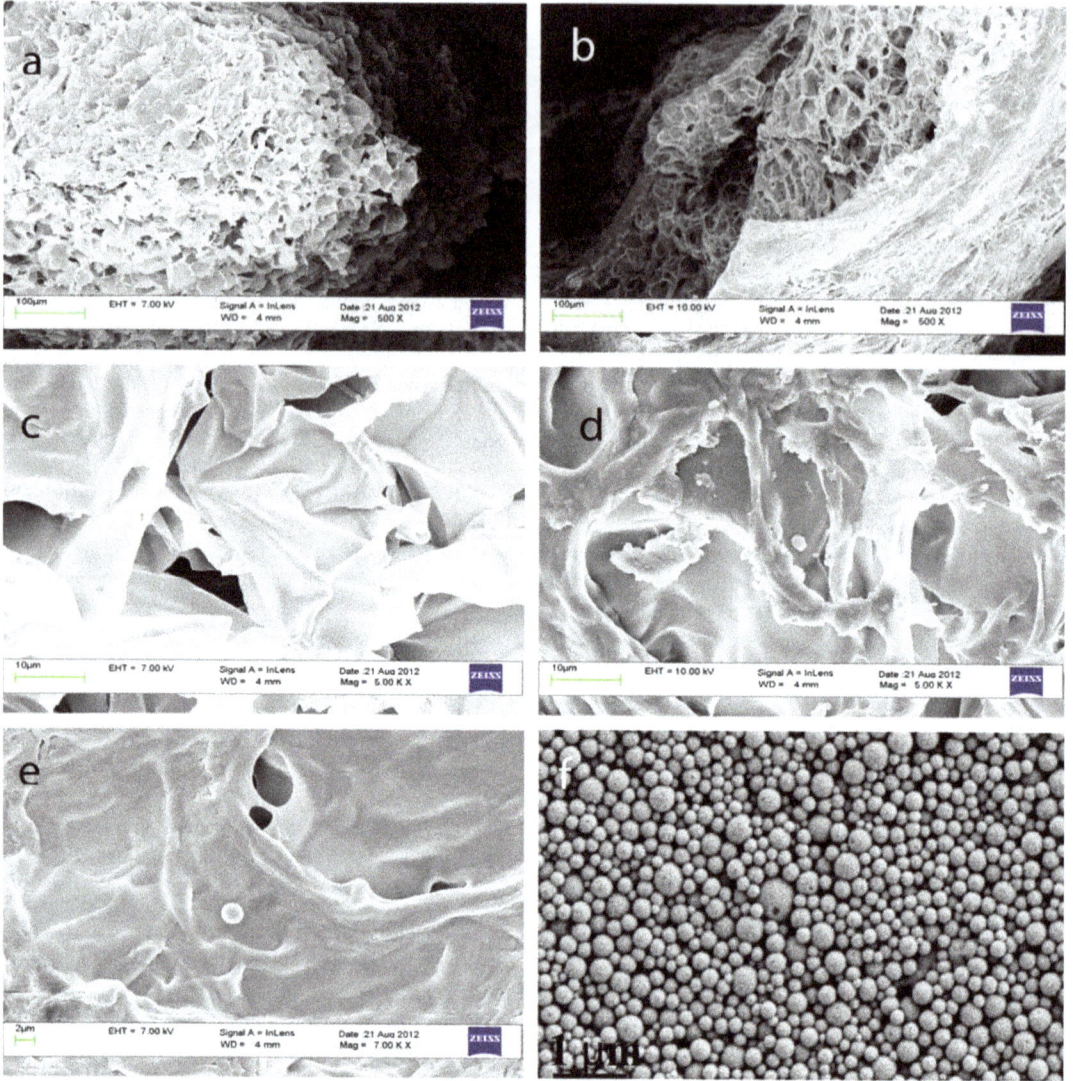

Figure 1 FESEM images of *a*, *c* gelatin gel 1% w/v and *b*, *d*, *e* gelatin gel with 5% PGLA NPs, *f* shows synthesized PLGA-NPs

cross-section of PLGA5-gelatin; several NP aggregates were embedded in the hydrogel matrix. Figure 1e shows a magnification image in which a single PLGA nanoparticle can be observed; however, the degree of dispersion in the gelatin matrix could not be assessed from this image.

Chemical characterization by ATR-FTIR

Figure 2 shows the ATR-FT-IR spectra of pure PLGA, gelatin (1% w/v) and two nanocomposite hydrogels prepared at a fixed gelatin concentration (1% w/v) and two different PLGA-NPs concentrations, 1% w/w and 5% w/w. The two nanocomposite hydrogels present the characteristic bands of gelatin (red spectra in Fig. 2), a band situated around $3288\ cm^{-1}$ corresponding to the NH-stretching coupled with hydrogen bonding of the water molecule O—H; C=O stretching around $1630\ cm^{-1}$ for the amide I; the bending vibration of N—H groups and stretching vibrations of C—N

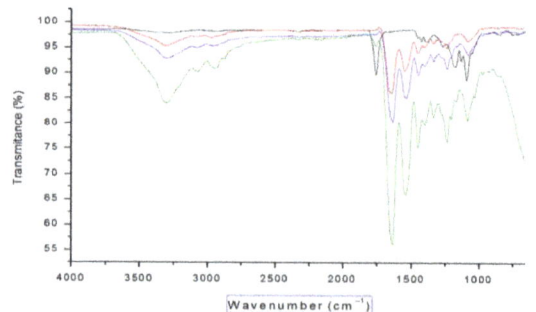

Figure 2 ATR-FTIR spectra corresponding to pure PLGA (black); gelatin 1% (w/v) (red); PLGA1-gelatin (blue) and PLGA5-gelatin (green) nanocomposite hydrogels

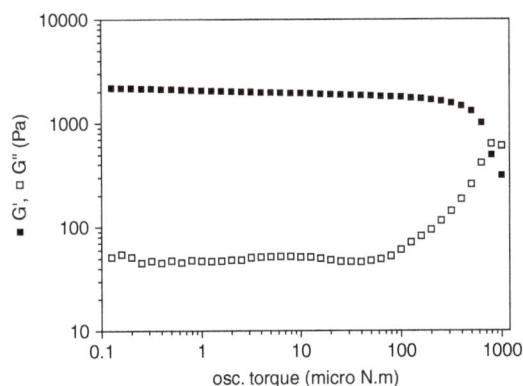

Figure 3 Evolution of G′(■) and G‴(□) as function of oscillatory torque for gelatin [10% (w/v)]

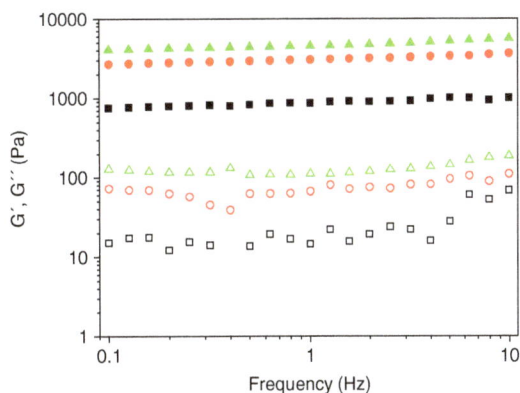

Figure 4 Evolution of G′ (closed symbols) and G″ (open symbols) as function of frequency for gelatin prepared at different concentrations (■) 5% w/v; (●) 10 w/v and (▲) 15% w/v

groups around 1539 cm^{-1} for the amide II; the vibrations in the plane of C—N and N—H groups of bound amide around 1237 cm^{-1}.[11]

Regarding the spectra corresponding to pure PLGA, the characteristic peak located at 1748 cm^{-1} can be attributed to the absorbance of the carbonyl group C=O.[12] This peak was visible only in the spectra corresponding to the nanocomposite hydrogel containing the highest concentration of PLGA nanoparticles (5% w/w). For the nanocomposite hydrogel containing 1% w/w of PLGA nanoparticles, it was not possible to ascertain the presence of PLGA NPs by this technique.

Rheological characterization

The linear viscoelastic region is observed in the low oscillation stress region where both, the elastic modulus and the loss modulus, are independent of the oscillation stress. As a representative example (Fig. 3), the evolution of the elastic and loss modulus with the oscillation torque is shown for gelatin at a 10% w/v concentration. The linear viscoelastic region extends from 0.1 and 100 Pa. A value of 10 Pa was chosen to carry out the rest of the rheological measurements.

As a first step, frequency sweeps were carried out for pristine gelatin gels at different concentrations (5, 10 and 15% w/v) and the results are shown in Fig. 4. It can be observed that, in all cases, the elastic modulus (G′) is higher than the loss modulus (G″) and both remain independent of frequency; this is characteristic of gel behavior.[13]

Figure 4 shows a representation in double logarithmic scale of the values of elastic modulus extracted from frequency sweeps at a frequency = 1 Hz as a function of gelatin concentration. It can be observed that the elastic moduli of gelatin hydrogels increases with the gelatin concentration. In previous studies,[14–16] the viscoelastic results obtained for polymer gels have been analyzed using a theoretical model based on a scaling approach which relates modulus to concentration through the following type of equation

$$G \approx C^n \tag{1}$$

where n is an exponent which depends upon the conformation of the chain linking junction points and is related to the fractal dimension ν^{-1} of the object between the

junctions through the following equation.[17]

$$G \approx C^{3\nu/(3\nu-1)} \tag{2}$$

Fitting the data represented in Fig. 5 for gelatin hydrogels to equation (2) yields a fractal dimension, $\nu^{-1} \sim 1.5$, which is characteristic of rod-like chains. This result is in agreement with published results regarding gelatin gels considered as entangled networks of rigid rods possibly connected by flexible links.[18]

For the preparation of nanocomposite hydrogels, a concentration of 1% w/v of gelatin was chosen. Figure 6 shows the elastic modulus of gelatin 1% w/v and its nanocomposite hydrogels PLGA1-gelatin and PLGA5-gelatin. It can be observed that, for both nanocomposite hydrogels, G′ is relatively independent from the frequency applied in the studied range so that the incorporation of PLGA nanospheres in gelatin gels does not change the typical gel behavior.

The inset in Fig. 6 shows a representation of the elastic modulus as a function of the PLGA nanoparticles. The introduction of PLGA nanospheres at a 1% w/w with respect to gelatin weight does not influence the elastic modulus of the pristine gelatin. However, an increase of the

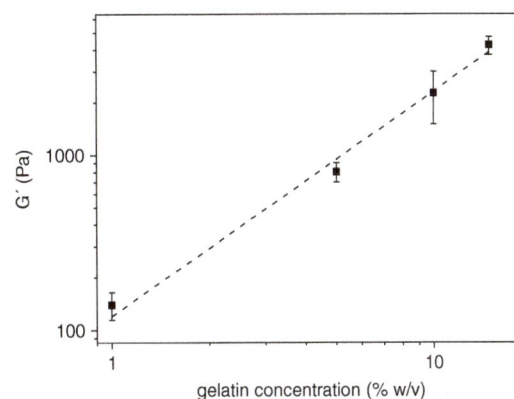

Figure 5 Elastic modulus G′ as function of gelatin concentration: dashed line represents fitting of data to equation (1)

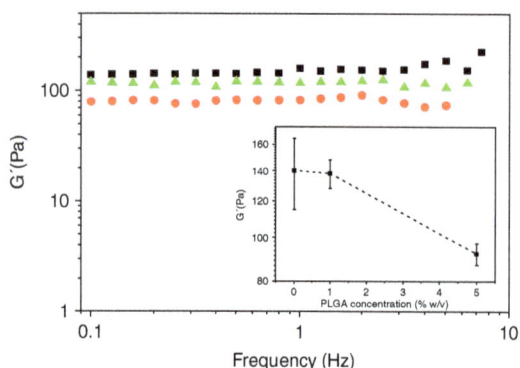

Figure 6 Evolution of G′ as function of frequency for gelatin 1% w/v (■); PLGA1-gelatin (▲) and PLGA5-gelatin (●): inset shows representation of elastic modulus as function of PLGA concentration

concentration of PLGA nanospheres in the gelatin gel to 5% w/w significantly decreases the elastic modulus of the nanocomposite hydrogel with respect to gelatin 1% w/v. This might indicate that the incorporation of PLGA nanoparticles inside gelatin prevents the formation of the gel above a certain concentration, and hence, the degree of crosslinking is lower resulting in a lower elastic modulus.

Conclusions

The inclusion of PLGA nanoparticles in gelatin hydrogels was successfully performed and characterized. The FESEM images revealed differences in the porous structure obtained. Specifically, the presence of PLGA NPs induces a more homogeneous pore structure of the hydrogel nanocomposite than in the case of the neat gel.

The introduction of PLGA nanospheres at 1% w/v with respect to gelatin weight does not influence the elastic modulus of the pristine gelatin but an increase of the amount of PLGA spheres (5% w/v) influences negatively the elastic modulus of the nanocomposite hydrogel with respect to the gelatin 1% w/v.

The successful introduction of PLGA NPs, in the gelatin hydrogel, provides a good tool for future applications in controlled drug release systems and in food technologies

since biodegradable polymeric nanoparticles are able to encapsulate different natural and biological systems and release them in a complex and controlled way.

Conflicts of Interest

The authors have no conflicts of interest to declare.

Acknowledgements

Authors would like to thank the European Genis Lab Project for the financial support.

References

1. J. Vandervoort and A. Ludwig: *Eur. J. Pharm. Biopharm.*, 2004, **57**, (2), 251–261.
2. S. Young, M. Wong, Y. Tabata and A. G. Mikos: *J. Controll. Rel.*, 2005, **109**, (1–3), 256–274.
3. Y. Fukunaka, K. Iwanaga, K. Morimoto, M. Kakemi and Y. Tabata: *J. Controll. Rel.*, 2002, **80**, (1–3), 333–343.
4. M. Yamamoto, Y. Takahashi and Y. Tabata: *Biomaterials*, 2003, **24**, (24), 4375–4383.
5. M. Yamamoto, Y. Ikada and Y. Tabata: *J. Biomater. Sci. Polym. Ed.*, 2001, **12**, (1), 77–88.
6. N. Rescignano, M. Amelia, A. Credi, J. M. Kenny and I. Armentano: *Eur. Polym. J.*, 2012, **48**, (7), 1152–1159.
7. J. M. Anderson and M. S. Shive: *Adv. Drug Deliv. Rev.*, 2012, **64**, (Suppl. 0), 72–82.
8. B. D. Ulery, L. S. Nair and C. T. Laurencin: *J. Polym. Sci. B: Polym. Phys.*, 2011, **49**, (12), 832–864.
9. N. Rescignano, R. Hernández, I. Armentano, D. Puglia, C. Mijangos and J. M. Kenny: *Polym. Degrad. Stabil.*, 2014, **108**, 280–287.
10. H. Laroui, G. Dalmasso, H. T. T. Nguyen, Y. Yan, S. V. Sitaraman and D. Merlin: *Gastroenterology*, 2010, **138**, (3), 843–853.e842.
11. Z. A. Nur Hanani, Y. H. Roos and J. P. Kerry: *Food Hydrocoll.*, 2012, **29**, (1), 144–151.
12. S. Manoochehri, B. Darvishi, G. Kamalinia, M. Amini, M. Fallah, S. N. Ostad, F. Atyabi and R. Dinarvand: *Daru*, 2013, **21**, (1), 58.
13. G. M. Kavanagh and S. B. Ross-Murphy: *Progr. Polym. Sci.*, 1998, **23**, (3), 533–562.
14. D. López, C. Mijangos, M. E. Muñoz and A. Santamaría: *Macromolecules*, 1996, **29**, (22), 7108–7115.
15. R. Hernández, V. Zamora-Mora, M. Sibaja-Ballestero, J. Vega-Baudrit, D. López and C. Mijangos: *J. Colloid Interf. Sci.*, 2009, **339**, (1), 53–59.
16. R. Hernández, J. Sacristán and C. Mijangos: *Macromol. Chem. Phys.*, 2010, **211**, (11), 1254–1260.
17. P. G. de Gennes: 'Scaling concepts in polymer physics'; 1979, New York, Cornell University Press.
18. C. Joly-Duhamel, D. Hellio, A. Ajdari and M. Djabourov: *Langmuir*, 2002, **18**, (19), 7158–7166.

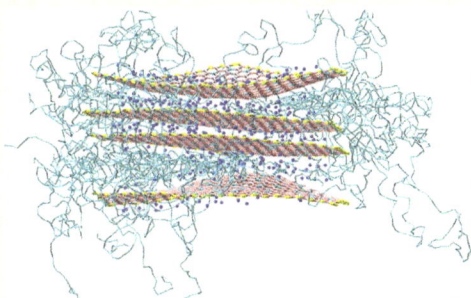

Preparation and characterization of poly(ethylene terephthalate)/hyperbranched polymer nanocomposites by melt blending

Mina Ahani, Marziyeh Khatibzadeh* and Mohsen Mohseni

Department of Polymer Engineering and Color Technology, Amirkabir University of Technology, Tehran, Iran

Abstract In this paper, nanocomposites of polyethylene terephthalate (PET) including polyesteramide-based hyperbranched polymer, or PET/Hyperbranched polymer nanocomposites, have been prepared via melt blending method with different hyperbranched polymer contents. In addition, morphology, surface structure, and thermal properties of these nanocomposites and virgin PET were studied by atomic force microscopy, attenuated total reflection fourier transform infrared spectroscopy and differential scanning calorimetry (DSC), respectively. Dynamic mechanical analysis experiments in solid state were carried out to follow the effect of hyperbranched polymer on the dynamic mechanical properties of these nanocomposites. The structure of the nanostructured hyperbranched polymer was also studied by small-angle X-ray scattering. The rheometric mechanical spectroscopy results showed that the hyperbranched polymer as a modifier decreased the complex viscosity and enhanced liquid-like behavior. This happened more significantly by increasing the content of hyperbranched polymer. The DSC analysis results revealed that crystallinity and glass transition temperature decreased by adding the amount of hyperbranched polymer.

Keywords Polyethylene terephthalate (PET), Hyperbranched polymer, Nanocomposite, Melt blending

Introduction

Polymer nanocomposites are a new class of engineering materials and a method of enhancing polymer properties using nanoscale modifiers, which have found many applications in various industrial fields such as automotive, construction, and packaging.[1] Nanocomposites are a combination of two or more phases containing different compositions or structures, where at least one of the phases is in the nanoscale regime. They show extraordinary advantages in mechanical, thermal, optical, and chemical properties in comparison with virgin polymers or conventional macro- and microcomposites. These enhancements have attributed to the small size and high surface-to-volume ratio of the nanoscale materials.[2,3] Melt intercalation (blending),[4] solgel,[5] and *in situ* polymerization[6] are the different methods to prepare polymer nanocomposites. The melt blending method is the most useful approach for industrial applications due to the absence of environmentally harmful solvents, and compatibility with current industrial

compounding and processing techniques. In this method, polymer and modifier are combined in internal batch mixer or in a twin screw extruder under shear and above the softening temperature of the polymer.[7]

Poly(ethylene terephthalate) (PET) is a linear polyester widely used in different applications such as food and beverage packing, textile fiber production, automotive components, and molded goods because of its relatively good physical, mechanical, barrier, and optical properties.[8,9] Despite many excellent properties of PET, the disadvantages such as the lack of functional groups, hydrophobic nature, highly compact molecular structure, and relatively high crystallinity have limited its application. These deficiencies result in low dyeability[10,11] as an important parameter in textile usages. Physical blending of PET with other polymers or additives containing functional groups with lower crystallinity has been the most frequent method of modification.[12] One of these approaches is incorporation of hyperbranched polymer as a modifier into the PET matrix in order to improve dyeability of PET.[12] Apart from the improvement in dyeability, the blend of this hyperbranched polymer with PET attained higher adsorption properties due to the formation of free volume between chains of

*Corresponding author, email khatib@aut.ac.ir

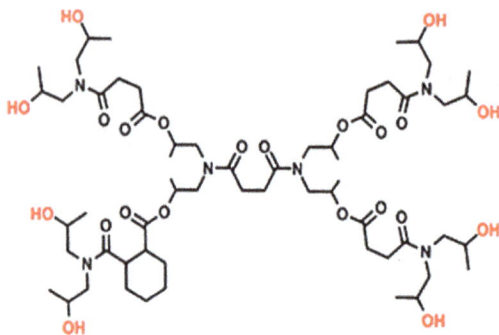

Figure 1   Chemical structure of Hybrane H1500

PET that is related to the presence of hyperbranched polymer as a modifier.[13,14] Hyperbranched polymers exhibit characteristics that are very different from those of linear polymers, for example, low viscosity, high solubility, and high density of the functional end-groups.[15] These particulate polymers provide various kinds of functional end groups, from hydrophilic to hydrophobic and from reactive to non-reactive.[16,17] The presence of numerous terminal groups is a typical characteristic of hyperbranched polymers. The terminal groups can affect the properties of the polymer and make such polymers attractive candidates for many material applications including targeted drug delivery, rheology modifiers, plasticizers, supporting material for catalysts, and versatile scaffolds for further synthesis.[18,19] The content of the terminal groups in the hyperbranched polymer is a key index for polymer grafting, drug conjugation and catalyst supports.[20] These polymers also have the ability of changing and disordering the crystalline structure of PET leading to change in physical and mechanical properties. Hence, it is a matter of great importance to study the effect of the modification process on the physical and mechanical properties of the final blends.

We have previously reported[13,14] the adsorption properties of a disperse dye toward PET sheets modified using various loads of a hyperbranched polymer in terms of thermodynamic and kinetic parameters. According to these findings, the adsorption of the dye on the modified PET sheets was much higher than that of the virgin PET one. Eventually, the results indicated that the color yield of the virgin PET was improved by incorporating hyperbranched polymer.

In this study, PET nanocomposites with hyperbranched polymer were prepared via melt blending method using an internal mixer and the viscoelastic behavior, thermal and dynamic mechanical properties of these nanocomposites and virgin PET were investigated. The aim was to demonstrate the effect of adding various amounts of hyperbranched polymer on the morphology, thermal and dynamic mechanical properties of PET/hyperbranched polymer nanocomposites.

Experimental

Materials

In this work, polyethylene terephthalate chips (Zimmer Specification, Germany) with an intrinsic viscosity of 0.60 dl g^{-1}, water content of 0.25 wt %, and melting point of 250 °C were

purchased from Shahid Tondgoyan Petrochemical Company, Mahshahr, Iran.

Hyperbranched polymer (Hybrane H1500, M_w = 1500, designated as H, possessing only hydrophilic hydroxyl end groups) was kindly supplied by DSM Company. Fig. 1 shows the schematic structure of this polymer.

All chemical reagents were of analytical grade and were used without further purification.

Preparation of nanocomposites via melt blending method

In this study, all the nanocomposites including 0.5, 1, 2, and 3 wt % of Hybrane H1500 with PET, which will be designated as PET + 0.5H, PET + 1H, PET + 2H, and PET + 3H were formed by melt blending method. Prior to any processing, all materials were dried for 24 h at 80 °C in an oven. Mixing was carried out using an internal mixer (Brabender Plasticorder W50, Germany), equipped with a Banbury type rotor design, which was maintained at a temperature of 250 °C, and a rotor speed of 60 rpm. Mixing continued until the rotor torque reached a constant value. The overall time of the mixing process was about five minutes. After having discharged, sheets with one mm in thickness were prepared by hot pressing the blended mixtures. Each sample was pressed for five minutes at a pressure of 140 bar and a temperature of 250 °C. The press was subsequently cooled to room temperature using cold water.

Characterization

Atomic force microscopy

Tapping mode atomic force microscopy (AFM) (Ambios Tech (USA)) was used to characterize the samples. A scanning probe microscope was operated under ambient conditions with commercial silicon microcantilever probes. Manufacturer's values for the probe tip radius and probe spring constant were in the ranges of 5–10 nm and 20–100 N/m, respectively. Phase images were obtained using a resonance frequency of approximately 300 kHz for the probe oscillation and a free-oscillation amplitude of 60 ± 5 nm. This technique was used to study the homogeneity of the PET/H nanocomposites.

Attenuated total reflectance

Attenuated total reflectance (ATR-FTIR) measurements on samples were carried out on a Bomem Hartman & Braun FTIR Spectrophotometer. The samples were analyzed in the reflectance mode in the range of 400–4000 cm^{-1} at room temperature.

Differential scanning calorimetry

Thermal properties and crystallinity of virgin PET and PET/H nanocomposites were determined by a Mettler Toledo differential scanning calorimeter instrument under N$_2$ atmosphere. The sample pan and the reference pan were heated from 0 to 300 °C at the rate of 10 °C/min. The glass transition temperature (T_g), the melting point (T_m), the cold crystallization (ΔH_c), and the heat of melting (ΔH_m) were determined from the differential scanning calorimetry (DSC) thermograms. The glass transition temperature and melting point were obtained from the midpoint of the transitions. The heat of melting, ΔH_m, and cold crystallization, ΔH_c, were determined by integrating the areas under the peaks. The percent crystallinity was then calculated from the following relation:

$$\text{Crystallinity\%} = (\Delta H_{\text{m}} - \Delta H_{\text{c}})/\Delta H_{100\%} \qquad (1)$$

where heat of melting and cold crystallization are in J g^{-1}. The term $\Delta H_{100\%}$ (135.8 J g^{-1}) was a reference value and represented the heat of melting of a 100% crystalline PET.[21]

Dynamic mechanical thermal analysis in the melt state (rheometric mechanical spectrometry)

A MCR300 rheometric mechanical spectrometer was utilized to perform the measurements in the melt state. The temperature was adjusted to 200 °C to equal the processing temperature. Parallel plate geometry was used and deformation was applied on the molten samples in the frequency range 0.1–500 s^{-1}. The strain was kept at 10% to keep the material in its linear viscoelastic region. Preliminary experiments confirmed the linearity.

DMA in the solid state (dynamic mechanical thermal analysis)

Dynamic mechanical thermal analysis (DMA) was carried out by means of a Tritec2000 in rotational mode, in order to study the thermal and mechanical properties of samples. Deformation was applied at a frequency of 1 Hz and strain of 0.1%. The preliminary tests confirmed that the selected strain was in the linear viscoelastic zone. The temperature was increased at a rate of 5 °C/min from 0 to 250 °C.

Small-angle X-ray scattering

Small-angle X-ray scattering (SAXS) was used in order to study the size and shape of nanoparticles. SAXS works on the basis of difference in electronic density in materials. On the basis of this difference, X-ray beam in passing through material scattered in small angles. SAXS measurements were carried out using a S3-MICRO diffractometer (Hecus, Austria) equipped with a copper radiation source ($\lambda = 0.154$ nm) operating at 40 kV and 49 mA. The beam was monochromatized by a silicon monochromator and collimated by a set of slits defining pinhole geometry. A one-dimensional position sensitive detector was used to record the SAXS intensity as function of the modulus of the scattering vector. The data were recorded in the reflection mode over a 2θ range of about 0–10° at a rate of 0.04 °C/min.

Results and discussion

Morphological analysis: atomic force microscopy results

AFM is one of the most powerful techniques for nanoscopic characterization of surfaces. Usually, obtaining information on the sample's morphology is straightforward, since AFM images primarily contain topographical information.[22] The AFM images which reported in Fig. 2 are the original scans without any filtering. The phase contrast image for virgin PET sheet showed the smooth and uniform surface without any nanoparticle. For nanocomposites using 1 and 2 wt % hyperbranched polymer, a few numbers of light-colored particles are observed indicative of the existence of hyperbranched

polymer. These observations confirm that no aggregates are formed and there seems a complete homogeneous surface. However, some aggregations are seen in nanocomposite containing 3 wt % hyperbranched polymer. Accordingly, the PET/H nanocomposites containing up to 2 wt % hyperbranched polymer are more homogenous.

Structural studies

The ATR-FTIR spectra was recorded to assess structural changes, if any, made in the blended samples of the alteration of existing functional groups as a consequence of the presence of hyperbranched polymer. Fig. 3 shows the spectra of the virgin PET sheet (Fig. 3b) and those of the pure hyperbranched polymer (Fig. 3a) as well as the blended samples. The ATR-FTIR results show that the presence of hyperbranched polymer is notable in all the nanocomposites. In almost all cases, the presence of the hyperbranched polymer is proved by the presence of the peaks at 1400–1500 and 1600–1700 cm^{-1} that is related to amide C=O stretching.

Moreover, virgin PET sheet shows a sharp peak for OH-stretching at 3400 cm^{-1} due to the presence of terminal hydroxyl groups in the chain. After blending, the intensity of this peak increased reflecting the increase in hydroxyl groups in the treated samples. Similarly, a broad peak at 3500–3600 and 3600–3770 cm^{-1} also increased in the treated samples, which indicates the enhanced hydrogen bonding among hydroxyl moieties existed in the samples containing hyperbranched polymer.

Crystallinity and glass transition temperature: differential scanning calorimetry results

The thermal properties of the blended PET samples may be changed when hyperbranched polymer as an additive was added to the virgin PET. This change can be possibly used to probe the existence of the hyperbranched polymer in the blended PET samples. Fig. 4 presents DSC diagrams for the virgin and blended PET samples during heating at a rate of 10 °C/min. It can be found from Fig. 5 that the glass transition temperature of virgin PET sheet was 80.07 °C and changed to 76.7, 73.96 and 72.49 °C when 0.5, 1 and 2 wt % hyperbranched polymer was added the virgin PET, respectively. It is concluded that as the content of hyperbranched polymer increased, the glass transition temperature decreased. This means that samples became more flexible upon addition of the amorphous hyperbranched polymer. For further investigations, the crystallinity for each sample was calculated from the heat of fusion during heating as shown in Fig. 5 and it decreased with the addition of hyperbranched polymer content in the virgin PET. As shown in Fig. 5, the crystallinity had reached 44% in virgin PET sheet. When 0.5, 1, and 2 wt % hyperbranched polymer was added to the virgin PET sheet, the crystallinity content of PET dropped to 41.65, 40, and 38%, respectively.

Dynamic mechanical thermal analysis in the melt state: RMS results

Fig. 6 shows diagrams of the complex viscosity η^* and storage modulus G' versus frequency of applied deformation, ω, for the virgin and blended PET samples. It is clear from the

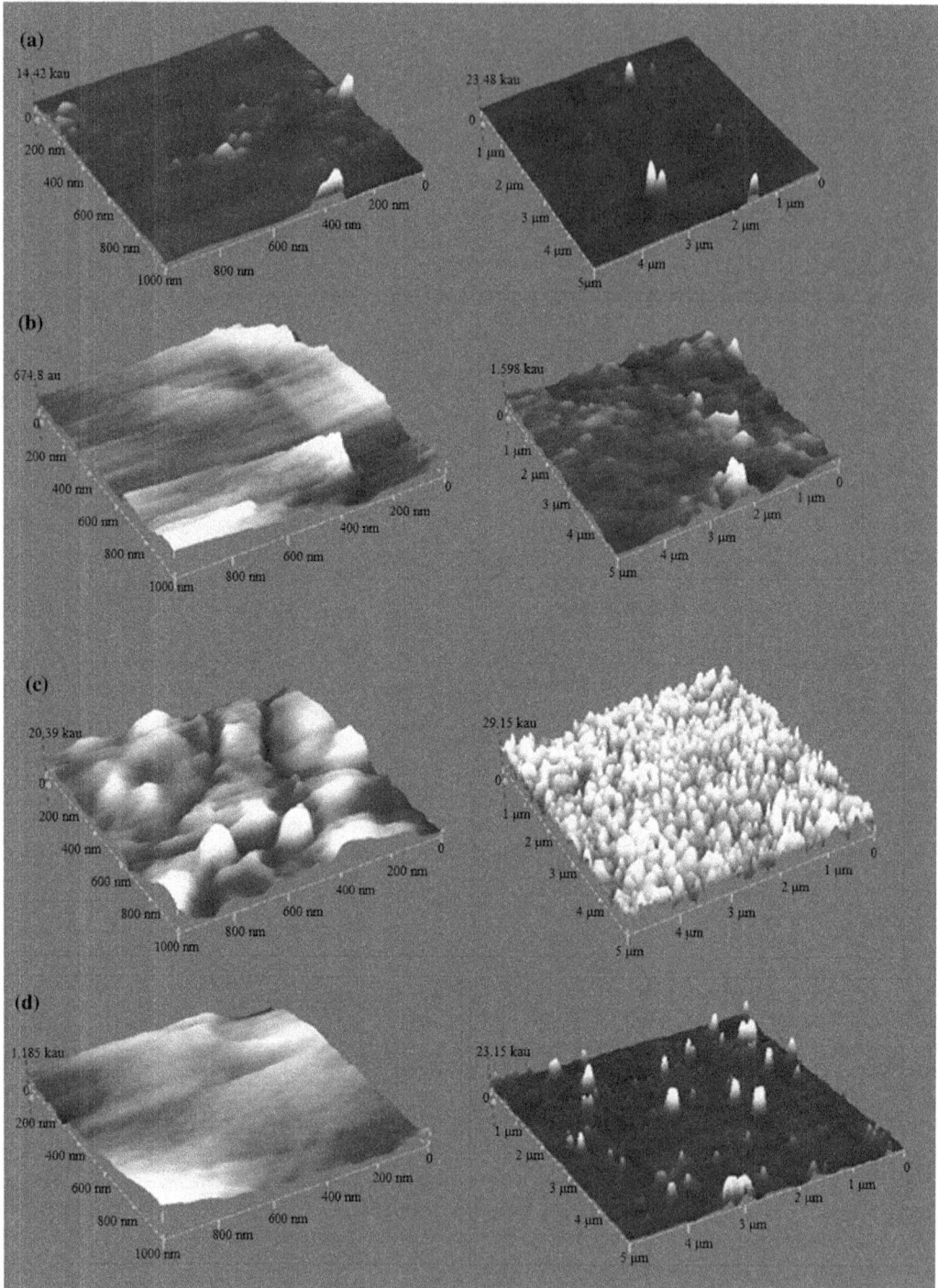

Figure 2   *a–d* AFM images *a* virgin PET, *b* PET + 1H, *c* PET + 2H, and *d* PET + 3H (right 1micron and left 5 micron)

diagrams that virgin PET showed higher viscosity and modulus values compared with blended samples. The hyperbranched polymers as a subgroup of dendritic polymers were very well known for their reducing effect on viscosity. Different mechanisms have been proposed by researchers among which the ball-bearing effect and reducing free volumes between the matrix chains can be mentioned.[23] In one study, the reduction of viscosity in blended samples with hyperbranched polymer has been attributed to two factors. The first factor involved the interference of physical entanglements between linear

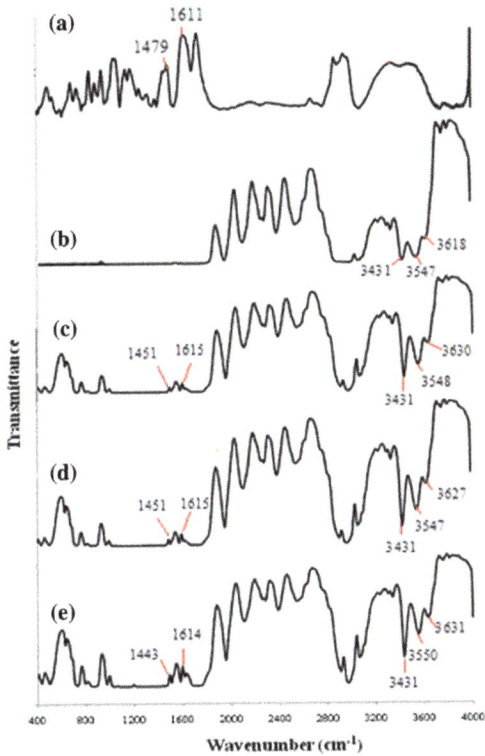

Figure 3 *a–e* ATR-FTIR spectra of virgin PET *b*, pure hyperbranched polymer *a*, PET + 0.5H *c*, PET + 1H *d*, and PET + 2H *e*.

polymers by the globular-shaped hyperbranched polymers. The second factor is related to the low intrinsic viscosity of

hyperbranched polymers due to their packed structure.[24] The hyperbranched polymers also exhibit Newtonian behavior, indicating the absence of physical entanglements in these systems.[25] Therefore, it is predicted that by adding hyperbranched polymer to PET, Newtonian behavior can be manifested. As can be seen in Fig. 6*a*, PET + 2H sample shows Newtonian behavior, that is, a constant viscosity, while virgin PET and PET + 1H samples show shear thinning trend.

Dynamic mechanical thermal analysis in the solid state: dynamic mechanical thermal analysis results

DMTA is one of the most versatile methods to simulate the conditions both on the production line and in operational use. Apparently, the viscoelastic behavior of the material obtained from the DMTA data originates from its nanoscale structure. Therefore, monitoring the nanostructure can provide valuable information to gain a better understanding of the mechanisms through which the properties have changed.[23,26,27] DMTA can be used to predict and evaluate application performance of the nanocomposites. It was shown in the rheological studies that the hyperbranched polymers reduce the viscosity. In other words, these molecules facilitate motions of the PET chains on each other in the melt state. Utilizing the same logic, they had the same characteristics in the solid state as deduced from the reduced T_g of the blended samples. This means that hyperbranched polymers were spread between PET chains and reduced entanglements. As a consequence, the degree of freedom of the chains increased and the chains were more capable of moving. The DMTA results of all samples are presented in Figs. 7 and 8 in terms of temperature dependence of the loss factor (tan delta) and the storage modulus (G'), respectively. Fig. 7 shows that blended samples had only one relaxation peak, the primary relaxation peak associated with

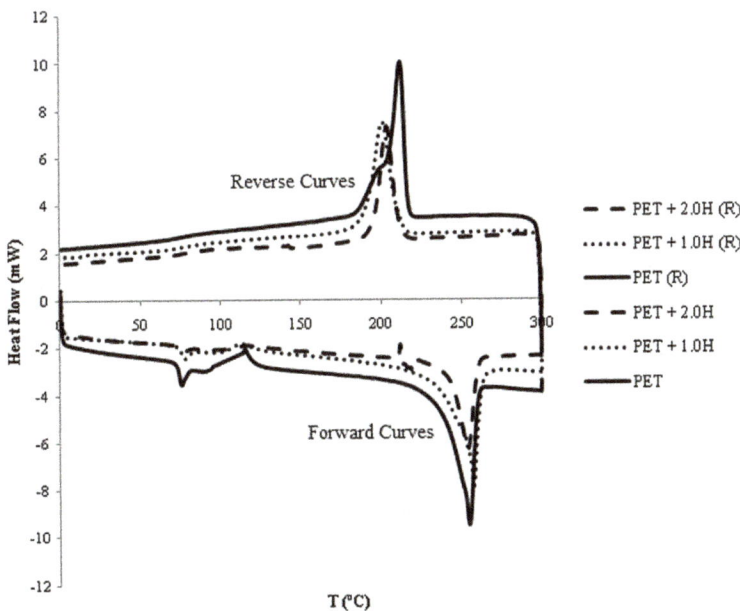

Figure 4   DSC diagrams of virgin and blended PET samples

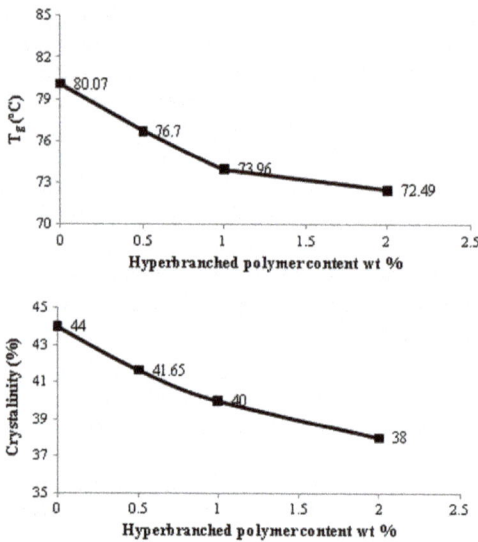

Figure 5   Glass transition temperature (T_g) and crystallinity (%) versus hyperbranched polymer wt % for virgin PET and PET/H nanocomposites

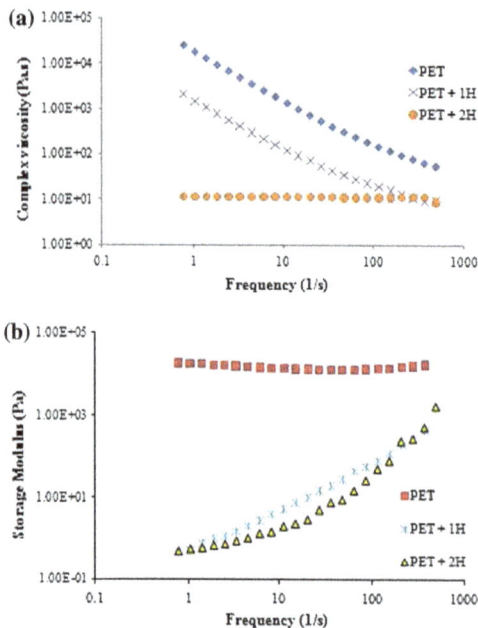

Figure 6   Complex viscosity η^* (a) and storage modulus G' (b) versus frequency ω of applied deformation for the virgin PET and PET/H nanocomposites

glass transition temperature. This represents that blended samples had only one maximum tan delta, indicating that a homogenous blending had occurred. In other words, the hyperbranched polymer had been dispersed properly in PET and was able to introduce a homogenous compound with a single glass transition temperature. These results indicated that the blends were partially miscible. Interestingly, with the

increase in the hyperbranched polymer content, the glass transition temperature of blended PET sheets tended to shift towards lower temperatures. Thus, the decreased value of T_g resulting on the addition of the hyperbranched polymer confirms that the mobility of polymeric chains greatly increased which in turn affected the softer and more flexibility of the samples.

Storage modulus (G') can be used to evaluate the hardness of polymer materials. Fig. 8 reports the storage modulus of samples versus temperature. The trends of storage modulus curves were qualitatively similar. All the storage modulus versus temperature curves experienced a gradual decline in storage modulus with temperature increasing from 0 to around 80 °C and then a decrease occurred in modulus for virgin PET and all the blended ones. Storage modulus indicates the strength of materials against deformation at different temperatures under oscillation forces. Therefore, higher storage modulus means that sheets can resist against deformation. As can be seen in Fig. 8, increasing hyperbranched polymer's content decreases the storage modulus. This can be attributed to a more compacted structure of virgin PET sheet in comparison with blended PET sheets.

Small-angle X-ray scattering data evaluation

According to the SAXS theory, the SAXS signal is produced by electron density inhomogeneity in the sample. The angle between transmitted beam and scattered beam is scattering angle.[28] The scattering intensity is proportional to the square of the density differences between the constituting phases. By plotting scattering intensity versus scattering vector in logarithmic scale, two regions are achieved. Very small angles are Guinier region and follows Equation (2):[28]

$$I(q) \; \alpha \; G \exp(-q^2 R_g^2 / 3) \tag{2}$$

where $I(q)$ is the scattering intensity, R_g is the gyration radius which indicates a measure of the mean square distance of the scattering centers within particle domains from the center of gravity and q, the scattering vector is defined as Equation (3):[28]

$$q = \frac{4\pi}{\lambda} \sin(\theta) \tag{3}$$

where λ is the X-ray wavelength and θ is the scattering angle. The R_g is related to the size of particle and could be calculated from the slope in the linear region of a plot of ln ($I(q)$) versus q^2.

A log–log plot of SAXS data, called Porod plot, comforted in other region and could give an indication of the fractal nature of particle, specifically if the system displays a mass or surface fractal behavior.[29] Porod formula[30] is shown in Equation (4):

$$I(q) \; \alpha \; q^{-\alpha} \tag{4}$$

where α is the power exponent obtained from the slope in the linear region of a double logarithmic plot of $I(q)$ versus q. The power exponent is simply related to the fractal characteristics; thus, it is a measure of the compactness or shape of materials such as a sharp and smooth surface between the two regions of different electron density. An increase in the fractal dimension implies more compact structure of domains. For $1 < \alpha < 3$, objects are considered as mass fractal in three-dimension and fractal dimension is determined by $D = -\alpha$. Mass fractal

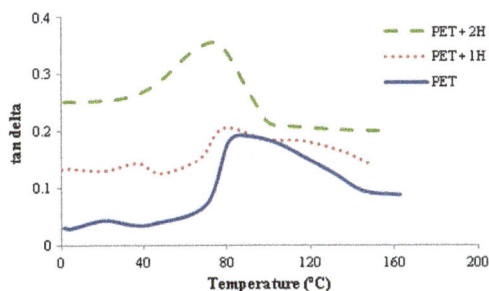

Figure 7   Temperature dependence of tan delta for the virgin PET and PET/H nanocomposites

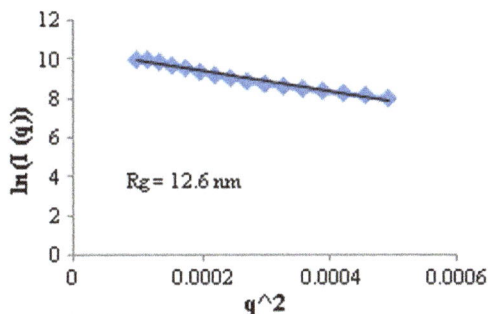

Figure 9   Guinier plot of SAXS for the blended PET with 2 wt % hyperbranched polymer

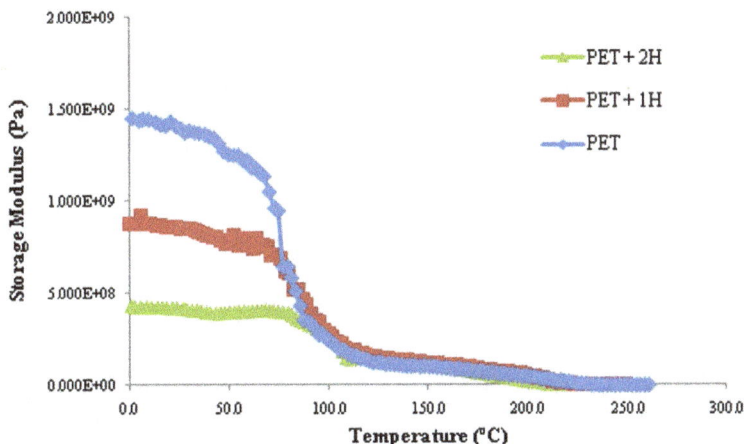

Figure 8   Temperature dependence of the storage modulus for the virgin PET and PET/H nanocomposites

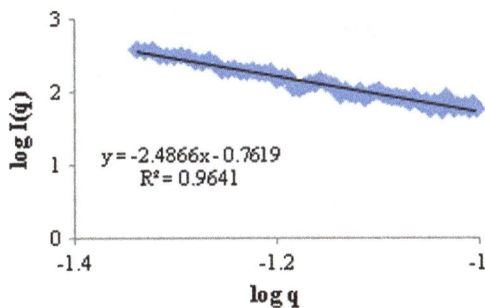

Figure 10   Porod plot of SAXS for the blended PET with 2 wt % hyperbranched polymer

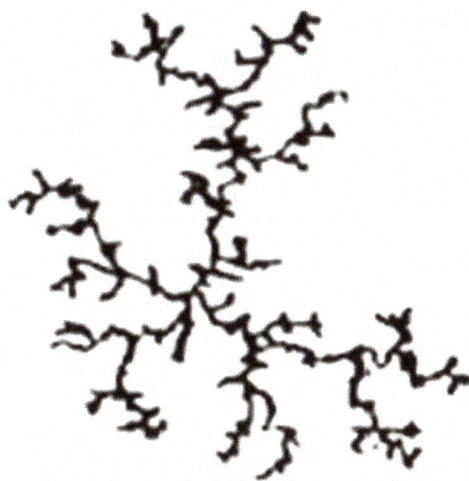

Figure 11   The schematic shape of hyperbranched polymer

dimensions are always less than the dimension of the space in which the fractal exists, which implies that the mass of the particulate increases less rapidly than the volume it occupies. For $3 < a < 4$, objects are surface fractal with fractal dimension equal to $D_s = 6 - a$.[31] Fractal surfaces, on the other hand, have the property that the surface area S varies as a non-integer power of the length. In other words, D_s is a measure of surface roughness of the scattering particle. SAXS profile of virgin PET exhibited a very low scattering intensity and this means that

no particles exist in virgin PET. The Guinier plot and Porod plot of SAXS data for the blended PET showed a linear region (Figs. 9 and 10). The R_g and D are obtained from the slope of the curves. The power law exponent (D) and R_g for blended PET were 2.48 and 12.6 nm, respectively. The power law exponent for the sample was lower than 3 and suggesting blended PET had mass fractal behavior.[32] So based on this result, the shape of hyperbranched polymer can be imagined to look similar to Fig. 11.

Conclusion

Successfully, PET/hyperbranched polymer nanocomposites have been prepared via melt blending method. AFM studies were performed to examine the surface morphology for virgin PET and nanocomposites. According to the observations made by AFM, dispersion of the hyperbranched polymer in the nanocomposites containing 0.5, 1, and 2 wt % hyperbranched polymer was uniform, but some aggregations were observed in sample containing 3 wt % hyperbranched polymer. Thermal analysis by DSC was used to probe the effect of hyperbranched polymer on crystallinity and glass transition temperature of nanocomposites. The presence of the hyperbranched polymer decreased the glass transition temperature as well as the crystallinity of the nanocomposites compared with the virgin PET. It was concluded from the rheological data that the flowability of nanocomposites was enhanced. Solid-state DMA results clarified that microscopic properties such as T_g and viscosity were changed as a result of the melt blending process. SAXS measurements determined the size and shape of hyperbranched polymer in nanocomposites.

Acknowledgments

Hybrane H1500 was kindly provided by DSM from the Netherlands.

Disclosure statement

No potential conflict of interest was reported by the authors.

References

1. K. Shahverdi-Shahraki, T. Ghosh, K. Mahajan, A. Ajji and P. J. Carreau: 'Morphology and thermal properties of poly (ethylene terephthalate)-modified kaolin nanocomposites', Polym. Compos., 2016, 37, 1443–1452.
2. H. Ishida, S. Campbell and J. Blackwell: 'General approach to nanocomposite preparation', Chem. Mater., 2000, 12, (5), 1260–1267.
3. M. C. Costache, M. J. Heidecker, E. Manias and C. A. Wilkie: 'Preparation and characterization of poly(ethylene terephthalate)/clay nanocomposites by melt blending using thermally stable surfactants', Polym. Adv. Technol., 2006, 17, 764–771.
4. S. G. Kim, E. A. Lofgren and S. A. Jabarin: 'Dispersion of nanoclays with poly(ethylene terephthalate) by melt blending and solid state polymerization', J. Appl. Polym. Sci., 2013, 127, (3), 2201–2212.
5. M. Catauro, F. Bollino and F. Papale: 'Preparation, characterization, and biological properties of organic–inorganic nanocomposite coatings on titanium substrates prepared by sol–gel', J. Biomed. Mater. Res. Part A, 2014, 102, (2), 392–399.
6. H. Yin, L. Ma, M. Gan, Z. Li, X. Shen, S. Xie, J. Zhang, J. Zheng, F. Xu, J. Hu and J. Yan: 'Preparation and properties of poly (2, 3-dimethylaniline)/organic–kaolinite nanocomposites via in situ intercalative polymerization', Compos. Sci. Technol., 2014, 94, 139–146.
7. N. Najafi, M. C. Heuzey and P. J. Carreau: 'Polylactide (PLA)-clay nanocomposites prepared by melt compounding in the presence of a chain extender', Compos. Sci. Technol., 2012, 72, (5), 608–615.
8. C. N. Barbosa, F. Gonçalves and J. C. Viana: 'Nano and hybrid composites based on poly (ethylene terephthalate): blending and characterization', Adv. Polym. Technol., 2014, 33, (2), 21397–21408.
9. E. M. Aizenshtein: 'Polyester fibres continue to dominate on the world textile raw materials balance sheet', Fibre Chem., 2009, 41, 1–8.
10. S. M. Burkinshaw, 'Chemical principles of synthetic fibre dyeing', 1st edn, 1–9; 1995, London, Blackie Academic & Professional.
11. R. M. Stinson and S. K. Obendorf: 'Simultaneous diffusion of a disperse dye and a solvent in PET film analyzed by Rutherford backscattering spectrometry', J. Appl. Polym. Sci., 1996, 62, 2121–2134.
12. M. Khatibzadeh, M. Mohseni and S. Moradian: 'Compounding fibre grade polyethylene terephthalate with a hyperbranched additive and studying its dyeability with a disperse dye', Color. Technol., 2010, 126, 269–274.
13. M. Ahani, M. Khatibzadeh and M. Mohseni: 'Studying the thermodynamic parameters of disperse dyeing of modified polyethylene terephthalate sheets using hyperbranched polymeric additive as a nanomaterial', J. Ind. Eng. Chem., 2013, 19, 1956–1962.
14. M. Ahani, M. Khatibzadeh and M. Mohseni: 'Investigating the effects of different loadings of a nanostructured hyperbranched polymer on the kinetic parameters of disperse dyeing of modified poly(ethylene terephthalate) sheets', Fiber. Polym., 2015, 16, (3), 606–613.
15. K. C. Cheng, S. C. Chang, Y. H. Lin and C. C. Wang: 'Mechanical and flame retardant properties of polylactide composites with hyperbranched polymers', Compos. Sci. Technol., 2015, 118, 186–192.
16. S. Wei, Y. Zhu, Y. Zhang and J. Xu: 'Preparation and characterization of hyperbranched aromatic polyamides/Fe₃O₄ magnetic nanocomposite', React. Funct. Polym., 2006, 66, 1272–1277.
17. J. H. Lee, K. Orfanou, P. Driva, H. Iatrou, N. Hadjichristidis and D. J. Lohse: 'Linear and nonlinear rheology of dendritic star polymers: experiment', Macromolecules, 2008, 41, 9165–9178.
18. D. Konkolewicz, M. J. Monteiro and S. Perrier: 'Dendritic and hyperbranched polymers from macromolecular units: elegant approaches to the synthesis of functional polymers', Macromolecules, 2011, 44, 7067–7087.
19. A. Tomuta, F. Ferrando, N. Serra and X. Ramis: 'New aromatic–aliphatic hyperbranched polyesters with vinylic end groups of different length as modifiers of epoxy/anhydride thermosets', React. Funct. Polym., 2012, 72, 556–563.
20. H. Chen, J. Jia, W. Zhang and J. Kong: 'A simple and general method for the determination of content of terminal groups in hyperbranched polymers derived from ABₙ monomers', Polym. Test., 2014, 35, 28–33.
21. H. W. Starkweather, P. Zoller and G. A. Jones: 'The heat of fusion of poly(ethylene terephthalate)', J. Polym. Sci., Polym. Phys. Ed., 1983, 21, 295–299.
22. B. Basnar, G. Friedbacher, H. Brunner, T. Vallant, U. Mayer and H. Hoffmann: 'Analytical evaluation of tapping mode atomic force microscopy for chemical imaging of surfaces', Appl. Surf. Sci., 2001, 171, 213–225.
23. C. M. Nunez, B. Chiou, A. L. Andrady and S. A. Khan: 'Solution rheology of hyperbranched polyesters and their blends with linear polymers', Macromolecules, 2000, 33, 1720–1726.
24. R. Mezzenga, L. Boogh and J. E. Manson: 'A review of dendritic hyperbranched polymer as modifiers in epoxy composites', Compos. Sci. Technol., 2001, 61, 787–795.
25. M. Ganjaee, N. Sari, S. Stribeck, A. Moradian, S. Zeinolebadi, S. Bastani and E. Bottaf: 'Dynamic mechanical behavior and nanostructure morphology of hyperbranched-modified polypropylene blends', Polym. Int., 2014, 63, 195–205.
26. A. Bartolotta, G. Di Marco, F. Farsaci, M. Lanza and M. Pieruccini: 'DSC and DMTA study of annealed cold-drawn PET: a three phase model interpretation', Polymer, 2003, 44, 5771–5777.
27. V. Pistor, F. G. Ornaghi, H. L. Ornaghi and A. J. Zattera: 'Dynamic mechanical characterization of epoxy/epoxycyclohexyl–POSS nanocomposites', Mater. Sci. Eng. A, 2012, 532, 339–345.
28. C. J. Brinker and G. W. Scherer: 'Sol–gel science: the physics and chemistry of sol–gel processing', 531–541; 1990, New York, Academic Press.
29. J. Habsuda, G. P. Simon and Y. B. Cheng: 'Organic–inorganic hybrids derived from 2-hydroxyethylmethacrylate and (3-methacryloyloxypropyl) trimethoxysilane', Polymer, 2002, 43, 4123–4136.
30. S. Yano, K. Iwata and K. Kurita: 'Physical properties and structure of organic-inorganic hybrid materials produced by sol-gel process', Mater. Sci. Eng. C, 1998, 6, 75–90.
31. K. J. West, B. F. Zhu, Y. C. Cheng and L. L. Hench: 'Quantum chemistry of sol-gel silica clusters', J. Non-Cryst. Solids, 1990, 121, 51–55.
32. H. Yahyaei and M. Mohseni: 'Mechanically controlled, morphologically determined sol–gel derived UV curable hybrid nanocomposites: SAXS and DMTA studies', J. Sol-Gel Sci. Technol., 2013, 66, 187–192.

5

The thermo-mechanical response of PP nanocomposites at high graphene loading

Kai Yang*[1], Maya Endoh[1], Rebecca Trojanowski[2], Radha P. Ramasamy[1], Molly M. Gentleman[1], Thomas A. Butcher[2] and Miriam H. Rafailovich*[1]

[1]Materials Science and Engineering, Stony Brook University, NY 11794, USA
[2]Energy Resources Division, Brookhaven National Laboratory, Upton NY 11973, USA

Abstract Authors have successfully fabricated polypropylene/graphene nanoplatelets (PP/GNPs), nanocomposites that are thermally conductive, processable, and flame resistant. Thermal conductivity measurements indicated that the thermal coefficient scaled linearly with GNP loading, where a value of $2.0\,W\,m^{-1}\,K^{-1}$ was achieved at 40 wt-% loading. Tensile measurements indicated that the modulus increased linearly with GNP loading, while the Izod impact, after an initial decrease, remained constant for loadings up to 50 wt-%. Small angle X-ray scattering (SAXS) showed a large decrease in the amount of lamellar structure relative to the neat PP, while wide angle X-ray scattering (WAXS) showed a high degree of crystallinity. These results are consistent with formation of a new type of layered nanocomposite, composed of crystalline PP chains oriented onto layered GNPs.

Keywords Flame retardant, Nanocomposites, Thermal conductivity

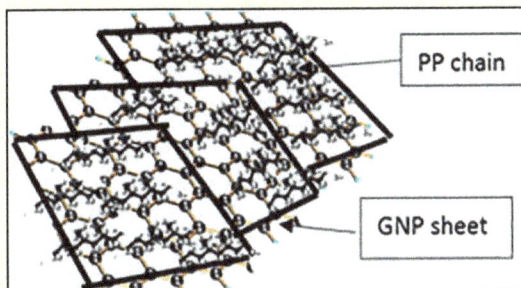

Introduction

Biofuels are currently being introduced as `green' replacements for fossil fuels in numerous applications including transportation and home heating. One major drawback of these fuels is their high acidity, which makes them corrosive and limits their usefulness as energy generators. Consequently, there has been intense research in finding a suitable, non-corrosive, heat-resistant, and yet, thermally conductive and malleable material for replacing the metals in heat exchange or fuel storage applications.[1–8] Polypropylene (PP), a crystalline polyolefin elastomer, has been proposed owing to its high corrosion resistance,[9] toughness, and ease of processing. On the other hand, PP is also extremely flammable, and has a poor thermal conductivity coefficient typical of most polymers, $k = 0.2\,W\,m^{-1}\,K^{-1}$.

Numerical simulations of the heat exchange parameters in a typical boiler found that the efficiency was always limited by the component with the lowest thermal coefficient, which in a metal heat exchanger is the gas-side convective coefficient.[10,11] In the case of metal heat exchangers, the thermal coefficient of the metal body was approximately $k \geq 10\,W\,m^{-1}\,K^{-1}$, but that of the flue gas was only $k \approx 2.0\,W\,m^{-1}\,K^{-1}$. Hence the materials of the heat exchange unit could easily be replaced by a nanocomposite with a coefficient slightly larger than $2.0\,W\,m^{-1}\,K^{-1}$ without loss of efficiency in the performance of the unit.

Despite the apparent simplicity, it still required nearly an order of magnitude increase of the polymer thermal coefficient. Numerous nanoparticles exist, i.e. nanotubes, carbon black, graphene, metallised graphene, or metallic particles, which have been shown to increase the electrical conductivity,[12–15] but increasing the thermal conductivity has proven far more difficult,[8,16] especially when considerations of corrosion and thermal resistance were also imposed, which ruled out water-soluble polymers such as PEI, where high conductivity has been reported.[17] For example, Dittrich et al., Shi et al., and Chatterjee et al. have demonstrated that it is possible to increase the thermal coefficient of PP nanocomposites with graphene,[18] metal/graphene combination,[19] or graphene and other carbon particle mixtures,[20] but at the concentrations reported only a value of $K < 0.3\,W\,m^{-1}\,K^{-1}$ was achieved. Electronic conductivity can be accomplished when the percolation threshold of the particle is reached and the conducting components are at least in point contact.[21] Point contact, on the other hand, is insufficient for phonon transfer, which is required for

*Corresponding author, email miriam.rafailovich@stonybrook.edu; dyclexr@hotmail.com

thermal conductivity. As pointed out previously by several groups, Phonon transfer is a more complicated process, which requires proper coupling between the nanoparticle structure and the polymer matrix over much larger contact areas.[22–24] Therefore, the large aspect ratio of graphene and large contact area allows for good coupling, making graphene an obvious choice for engineering high thermal conductivity nanocomposites.

The tensile properties are another essential component in choosing the appropriate material. Heat exchange units need large surface areas in order to increase their efficiency in interacting with their environment. These can be achieved by forming heat exchange coils, which require good extrusion characteristics and processability. Addition of even small volume fractions of nanoparticles, such as clays, have been shown to decrease impact strength, leading to embrittlement and making it very difficult to process.[24–26] Kalaitzidou et al. compared the mechanical properties of PP with different layered fillers and showed that for loading up to 20% volume PP/GNP outperformed clays or nanotubes.[27]

Since the heat exchange applications often require the unit to be in close proximity to electronic components or open flame, the materials must also be flame resistant. Several groups have shown that Polystryene (PS)[28] and Polylactic Acid (PLA)[29] can be rendered somewhat flame retardant, when graphene platelets were used in combination with other particles. Applying the UL-94 standardised test indicated that composites with up to 12 wt-% loading obtained a V2 rating, or these compounds extinguished within a time $t < 30$ s but dripped and ignited an underlying cotton. For use in heat exchangers, a better rating, preferably V0 is desirable. Recently, Dittrich et al.[30] compared the response to an approaching heat front of PP compound with graphene, carbon nanotubes, and carbon black, where they reported that even though all three nanoparticles were able to affect a significant reduction in the heat release rate, the graphene/PP nanocomposites yielded the best results at a fixed loading of 5 wt-%. Since PP is one of the most corrosion-resistant polymers, with better performance than either PS or PLA, we proposed to build on these results and investigate the properties of PP compounds as a function of graphene loading in order to determine whether a concentration could be found, which would yield a composite that would have K ≈ 2.0 W m^{-1} K^{-1}, while at the same time have an even lower peak heat release rate (PHRR), possibly achieve UL-94-V0 designation, while still maintaining sufficient processability.

Recently Zhao et al.[31,32] and Milani et al.[33] performed a series of elegant experiments where PP was synthesised from the monomer directly on the surface of graphene platelets. In this manner, they were able to ensure complete exfoliation in their nanocomposites. Using SEM, Zhao et al.[31] studied the growth of crystalline structures and found that at constant temperature, the onset of crystallinity was faster, while the degree of crystallinity was higher than the bulk for structure nucleated directly on the graphene platelets. In this paper, we study the crystallisation properties of PP/graphene samples as a function of filler concentration, where, in order to achieve higher loading we use melt blending rather than direct synthesis. Using SEM together

with other complementary techniques, i.e. small angle and wide angle X-ray scattering (SAXS, WAXS), dielectric spectroscopy, and thermal conductivity, we find that the layered structures are also formed at higher graphene concentrations, where the PP polymer is adsorbed as thin layers on the graphene, and forms surface-nucleated highly crystalline structures, commensurate with the underlying graphene platelets. This coupling between the polymer and the graphene may then provide an explanation for the enhanced phonon coupling leading to relatively high thermal coefficient and excellent flame retardant properties while maintaining ductility even at very high loading.

Experimental section

Materials

Graphene nanoplatelets: The Graphene Nanoplatelets (GNPs) were purchased from XG Sciences (Lansing, MI). The GNPs have an average thickness of approximately 15 nm and a typical surface area of 50–80 m^2 g^{-1}. GNPs (H5) were tested, which had nominal diameters of 5 μm.

Particles: The multi-walled carbon nanotubes (MWCNTs) were provided by Cheap Tubes Inc (Grafton, VT). The lengths of the MWCNTs used in this study were 0.5–2.0 μm, which have a diameter of 30–50 nm. Copper Microparticles (Copper powder, <75 μm, 99%; Sigma Aldrich, St. Louis, MO)) and Carbon black was (Cabot corporation, Vulcan 3 N330, Boston, MA) used as an additive to improve the thermal and electrical conductivity.

Polymer: The matrix polymer used was Polypropylene (Resin 3825, Total Petrochemical, Houston, TX, USA), the melt flow of which is 30 g 10 min^{-1} and density is 0.905 g cc^{-1}.

Methods

Compounding

A C.W. Brabender instrument, type Intelli-Torque Plasti-Corder was used to blend the nanocomposites. The blender was equipped with two screw-type roller blades in a heating chamber. The polymer pellets were first added to the chamber at a rotation speed of 20 rev min^{-1} and temperature of 180°C. The GNPs, MWCNT, carbon black, and copper micro particles were then inserted and mixed at the same revolutions per minute for 2 min. Either the GNPs or the MWCNTs were gradually added into the chamber, while blending. The entire mixture was blended at 100 rev min^{-1} for 15 min under nitrogen gas flow, which prevented degradation of the mixture from heat-induced oxidation. The mixing torque at 15 min was recorded and the mixture was allowed to cool at room temperature and then pelletised and moulded in a hot press at $T = 180$°C into the different shapes required for the thermal conductivity, mechanical tests and flame tests.

Thermal Conductivity tests were measured using Quick Thermal Conductivity Meter (QTM500) at 23°C. Moulded samples of dimension 64 × 64 × 10 mm were tested and the results quoted represent the statistical average of the measurements from five identical specimens.

Tensile properties

Nanocomposite samples were moulded into dog bone shapes, 1.5 mm thick and 6.4 cm long, for use with the

Instron Tensile Tester (Model: 5542) equipped with a 500N load cell and hydraulic pressure grip. The measurement was performed at a rate of $1\,mm\,min^{-1}$ and the Young's modulus, tensile strength, and tensile strain were analysed from the acquired data. Specimens were moulded from the hot press and tested according ASTM D638.

Izod impact

Notched Izod impact tests were measured using TMI Impact tester (model 43-02) at 23°C, 50% RH with 5.5 J pendulum. The samples with dimensions $64 \times 12.7 \times 3.2\,mm$ were used according to ASTM D256. The statistical average of the measurements of 10 specimens was taken to obtain a reliable data with appropriate standard deviation.

Scanning electron microscopy

The distribution of GNPs, CNT was imaged using JEOL 7600F Analytical high resolution SEM at the Center for Functional Nanomaterials (CFN), Brookhaven National Laboratory, Upton, NY, USA. Cross-sectioned PP/GNPs specimens were placed on a conductive, double-sided, carbon adhesive tab and imaged at 5 kV accelerating voltages using a secondary electron imaging (SEI) detector. Ten nanometres of gold were coated on the surface of the specimens in order to make the specimens conduct.

Dynamic mechanical analysis

The DMA test was performed on a DMA Q800 from TA Instruments in the single cantilever mode at a fixed frequency of 1 Hz. The samples were moulded in the dimensions $40 \times 10 \times 2\,mm$. The measurement temperature range was from -30 to 110°C with a heating rate of $3°C\,min^{-1}$; the storage modulus and tan δ were recorded as a function of temperature.

Synchrotron X-ray characterization

Wide angle X-ray scattering and SAXS measurements were carried out at the Advanced Polymers Beamline (X27C, $\lambda = 0.1371\,nm$) in the National Synchrotron Light Source (NSLS), Brookhaven National Laboratory (BNL). Aluminium oxide (Al_2O_3) and a silver behenate (AgBe) standard were used to calibrate the detector distance for WAXS and SAXS, respectively. A 2D MAR CCD X-ray detector (MAR-USA) with a resolution of 1024×1024 pixels (pixel size $= 58.44\,\mu m$) was used to acquire 2D-WAXS and 2D-SAXS images. For SAXS, the exposure time was 120 s and the sample to the detector length was 1689 mm. For WAXS, the exposure time was 300 s and the sample to the detector length was 169 mm. WAXD has been changed to WAXS.

Two-Dimensional-WAXS and SAXS patterns were integrated with proper orientation correction to obtain their corresponding scattering profiles as a function of $|q| = 4\pi\sin\theta/\lambda$, where q is the first-order scattering vector, λ is the wavelength of X-ray beamline (0.137 nm), and 2θ is the scattering angle. For integrated 1D-WAXD curves, a multi-peaks Gaussian fitting was used to obtain the areas of crystalline and amorphous peaks.

The overall crystallinity X_C was calculated by

$$X_C = \frac{\sum A_{cryst}}{\sum A_{cryst} + \sum A_{amorp}}$$

Where $\sum A_{cryst}$ and $\sum A_{amorp}$ are crystalline and amorphous phases, respectively.

Vertical oxidative flame test

A vertical burning chamber purchased from Underwriters Laboratories Inc (Melville, NY). was used to assess flammability under oxidative atmosphere. Samples of dimensions $125 \times 13 \times 5\,mm$ were moulded and tested using the protocol established by ASTM D3801 UL94 V0. Ten specimens were preconditioned with 50% relative humidity for 48 h at 23°C. A flame of $20 \pm 2\,mm$ was applied to the sample for 10 s and then t_1, the time the sample took to self-extinguish was measured. The flame was reapplied for another 10 s and t_2 the time to extinguish the flame was measured. The criteria for achieving V0 could require that either t_1 or t_2 be $< 10\,s$. The results reported are the average of five samples for each type of sample.

Limiting oxygen index

Limiting Oxygen Index (LOI) was conducted by Stanton Redcroft FTA Flammability Unit. Sample of dimensions $100 \times 6.5 \times 3\,mm$ were moulded and tested according to ASTM D2863.

Heat and mass loss rates

The heat release rate (HRR) and mass loss rate (MLR) of each sample were tested using cone calorimetry at Israel Chemistry Limited Industrial Products (ICL-IP, Ardsley, NY, USA). The samples were made by moulding the composite into a $75 \times 75 \times 5\,mm$ square. Thin aluminium foil was used to wrap all sides of samples except for the upper face to avoid the burning splash. The samples were then exposed in ambient atmosphere, to an external radiant flux of $50\,kW\,m^{-2}$, perpendicular to the sample surface and the HRR and MLR were measured as a function of exposure time. The standard uncertainty of the measured heat release rate (HRR) was $\pm 10\%$. The cone calorimetry test was conducted according to the ISO 5660 protocol.

Results

Thermal conductivity of nanocomposites

Authors first produced a set of PP nanocomposites, with the same particle loading of 20% weight ratio, but with different nanoparticles such as copper microparticles, CNT, and carbon black, which are all known to have a high thermal and electrical conductivity. The thermal conductivity measured for nanocomposites samples, 10 mm thick, is tabulated below. From the table, we can see that the spherical particles, such as CB and Cu, have only a minimal effect on the thermal conductivity. In the case of MWCNT, from the Table 1, we can see that even for the 20 wt-% which is far above the percolation threshold, the coefficient that can be reached is $k = 0.57\,W\,m^{-1}\,K^{-1}$, which is still far below the required minimum value of $k = 2.0\,W\,m^{-1}\,K^{-1}$. From the table we can see that the compounds with GNPs particles had the highest thermal coefficients is GNPs who achieved a value of $k = 1.12\,W\,m^{-1}\,K^{-1}$ at a loading of 20%. Hence GNPs appeared to be the most promising approach to increase the thermal coefficient.

Table 1   Thermal conductivity test of polypropylene nanocomposites with graphene nanoplatelets (GNPs), carbon black, carbon nanotube, and copper micro particles

Sample	(Filler concentration, wt-%)	Thermal conductivity $W\,m^{-1}\,K^{-1}$
PP	100	0.23
PP/Cu	80/20	0.29
PP/CB	80/20	0.36
PP/CNT	80/20	0.57
PP/GNPs	80/20	1.12

In order to determine whether it was possible to achieve the minimum required value of $k = 2.0\,W\,m^{-1}\,K^{-1}$, and determine the functional dependence of the thermal coefficient on particle loading, we measured the conductivity as a function of particle weight fraction.

In Fig. 1 we plot the thermal conductivity for GNPs as a function of loading, Φ, where we find a direct linear relationship given by the following equation

$$y = 4.77\Phi + 0.16\,(\text{for GNPs})\,R^2 = 0.999 \qquad (1)$$

Zhang et al.[34] have studied the dependence of the thermal conductivity on the loading of thermally conductive nanoparticles, and identified three regimes corresponding to the internal organisation of the nanoparticles. The first, persisting up to approximately 30 vol.-%, is linear, where the increase is directly correlated to the increase in interfacial area between the particles and the matrix. Large discontinuities in the slope were observed at higher concentrations and were associated with changes in the conductivity pathway owing to particle agglomeration. The high degree of linearity shown in Fig. 1 is consistent with a uniform distribution of the nanoparticles, which does not change with increase in concentration.

From the figure, we can also see that the thermal conductivity, equivalent to that of the carrier gas in the heat exchange unit, $k = 2.0\,W\,m^{-1}\,K^{-1}$, can be obtained with a minimum filler concentration of 40 wt-% for GNPs. This concentration is significantly higher than that previously reported in the literature, and hence performed a series of thermomechanical measurements in order to determine the structure property relationships of the composite at high loading.

Tensile/rheological response at high loading

Torque measurement

A direct indicator of the processing ability of a compound can be determined by the torque used in the blending and extrusion process. The torque is initially high but decreases as the dispersion increases, till a steady state is reached after 15 min. In Fig. 2, we plot the torque versus the GNPs weight fraction, where we can see, that as expected for a filled system, the torque increases monotonically with filler concentration. Authors find that in contrast to other filled polymer systems, the increase is relatively small, being only 57%, even for a filler concentration of 50 wt-%.[35] Since the torque is proportional to the low shear rate viscosity, we attempted to fit the data to a modified Mooney relation,[36] but were not successful. In contrast to other fillers, such as clays or carbon nanoparticles and tubes, the density of graphene is rather small as $0.2-0.3\,g\,cc^{-1}$, yielding values for the volume fraction in excess of 50%, which was well out of range for the Mooney approximation.

Tensile properties

In order to determine the influence of high GNPs loading on the mechanical properties, tensile tests were performed as a function of filler concentration (Fig. 3a). From the figure we find that, even though the initial slope increases with filler concentration, a large decrease in tensile toughness, as defined by the integral under the stress strain curve, occurs immediately at a GNPs concentration of 1%. From the figure

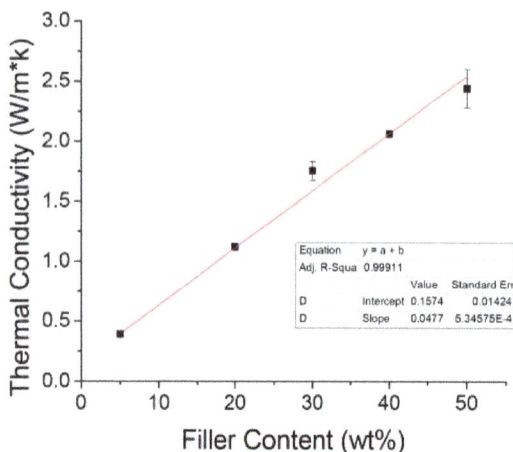

Figure 1   Thermal conductivity of polypropylene/graphene nanoplatelets (PP/GNPs) nanocomposites as a function of GNPs weight percentage

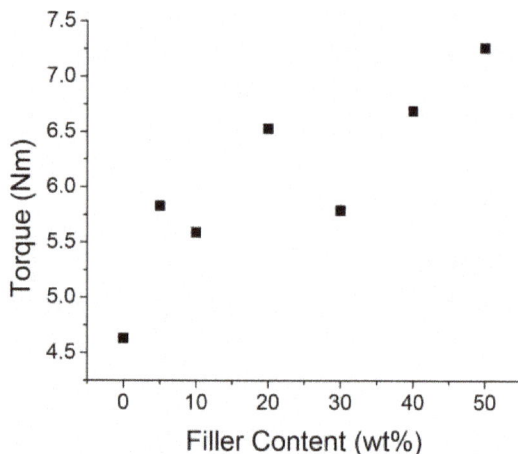

Figure 2   Torque measurement in blending process for polypropylene/graphene nanoplatelets (PP/GNPs) nanocomposites as a function of GNPs weight percentage

we find that the tensile modulus, M, increases linearly with filler concentration, Φ, according to the relationship, $M = M_o + 0.067\ \Phi$. Reasonable mechanical properties and processability could be obtained at a loading of 40%, which is the minimum required amount to obtain the desired thermal coefficient of $k = 2.0\,\mathrm{W\,m^{-1}\,K^{-1}}$.

Izod impact test

The Izod impact parameter was also measured and the results are shown in Fig. 4. The Izod impact parameter initially decreases linearly, to 40% of the PP homopolymer value at a GNPs concentration of 10 wt-%, and then the value plateaus and remains constant for loading up to 50 wt-%. Hence the filler concentration, $\Phi \approx 0.40$, which produces the PP nancomposites with the desired thermal coefficient, has nearly the same Izod factor and toughness as the formulation containing only $\Phi \approx 0.10$ filler. This is consistent with the results shown in Fig. 2 where the torque required to mix this formulation is 50% higher than that required for the $\Phi \approx 0.10$ mixture.

A possible explanation for the observed mechanical response with increasing filler concentration can be obtained by examining the microstructures of the fractured sample surfaces. In Fig. 5, we show the SEM images of the fracture surfaces following the Izod impact test of the unfilled PP and those containing 20 and 40 wt-% GNPs. From the figure, we can immediately see that in contrast to the unfilled polymer,

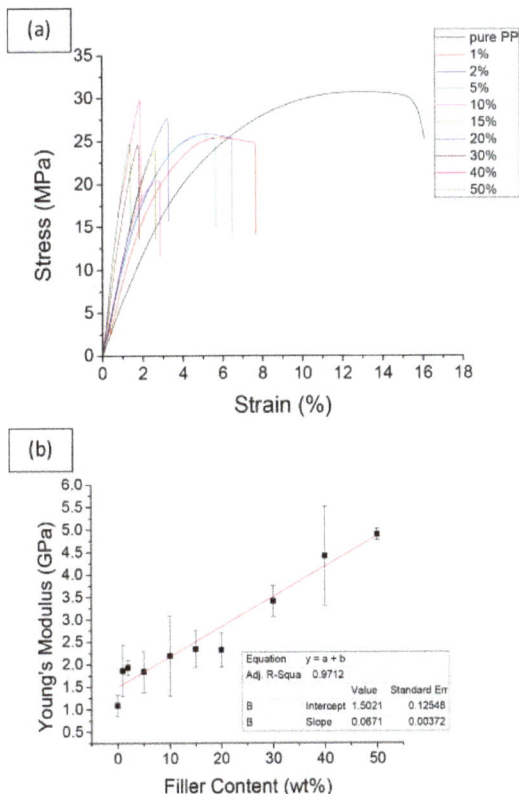

Figure 4 Izod impact test for polypropylene/graphene nanoplatelets (PP/GNPs) nanocomposites as a function of GNPs weight percentage

nancomposites have become a layered type of material, composed mostly of graphene sheets, covered by a thin polymer coating (lighter material). The sheets appear slightly separated by micro-cracks visible between the layers, which are likely responsible for the large decrease of the Izod impact and tensile toughness relative to the solid unfilled polymer. The high volume fraction of the layer is consistent with the graphene volume fractions of 55 and 77% corresponding to the weight fractions of 20 and 40% shown in the figure.

The decrease in the Izod impact and tensile toughness is directly related to the concentration of the defects, which allow cracks to propagate and dissipate internal energy. The low density of GNPs allows for the introduction of a large volume fraction of particles, reaching a value of 5 vol.-%, even at a loading of 1 wt-%. In the neat entangled polymer, propagation of cracks and dissipation of impact energy occurs primarily through stretching or disentangling of polymer chains, followed by chain scission.[37] When particles are introduced, chain entanglements are reduced, and micro-cracks around the filler interfaces form following impact, which allow for faster crack propagation and dissipation of energy. The Izod Impact and Tensile toughness are therefore most sensitive to small volume fractions of nanofillers, explaining the rapid decrease in mechanical toughness at low filler concentrations. For filler concentrations in excess of 10 wt-%, saturation of micro-cracks occurs, stabilising the nanocomposites against further decrease in mechanical properties.

Thermo-mechanical analysis

In order to determine whether the graphene fillers affect the thermal properties, DMA was performed. No change in either glass transition or melting temperatures were observed for concentrations up to 40 wt-%. In Fig. 6, we show the storage moduli of the PP/GNPs nanocomposites as a function of temperature for different GNPs concentrations. From the figure we see that the storage modulus decreases rapidly with increasing temperature. The storage modulus at room

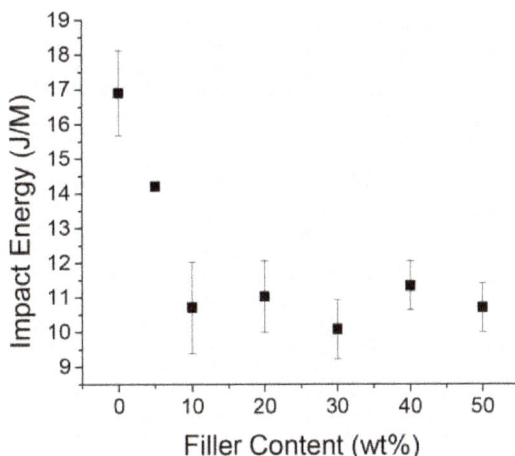

Figure 3 a Stress versus strain curve; b Young's modulus of polypropylene/graphene nanoplatelets (PP/GNPs) nanocomposites as function of graphene weight percentage from tensile test at room temperature

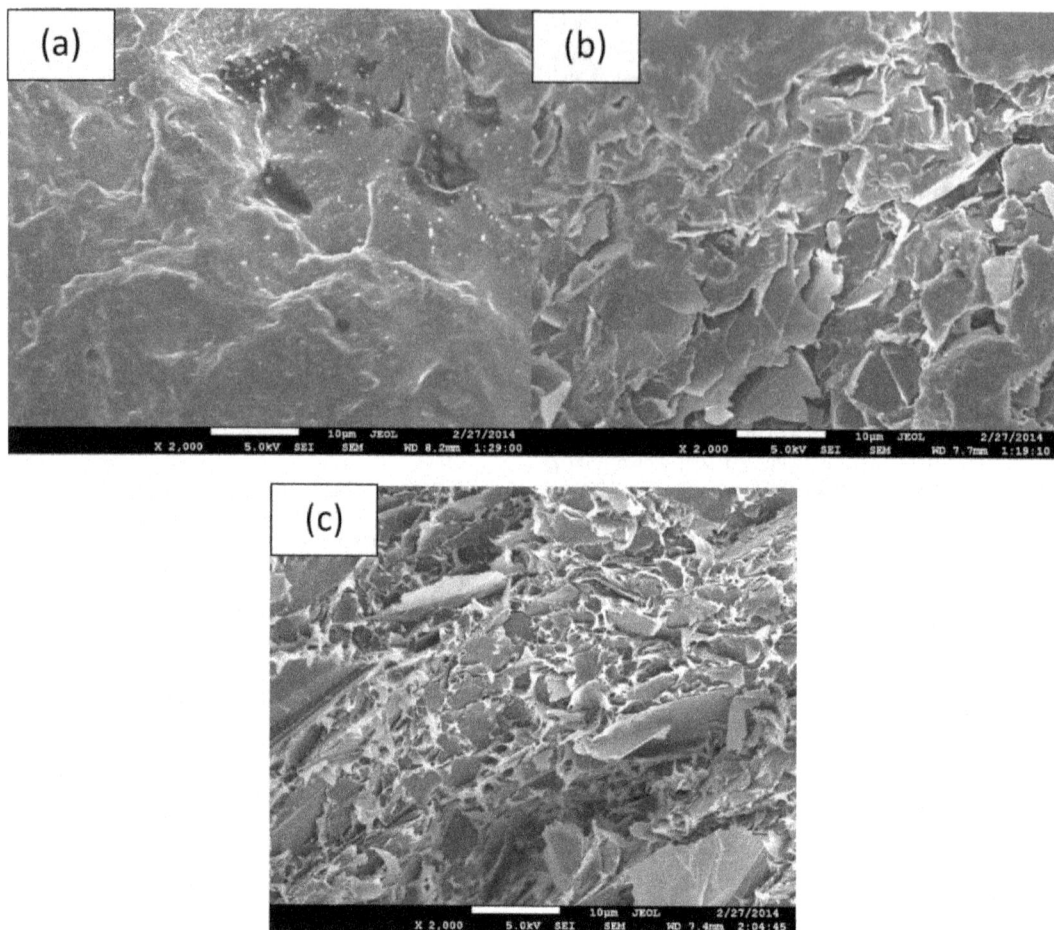

Figure 5 The SEM images of *a* pure polypropylene (PP); *b* PP/graphene nanoplatelets (GNPs) (80/20 wt-%); *c* PP/GNPs (60/40 wt-%)

temperature is plotted in Fig. 6*b* where we find that it observes the same linear increase with GNPs loading as the tensile modulus, with the two quantities having a relative same result. In Fig. 6*c*, we also plot the storage modulus at $T = 100°C$ as a function of filler concentration. From the figure we find that in this case the modulus increases more rapidly, or as a power law in concentration. Hence the addition of the GNPs is more effective of improving the mechanical properties at elevated temperature than in ambient. This factor can be beneficial in preventing dripping when the material is exposed to flame, as is discussed below.

Influence of GNPs on PP crystallinity

SAXS/WAXS

It is known that the crystal structure of the matrix has a key role in determining the properties of polymer-based composites. WAXS spectra were obtained on all the samples in order to investigate the crystal structure of the bulk samples. From Fig. 7, we can see that the peaks corresponding to the alpha structure are clearly visible, but have different relative peak area. A fitting procedure was applied to all the WAXS patterns in order to measure the degree of crystallinity. According to the method proposed

by Hindeleh and Johnson,[38] experimental WAXS spectra were produced by convolution of Lorentzian functions. In Table 2, we showed the degree of crystallinity as a function of GNPs weight ratio from $0.7 Å − 1 < q < 2.1 Å − 1$, where we find that the degree of crystallinity remains high and relatively constant for all the samples. Hence the addition of GNPs does not affect the crystallinity of PP even at loadings as high as 40 wt-%. The peaks observed in Fig. 7 correspond primarily to the α-phase, which was previously shown to nucleate readily on the platelet surfaces at lower concentrations. No peak was observed, even at the highest loading for β crystal (300), to indicate the presence of a β-phase, as reported by Xu *et al.*[39] for a PP nanocomposite with low GNP volume fraction subjected to high shear rates. From the figure we can see though that the relative amplitudes of the peaks area differ significantly from the spectra of PP homopolymer, where the ratio of intensities between the (110) and (111) orientation are 18.5 and 11%, respectively.

In Table 2 and Fig. 7, when GNPs are added, we can see that the intensity of the (110) and (111) peaks increases to 28 and 17% compare to 18.5 and 11%, respectively for virgin polymer, while that of the (040), (130), and (041) peaks decreases

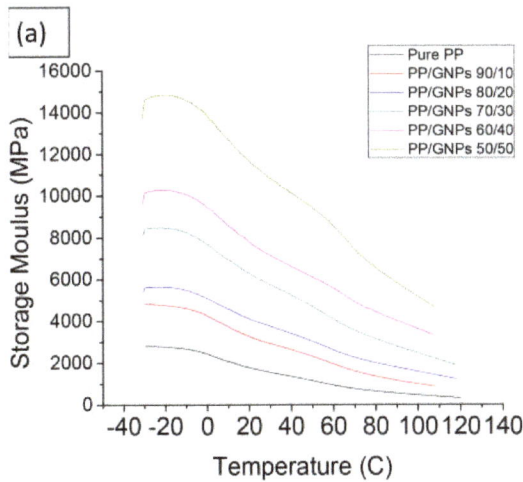

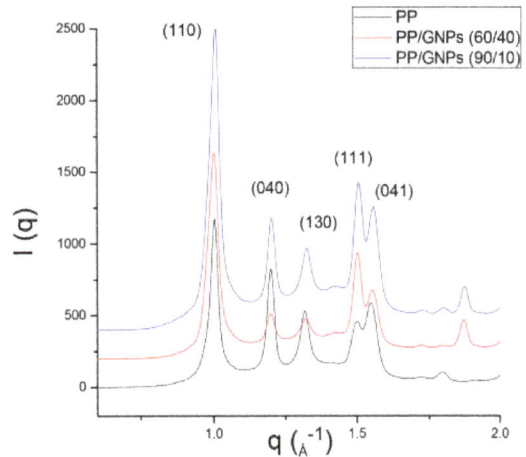

Figure 7 Wide angle X-ray scattering data of poly-propylene/graphene nanoplatelets (PP/GNPs) nanocomposites

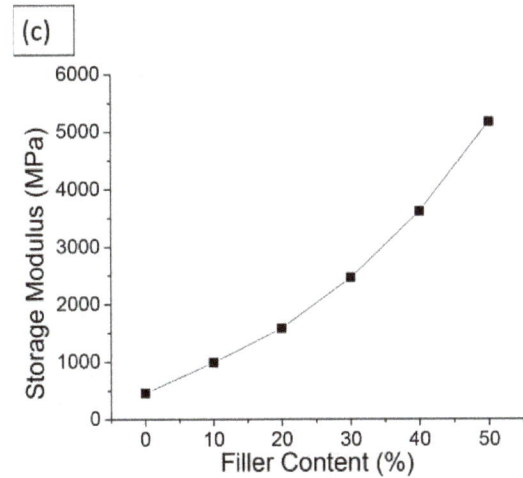

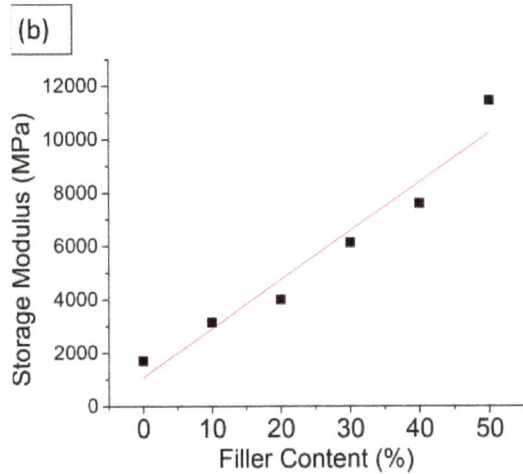

Figure 6 *a* Storage modulus of polypropylene/graphene nanoplatelets (PP/GNPs) nanocomposites as a function of temperature. *b* Storage modulus at 23°C of the PP/GNPs nanocomposites as a function of GNPs weight ratio; *c* Storage modulus of nanocomposites at 100°C as a function of GNPs weight ratio

with increasing GNPs concentration, such that for the sample with 40 wt-% loading we find values of 6, 9, and 10.5%, as opposed to 16,13.5, and 16% for the unfilled sample. The (311) of the β-phase configuration overlaps with the (111) peak of the alpha configuration, but since we do not see the pre-dominant peak at (300), this orientation is not probable. Hence from this data we can conclude that the GNP surfaces nucleate a somewhat different orientation of α phase. This feature becomes more pronounced as the GNP concentration increases and a larger volume fraction of the PP matrix chains are adsorbed on the GNP where they are crystalized with the orientation that is nucleated on the platelet surfaces, rather than the bulk structure. Hence this effect was not apparent in previous studies performed at concentrations below 10% where the predominant scattering originated from bulk PP. It is interesting to note though that the difference in orientation of the surface nucleated films is not seen when the chains are synthesised directly from the GNP surfaces.[31,32] Surface syn-thesis of chains pins them onto the surface, whereas chains that are adsorbed can orient on the surface as assumed configuration which are commensurate with the surface lat-tice structure. This will be investigated in a subsequent pub-lication where the scattering structure factors of different orientations are being calculated.

Small angle X-ray scattering spectra, as seen in Fig. 8, were obtained in order to determine the effect of the GNPs on the lamellar orientation. From the figure we can see that for the unfilled system a clear peak is observed corresponding to a lamellar spacing of $q = 0.04 \, \text{Å}^{-1}$. Xu et al.[39] reported on the presence of this peak at a loading of $< 10\%$, but from our data we find that when the loading increases to 10 wt-%, the amplitude of this peak is greatly decreased and at concentrations of 30% or higher it is nearly undetectable. These results indicate that the internal crystal structure can be very different for blends with a high degree of loading. In this case, most of the polymer is adsorbed in thin layers onto the platelet surfaces, which now constitute a significant fraction of the total volume. As has been shown previously, the platelet surfaces are very effective at nucleation in plane crystal structure of the adsorbed polymer chains, and hence the high degree of crystallinity measured by

Table 2 The crystallinity and the area of the corresponding peak for each individual orientation of polypropylene (PP)/graphene nanoplatelets (GNPs) nanocomposites

Sample	Weight ratio/%	Degree of crystallinity (%)	Area of the peak (%)				
			(110)	(040)	(130)	(111)	(041)
PP	100	78	18.5	16	13.5	11	16
PP/GNPs	90/10	73	23.5	10	10.5	14.5	12
PP/GNPs	60/40	75	28	6	9	17	10.5

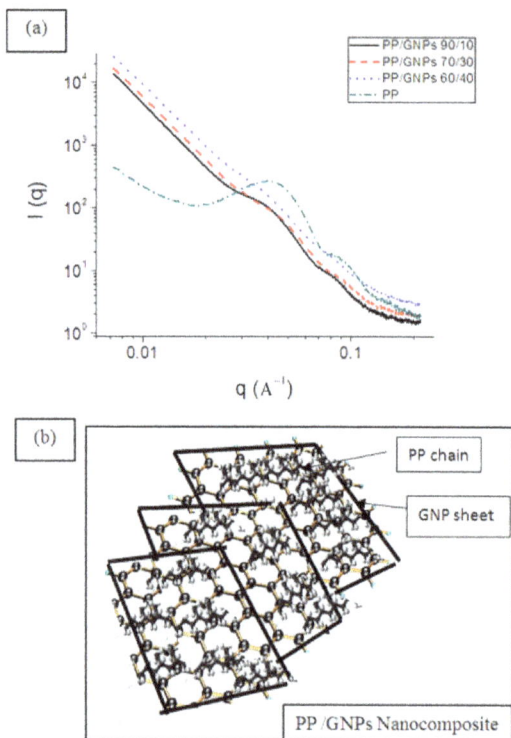

Figure 8 *a* Small angle X-ray scattering data of polypropylene/graphene nanoplatelets (PP/GNPs) nanocomposites. *b* Schematic representation of the internal structure of the high filler concentration PP/GNPs nanocomposite, where PP chains with crystalline order are adsorbed on the graphene platelet surface

WAXS. On the other hand, the lamellar structure, which requires chain folding and long-ranged order, is disturbed by the presence of the GNPs. Since the platelets do not have a preferred orientation, in the absence of shear, within the matrix, no long-ranged correlations are present that would produce a peak in the SAXS spectrum, as shown in Fig. 8*a*. A schematic representation of the nanocomposite based on the SAXS/WAXS spectra is sketched in Fig. 8(b), where we postulate that the nanocomposite is now composed of separated graphene sheets upon which PP chains with a high degree of in-plane crystalline order are adsorbed. The lack of peaks in the SAXS spectra indicates that the polymer appears to coat exfoliated graphene sheets, as shown, rather than be intercalated within multilayered graphene structures. The large number of graphene sheets also appears to have disrupted the lamellar ordering of the PP chains, since the lamellar peak observed for the pure PP sample is greatly

decreased at a filler loading of 10 wt-% and completely disappears at a loading of 30 or higher wt-%. This transition from a polymeric materials with internal lamellar structure to a separated platelet type composite is also consistent with the SEM images shown in Fig. 5, where the fracture surfaces are composed almost entirely of platelets at high loading fractions.

These discontinuities in the internal structure are not reflected in the thermal conductivity, which is well fitted by equation (1) throughout the entire concentration range. This is an illustration of the fact that electrical and thermal conductivity arises from different physical processes. Electrical conductivity is a property mostly dependent on the nature of the GNP particles, and occurs via electron transport when the GNPs are in direct contact. The properties of the matrix enter peripherally when electron tunnelling occurs as the platelets come in increasingly closer proximity approaching percolation. On the other hand, the thermal conductivity is a function of the phonon transfer between the GNPs platelets and the polymer matrix, which is dependent on the GNP/polymer coupling as opposed to the GNP/GNP coupling. It has been shown that in the case of GNPs, the propagation of phonons may be affected by the surrounding medium.[30] They showed that the open 2D lattice structure couples strongly to its environment, and the nature of the substrate affects interfacial propagation properties. In the case of PP, the WAXS data are consistent with adsorption of single chains onto the GNP platelets, where the crystalline orientation of the chains differs from that in the neat polymer. Time-dependent data from Xu *et al.*[39] of the crystallisation process indicate that crystallisation on the platelets occurs before the chains are organised into the lamellar structure, hence the coupling between the PP chains and the GNP surfaces is strong, consistent with good phonon coupling.

Heat flux response

Deformation and oxygen consumption

The thermal conductivity of a material involves the steady-state process of heat energy transfer, while flame retardance involves the ability of the system to dissipate heat rapidly from an approaching intense heat source. Since the process may involve different mechanisms, it is important to understand the response of the materials to both steady state and intense pulse heat sources. Flame retardance can be quantified in terms of the time that it takes to extinguish a flame, thus it requires the ability to both dissipate an intense heat pulse quickly in order to reduce temperatures below combustion, while at the same time decreasing the reduction in modulus with temperature to prevent dripping.

The results for nanocomposites with different GNPs/polymer ratios exposed to an open flame for 10 s are tabulated in Table 3. From the table we find that even though PP drips profusely, dripping stops very quickly at concentrations of 10% or higher, even though concentrations of 40% or higher are required for rapid extinction of the flame.

From Fig. 3, we find that the major effect of the graphene filler is to increase the viscosity at higher temperatures, which reduces dripping. The filler concentration of 10% where dripping stops also corresponds to the value where the increase in modulus at 120°C relative to that of pure PP, becomes significant. In order to further explore the mechanism for reducing the flameout time we measured the Limiting Oxygen Index, which determines the rate of oxygen consumption during burning. The results are tabulated in Table 4, where we find a small but steady increase in the amount of oxygen required for burning, which is consistent with a slower rate of combustion, but no dramatic change in the combustion reaction. Hence, GNPs do not appear to be catalysing a different reaction pathway. Rather, as loading increases, GNPs provide a barrier to oxygen penetration, while increasing the thermal dissipation of the heat, thereby reducing the rate of burning.

Cone calorimetry

The reaction of a material to a rapidly approaching heat flux is best measured using cone calorimetry. Here, we examined the rate of gasification in air of PP/GNPs nanocomposites as a function of weight ratio. In Fig. 9, video images of the combustion process reveal that for the higher loading samples, a char is observed, which forms an elastic shell surrounding the sample that inflates during combustion. In the video of the burning process submitted as part of the supplementary materials, one can see chars with greater elasticity can expand more and are better at containing the hot gases, which are responsible for combustion. Images of the chars are shown in Fig. 10, where we can see that the PP homopolymer burns completely without char formation. Addition of GNPs initiates char, which forms a solid shell for

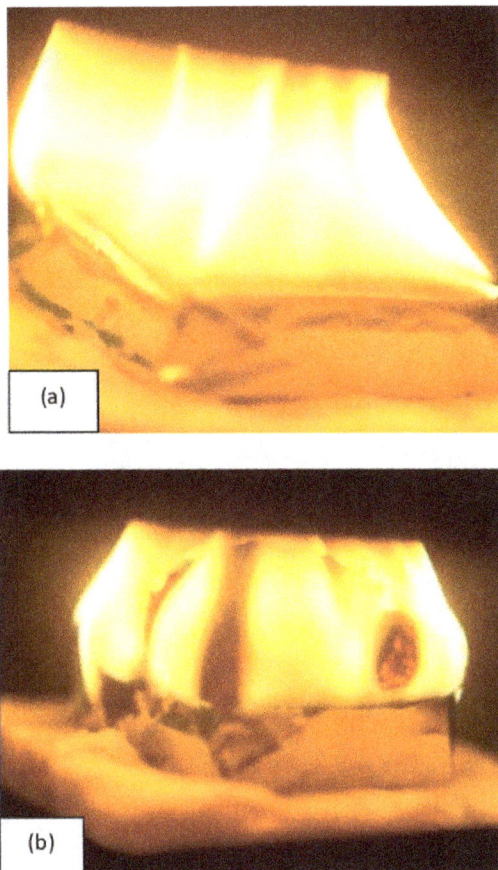

Figure 9 **Images of burning process in Cone calorimetry of *a* pure polypropylene (PP) and *b* (PP/graphene nanoplatelets (GNPs)) nanocomposites**

Table 3 **Result of UL-94 vertical burning test of polypropylene/graphene nanoplatelets (PP/GNPs) nanocomposites**

Sample	Weight ratio/%	Dripping	$T_1(s)$	$T_2(s)$	UL-94 V0
PP	100	Yes	≥30	–	NG
PP/GNPs	90/10	No	≥30	–	NG
PP/GNPs	80/20	No	≥30	–	NG
PP/GNPs	70/30	No	≥30	–	NG
PP/GNPs	60/40	No	2	2	V0
PP/GNPs	50/50	No	1	3	V0

Table 4 **Limiting oxygen index (LOI) of polypropylene/graphene nanoplatelets (PP/GNPs) nanocomposites as a function of GNPs weight percentage**

Sample	Weight ratio/%	LOI (%)
PP	100	18.3
PP/GNPs	90/10	19.5
PP/GNPs	80/20	20.6
PP/GNPs	70/30	22.6
PP/GNPs	60/40	23.4

composites with loading of 10% or higher. The mass of the char residue in each case is consistent with the GNP fraction, hence the chars are mostly likely composed almost entirely of GNPs. Graphene nanoplatelets are known to absorb and re-radiate the heat in the approaching front, thereby reducing the effective flux reaching the sample, and decreasing the combustion rate.[10] This can be seen from the effect of the char on the heat and mass loss rates of the samples, plotted in Fig. 11. The mass of char residue after the cone calorimetry is exactly same with the amount of GNPs, which is consistent with TGA result that confirmed the char is mainly GNPs. The peak heat release rate (PHRR) of Polypropylene homopolymer is as high as 2045 kW, the heat release rate (HRR) reaches a value around 659 kW m^{-2}, while the mass loss rate is 15 g s^{-1}. Addition of GNPs reduces the PHRR continuously to nearly half its original value at a concentration of 5%, but only a minimal decrease in the Heat Release Rate (HRR) or mass loss (MLR) rates is observed. A significant change in the HRR and MLR profiles is observed at a concentration of 10%, where the HLR and MLR are reduced to 400 kW m^{-2} and 10 g s^{-1}.

The largest reduction occurs in the PHRR, indicating that combustion is greatly reduced. This is consistent with the

Figure 10   Photographs of the char residue after the cone calorimetry test. *a* Pure Polypropylene (PP); *b* PP/graphene nano-platelets (GNPs) (99/1 wt-%); *c* PP/GNPs (98/2 wt-%); *d* PP/GNPs (95/5 wt-%); *e* PP/GNPs (90/10 wt-%); *f* PP/GNPs (80/20 wt-%); *g* PP/GNPs (70/30 wt-%); *h* PP/GNPs (60/40 wt-%)

decrease in oxygen consumption from 18.3 to 19.5%, which is also occurs at this concentration. Inspection of the morphology of the chars in Fig. 10 shows that for samples containing <10% GNP filler, the char is porous, while the char containing higher filler concentrations has a uniform, solid appearance. Hence, in addition to improved thermal conduction, the solid char surface also blocks diffusion of oxygen to the hot gases in the interior, impeding the rate of combustion.

The flame out time is strongly correlated to the mass loss rate. From Fig. 11, authors see that an abrupt increase from 184 s for pure PP to 350 s for samples with 10% or higher loading of GNPs. The flame out time, as well as the HLR and MLR do not change significantly with further increase in GNPs loading, as would be expected if the phenomena were a result of a percolated structure formed by the filler particles. This structure would at once increase heat conduction, while decreasing gas flow and hence reducing the rate of mass loss. Both of these effects would contribute strongly to the flame-retardant behaviour.

Cone calorimetry also measures the time to ignition of the sample from an approaching heat front. The time to ignition is an additional measure of the ability of the sample to dissipate heat before the combustion temperature is reached. In contrast to the MLR or HLR, ignition occurs before char formation, and hence, is not related to the char structure. Therefore, decreased ignition times are known to occur in nancomposites[40] even when flame retardancy is increased. The time to ignition for PP/GNPs nanocomposites obtained from cone calorimetry are plotted in Fig. 12 as a function of GNPs concentration. From the figure we can see that the time to ignition is initially decreased relative to that of the PP homopolymer for GNPs filler concentrations of 5% or lower. A large increase occurs at a filler concentration of 10%, where the value becomes slightly larger than that of pure PP, and in contrast to the MLR and HLR values, the

times to ignition continue to increase sharply up to about 30% after which a plateau is reached.

Figure 12 is a plot of the time to ignition as a function of filler concentration. From the figure, we find that initially the time to ignition decreases at the lowest filler concentration, and begins to increase slowly for increased loading. The value for pure PP is recovered at a loading of 10% and increases continuously thereafter. The decreased time to ignition is a common phenomenon in filled systems, and there are problems even when the filler decreases the heat release and mass loss rates, or otherwise improves other flame retardant properties. The phenomenon has been explained as being owing to a rapid increase in the local temperature near the filler as the heat front approaches and the lower heat capacity of the filler causes a more rapid increase in the local temperature. The poor con-ductivity of the matrix does not allow the heat to dissipate, leading to the decrease in time to ignition. In this case, restoring, and even increasing the time to ignition, does seem to correlate well with the electrical percolation concentration. This indicates that thermal conductivity within a filled system may be subject to two time constants. When the system is subjected to small thermal gradients (i.e. $1°C\,min^{-1}$), the thermal conductivity increases linearly with filler concentration, or with the number of filler polymer interfaces indicating that thermal conduction is moderated only by phonon conduction. On the other hand, when the sample is subjected to a rapidly approaching heat front, as is the case when the material is exposed to a flame, thermal conduction must occur on a timescale that is much larger than the phonon frequency in the matrix. In this case, the much more rapid conductivity properties of the electrically conducting particles dominate and actual contact between the particles is required to dissipate the heat. For low GNP loading, heat will accumulate within the vicinity of the particle, accel-erating combustion, till some contact between particles is reached as shown[41] and a conducting pathway is created with

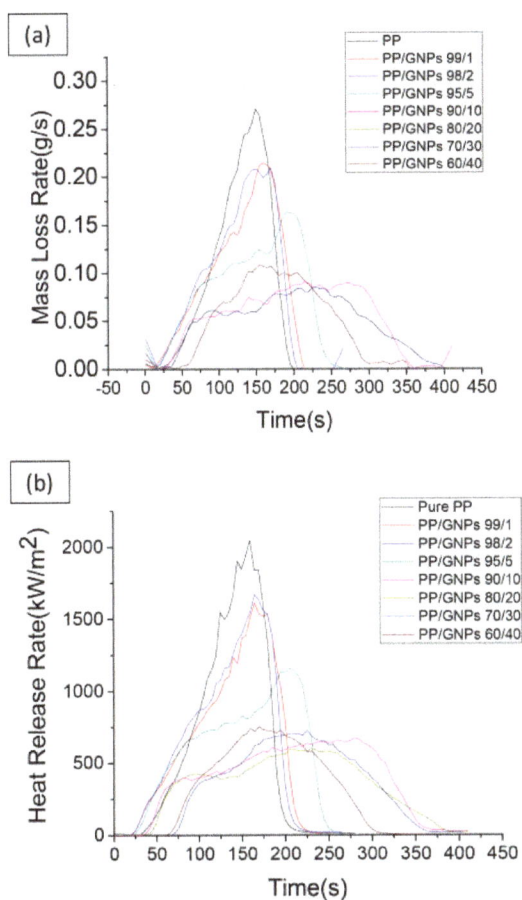

Figure 11 Overlays of mass loss curves *a* and *b* heat release curves of polypropylene/graphene nanoplatelets (PP/GNPs) nanocomposites performed in cone calorimeter as a function of GNPs weight concentration at a heat flux of 50 kW m^{-2} in air atmosphere

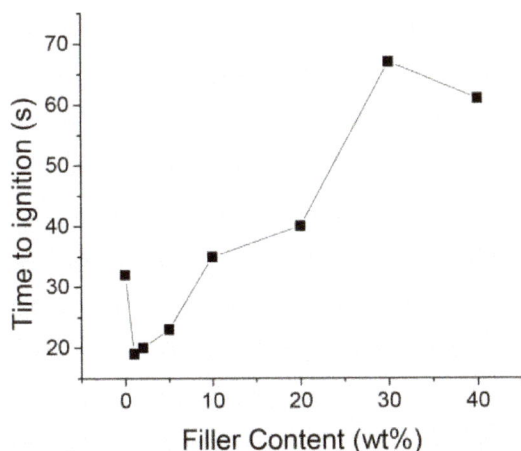

Figure 12 The time to ignition (TTI) of polypropylene/graphene nanoplatelets (PP/GNPs) nanocomposites in Cone calorimetry test as a function of GNPs concentration

particle/particle contacts. Furthermore, we can see that the maximum effect is reached at a concentration of 40 wt-%, when the PP/GNP nanocomposites achieves a UL-94-V0 rating also coincides with the percolation threshold reported and heat dissipation is now dominated by the GNP particles rather than the matrix.

Conclusion

Authors have successfully fabricated the PP/GNP nanocomposites with well-dispersed GNPs via melt blending. Thermal conductivity measurements on the nanocomposites indicated that the thermal coefficient scaled linearly with GNP loading. A value of 2.0 W m^{-1}K^{-1} was achieved at a loading of 40 wt-%. This value, could not be achieved with conductive particles or carbon nanotubes, and hence opens multiple uses for these compounds in heat exchange units or storage tanks. For these applications, the corrosion resistance of PP is ideal for promoting the use of bio-fuels, whose high acidity easily corrodes metal enclosures.

Tensile measurements indicated that the modulus increased, linearly with GNP loading throughout the range tested (1–50 wt-%), while the Izod impact and toughness decreased initially up to a loading of 10% and then remained constant for up to 50%. The torque required for extrusion increased only by 40% over this range indicating that the compound remained processable despite the high loading content.

Small angle X-ray scattering data, together with SEM analysis, indicated that these unusual properties occurred as a result of the ability of PP to incorporate very large volume fractions of GNPs, causing a transformation from a filled polymer matrix with GNP inclusions to a Nacre-like layered nanocomposite, consisting mostly of GNPs coated with polymer chains. The WAXS spectra indicated that the polymer chains had a high degree of crystallinity and were templated on the GNP layer surface in the α-phase, but favouring a different orientation than the chains within the folded lamellae of the neat PP polymer. This internal structure provides a possible explanation for the relatively large impact toughness at high loading, good phonon coupling, and the unusually high thermal coefficients achievable with the GNPs.

The response of the compound to a rapidly approaching heat front was also probed using cone calorimetry, where large reductions in the heat release rate and mass loss rates were observed for concentrations larger than 10 wt-%. Minimal changes were observed for larger concentrations, though a concentration of 40 wt-% was required to render the composite flame retardant according to UL-94-V0 convention. This response was attributed to good thermal conductivity coupled with char forming properties which hindered the oxidation process, as evidenced by the increase in the LOI. The time to ignition though had a parabolic function, initially decreasing and not returning to baseline till loadings higher than 10 wt-% were achieved.

Conflicts of interest

None.

Acknowledgements

Authors acknowledge partial support for this work from National Oilheat Research Alliance (NORA), The New York State Energy Research and Development Authority (NYS-ERDA) and National Science Foundation – Chemical, Bioengineering, Environmental and Transport Systems (NSF-CBET). Research carried out in part at the Center for Functional Nanomaterials and NSLS, Brookhaven National Laboratory, which is supported by the U.S. Department of Energy, Office of Basic Energy Sciences, under Contract No. DE-AC02-98CH10886. Authors would like to thank Israel Chemistry Limited (ICL) for providing the Cone Calorimetry and Limiting Oxygen Index test. One of the authors (Radha Perumal Ramasamy) gratefully thanks Government of India for offering Raman fellowship to support his stay in USA.

References

1. R. Haggenmueller, C. Guthy, J. R. Lukes, J. E. Fischer and K. I. Winey: *Macromolecules*, 2007, **40**, (7), 2417–2421.
2. J. E. Peters, D. V. Papavassiliou and B. P. Grady: *Macromolecules*, 2008, **41**, (20), 7274–7277.
3. P. Song, Z. Caob, Y. Caib, L. Zhaob, Z. Fangb and S. Fua: *Polymer*, 2011, **52**, (18), 4001–4010.
4. M. A. Raza, A. V. K. Westwood and C. Stirling: *Mater. Chem. Phys.*, 2012, **132**, (1), 63–73.
5. K. Kalaitzidou, H. Fukushima and L. T. Drzal: *Carbon*, 2007, **45**, (7), 1446–1452.
6. H. Kim, A. A. Abdala and C. W. Macosko: *Macromolecules*, 2010, **43**, (16), 6515–6530.
7. Z. D. Han and A. Fina: *Prog. Polym. Sci.*, 2011, **36**, (7), 914–944.
8. J. R. Potts, D. R. Dreyerb, C. W. Bielawskib and R. S. Ruoff: *Polymer*, 2011, **52**, (1), 5–25.
9. R. Wake, K. Takano and R. Yoshihara: *J. Iron Steel Inst. Jpn.*, 1995, **81**, (10), 983–988.
10. B. Weidenfeller, M. Hofer and F. R. Schilling: *Compos. A Appl. Sci. Manuf.*, 2004, **35**, (4), 423–429.
11. K. Enomoto, S. Fujiwara, T. Yasuhara, H. Murakami, J. Teraki, N. Ohtake and Jpn: *J. Appl. Phys.*, 2005, **44**, (24-27), L888–L891.
12. X. Shi, J. L. Hudson, P. P. Spicer, J. M. Tour, R. Krishnamoorti and A. G. Mikos: *Macromolecules*, 2004, **37**, (24), 9048–9055.
13. O. Meincke, D. Kaempfera, H. Weickmanna, C. Friedricha, M. Vathauerb and H. Warthb: *Polymer*, 2004, **45**, (3), 739–748.
14. J. Sandler, M. S. P. Shaffera, T. Prasseb, W. Bauhoferb, K. Schulte and A. H. Windlea: *Polymer*, 1999, **40**, (21), 5967–5971.
15. F. M. Du, J. E. Fischer and K. I. Winey: *J. Polym. Sci., Part B: Polym. Phys.*, 2003, **41**, (24), 3333–3338.
16. B. Li and W. H. Zhong: *J. Mater. Sci.*, 2011, **46**, (17), 5595–5614.

17. S. Ghose, K. A. Watson, D. M. Delozier, D. C. Working, J. W. Connell, J. G. Smith Jr, Y. P. Sun and Y. Lin: *High Perform. Polym.*, 2006, **18**, (6), 961–977.
18. B. Dittrich, K. -A. Wartig, D. Hofmann, R. Mülhaupt and B. Schartel: *Polym. Adv. Technol.*, 2013, **24**, (10), 916–926.
19. Y. Q. Shi, X. Qian, K. Zhou, Q. Tang, S. Jiang, B. Wang, B. Wang, B. Yu, Y. Hu and R. K. K. Yuen: *Ind. Eng. Chem. Res.*, 2013, **52**, (38), 13654–13660.
20. S. Chatterjee, F. Nafezarefia, N. H. Taib, L. Schlagenhaufa, F. A. Nüescha and B. T. T. Chua: *Carbon*, 2012, **50**, (15), 5380–5386.
21. K. Kalaitzidou, H. Fukushima, P. Askeland and L. T. Drzal: *J. Mater. Sci.*, 2008, **43**, (8), 2895–2907.
22. J. H. Seol, I. Jo, A. L. Moore, L. Lindsay, Z. H. Aitken, M. T. Pettes, X. Li, Z. Yao, R. Huang, D. Broido, N. Mingo, R. S. Ruoff and L. Shi: *Science*, 2010, **328**, (5975), 213–216.
23. A. Ghatak, K. Vorvolakos, H. She, D. L. Malotky and K. Manoj: *J. Phys. Chem. B*, 2000, **104**, (17), 4018–4030.
24. P. C. Ma, N. A. Siddiquia, G. Maromb and J. -K. Kim: *Compos. A*, 2010, **41**, (10), 1345–1367.
25. I. Y. Phang, T. Liu, A. Mohamed, K. P. Pramoda, L. Chen, L. Shen, S. Y. Chow, C. He, X. Lu and X. Hu: *Polym. Int.*, 2005, **54**, (2), 456–464.
26. I. Zaman, B. Manshoor, A. Khalid and S. Araby: *J. Polym. Res.*, 2014, **21**, (5), 11.
27. K. Kalaitzidou, H. Fukushima and L. T. Drzal: *Compos A*, 2007, **38**, (7), 1675–1682.
28. C. L. Bao, L. Song, C. A. Wilkie, B. Yuan, Y. Guo, Y. Hu and X. Gong: *J. Mater. Chem.*, 2012, **22**, (32), 16399–16406.
29. M. Murariu, A. L. Dechiefa, L. Bonnauda, Y. Painta, A. Gallosb, G. Fontaineb, S. Bourbigotb and P. Duboisa: *Polym. Degrad. Stab.*, 2010, **95**, (5), 889–900.
30. B. Dittrich, K. -A. Wartigb, D. Hofmannb, R. Mülhauptb and B. Schartel: *Polym. Degrad. Stab.*, 2013, **98**, (8), 1495–1505.
31. S. M. Zhao, F. Chena, Y. Huangc, J. -Y. Dongc and C. C. Han: *Polymer*, 2014, **55**, (16), 4125–4135.
32. S. M. Zhao, F. Chena, C. Zhaoa, Y. Huangb, J. -Y. Dongb and C. C. Han: *Polymer*, 2013, **54**, (14), 3680–3690.
33. M. A. Milani, D. Gonzálezb, R. Quijadab, N. R. S. Bassoc, M. L. Cerradad, D. S. Azambujaa and G. B. Galland: *Compos. Sci. Technol.*, 2013, **84**, 1–7.
34. S. Zhang, X. Y. Cao, Y. M. Ma, Y. C. Ke, J. K. Zhang and F. S. Wang: *Exp. Polym. Lett.*, 2011, **5**, (7), 581–590.
35. K. L. Lim, Z. A. Mohd Ishak, U. S. Ishiaku, A. M. Y. Fuad, A. H. Yusof, T. Czigany, B. Pukanszky and D. S. Ogunniyi: *J. Appl. Polym. Sci.*, 2005, **97**, (1), 413–425.
36. M. Mooney: *J. Colloid. Sci.*, 1951, **6**, (2), 162–170.
37. C. Creton, E. J. Kramer, C. Y. Hui and H. R. Brown: *Macromolecules*, 1992, **25**, (12), 3075–3088.
38. A. M. Hindeleh and D. J. Johnson: *J. Phy. D*, 1971, **4**, (2), 259.
39. J. Z. Xu, J. Chen, Y. Wang, H. Tang, Z. -M. Li and B. S. Hsiao: *Macromolecules*, 2011, **44**, (8), 2808–2818.
40. F. Laoutid, L. Bonnauda, M. Alexandreb, J. -M. Lopez-Cuestac and P. H. Dubois: *Mater. Sci. Eng. R Rep.*, 2009, **63**, (3), 100–125.
41. R. P. Ramasamy, K. Yang and M. H. Rafailovich: *RSC Adv.*, 2014, **4**, (85), 44888–44895.

6

Morphology related bulk and surface mechanical properties of ultralow diameter VGCF–iPP monofilament nanocomposites under potential confinement conditions

Golan Abraham Tanami*[1], Liliya Kovalenko[1], Chung Chueh Chang[2], Miriam Rafailovich[3] and Gad Marom[1]

[1]Casali Center for Applied Chemistry, The Institute of Chemistry and the Center for Nanoscience and Nanotechnology, The Hebrew University of Jerusalem, Jerusalem 91904, Israel

[2]Advanced Energy Research and Technology Center (AERTC), State University of New York, Stony Brook, NY 11794, USA

[3]Department of Materials Science and Engineering, State University of New York, Stony Brook, NY 11794-2275, USA

Abstract This study looks at the effects of imposed confinement on morphology related properties in vapor grown carbon nanofibers (VGCF) polypropylene nanocomposite filaments. Different levels of imposed confinement are achieved through varying the content of the VGCF and the dimensions of the nanocomposite filaments. The resulting molecular confinement is apparent through a sharp increase (termed the *inversion point*) in a number of properties at low filament diameters, which is associated with a formation of a 'shish kebab' structure and a transition from a skin/core morphology to an axially oriented lamellar structure.

Keywords Nanocomposites, Particle reinforced composites, Mechanical properties, Extrusion, Surface morphology, Confinement

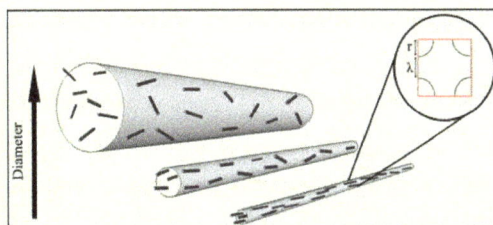

Introduction

Polymer nanocomposites have increasingly gained attention over the past few years in view of their potential to serve as engineering materials in high property and fracture toughness applications. Furthermore, nanocomposites have several advantages over their classical fiber composite counterparts as they can be tailored for diverse biomimetic applications, which aim to synthesize functional nanocomposites with sophisticated hierarchical structures inspired by nature.[1–3] In addition to their reinforcing potential, nanofillers may affect properties indirectly by acting on the polymer matrix of the nanocomposite by either or both mechanisms, as follows: (i) due to their large interfacial area they can induce significant nucleation, to form materials of different properties (termed 'interface materials');[4–6] and (ii) even minute quantities of nanofillers result in small interparticle spacing with potential molecular confinement of the matrix.[7–9] Many types of nanofiller, such as nanoclay, carbon black, metallic and ceramic nanoparticles, carbon nanotubes (CNT) and vapor grown carbon fibers (VGCF), have been incorporated with different polymers in order to develop materials with better barrier, electrical, thermal or mechanical properties.[10–14] Currently, different methods such as melt mixing (extrusion), *in situ* polymerization and electrospinning are widely used in the preparation of polymer nanocomposites.[15–17] The recent scientific literature already offers a range of examples of how nanoparticles, by surface interactions and confinement under different conditions, affect the morphology of the polymer matrix and in turn the properties of the nanocomposites.[9,18–24]

A correlation between molecular orientation in polymer nanocomposite structures to the confined morphology and structure that link to axially enhanced properties has been established recently. The extent of molecular confinement can be manipulated by filaments draw speed and by introducing nanofillers to the system (e.g. CNT or VGCF, etc.). For instance, micron sized thin monofilaments of isotactic polypropylene (iPP) generated under extension flow conditions in the presence of VGCF and CNT showed higher

*Corresponding author, email gad.marom@mail.huji.ac.il

molecular orientation coupled with improved mechanical properties.[6,25,26] Another study of iPP–multiwall CNTs (MWNT) films revealed that the MWNT acted as α-nucleating agents and that fibrillar morphology was observed instead of the common spherulitic one.[27] In addition, extruded monofilaments of iPP–VGCF showed, besides an improvement of mechanical properties, an ordered and orientated lamellar morphology. The proportion of the orientated crystals and of aligned VGCF increased with VGCF addition from 5 to 10%, and from 10 to 35% for the drawn and VGCF reinforced filaments.[6]

The confinement effect in nanocomposites has also been studied recently where for example a Monte Carlo simulation of the conformational properties of polymer chains in platelet nanocomposites was conducted. It was shown that highly orientated layered nanocomposites, as obtained through melt extrusion, lead to an increase in confinement of the polymer chains and to a substantial decrease in their entropy.[28] In another paper, a series of two blends (isotactic polypropylene with silsesquioxane (POSS) and another of nylon 6 with POSS) were produced, and then layered to nanometer thicknesses to test the effects of confinement upon polymer property modification. Mechanical properties increased with respect to decreasing film and layer thickness, e.g. the increase in modulus was from 310 MPa for the neat and thick polymer (305 μm) to 1.95 GPa to the thin and reinforced polymer (25 μm, 3 wt-%POSS, 0.3 wt-%DBS), and that was attributed to changes in the crystal size and structure via confinement of the individual layers.[29] A different study based on POSS and nylon-6 as the matrix confirmed that nanomechanical and bulk thermomechanical properties are related to the observed composite morphology. Studies of nano-indentation of POSS/nylon 6 composites revealed significantly elevated nanoscale properties as compared with the neat polymer, e.g. increase in surface modulus, – from 137 MPa to 2.38 GPa – and hardness – from 24 to 157 MPa upon 10 wt-% addition of POSS.[30] In a number of studies confinement has been achieved through electrospinning, e.g. of PS-MWNT, showing a remarkable increase in the modulus of the confined system, e.g. 2–4 GPa for diameters of the order of 3.0 μm compared with 48 GPa for 250 nm composite fibers (about 13 GPa for the unfilled system).[9] Liu et al. reported that in semicrystalline polymers such as PEVA and LDPE, confinement imposed by a reduction of about 8 μm in diameter for the electrospun fibers decreased the melting point by almost 20%, while at the same time improved the modulus by almost 60%.[31] Hershkovits-Mezuman et al. investigated the transition from edge on to flat on lamellar morphology under confinement in spin cast pristine and carbon nanofiber or nanoclay filled linear low density polyethylene thin films. They concluded that the addition of nanoparticles generates new interactions such as polymer nucleation at the nanoparticle surface, which disrupts the thickness uniformity and the pristine morphologies.[20] Bartczak et al. studied the crystallization of high density polyethylene either on the calcite crystals or between the layers of ethylene octene rubber, and found a strong dependence of the morphology on the film thickness: for films thicker than 0.4 μm, the crystal resembles the conventional spherulite, for films with smaller thickness the

morphology is in sheaf-like aggregates within which the lamellae are oriented edge-on with respect to the substrates.[32]

The aim of the present study was to look into variable imposed confinement states and to understand how they affect the morphology and – in turn – the properties of VGCF–iPP nanocomposites. The overall confinement state would be produced by employing small structural dimensions, in analogy to our previous study where confinement was achieved by reducing the thickness of polymer films.[20] Here, confinement was to be induced through employing an extrusion drawing process to form ultralow diameter monofilaments. In addition, relatively high loadings of VGCF (up to 10 wt-%) were expected to increase the confinement level further by decreasing the size of continuous polymer domains.

Material and methods

Materials

For the preparation of extruded monofilament nanocomposites, we used iPP high flow nucleated homopolymer (Mw=135 000 g mol^{-1}; Capilen U77; Carmel Olefins, Israel) as the matrix. Acicular nano-fibers were used as the reinforcing phase: VGCF (graphitized up to 2800°C, average diameter 150 nm, length 10–20 μm, density 2.0 g cm^{-1}, without surface treatment, i.e. as is), supplied by Showa Denko KK, Japan. In addition, 0.1 wt-% of the antioxidant Irganox B-225 (Ciba) was added in order to prevent polymer degradation at high temperatures.

Fabrication of nanocomposite monofilaments

Nanocomposites of 1–10 wt-%VGCF were prepared from precompounded iPP/VGCF flakes, as follows. First, a weighed quantity of VGCF was sandwiched between two iPP films with 0.1 wt-% antioxidant (Irganox), then hot pressed and caged in the polymer (to avoid health hazards when transferring VGCF directly from the balance to the extruder and to keep the original weight) and finally cut to flakes by scissors. The flakes were then transferred to a twin screw microcompounder (DSM Xplore, 15 cc Micro Extruder; Geleen, The Netherlands). The melt blending process was carried out at 195°C (under constant flow of N$_2$ gas, to prevent oxidation) for a period of 10 min at a rotational speed of 100 rpm. Subsequently, monofilaments were produced from the melt, by drawing through a die of 300 μm at a speed of 50–2200 rev min^{-1}. They were collected onto a winding cylinder with a diameter of 10–15 cm. The drawing speed, defined as $v=\omega R$, dictates the final monofilament diameter, where ω is the angular speed (rev min^{-1}) and R is the radios of the collecting cylinder (cm).

Characterization

Optical microscopy

Sorting each monofilament according to size was conducted using optical microscopy (Nikon Eclipse 80i). The monofilament diameter was measured at more than 10 points along a about 45 cm section, using analysis software (NIS, Elements BR 3.0).

Bulk mechanical properties

Tensile testing measurements were carried out using a universal testing machine (Model 4502; Instron, UK) with a static load cell of ±10 N. The gage length was 20 mm and the loading rate was 20 mm min^{-1}. The data presented in the graphs are based on at least 40 specimens per diameter, and more than 1200 samples overall. Statistical processing was carried out using Origin 8.0 (Northampton, MA, USA) and Excel (Microsoft).

Surface mechanical properties

Two different atomic force microscope (AFM) instruments and methods were employed to characterize the surface mechanical properties. Stiffness measurements at room temperature of the relative modulus were conducted by shear modulation using an AFM (Dimension 3000; Veeco/ Digital Instruments, USA) in contact mode. The principles of shear modulation force microscopy (SMFM) method and the set-up of the experiment were described in an earlier paper.[33] A drive signal that generated shear was applied on the monofilament along its main axis (z direction). The signal amplitude was varied from 7.5 to 82.5 mV, which corresponded to an x piezo displacement of 1.5–16.5 nm. The cantilever response (i.e. the tip deflection, Δx) was recorded to estimate the stiffness of the fiber surface. At least three monofilaments were measured for the same diameter of certain VGCF loading. Each monofilament was measured at three different locations. Three measurements were done at each location. Hence, each point in the graph represented an average of 27 measurements. According to the mathematical model proposed in,[33] the deflection of the cantilever (i.e. the response signal) is proportional to shear modulus of the specimen: $\Delta x \propto G^{-2/3}$; since shear and elastic moduli (E) are also proportional to each other: $\Delta x \propto E^{-2/3}$. The relative surface modulus is essentially the surface modulus of certain monofilament with specific VGCF loading, divided by the surface modulus of the reference specimen, i.e. pristine monofilaments. The mathematical expression is as follows

$$\frac{\Delta x_1}{\Delta x_2} \propto \left(\frac{E_2}{E_1}\right)^{2/3} \Rightarrow \frac{E_1}{E_2} \propto \left(\frac{\Delta x_1}{\Delta x_2}\right)^{3/2}$$

The same drive response AFM traces (Fig. 2a) were used to calculate the energy to yield (termed surface Y. Energy) by integrating the area under the elastic region (the initial linear section) of the traces.

Temperature dependent relative modulus measurements were carried out by nano-indentation, using the AFM (Dimension Icon; Bruker, USA) in the dynamic penetration mode. Monofilaments with varied VGCF content and diameter were mounted on Si wafer (1×1 cm) using thermal glue (solvent free epoxy, Double/Bubble; Hardman Adhesives, USA). The AFM tip was placed on top of the filament and calibrated for modulus measurements for a scan area of 500×500 nm in a temperature range of 25–130°C. Considering the nanometric diameter of the tip, the surface of the micrometer diameter filament can be regarded as flat. The measured modulus at a specific temperature was divided by the room temperature modulus of the specific monofilament and termed relative modulus.

Morphology of monofilaments

The samples were dipped in an etching solution according to the standard procedure for highlighting crystalline structure.[34] Etching time was varied between 0.5 and 4 h,

according to monofilament thickness. Specimens were coated with a Au–Pd nanolayer using a SC7640 Sputter. Electron micrographs were acquired with high resolution SEM (FEI Sirion) operated at 3–5 kV.

Development of supramolecular structure

In order to follow the changes in crystalline structure, e.g. shish kebab appearance and orientation, small angle X-ray scattering (SAXS) technique was employed. The pristine and 5 wt-%VGCF loaded filaments were measured. Each sample was measured twice. The X-ray generator used, MicroMax-007HF (Rigaku Corporation), has a rotating anode operating at 40 kV and 30 mA with a copper target producing K_α photons with an energy of 8 keV (wavelength of 1.54 Å). The focal spot size on the anode is 70×70 µm. A focused monochromatic beam is obtained using Confocal Max-Flux optics consisting of a CMF-12-100Cu8 focusing unit (Osmic Inc., a Rigaku company). The beam continues into a vacuum flight path (about 15 torr), which is 2.54 cm diameter and contains two slits. The scatterless hybrid metal/Ge single-crystal slits (Forvis Technologies, Inc.) are fully motorized. Performance is optimized when the first slit (after the optics) is set to 1×1 mm and the beam spot size is set by the second slit (close to the sample) at 0.7×0.7 mm. The motorized sample stage consists of a Huber goniometer (to control the rotational angle) combined with X-Y-Z translation stages (Forvis Technologies, Inc.). The scattered beam enters a large He filled flight path (about 36 cm in diameter). A MAR345 image plate detector (Marresearch GmbH) is stationed at the end of this flight path on a motorized plate. The sample to detector distance is about 1850 mm. The formation of a shish kebab structure was identified by qualitative observation of the 2D pattern. Appearance of layered lamellar assemblies (termed kebabs) generates clear meridional maxima, while the microfibrillar structure (termed shish) sometimes gives rise to an equatorial streak.[35,36] Since the kebab formation has a clear intensity maximum on the 2-D SAXS diffraction pattern, its orientation can be calculated using the full width at half maximum (FWHM) determined from the diffraction intensity as a function of the azimuthal angle.

Results

In this research, variable confinement levels were achieved by changing the diameter of the monofilaments, thus the studied properties presented below are evaluated as a function of the diameter. Bulk properties of the monofilaments – measured by universal tensile testing – and surface properties – measured by atomic force microscopy, are also presented in separate sections.

Bulk mechanical properties

Figure 1a presents the filament stiffness as a function of the diameter. In the region between 250 and 50 µm, monofilaments with 10 and 5 wt-%VGCF possess ∼2.5 and 1.5 times higher initial stiffness values respectively, than the pristine ones. The monofilaments maintain a fairly constant stiffness values within error with a slight increase when reaching to 100 µm. At 100–50 µm, the values decline, followed by a sharp increase that appears as an inversion point below the

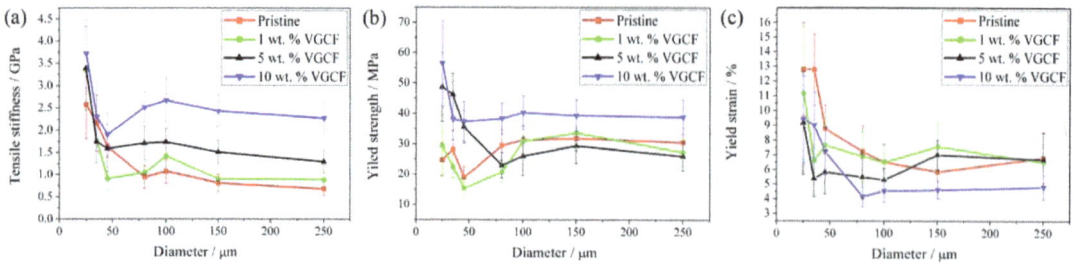

Figure 1 Bulk mechanical properties as function of filament diameter, presenting *a* tensile stiffness, *b* yield strength and *c* yield strain

diameter of 50 μm, with stiffness values approaching 4 GPa for 5–10 wt-%VGCF, and 2 GPa for the pristine and 1 wt-% monofilaments. Figure 1b exhibits the strength at yield point, where it is seen that the 10 wt-%VGCF nanocomposite is the strongest monofilament throughout the entire range, with yield strength values of about 60 MPa. The yield strength exhibits an inversion point for the same diameters as the stiffness results, and its values at diameters of 25 μm are 50 MPa for the 5 wt-%VGCF, 30 MPa for the 1 wt-%VGCF and 25 MPa for the pristine sample. The nanocomposite mono-filaments strain values at yield point can be viewed in Fig. 1c. For 10 wt-%VGCF and diameters in the range between 250 and 100 μm, the strain is lower than the strain values of the pristine and the 1–5 wt-%VGCF monofilaments. The inversion point noticed earlier presents as well, essentially in the same diameter region as before. Thereafter, elevated yield strain values up to 11–13% for 1 wt-% and pristine respectively, and 9% for the 5–10 wt-%VGCF specimens can be found.

Surface mechanical properties

Figure 2a shows the drive versus response amplitudes for various monofilaments, as recorded by the AFM–SMFM method. It is noted that the drive force of the AFM tip was directed along the monofilaments axis. Lower modulus materials are more responsive to an applied force (drive amplitude), thus having lower slopes. As can be seen from the linear section of the 250 μm^{-1} wt-%VGCF monofila-ment, it has the lowest slope, i.e. its responsiveness with the applied load which originates in the AFM tip is fairly high, reflecting its substantially lower surface shear modulus. For instance, at an applied load of 30 mV, the 25, 50 and 80 μm monofilaments have response values of 80, 100 and 150 mV respectively, while the 250 μm has a response value of ca. 250 mV. Figure 2b presents the relative surface elastic modulus measured at room temperature (using the SMFM

method), which is essentially the modulus ratio of the iPP–VGCF to the iPP (pristine) monofilaments. It is apparent that the surface elastic modulus decreases with the VGCF content throughout the entire studied diameter range, implying that the highly loaded VGCF monofilaments have softer skins. Here too an inversion point is observed at around 50 μm. Figure 2c presents the relative yield energy at the skin. Apparently, the behavior is similar to that of the elastic modulus, namely, the yield energy for 10 wt-%VGCF monofilaments is lower than those of the 1–5 wt-%, where below 100 μm diameters there is a drop of approximately 80% in the energy for all the VGCF contents.

A different AFM procedure utilizing the dynamic penetra-tion mode was employed to measure the relative modulus (with respect to room temperature value) at the surface of the monofilaments as a function of the temperature in the range of 25–130°C. Figure 3 is an example of a 3D presenta-tion of the modulus in an area of 500×500 nm^2 at 110°C, for the 25 μm pristine monofilaments (Fig. 3a) and for the 5 wt.-% VGCF monofilaments (Fig. 3b). It can be seen (Fig. 3a) that whereas no readable signal above the noise level could be extracted from the pristine sample (the S/N ratio is very low), the reinforced sample retains a readable S/N ratio even at this elevated temperature (110°C), reflecting a significantly higher surface modulus. The normalized moduli with respect to the room temperature values (E_{T*}/E_{RT}) are depicted in Fig. 4 as a function of temperature in the range of 25–130°C, comparing the pristine and the 5 wt-%VGCF monofilaments for three filament diameters, namely, 250 μm (Fig. 4a), 50 μm (Fig. 4b) and 25 μm (Fig. 4c). A common characteristic of the graphs in Fig 4 is that they all exhibit a sharp softening temperature – expressed by an abrupt drop in E_{T*}/E_{RT} – that varies with the diameter and composition of the monofila-ments. It is apparent that the softening temperatures of

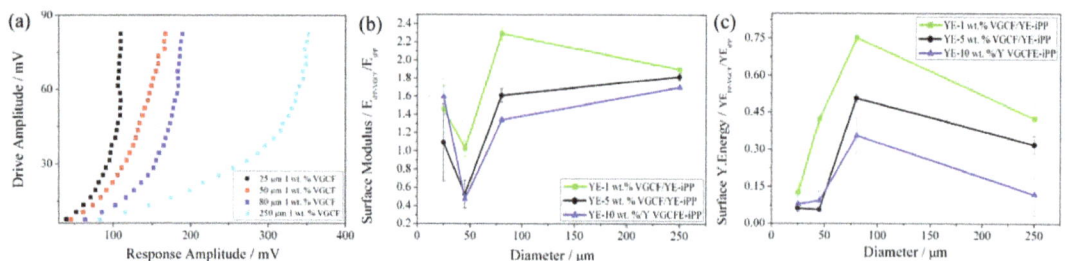

Figure 2 Surface mechanical properties as function of filament diameter, presenting *a* drive response amplitude, *b* sur-face modulus of loaded filaments relative to pristine and *c* surface yield energy of loaded filaments relative to pristine

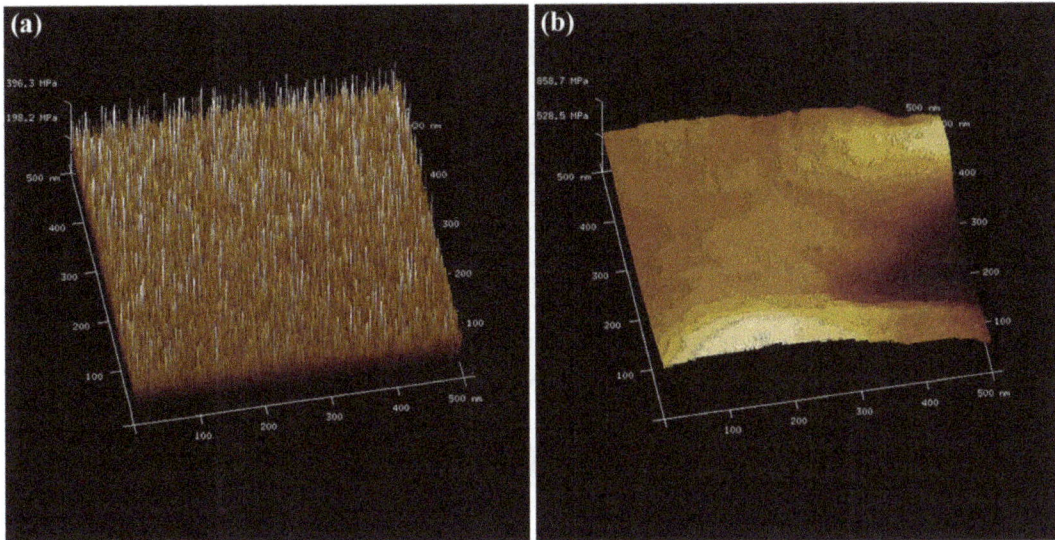

Figure 3   3D illustration of surface modulus in scan area of 500 × 500 nm at 110°C of *a* pristine and *b* loaded filaments

the 250 µm monofilaments are identical at around 100°C regardless of the VGCF (Fig. 4*a*). At lower filament diameters of 50 and 25 µm, the 5 wt-%VGCF monofilaments retain their temperature stability considerably, while that of the unreinforced monofilaments is reduced by some 30°C. Moreover, the 50 and 25 µm, 5 wt-%VGCF monofilaments retain more than 20 % of the room temperature surface stiffness even beyond the softening point up to at least 140°C.

Morphology of monofilaments

In order to gain a comprehensive understanding of the monofilaments behavior over a range of varied diameters and increased potential confinement, SEM analysis was carried out on mildly etched samples (see the section on 'Morphology of monofilaments'). Fig. 5 presents SEM images of a number of filaments according to their diameter and VGCF content, as specified. It is observed that 250 µm thick monofilaments of pristine-iPP (Fig. 5*a*) and 1 wt-%VGCF (Fig. 5*b*) have nearly smooth surfaces with marks of a circumferential morphology in the skin, under which the core is completely smooth (Fig. 5*a*). A similar observation applies to the pristine 50 µm (Fig. 5*c*) and the 50 µm 1 wt-%VGCF (Fig. 5*d*) – the latter revealing a distinctive circumferential morphology. For smaller filament diameters of similar respective compositions, i.e. 25 µm of pristine (Fig. 5*e*) and 1 wt-%VGCF (Fig. 5*f*), the

surfaces are completely smooth. Then, as either the diameter of the monofilament is decreased further or the content of VGCF is increased, the morphology becomes distinctively axial. This occurs through a transition state at which a circumferential skin and axial core morphologies exist simultaneously, forming a shell/core structure (Fig. 5*g–i*, see legend for details) – eventually reaching an all axial morphology for the 25 µm 5 wt-%VGCF (Fig. 5*j*) and 10 wt-%VGCF (Fig. 5*k* and *l*). Table 1 summarizes the main SEM morphological observations. It is noted, as discussed below, that the transition state comprising a shell/core structure, may coincide with the property inversion point observed for the various properties presented in Figs. 1 and 2.

This table presents changes of VGCF loadings (left–right) and diameter (up–down).

Development of supramolecular structure

In view of the SEM observation, a structural analysis was performed by small angle X-ray scattering. Figure 6 presents two series of 2D SAXS diffraction photos for pristine and for the 5 wt-%VGCF monofilaments over the whole tested diameter range. Moving from high to low diameters, it is apparent for the two photo series that for high diameters the scattering patterns comprise equatorial streaks that are indicative of axial molecular alignment. Then, as the

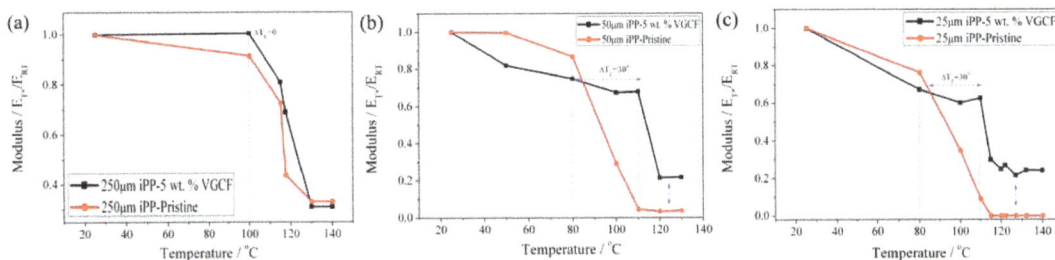

Figure 4   Surface moduli of pristine and loaded filaments at different diameters: *a* 250 µm, *b* 50 µm and *c* 25 µm, as function of temperature

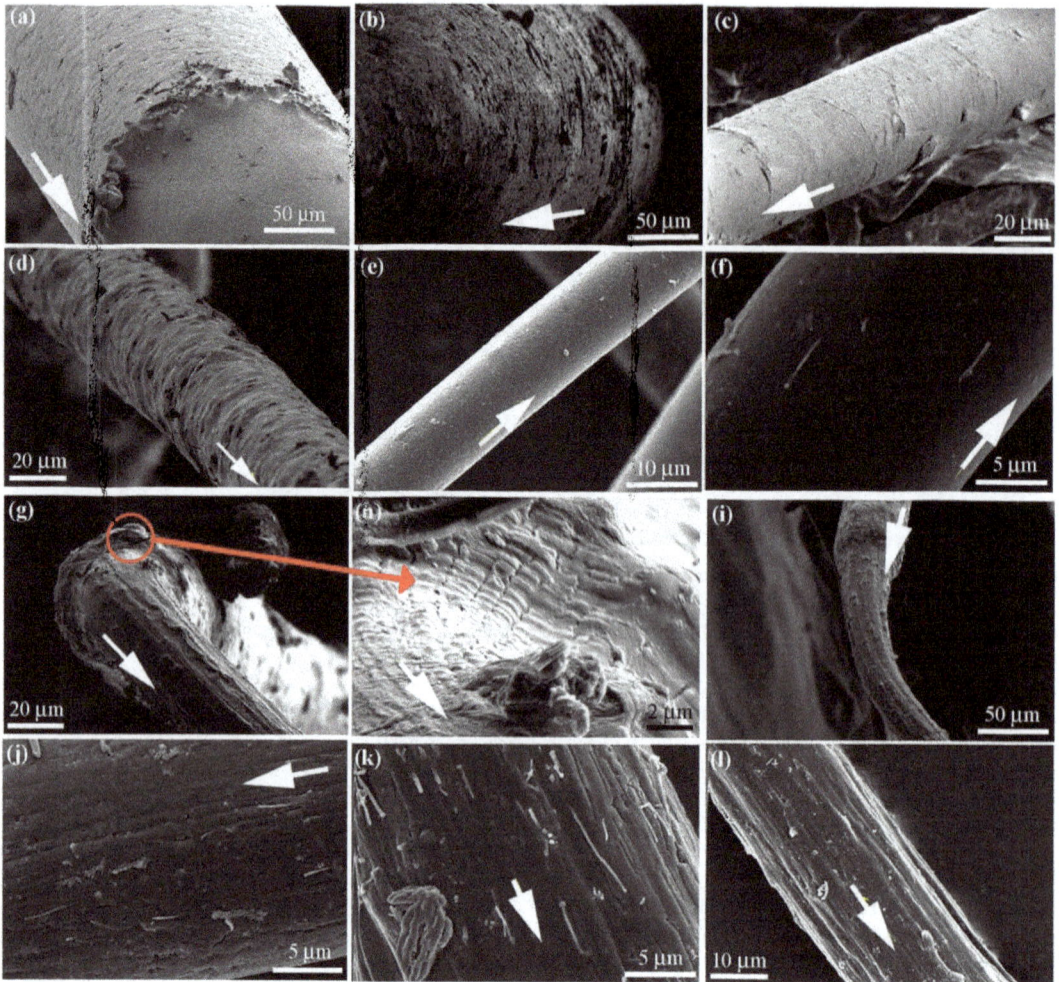

Figure 5 Scanning electron micrographs presenting filaments with different VGCF loadings and diameters: arrows mark longitudinal axis of filaments. *a* 250 μm pristine; *b* 250 μm 1 wt-%; *c* 50 μm pristine; *d* 50 μm 1 wt-%; *e* 25 μm pristine; *f* 25 μm 1 wt-%; *g, h* 50 μm 5 wt-%; *i* 50 μm 10 wt-% ; *j* 25 μm 5 wt-%; *k, l* 25 μm 10 wt-%

diameter decreases meridional scattering rings appear progressively, reflecting the formation of a lateral lamellar structure on top of the axial one, which together result in a 'shish kebab' morphology.[35] An important observation is that the presence of VGCF facilitates the formation of ordered 'shish kebab' morphology, so that clear scattering from an ordered lamellar structure is observed already at a filament diameter of 100 μm in comparison to 35 μm for the pristine filaments. Figure 7 presents a quantitative estimate of the orientation of the lateral lamellar structure expressed by the FWHM of the SAXS Gaussian fitted meridional peaks. It is seen that particularly at the higher diameter range the, VGCF monofilaments exhibit significantly higher levels of

orientation (sharper peaks) – then, at lower diameters, the effect of the VGCF becomes insignificant as it is being overshadowed by that of the high draw ratios.

Discussion

The results by and large portray consistent trends of properties as a function of the diameter of the monofilaments. Moreover, there seems to be a substantial structure–property correlation so that morphological changes that occur as the filament diameter is reduced (e.g. the transition from shell/core to axial morphology) are expressed directly in the physical property trends. A number of parameters

Table 1   Morphology as function of VGCF loadings and diameter

Diameter/μm	Pristine	1 wt-%VGCF	5 wt-%VGCF	10 wt-%VGCF
250	Slightly circumferential	Slightly circumferential	Moderately circumferential	Moderately circumferential
50	Slightly circumferential	Definite circumferential	Shell/core	Shell/core
25	Smooth	Smooth	Axial	Definite axial

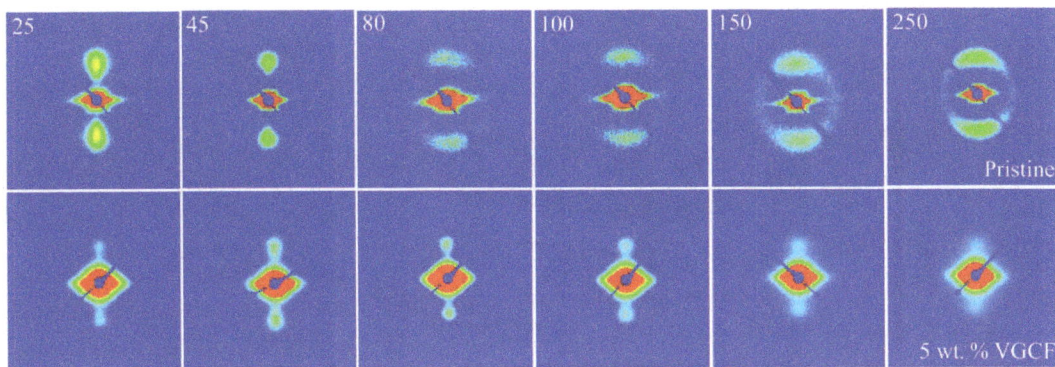

Figure 6   2D SAXS diffraction pattern of pristine and VGCF reinforced filaments with diameters as noted

that affect the structural changes can be considered, as published elsewhere,[39] while here, the focus is on the property trends. Examining the details of those trends, a number of features are predictable by composites theory considerations. For example, that the strength and stiffness of the monofilaments increase as the volume fraction of VGCF is increased or that the thermal stability of the filament surface is higher with VGCF. The slight increase of the stiffness and yield strength in Fig. 1a and b may possibly be correlated with better molecular orientation which is a direct product of the drawing process. Higher drawing speeds facilitate promotion of disentanglements of polymer chains; hence, longer sequences of crystalline regions and improvement of mechanical properties are feasible.

Two outstanding observations, however, are underlined as follows. The first is a gradual property decrease that starts around a diameter of 100 μm. The second is a sharp inversion that occurs at low filament diameters and for higher VGCF concentrations, and which is expressed throughout the experimental results. Seemingly, these observations correlate with a structural transformation from a circumferential to an axial morphology via a transition state of a shell/core structure and with the appearance of 'shish kebab' morphology. The SEM images are consistent with different crystalline orientations due to nucleation from different surfaces. One orientation arises from nucleation and orientated crystallization by the bounding surface of the

extruder die, forming a thin shell around the filament. The shell comprises an oriented crystalline structure and since this layer follows the outline of the extrusion die, it produces circumferential order, which also propagates several lamellae into the bulk, estimated from the SEM image to be approximately 5–10 μm thick. This can be seen in Fig. 5a where the shell is approximately 10 μm thick and constitutes only a small component of the much thicker extruded filament. The nanofibers provide an additional surface for orientation within the bulk of the filament, which, as discussed by Larin et al,[37] induces nucleation somewhat earlier than the bulk. The exact mechanism is not yet understood, but when the VGCF nanofibers are present, a core phase is observed with axial orientation in the thinnest filaments (25 μm) studied and shown in Fig. 5i and j. For filaments around 50 μm, two well defined orientations are observed, which are separated by an apparent interface – expected to introduce some mechanical weakness into the system. We then postulate that the formation of this interface structure and its relative proportion with respect to the two bulk phases may also explain the features in Figs. 1 and 2, where a minimum is observed in both the tensile strength and the surface moduli for filaments around 50 μm. When the filament diameter becomes even smaller, the shish structure seems to win out over the circumferential one, and the whole filament now consists of a shish structure oriented along the axis of the filament. The sharp property increase beyond the inversion point reflects a formation of a highly oriented lamellar structure obtained by either shear[38] or draw[6] process.

Two additional factors might be considered regarding the nanocomposites, namely, the effects of the nanofibers on nucleation and orientated crystallization and on confinement. Considering the first factor, the SAXS results show that the 'shish kebab' morphology of the 5 wt-%VGCF form already at higher filament diameters than the pristine. In fact, the SAXS results based on width analyses of the meridional scattering peaks (Fig. 7) also show that the VGCF generate a highly orientated lamellar structure that acquires its order at relatively higher diameters. Thicker monofilaments posses a gradient alignment of polymer molecules located in the core of the monofilament compared to those in its surface,[9,39] whereas smaller diameters exhibit larger fractions of core molecules aligned in the flow direction.

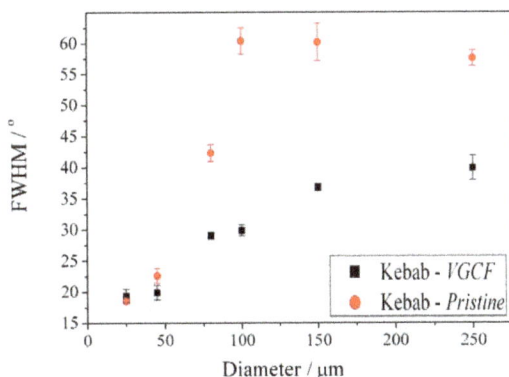

Figure 7   FWHM (i.e. molecular orientation) of iPP pristine and iPP–5 wt-%VGCF kebab as function of filament diameter

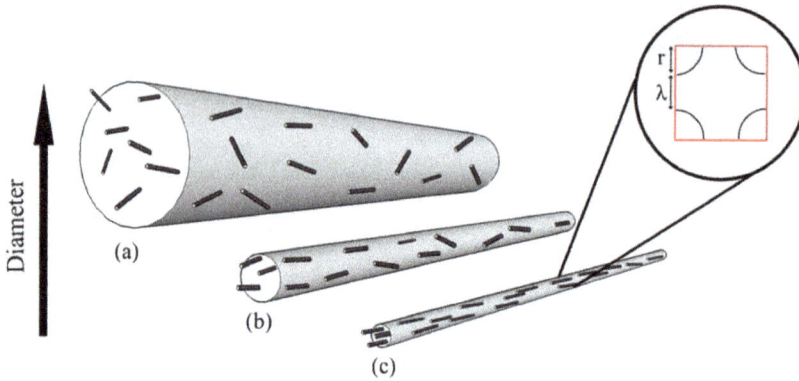

Figure 8 Illustration of nanofillers (their radius marked with '*r*') dispersed inside filaments, presenting their better alignment and overlapping due to decreased interfiber distance (marked as 'λ') as consequence of lowering filaments diameter, starting from *a* thick to *b* thin and ending with *c* ultra low diameter

Obviously, the better orientated structure of the low diameter nanocomposites is responsible for their significantly higher mechanical properties.

Considering the second factor – confinement – in view of our published results on confinement effects in thin films that showed how radical structural changes occur over constricted nanometric dimensional variations,[20] we suggest that confinement is responsible for the sharp property change. The reason why confinement appears to generate a sharp rather than a gradual change as a function of diameter could be – as claimed previously in the cited paper – that a specific molecular dimension is attained, e.g. the radius of gyration of the polymer chain. For instance, significant improvements in bulk stiffness and tensile strength have been achieved in this work in comparison to a previous study of iPP–VGCF 120 μm filaments,[6] namely, from 1.3 to 3.5–4.0 GPa, and from 40 MPa to almost 60 MPa respectively. Other confined systems also reported remarkable improvements. For example, stiffness values of Polyethylene terephthalate fibers increased from 2.0–3.0 to 15.0 GPa and the yield strength increased as well, from 50–150 MPa to 1.1 GPa for the electrospun (melt spinning) fibers. Also, nylon 6 and 6,6 electrospun fibers were reported to exhibit

an increase of about 4.0 and 900 MPa in stiffness and yield strength respectively, relative to the bulk system.[40]

The source of potential confinement is the crowding of the nanofibers as the diameter of the monofilament is decreased. The crowding phenomenon is depicted schematically in Fig. 8 that presents high and low diameter monofilaments with their corresponding arrays of nanofibers. The graphic presentation demonstrates how a disoriented open array in a thick monofilament turns into a highly orientated dense one in a thin monofilament. Because of the high aspect ratio (length to diameter ratio) of the VGCF, there will be overlapping sections of the VGCF so that cross-sections of the monofilaments, as shown in Fig. 8c, will exhibit gradually higher and higher two dimensional volume contents of VGCF. Assuming that the nanofibers are organized in a square array, as illustrated in the magnified view, the interfiber distance λ is related to the nanofibers volume fraction ϕ_f and their radius r by $\phi_f = \pi r^2/(\lambda + 2r)^2$. At $\phi_f = 0.79$ (the maximum fiber volume fraction for square packing), $\lambda = 0$. It can be easily calculated that as the volume fraction of the overlapping nanofibers increases, interfiber distances of nanometric orders prevail, as demonstrated in Fig. 9 that depicts the interfiber spacing as a function of the fiber volume fraction.

Conclusions

The formation of a highly orientated lateral lamellar structure as the monofilament diameter is decreased, which is responsible for an abrupt jump in the mechanical properties, is facilitated markedly by nanofibers (VGCF). Two factors – either or both – might potentially be active, namely, the effects of the nanofibers on nucleation and orientated crystallization and on confinement at interoverlapping nanofiber regions. The argument for confinement is that it appears to generate a sharp rather than a gradual change as a function of diameter, possibly indicating that a specific polymer chain dimension is attained, e.g. the lamellar length and radios of gyration related length scales.

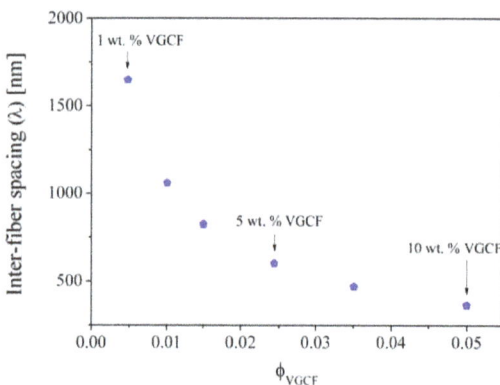

Figure 9 Interfiber distance as function of volume fractions: weight fractions are noted as well

Conflicts of interest

The authors have no conflicts of interest to declare.

Acknowledgements

The authors are indebted to Mrs Carmen Tamburu and Dr Yaelle Schilt from the X-Ray Scattering Laboratory of Professor Uri Raviv at the Hebrew University of Jerusalem for their assistance in the analysis of the SAXS results. MR gratefully acknowledges the NSF Inspire program.

References

1. K. Shanmuganathan, J. R. Capadona, S. J. Rowan and C. Weder: *Progress in Polymer Science*, 2010, **35**, (1–2), 212–222.
2. C. K. Chan, T. S. S. Kumar, S. Liao, R. Murugan, M. Ngiam and S. Ramakrishman: *Nanomedicine*, 2006, **1**, (2), 177–188.
3. C. S. S. R.Kumar: 'Biomimetic and Bioinspired Nanomaterials'; 2010, New York, Wiley & Sons.
4. L. Schadler: *Nat. Mater.*, 2007, **6**, (4), 257–258.
5. H. D. Wagner and R. A. Vaia: *Materials Today*, 2004, **7**, (11), 38–42.
6. B. Larin, T. Lyashenko, H. Harel and G. Marom: *Composites Science and Technology*, 2011, **71**, (2), 177–182.
7. S. Srivastava and J. K. Basu: *Physical Review Letters*, 2007, **98**, (16), 165701.
8. J. Choi, M. J. A. Hore, J. S. Meth, N. Clarke, K. I. Winey and R. J. Composto: *ACS Macro Letters*, 2013, **2**, (6), 485–490.
9. Y. Ji, C. Li, G. Wang, J. Koo, S. Ge, B. Li, J. Jiang, B. Herzberg, T. Klein, S. Chen, J. C. Sokolov and M. H. Rafailovich: *Epl*, 2008, **84**, (5) 56002.
10. L. S. S. Pulickel M. Ajayan, Paul V. Braun: 'Nanocomposites Science and Technology'. 2003, New York, Wiley & Sons.
11. H. Miyagawa, M. Misra and A. K. Mohanty: *Journal of Nanoscience and Nanotechnology*, 2005, **5**, (10), 1593–1615.
12. F. Hussain, M. Hojjati, M. Okamoto and R. E. Gorga: *Journal of Composite Materials*, 2006, **40**, (17), 1511–1575.
13. L. Vaisman, H. D. Wagner and G. Marom: *Adv. Colloid Interface Sci.*, 2006, **128**, (21), 37–46.
14. P. C. Ma, N. A. Siddiqui, G. Marom and J. K. Kim: *Compos. Pt. A-Appl. Sci. Manuf.*, 2010, **41**, (10), 1345–1367.
15. Z. M. Huang, Y. Z. Zhang, M. Kotaki and S. Ramakrishna: *Composites Science and Technology*, 2003, **63**, (15), 2223–2253.
16. J. Jordan, K. I. Jacob, R. Tannenbaum, M. A. Sharaf and I. Jasiuk: *Materials Science and Engineering: A*, 2005, **393**, (1–2), 1–11.
17. E. T. Thostenson, C. Y. Li and T. W. Chou: *Composites Science and Technology*, 2005, **65**, (3–4), 491–516.
18. A. Mejia, N. Garcia, J. Guzman and P. Tiemblo: *Eur. Polym. J.*, 2013, **49**, (1), 118–129.
19. A. J. Muller, M. L. Arnal, M. Trujillo and A. T. Lorenzo: *Eur. Polym. J.*, 2011, **47**, (4), 614–629.
20. A. Hershkovits-Mezuman, H. Harel, Y. T. Wang, C. H. Li, J. C. Sokolov, M. H. Rafailovich and G. Marom: *Compos. Pt. A-Appl. Sci. Manuf.*, 2010, **41**, (9), 1066–1071.
21. A. Baji, Y. W. Mai, S. C. Wong, M. Abtahi and X. S. Du: *Composites Science and Technology*, 2010, **70**, (9), 1401–1409.
22. N. Muksing, S. Coiai, L. Conzatti, E. Passaglia, R. Magaraphan and F. Ciardelli: *Journal of Nanoscience and Nanotechnology*, 2010, **10**, (9), 5814–5825.
23. Y. Shen, Z. H. Guo, J. Cheng and Z. P. Fang: *J. Appl. Polym. Sci.*, 2010, **116**, (3), 1322–1328.
24. L. Chen, K. Liu, T. X. Jin, F. Chen and Q. Fu: *Express Polym. Lett.*, 2012, **6**, (8), 629–638.
25. A. Y. Feldman, B. Larin, N. Berestetsky, G. Marom and A. Weinberg: *Journal of Macromolecular Science, Part B: Physics*, 2007, **46**, (1), 111–117.
26. G. A. Tanami, E. Wachtel and G. Marom: *Polymer Composites*, 2013, **34**, (3), 382–389.
27. E. Assouline, A. Lustiger, A. H. Barber, C. A. Cooper, E. Klein, E. Wachtel and H. D. Wagner: *Journal of Polymer Science Part B: Polymer Physics*, 2003, **41**, (5), 520–527.
28. Y. Termonia: *Phys. Rev. E*, 2013, **88**, (01263), 1–5.
29. M. M. Herbert, R. Andrade, H. Ishida, J. Maia and D. A. Schiraldi: *Polymer*, 2013, **54**, (26), 6992–7003.
30. R. Misra, B. X. Fu, A. Plagge and S. E. Morgan: *J. Polym. Sci. Pt. B-Polym. Phys.*, 2009, **47**, (11), 1088–1102.
31. Y. Liu, S. Chen, E. Zussman, C. S. Korach, W. Zhao and M. Rafailovich: *Macromolecules*, 2011, **44**, (11), 4439–4444.
32. Z. Bartczak, A. S. Argon, R. E. Cohen and M. Weinberg: *Polymer*, 1999, **40**, (9), 2331–2346.
33. S. Ge, Y. Pu, W. Zhang, M. Rafailovich, J. Sokolov, C. Buenviaje, R. Buckmaster and R. M. Overney: *Physical Review Letters*, 2000, **85**, (11), 2340–2343.
34. R. H. Olley and D. C. Bassett: *Polymer*, 1982, **23**, (12), 1707–1710.
35. R. H. Somani, L. Yang, B. S. Hsiao, P. K. Agarwal, H. A. Fruitwala and A. H. Tsou: *Macromolecules*, 2002, **35** (24), 9096–9104.
36. S. Kimata, T. Sakurai, Y. Nozue, T. Kasahara, N. Yamaguchi, T. Karino, M. Shibayama and J. A. Kornfield: *Science*, 2007, **316**, (5827), 1014–1017.
37. B. Larin, C. A. Avila-Orta, R. H. Somani, B. S. Hsiao and G. Marom: *Polymer*, 2008, **49**, (1), 295–302.
38. Y. H. Chen, G. J. Zhong, J. Lei, Z. M. Li and B. S. Hsiao: *Macromolecules*, 2011, **44**, (20), 8080–8092.
39. M. J. Folkes and D. A. M. Russell: *Polymer*, 1980, **21**, (11), 1252–1258.
40. J. Yao, C.W. M. Bastiaansen and T. Peijs: *Fibers*, 2014, **2**, (2), 158–187.

7

Melt flow behavior of high density polyethylene nanocomposites with 1D, 2D and 3D nanofillers

Y. Q. Gill, J. Jin and M. Song*

Department of Materials, Loughborough University, Loughborough LE11 3TU, UK

Abstract The melt flow behavior of high density polyethylene (HDPE) and its nanocomposites with different kinds of nanofillers (kaolin, bentone, sodium montmorillonite (Na-MMT) clay, carbon nanotubes, graphene oxide (GO) and carbon black) is assessed. The dimensional effect (1D, 2D and 3D) of nanofillers on the melt flow singularity of HDPE is investigated. Results revealed that 2D nanofillers in the polymer matrix have significant improvements on the extrusion window of HDPE. The 2D nanofillers: exfoliated Na-MMT clay and GO showed the most enhanced melt flow behavior and processing window for HDPE up to 6.5°C, the best result reported so far. In comparison to the 2D nanofillers, 1D and 3D nanofillers, and micro-size clays also can broaden the processing window.

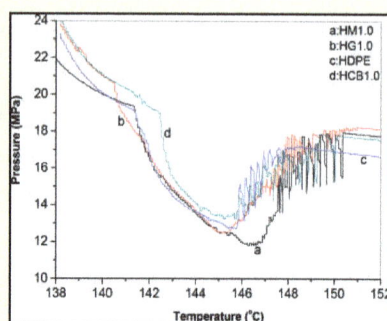

The addition of nanofillers resulted in an expansion for the extrusion window of polyethylene. 2D nanofillers (exfoliated clay and graphene oxide) showed the most enhanced window up to 6.5°C, the best result reported so far.

Keywords Polymer nanocomposite, Melt flow singularity, Carbon Nanotubes, Graphene

Introduction

In polymer processing, polyethylene (PE) is usually processed at temperatures between 160 and 200°C, which are much higher than the equilibrium melting point of PE and are mainly adopted to avoid any interference of melt flow instabilities like high die swell, shark skin etc. Recent studies[1–7] have shown the presence of an extrusion window for linear PE in between the flow instability and flow-induced solidification regions. The extrusion window effect is a specific melt flow singularity in extrusion and is described by the appearance of a pressure reduction occurring at and around 150°C, while maintaining a constant material throughput without distortion. The existence of an extrusion window for linear PE was first reported by Keller.[1] For polymer processing, the extrusion window has the advantages of an energy efficient processing route (low extrusion pressure and temperature) that can provide the usual continuous flow without any flow instabilities and a smooth extrudate with minimum die swell. On the other hand, the research conducted so far on most of the commercial PE grades has shown that the window effect is conspicuously too narrow about 1–2°C to be adopted in polymer processing in an industrial scale extruder where

temperature variation in various zones can be ± 2–3°C, i.e. the width of window has to be about 5–6°C.[1–12] In order to make the processing industrially viable, the extrusion window has to be broadened to several degrees based on Keller's discovery,[1] so that the thermal fluctuations in the polymer processing can be tolerated.

Conventional rheological theories of extrusion instabilities cannot predict or even explain the observed melt flow singularity phenomenon and its flow criticalities. Lately, the molecular origin of melt flow singularity was studied in details by Xu et al.[11] Previous research carried out by Keller[5–7] had concluded that the capillary flow gives rise to the window effect and the convergence flow at the die entry, which can induce a transient mobile hexagonal mesophase into the aligned chains at the die wall surface. The mesophase structure is basically liquid crystals, which are produced during the transformation of PE melt into fibrous crystals during dynamic cooling inside a capillary experiment. Although Keller et al.[8] carried out in situ wide-angle X-ray analysis during flow in a specially designed capillary rheometer to confirm the formation of such a mesophase, but later studies[11] indicated that window effect appeared owing to shear flow rather than convergence flow. One of the primary factors in enhancing the convergence flow is the die surface entry angle. Xu et al.[11] carried out detailed flow analysis to investigate the effect of

*Corresponding author, email m.song@lboro.ac.uk

entry angle of die surface on the window and concluded that the extrusion window remains unaffected by the die entry angle. Keller in his extensive work[5–7] also suggested that the extrusion window can only appear in samples with a molecular weight in the range of from 1.3×10^5 to 1.0×10^6 gmol[1]. Although work carried out on low molecular weight grades of high density polyethylene (HDPE)[11] and linear low density polyethylene (LLDPE)[12] have reported the appearance of extrusion window for these polymers. The molecular origin of melt flow singularity phenomenon of linear PE has been attributed to slip flow which can usually promote smooth extrudate production with minimum die swell. The slip flow is produced owing to the disengagement of adsorbed chains from the anchored chains at the melt wall interface. This disengagement is also supported by the production of a flow induced mesophase structure.

Recent studies have shown that clay such as sodium montmorillonite (Na-MMT), bentone etc. can be exfoliated in polymers.[13–16] The nanoscale dispersion of individual clay layers achieved by exfoliation can increase the interaction between the polymer chains and the clay layers, which can influence chain dynamics during melt flow in different geometries. The aspect ratio, surface to volume ratio and the relaxation time of clay layers is generally very high and hence they can promote the stretched chain conformation of linear PE at high temperature. However, the effect of exfoliation on the melt flow singularity phenomenon is not clear. In order to clarify the matter, two grades of kaolin with different aspect ratios, exfoliated Na-MMT, intercalated bentone clays were selected for the research. The melt flow singularity behavior in Na-MMT, bentone and kaolin clay/HDPE systems was compared. The inclusion of different shaped fillers can have distinct effect on the rheological properties of polymers and hence alter their flow behavior e.g. the inclusion of low quantity of Multi-walled carbon nanotubes (MWCNT) in different polymers can reduce the overall viscosity of the polymer melt because their log rolling effect will influence the dynamics of the polymer chains and hence alter the interaction of the polymer chains at the capillary wall. The modification of chain dynamics of polymers at the wall will also be dependent on the shape of the nanofiller being incorporated into the polymer e.g. carbon nanotubes owing to their high aspect ratio and tube like structure will influence the engagement of the chains on the capillary wall from the bulk polymer chains in a different way in comparison to the spherical particles of carbon black or the layered structure of graphene.

The research will lead us to know how different shape fillers (1D, 2D and 3D) and their morphologies such as exfoliated and intercalated structures influence the melt flow singularity of HDPE.

Experimental

Materials and sample preparation

High density polyethylene (HDPE) (Density: 0.96gcm^{-3}, MFI: 4.0g/10min) powder with a particle size on average 850μm in length was supplied by Exxon Mobil Corporation, UK. Two type of kaolin clay from IMERYS Minerals Ltd were used as nanofillers, i.e., Barrisurf™ LX (BLX) with a shape factor of 60

and Barrisurf™ HX (BHX) with a shape factor of 100. Na-MMT (Cloisite®Na +) was supplied by Southern Clay Products, Inc (Gonzales, TX, USA). Bentone was supplied by Elementis Specialties Inc (Osaka Hightstown, NJ, USA).

Carbon black (B4040) was purchased from Cobalt Chemicals Company, Belgium. Multi-walled carbon nanotubes (MWCNT) chemically modified with hydroxyl group ($-$OH) about 3.5% was purchased from Chengdu Institute of Organic Chemistry, Chinese Academy of Science. Ultrafine grinding Graphite (UF4) with a particle size of 4–7μm was purchased from Graphite Kropfmühl AG, Germany. In order to prepare graphite oxide (GO), 2.5g of UF4 was mixed with 57.5mL of concentrated H_2SO_4 in an ice bath for 30min. In order to keep the temperature of the mixture below 20°C, $KMnO_4$ was added slowly to the mixture. Using a water bath the mixture was then heated to 35 $\pm$ 3°C with continuous stirring for 30min. 115mL of distilled water was added drop wise into the mixture which increased the temperature of the mixture to 98°C. The mixture was stirred for 15min at this temperature. Later 350mL of distilled water and 25mL of 30% H_2O_2 solution was added to terminate the oxidation reaction. Graphite oxide was collected by filtering and was successively washed with 5% HCl aqueous solution. HCl washing was repeated three times until there was no sulfate detected by $BaCl_2$ solution. Graphite oxide collected from the mixture was dried at 50°C under vacuum for 1week. Graphite oxide was prepared by carrying out ultrasonication of a 1mgml^{-1} concentrated dispersion of GO in water with a power of 300W for 1h at room temperature.

High density polyethylene (HDPE) with 0.2, 0.6, 1 and 2wt-% of BHX (HH), and with 1wt-% of BLX (HL1.0), Na-MMT (HMMT1.0), bentone (HB1.0), carbon black (HCB1.0), GO (HG1.0), MWCNT (HM1.0) composites were prepared by melt blending inside a lab-scale twin screw extruder. Premixing of all the samples was carried out in an aqueous medium according to patent EB2008/003130. The samples received from pre-mixing were then mixed on a lab scale twin screw extruder operating at 110revmin^{-1}. The temperature profile along the length of extruder was maintained at 160, 170, 180, 190 and 200°C, respectively.

Characterization

Wide angle X-ray diffraction (WAXD) analysis for the fillers and their composites was carried out using a Brüker AXS, D8 Advance X-ray diffractometer at a generator voltage of 40kV and a current of 40mA. JEOL 2100 FX Transmission Electron Microscope (TEM) was used to analyze the nanofiller structure in the HDPE matrix.

Dynamic cooling rheological analysis of HDPE and its composites was carried out using a twin bore Rosand RH7 capillary rheometer (Rosand Precision Ltd., UK). For rheological analysis, the barrels were charged with the sample using a funnel. During charging, the sample was compressed thoroughly to avoid any air bubbles build up in the melt. The sample was then heated inside the barrel at 160°C for 10min to relax the chains, remove the grain boundary and produce a uniform melt inside the barrel. The melt was then extruded from the capillary rheometer at constant shear rate and a constant cooling rate of 1.5°Cmin^{-1}. The extrudate emerging from the bottom of the rheometer was collected for visual analysis and die swell measurements. The die swell of extrudate was measured by an *in situ* laser detector situated at 2cm under the capillary die. During the measurement, the extrudates were cutoff manually to maintain their length at no more than 15cm so that the sagging effect can be avoided. The apparent shear stress and apparent shear rate are calculated by using the following equations

$$(\tau_w)_{app} = \frac{R\nabla P}{2L} \tag{1}$$

$$\gamma_{app} = \frac{4Q}{\pi R^3} \tag{2}$$

where R is the radius of the capillary radius, L is the die length, Q is the volumetric flowrate, ∇P is the pressure difference.

Results and discussion

Wide angle X-ray diffraction analysis can provide information about the interlayer d-spacing of the layered fillers upon the formation of composites. The formation of an exfoliated or intercalated structure can be observed from the changes in the position and intensity of the basal peak of the nanofiller. The WAXD pattern of BHX, HDPE and their composites are shown in Fig. 1a. The diffraction peaks which appear at 12.4° and 8.9° (2θ) are attributed to the basal peak (001) of natural kaolinite. All the diffraction patterns of the composites indicate the presence of kaolinite basal plane, though the peak intensity appears to decrease in low weight percentage filler systems, but for high weight percentage filler systems an increase in the peak intensity is observed. These results suggest that most of the kaolinite retains its natural state and chances of any intercalation can only be observed in low weight percentage filler systems. A number of studies,[17-20] had also reported the absence of an exfoliated or intercalated structure for kaolinite based composites and the main reason is always attributed to the internal structure of kaolinite. Figure 1b shows the WAXD patterns of HDPE, Na-MMT, bentone and their composites. The (001) diffraction peak for Na-MMT and bentone appear at 7.2° and 5.5° (2θ), respectively. In both the composites no peak is visible, which indicates that the layered structure of these two clays in the HDPE matrix is completely destroyed and an exfoliated composite structure was formed.

TEM micrographs for BLX/HDPE, BHX/HDPE, Na-MMT/HDPE and bentone/HDPE composites thin films are shown in Fig. 2. The bright region in the TEM micrographs shows the matrix phase and the dark dispersed entities represent the clay phase. The TEM images show that Na-MMT clay was fully exfoliated into individual platelets in the HDPE matrix with very small number of tactoids formation. Both kaolin clays remain in agglomerated state without sign of individual platelets formed and the kaolin tactoids have thickness in micrometres. In case of HB1.0, an intercalated structure for bentone clay is observed. The thickness for Na-MMT clay tactoids lies in the range of 40–65nm whereas the tactoids for bentone are much thicker about 70–100 nm and number of individual bentone platelets visible is also very small. The TEM micrographs also reveal the uniform distribution of Na-MMT and bentone clay in the HDPE matrix at a very high resolution while the dispersion of clay is much more uneven for the two kaolin clay samples. These results also confirm the results obtained from WAXD analysis, which showed a fully exfoliated structure for HMMT1.0, an intercalated structure for HB1.0 and a traditional filled composite structure for HL1.0 and HH1.0.

In order to understand the effect of carbon-based nanofillers on the rheological behavior of HDPE nanocomposites, TEM analysis of MWCNT (Fig. 3a), CB (Fig. 3c), and graphene (Fig. 3b, d) nanocomposites was also carried out. The micrographs indicate agglomeration for HM1.0, but also show the uniform distribution of MWCNT throughout the matrix which could have led to better barrier properties and enhanced extrusion window of HM1.0. The distribution of CB inside HDPE was not even visible at higher magnification, which shows the nanoscale distribution of spherical CB particles inside HDPE. The distribution of graphene (Fig. 3b, d) shows the unique characteristics of graphene sheets, which can only be visualized if the 2D sheets are individually separated from each other. Graphene layers are known to be flexible, transparent and very soft. Because of its soft nature the graphene layers can be folded over by the shearing action of extruder to produce a spool of graphene which is clearly shown in (Fig. 3d). This folding could cause a reduction in the aspect ratio of graphene sheets and hence result in diminishing their reinforcing capability. But such spools of graphene have minor visibility throughout the

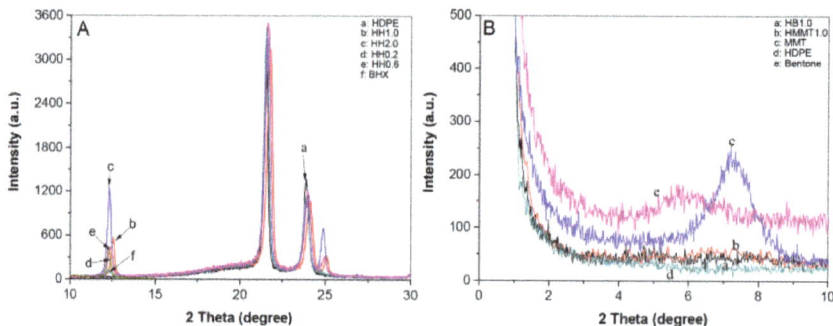

Figure 1   X-ray diffractograms for *a* high density polyethylene (HDPE), Barrisurf™ HX (BHX) and their composites containing various amounts of BHX clay and *b* sodium montmorillonite (Na-MMT), bentone, HDPE and their composites (HH: BHX/HDPE composite; BHX: Barrisurf™ HX kaolin clay; MMT: sodium montmorillonite clay; HMMT: Na-MMT/HDPE composite; HB: bentone/HDPE composite)

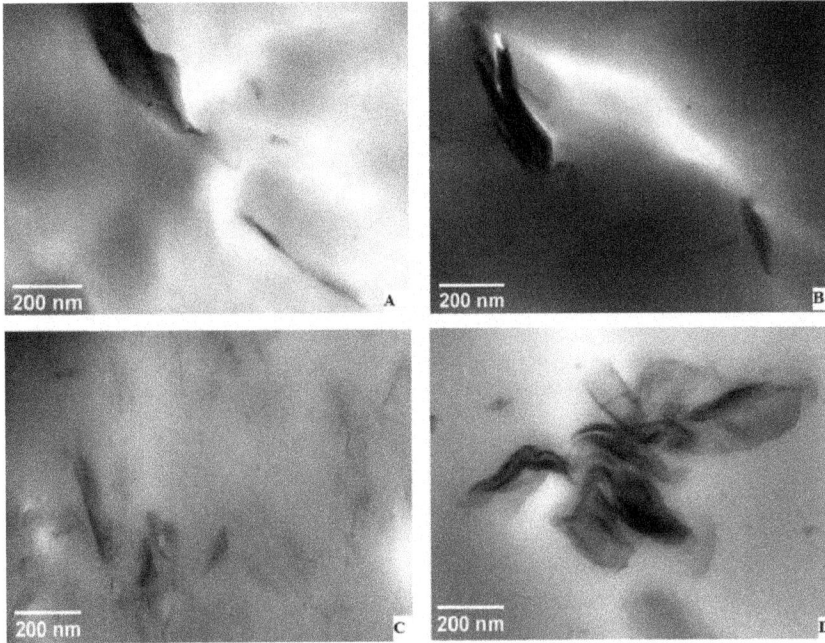

Figure 2 TEM micrographs for *a* BLX/high density polyethylene (HDPE), *b* BHX/HDPE, *c* sodium montmorillonite (Na-MMT)/HDPE and *d* bentone/HDPE composites (1wt-% fillers)

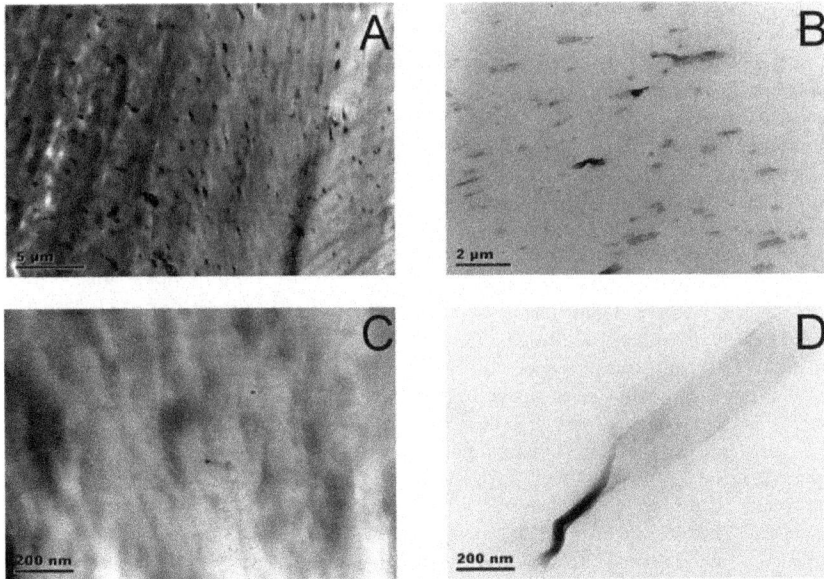

Figure 3 TEM micrographs of *a* Multi-walled carbon nanotubes (MWCNT)/high density polyethylene (HDPE), *b, d* graphene oxide (GO)/HDPE and *c* carbon black/HDPE composites (1wt-% fillers)

HDPE matrix which shows that at the nanoscale most of the graphene sheets had retained their platy structure and hence can contribute to the rheological properties of HDPE.

Figure 4*a* shows the extrusion pressure versus temperature data recorded under dynamic cooling conditions for the HDPE with different shear rates. The results show that the melt flow singularity does not appear at low shear rates of 200–250s^{-1} and the extrusion pressure at these shear rates

remains stable. These rates are too slow for the coil-stretch transformation to set in in the polymer melt at the die entry region. At 200s^{-1}, pressure increases steadily with the decrease in temperature, this increase in pressure is attributed to the increase in polymer melt viscosity with cooling. Similar behavior was observed at 250s^{-1}. This sort of viscosity behavior shows a typical Arrhenius type of relationship between viscosity and temperature. Gradual increase in

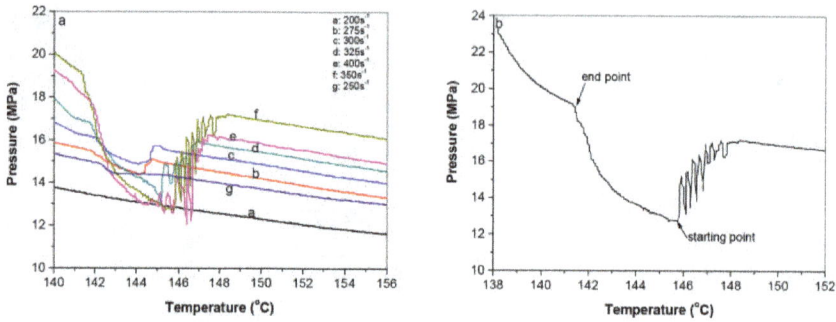

Figure 4 *a* Extrusion pressure versus temperature curves for the pure high density polyethylene (HDPE) at different strain rates recorded during the dynamic cooling experiment at a constant cooling rate of 1.5°Cmin^{-1} and using a capillary die with geometry: L–D–A:16–1–π. *b* Illustration of the definition and measurement of starting (T_w^s) and end (T_w^e) points

shear rate revealed the window at a shear rate of 275s^{-1}. At this rate, the decrease in pressure during cooling at a temperature of 144.24°C was observed because of the high degree of alignment of polymer melt chains along the flow direction in the rheometer die. This is the critical apparent shear rate γ_c^w, which corresponds to the onset of extrusion window effect. At 275 and 300s^{-1}, the window starts at 144.24 and 144.55°C (T_w^s = Extrusion Window Starting temperature) and ends at 141.89 and 141.92°C (T_w^e = Extrusion Window Ending temperature) and the minimum pressure (P_{min}) observed in the window are 14.44 and 14.98MPa. T_w^e also corresponds to the onset of flow-induced solidification region where a sudden increase in the pressure is observed. With the increase in shear rate to 325s^{-1}, the melt flow singularity phenomenon was also observed in between a stick-slip flow region[1-7,11] in which the pressure oscillates with the decrease in temperature and a flow-induced solidification region. Figure 4*b* illustrates the definition and measurement of starting (T_w^s) and end (T_w^e) points.

Figure 5 shows the rheological curves for the pure HDPE and its composites with different concentrations of BHX at a constant strain rate of 350s^{-1} recorded during the dynamic cooling experiment. Results show that the addition of BHX clay in the HDPE can influence the extrusion window and the enhancement is much more apparent than increase in strain rates. As shown in Fig. 4, that the addition of BHX clay influenced not only T_w^s and T_w^e temperatures of extrusion window but also P_{min}.

With the addition of 0.2 and 0.6wt-% of BHX clay, T_w^s and T_w^e temperatures remains mostly unaffected but an increasing trend in the window over the pure HDPE is observed. With 1wt-% BHX, however, both T_w^s and T_w^e temperatures shift and the broadening of window is more pronounced than in 0.2 and 0.6wt-% BHX clay composites. With 2wt-% BHX clay, a reduction in the extrusion window and P_{min} is observed with T_w^s shifting to a lower temperature as compared to other composites, while the T_w^e remains the same as the pure HDPE. The addition of high weight percentage of agglomerated kaolin clay particles will offer resistance to flow because of their ability to behave like irregular pentagon structures. Such structures will also assist in the entanglement of the polymer chains and hence reduce the polymer melt flow.

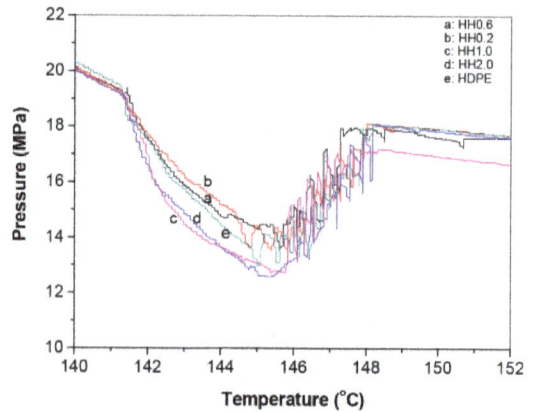

Figure 5 Extrusion pressure versus temperature curves for the pure high density polyethylene (HDPE) and its composites with different concentrations of Barrisurf™ HX (BHX) at a constant strain rate of 350s^{-1} recorded during the dynamic cooling experiment at a constant cooling rate of 1.5°Cmin^{-1} and using a capillary die with geometry: L–D–A:16–1–π (HH0.2 = HDPE with 0.2wt-% BHX clay; HH0.6 = HDPE with 0.6wt-% BHX clay and so on)

Figure 6 shows a comparison of the rheological curves for the HDPE (Fig. 6*d*) and its composites with Na-MMT (Fig. 6*c*), bentone (Fig. 6*a*), BLX (Fig. 6*b*) and BHX (Fig. 6*e*) clay. All the composites contain 1wt-% of clay and were extruded under the same conditions as all BHX composites. The addition of all four different types of clays shows their own individual effects on the extrusion window interval and P_{min}. With regards to BLX clay, a reduction in the extrusion window is observed owing to the movement of both T_w^s and T_w^e to higher temperatures in comparison to the pure HDPE. Also P_{min} for HL1.0 is only slightly less than that of the pure HDPE. These results indicate that the BLX clay is not uniformly dispersed and its hexagonal shaped agglomerates actually enhance the interaction between the adhered chains on the die wall and the bulk free polymer melt chains owing to their tumbling effect. In comparison to the two kaolin clay composites, the effect of intercalated bentone and exfoliated Na-MMT clay on the melt flow singularity is much more pronounced as depicted in Fig. 6*a, b*. In the presence of bentone and Na-MMT clay, T_w^s moves to higher

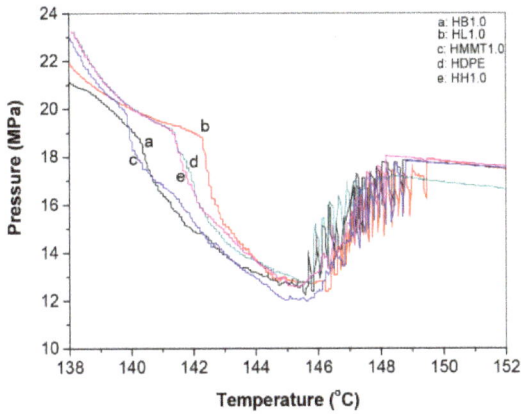

Figure 6 **Extrusion Pressure versus temperature curves for the pure high density polyethylene (HDPE) (e) and its composites with (a) sodium montmorillonite (Na-MMT), (b) bentone, (c) BLX, and (d) BHX clay at a constant strain rate of 350s^{-1} recorded during the dynamic cooling experiment at a constant cooling rate of 1.5°Cmin^{-1} and using a capillary die with geometry: L-D-A:16-1-π (HB1.0 = HDPE with 1wt-% bentone clay; HL1.0 = HDPE with1wt-% BLX clay; HMMT1.0 = HDPE with 1wt-% Na-MMT clay; HH1.0 = HDPE with 1wt-% BHX clay)**

Figure 7 **Extrusion pressure versus temperature curves for the pure high density polyethylene (HDPE) (c) and its composites with (a) multi-walled carbon nanotubes (MWCNT), (b) graphene oxide (GO) and (d) CB at a constant strain rate of 350s^{-1} recorded during the dynamic cooling experiment at a constant cooling rate of 1.5°Cmin^{-1} and using a capillary die with geometry: L–D–A:16–1–π (HM1.0 = HDPE with 1wt-% MWCNTs; HG1.0 = HDPE with 1wt-% graphene; HCB1.0 = HDPE with 1wt-% carbon black)**

temperatures while T_w^e moves to lower temperatures, which leads to the broadening of the extrusion window. The increase in the extrusion window interval for HM1.0 to 6.5°C is the highest yet reported so far. The observation regarding P_{min} is that with increasing Na-MMT loading, the pressure decreases within the extrusion window. However, the decrease in P_{min} and the increase in pressure in the flow-induced solidification region for all other composites are minor.

The extrusion temperature–pressure profiles recorded during the MFS phenomenon analysis for HDPE and its nanocomposites with CB, MWCNT and graphene are shown in Fig. 7. The results show the variation in the extrusion pressure recorded during the dynamic cooling process on capillary rheometer at a strain rate of 350s^{-1}. The duration of the extrusion window for pure HDPE is 4.3°C and the minimum extrusion pressure recorded during the window is 12.78MPa. The extrusion window observed for HM1.0 and HG1.0 (5.6 and 6.4°C) are broader than HDPE, whereas in case of HCB, a reduction (3.3°C) in the extrusion is observed. This reduction explains the effect of particle shape in extending and aligning the polymer melt chains tethered at the die wall which ultimately leads to enhanced slip flow of the disengaged bulk polymer chains on the tethered polymer melt chains.

Becuase of their circular shape CB enhance the engagement between the tethered and the bulk polymer chains and hence reduce the overall slip flow for the bulk polymer chains. On the other hand, the elongated and platy nature of MWCNT and graphene can align the tethered chains and also provide more slippage on surface to the bulk polymer chains. As detailed earlier on, that the spherical CB particles provide more friction to the polymer chains by promoting entanglements and hence the flow of CB composites through the die will be resisted more in comparison to the graphene and carbon nanotubes nanocomposites.

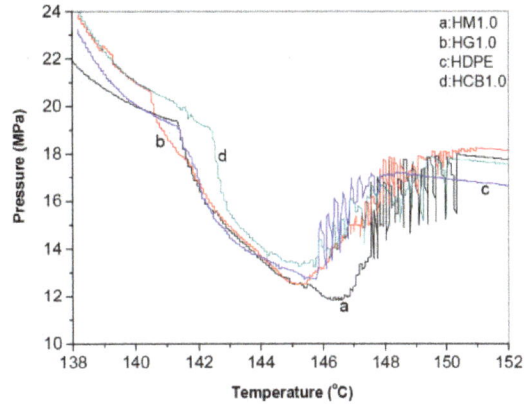

Another proof of the resistance offered by the CB particles to appearance of slip flow is the minimum extrusion pressure drop of 13.24MPa, this value is even higher than that of pure HDPE and is an indication of higher shear stress which is required to transform the stick-slip flow to slip flow of extrusion window. In case of HCB1.0, the extrudate diameter observed is slightly larger than pure HDPE. Table 1 shows the rheological data obtained from the dynamic cooling experiment of the HDPE and all its composites.

A schematic of extrusion pressure versus temperature for the HDPE flowing in capillary rheometer under dynamic cooling at a constant shear rate is shown in Fig. 8. The plot shows the four different flow regions encountered at or around the typical melt flow singularity phenomenon: (1) stick flow, (2) stick-slip flow, (3) slip flow and (4) flow-induced solidification. [5–7,11] In the stick flow region, extrusion pressure increases with decrease in the temperature and we have a typical liquid polymer melt inside the capillary die whose viscosity increases with decrease in temperature and in this region the viscosity and the temperature obey an Arrhenius type relationship. [5–7,11]

$$\eta = Ae^{E_a/RT} \tag{3}$$

In the above equation, γ is the shear viscosity, E_a is the melt/solid state flow activation energy, R is the universal gas constant and T is the flow temperature. In the stick flow region, the extrudate surface becomes rough and it shows surface melt fracture with loss of gloss and the die swell also increases with the decrease in temperature. In the stick-slip flow region, pressure oscillations with the decrease in temperature are observed and extrudate obtained consists of alternating regions of melt fracture and smooth extrudate surface. In the slip flow region, the phenomenon of melt flow singularity is observed and this region is the extrusion window where a reduction in pressure is observed along

Table 1 Rheological data obtained from the dynamic cooling experiment of HDPE and all its composites (HH0.2 = high density polyethylene (HDPE) with 0.2wt-% BHX clay; HL1.0 = HDPE with1wt-% BLX clay; HMMT1.0 = HDPE with 1wt-% Na-MMT clay; HB1.0 = HDPE with 1wt-% bentone clay; HG1.0 = HDPE with 1wt-% graphene; HCB1.0 = HDPE with 1wt-% carbon black; HM1.0 = HDPE with 1wt-% MWCNTs)

Sample	Strain rate ($1s^{-1}$)	T_w^s (°C) (starting point)	T_w^e (°C) (end point)	Processing window (°C)	P_{min} (MPa)
HDPE	200	–	–	–	–
	250	–	–	–	–
	275	144.24	141.89	2.35	14.44
	300	144.55	141.92	2.63	14.92
	325	145.12	141.6	2.82	13.43
	350	145.75	141.42	4.33	12.78
	400	145.02	142.07	2.95	12.94
HH0.2	350	144.75	141.29	3.46	14.92
HH0.6	350	145.54	141.42	4.12	14.33
HH1.0	350	146.01	141.32	4.69	12.59
HH2.0	350	145.01	141.36	3.65	12.96
HL1.0	350	146.2	142.31	3.89	12.72
HMMT1.0	350	146.4	139.87	6.53	11.97
HB1.0	350	145.5	140.29	5.21	12.66
HG1.0	350	146.72	140.36	6.36	12.48
HCB1.0	350	145.76	142.45	3.31	13.24
HM1.0	350	147.56	141.25	6.31	11.81

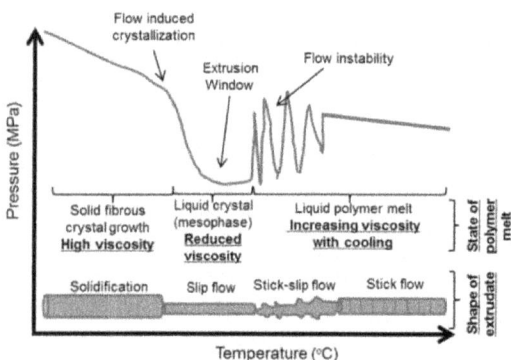

Figure 8 Extrusion pressure versus temperature schematic plot demonstrating melt flow singularity phenomenon of HDPE in capillary rheometer. The bottom of the graph shows the four different flow regions encountered at or around the MFS phenomenon i.e. (1) stick flow, (2) stick-slip flow, (3) slip flow and (4) flow induced solidification, along with the state of polymer melt and the shapes of the extrudate achieved in the different stages of extrusion

with a smooth and glossy extrudate with minimum die swell is obtained. The extrusion window is observed in a small region with the transformation of the polymer melt into liquid crystal mesophase. Molecular origin of the singularity is associated with the slip flow phenomenon,[11] which is influenced by the chain attachment to the die surface and its detachment from the bulk polymer melt. Thus, the singularity is influenced by the interfacial energy between the molecular chains and the wall. Brochard and de Gennes[21] and Drda and Wang[22] have proposed theories on the possibility of chain attachment to the wall surface and its disengagement from the polymer melt above a critical shear rate or shear stress. The disengagement process of the chains will be strongly dependent of the topological constraints between the adsorbed chains and the surrounding chains in the bulk, which would be dependent on molecular characteristics and flow conditions. Out of the extrusion

window a rapid increase in pressure is observed owing to the formation of solid fibrous crystal, which leads to flow induced solidification of the polymer melt and causes the viscosity and extrudate diameter increases rapidly. The solidification takes place because the temperature reaches the static solidification temperature during cooling and orientation-induced crystallization causes the viscosity to increase because of fibrous crystal formed inside the melt.

When the polymer chains are aligned, the small amount of filler particles added will act as lubricant surface that can effectively reduce the friction between the polymer chains and hence assist in the flow of polymer melt through the die (Fig. 9a). The introduction of individual clay and graphene layers produced by exfoliation or intercalation between the polymer chains can effectively decrease the overall friction in the system and hence the polymer chains will flow more easily during the extrusion window. If the polymer chains are entangled (Fig. 9b) the addition of exfoliated clay and graphene layers will produce large internal friction, which leads to energy dissipation in composites and can restrict the movement of polymer chains.[23] Also the filler agglomerates act like hexagonal particles (Fig. 9c), which perturb the orientation of the adsorbed chains and also favor their entanglement with the bulk polymer chains and hence cause a reduction in the extrusion window.

The polymer melt owing to their entangled structure has a short relaxation time[11,24], which can be altered significantly by the addition of nanofillers. The large platy structure of clay and graphene layers with high aspect ratio and high relaxation time[11,24] can help with alignment of polymer chains during flow and also help maintain the stretched chain confirmation of the adsorbed chains at the die wall. Owing to their structure, the exfoliated clay and graphene layers could also help in the orientation of the adsorbed chains at the capillary wall because these adsorbed chains can easily wrap themselves around the clay and graphene layers and that would ultimately stop their entanglements with the bulk polymer melt chains, which would lead to better flow and minimum die swell for longer temperature intervals.

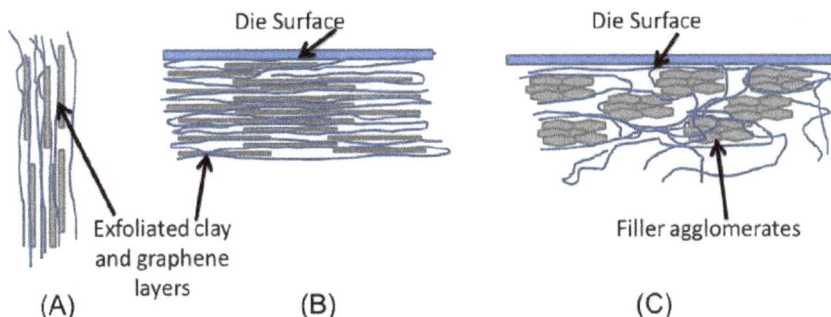

Figure 9 Schematic of the interaction: *a* exfoliated clay and graphene layers and the bulk polymer melt chains, *b* exfoliated clay and graphene layers along with adsorbed polymer melt chains on the die surface, *c* hexagonal shaped filler agglomerates and the adsorbed polymer melt chains on the die surface

The critical stress σ_c for wall slip is presented by Brochard and de Gennes theory on wall-slip at the metal-solid interface,[21] as

$$\sigma_c = \upsilon K_B T / \left(N_e^{\frac{1}{2}} \alpha \right) \qquad (4)$$

where υ is the number density of molecular chains adsorbed at the interface, $N_e^{\frac{1}{2}} \alpha$ is the entanglement distance.

The percentage area covered by the adsorbed chains at the metal wall and melt interface (θ) is obtained using the single layered adsorption model: [21]

$$\theta = b v_0 / (1 + b v_0) \qquad (5)$$

where v_0 is the number density of molecular chains in the bulk polymer melt and b indicates the intensity of adsorption. In case of polymer melt, strong interface adhesion on capillary wall is observed, which shows that $b v_0 >> 1$, and hence $\theta = 1$, which basically means saturated adsorption of molecular chains on capillary wall. So for the HDPE melt υ would remain same even under the influence of various nanofillers and varying temperature. Also since the amount of fillers used is very low so that could also not affect the HDPE melt υ. So in case of HM1.0, the lower value of P_{min} obtained, which indicates lower wall shear stress for slip flow is owing to increasing entanglement distance, which will help in the flow of bulk polymer chains by separating them from the adsorbed chains on the capillary wall.

As the temperature is reduced below the processing window temperature ($<142°C$), the extrusion pressure shoots up dramatically and a considerable increase in the die swell of the extrudate is observed. In Fig. 10, the die swell ratio for the pure HDPE at three different temperatures (140, 143, 160°C) is plotted against the strain rate (extrusion speed) maintained during the cooling process. The curve for 160°C represents the normal processing behavior for the HDPE melt in a capillary rheometer where the die swell gradually increases with increase in shear rates. The die swell (B) and the temperature of extrusion in a capillary rheometer are correlated by the following linear relationship:[16]

$$B = \alpha_1 - T\beta_1 \qquad (6)$$

Where α_1 and β_1 are coefficients related to material properties. It is also clear from Fig. 10 that in the flow-induced solidification temperature range (140°C) the die swell

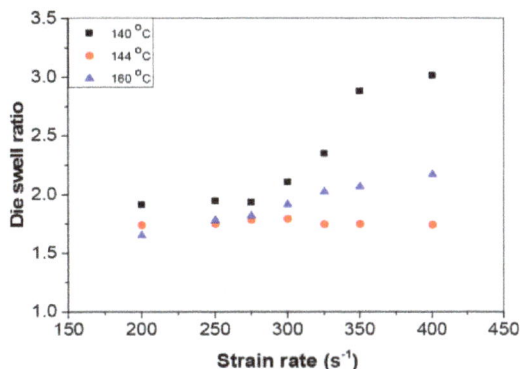

Figure 10 Plot of die swell ratio as a function of the strain rate maintained during dynamic cooling experiment for high density polyethylene (HDPE) at three different extrusion temperatures

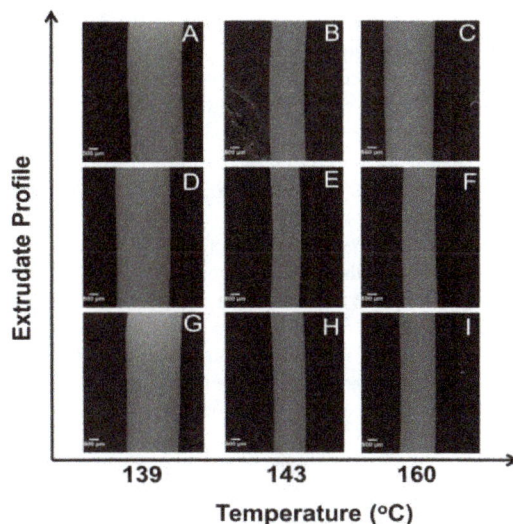

Figure 11 Extrudate profiles of the pure HDPE (*a–c*), HM1.0 (*d–f*) and HH1.0 (*g–i*) at three different temperatures 139, 143 and 160°C with a strain rate of 350s⁻¹

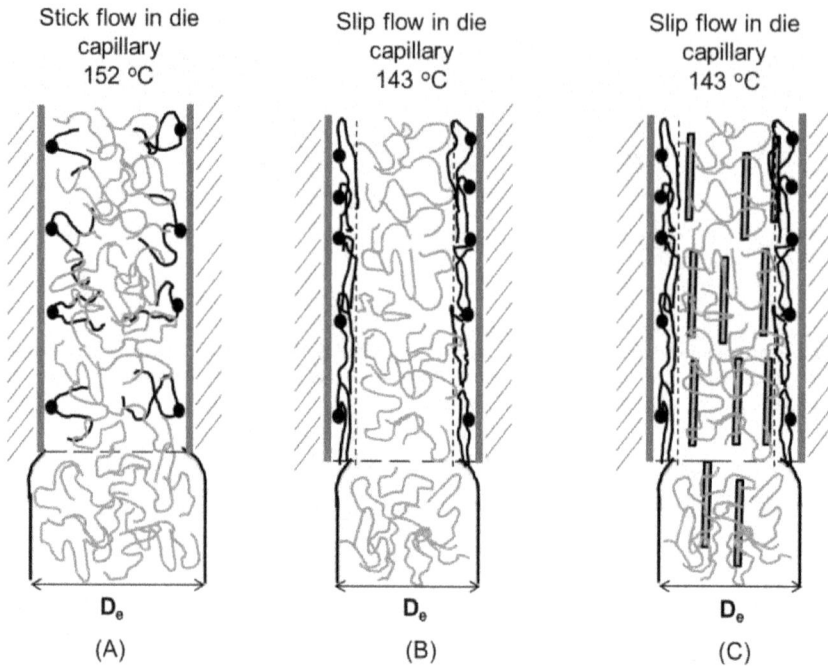

Figure 12 Schematic of the difference in the extrudate diameters (D_e) obtained in *a* the stick flow at 152°C, *b* slip flow at 143°C and *c* slip flow at 143°C in the presence of clay and graphene layers with a strain rate of 350s^{-1}. The black dots represent the end tethered chains on the die wall, while the entangled lines represent the polymer chains

increases quite steeply with increase in strain rates, whereas in the extrusion window temperature range (144°C) the die swell remains more or less constant. This result provides another proof for the liquid crystalline nature of the polymeric melt in the processing window because the liquid crystal polymers because their chain extension characteristics do not show any die swell regardless of the strain rate employed.

Figure 11 shows the extrudate profile for the pure HDPE, HM1.0 and HH1.0 at 139, 143 and 160°C. The three temperatures selected corresponds to the three distinct flow regimes: flow induced solidification, extrusion window and stable melt flow respectively. It is generally accepted [13,16] that the die swell occurs owing to the elastic recovery of the polymer melt after it emerges from the die.

The inclusion of clay and graphene layers with high surface area and ease of movement in comparison to the entangled polymeric chains can increase the energy dissipation of the composites at a given shear rate. [23] Also in the extrusion window, slip flow of polymeric melt takes place and most of the adsorbed chains remains tethered on the die wall and free chains slip over these tethered chains which are extruded out with minimum die swell as shown in Fig. 12*a–c*. The die swell of extrudate could depend on the disentanglement degree of the free chains. The addition of oriented high surface area clay and graphene layers will limit the elastic recovery of the stretched free polymeric chains.

In case of HCB1.0, the higher value of P_{min} obtained indicates higher wall shear stress for slip flow and is owing to the decreasing entanglement distance, which will resist in the flow of bulk polymer chains by entangling them with the adsorbed chains on the capillary wall. In case of MWCNT and graphene-based composites, the increase in the extrusion

window is also attributed to their high aspect ratio and relaxation time, which helps in stretched confirmation and alignment of chains tethered to the metal wall. Also the aligned chains can be easily wrapped up around graphene layers and the MWCNT, making the tethered chains unavailable for entanglements with the bulk of the polymer chains. The results indicate that die swell is minimized in the extrusion window temperature interval and it increases dramatically during the flow induced solidification. The difference between the die swell in the processing window and the stable melt flow regime indicate that the polymer melt extrudes more easily than at high processing temperature in the processing window, and thus validates the applicability of the processing window as an authentic method of processing.

Conclusions

The melt flow singularity of HDPE and its composites with different kinds of clay and carbon based nanofillers was investigated. The mechanism of melt flow singularity was analyzed with respect to the strain rate and filler content dependence of the critical points of the extrusion window. The addition of nanofillers resulted in the expansion of extrusion window of HDPE. When the polymer chains are aligned, the small amount of filler particles added will act as lubricant surface that can effectively reduce the friction between the polymer chains and hence assist in the flow of polymer melt through the die. Especially, the introduction of individual clay and graphene layers produced by exfoliation or intercalation between the polymer chains can effectively decrease the overall friction in the system and hence the polymer chains will flow more easily during the extrusion window. If the polymer chains are entangled the addition of exfoliated clay and graphene layers will produce large internal friction which leads to energy dissipation in composites and can restrict the movement of polymer chains. The exfoliated Na-MMT clay and graphene/HDPE composites showed the most enhanced window up to 6.5°C, the best result reported so far, owing to the larger interfacial contact area between the polymer chains and the surface of clay and graphene layers in the matrix.

Conflict of interest

The authors declare that they have no conflict of interest.

Acknowledgement

Authors thank EPSRC (UK) for providing funding (EP/G042756) for this work.

References

1. A. J. Waddon and A. J. Keller: *Polym. Sci. Polym. Phys.*, 1990, **28**, 1063–1073.
2. K. A. Narh and A. J. Keller: *Polymer*, 1991, **32**, 2512–2518.
3. K. A. Narh and A. J. Keller: *J. Mater. Sci. Lett.*, 1991, **10**, (22), 1301–1303.
4. A. J. Waddon and A. J. Keller: *Polym. Sci. Polym. Phy*, 1992, **30**, 923–929.
5. J. W. H. Kolnaar and A. J. Keller: *Polymer*, 1994, **35**, 3863–3874.
6. J. W. H. Kolnaar and A. J. Keller: *Polymer*, 1995, **36**, 821–836.
7. J. W. H. Kolnaar and A. J. Keller: *Polymer*, 1997, **38**, 1817–1833.
8. J. W. H. Kolnaar and A. J. Keller: *J. Non-Newtonian Fluid Mech.*, 1996, **67**, 213–240.
9. J. W. H. Kolnaar, A. J. Keller, S. Seifert, C. Zschunke and H. G. Zachmann: *Polymer*, 1995, **36**, 3969–3974.
10. H. M. M. van Bilsen, H. Fischer, J. W. H. Kolnaar and A. J. Keller: *Macromolecules*, 1995, **28**, 8523–8527.
11. H. Xu, A. Lele and S. Rastogi: *Polymer*, 2011, **52**, 3163–3174.
12. S. Pudjijanto and M. Denn: *J. Rheol.*, 1994, **38**, 1735–1744.
13. C. Dazhu, Y. Haiyang, H. Pingsheng and Z. Weian: *Compos. Sci. Technol.*, 2005, **65**, 1593–1600.
14. J. Aalaie, S. Malmir and M. Hemmati: *J. Macromol. Sci. Phys.*, 2012, **52**, 1–12.
15. G. Galgali, C. Ramesh and A. Lele: *Macromolecules*, 2001, **34**, 852–858.
16. N. Muksing, M. Nithitanakul, B. P. Grady and R. Magaraphan: *Polym. Test.*, 2008, **27**, 470–479.
17. L. Cabedo, E. Giménez, J. M. Lagaron, R. Gavara and J. J. Saura: *Polymer*, 2004, **45**, 5233–5238.
18. L. Domka, A. Malicka, N. Stachowiak and Pol: *J. Chem. Tech.*, 2008, **10**, 05–10.
19. L. Feijoo, P. Villanueva and E. Gime: *Macromol. Symp.*, 2006, **233**, 191–197.
20. M. P. Villanueva, L. Cabedo, J. M. Lagarón and E. Giménez: *J. App. Polym. Sci.*, 2010, **115**, 1325–1335.
21. F. Brochard and P. G. de Gennes: *Langmuir*, 1992, **8**, 3033–3037.
22. P. P. Drda and S. Q. Wang: *Phys. Rev. Lett.*, 1995, **75**, 2698–2701.
23. J. Jin, L. Chen and M. Song: *J. Nanosci. Nanotechnol.*, 2009, **9**, 6453–6459.

8

Key factors for tuning hydrolytic degradation of polylactide/zinc oxide nanocomposites

Samira Benali*, Sabrina Aouadi, Anne-Laure Dechief, Marius Murariu and Philippe Dubois*

Center of Innovation and Research in Materials and Polymers (CIRMAP), Laboratory of Polymeric and Composite Materials (LPCM), University of Mons & Materia Nova Research Center, Place du Parc 20, 7000 Mons, Belgium

Abstract The hydrolytic degradation of thin films of polylactide/surface treated zinc oxide [poly(lactic acid) (PLA)/ZnO$_s$] nanocomposites was investigated in phosphate buffer solution at the temperature of 37°C for more than 10 months. To produce PLA/ZnO$_s$ nanocomposites, the previously silanized metal oxide nanofiller has been dispersed into PLA by melt blending using twin screw extruders and the resulting dried pellets were shaped into thin films of about 70 μm thickness. For sake of comparison, pristine PLA was processed and investigated under similar conditions. The evolution of molecular weights of the PLA matrix, as well as of crystallinity and thermal parameters of interest, with the hydrolysis time, has been recorded by size exclusion chromatography (SEC) and differential scanning calorimetry (DSC), respectively. Accordingly, at longer hydrolysis time, the nanocomposites revealed better resistence to the hydrolytic degradation (lower weight loss, smaller decrease of molecular mass, no dramatic increase in dispersity), data that were also associated with the changes in the morphology of specimens over time as evidenced by visual analysis or by microscopy. The results show the possibility to tune the hydrolytic degradation and prolonging the service life of PLA throught the incorporation of a small amount of hydrophobic silanized nanofiller (ZnO$_s$). A bulk degradation mechanism was assumed, whereas the delayed degradation of nanocomposites was ascribed to a slowdown of the water diffusion into PLA matrix thanks to important increases of the crystallinity and especially to the hydrophobic properties of ZnO nanofiller treated with ~3 wt-% triethoxycaprylylsilane. Accordingly, the rate of hydrolytic degradation of PLA/ZnO$_s$ nanocomposite films can be reduced by increasing the loading of nanofiller and PLA crystallinity.

Keywords Polylactide, Zinc oxide nanoparticles, Hydrolysis, Degradation, Crystallization

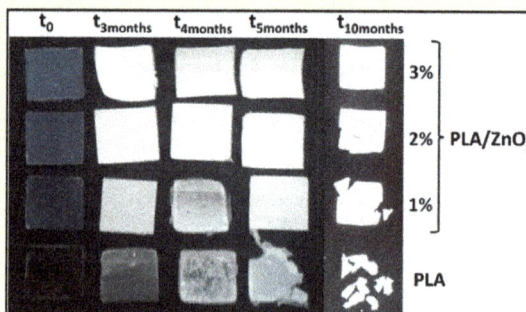

Introduction

High molecular weight poly(lactic acid) (PLA) is industrially obtained either through the direct condensation polymerization of lactic acid or by ring opening polymerization of lactide (the cyclic dimer of lactic acid, as an intermediate).[1,2] The raw material of PLA production, i.e. the lactic acid, can be produced at large industrial scale by the microbial fermentation of agricultural byproducts, mainly the carbohydrate rich substances. It just so happens that PLA is considered as a sustainable alternative to petrochemical derived products due to promising physical properties, low carbon footprint, broad possibility of processing using traditional equipment, etc. Furthermore, between the different end of life options, one of the main characteristics of PLA stands in its easiness to degrade hydrolytically, the final products of degradation being lactic acid, oligomers[3] that can then be consumed by microorganisms to produce carbon dioxide, water and solid biomass, while the chemical recycling can be considered as well. Therefore, PLA is actually a hydrolytically degradable and biobased aliphatic polyester that finds applications in medicine (due to its biocompatibility and bioresorbability), whereas the domains of fiber technology and packaging represent actually the principal sectors of utilization.[4] However, by considering the last trends and forecasts, PLA is more and more interesting

*Corresponding author, email samira.benali@umons.ac.be, philippe.dubois@umons.ac.be

for "durable" applications requiring high performance materials (automotive, electrical and electronic products). Nevertheless, PLA is not able to compete with the traditional polymers if (1) some required properties (heat and hydrolysis resistance, ductility, impact strength, crystallization rate, etc.) are not improved and if (2) the (bio)degradation is not controlled. Indeed, considerable efforts have been made to tailor the hydrolytic degradation rate of PLA in order to allow wider ecological applications without a compromise of the product properties before the onset of degradation.[5] On the other hand, it is important to design the end use properties of PLA by considering the requirements of application. Very recently, the antibacterial and anti-UV properties of PLA have been enhanced by the incorporation of a low amount of nanosized zinc oxide (1–3 wt-%).[5,6] Indeed, surface treated zinc oxide (ZnO$_s$) nanoparticles with a selected silane (i.e. triethoxycaprylylsilane) have been successfully used to avoid PLA degradation during production and melt processing of PLA/ZnO$_s$ nanocomposites.[5,6]

However, metallic oxides such as zinc oxide are also known as efficient catalysts for the recycling of PLA via catalyzed unzipping depolymerization at high temperature.[7,8] Very recently, Qu et al.[9] have reported a simple chemistry of using zinc oxide nanoparticles to catalyse the hydrolytic degradation of PLA at the temperature well below its glass transition temperature. They conclude that PLA can adsorb on the surface of ZnO. These nanoparticles function as the heterogeneous catalysts that accelerate the hydrolytic degradation of PLA in water at temperatures below its T_g. The activation energy for ZnO catalyzed PLA hydrolysis is about 38% lower than of pure PLA hydrolysis. Thus, by considering the potential of Zn based products (such as untreated ZnO) in triggering PLA degradation, other researchers have preferred to use the solvent casting method for the production of PLA–ZnO nanocomposite films,[10] in our opinion, rather with limited applicability with respect to the melt blending approach developed by our laboratory,[6] and which creates new opportunities for the application of these nanocomposites at larger scale. Still, for the implementation of these new products, it is strongly required to know their behavior under some conditions (i.e. under moisture, under the light, or under other aging factors).[5,11] Accordingly, it was of interest to verify whether the hydrolytic degradation of PLA chains from PLA–ZnO$_s$ nanocomposites can be affected by the presence of ZnO$_s$ nanoparticles.

However, the degradation mechanism of PLA triggered by hydrolysis has been largely investigated. Degradation of high molecular weight PLA is autocatalyzed by the carboxyl end groups initially present or generated upon ester bond cleavage.[7,8,12–15] For material thicknesses lower than 0.5 mm, PLA hydrolytic degradation has been reported to take place mainly in the material bulk rather than on the surface (bulk degradation mechanism).[8,16,17] This surface–interior differentiation or heterogeneous degradation of large sized PLA has been evidenced by the study of Li.[8] In parallel, the physical structure of PLA has been found to affect the hydrolytic degradation mechanism, as the hydrolytic chain cleavage proceeds preferentially in the amorphous regions,

leading therefore to an increase of the polymer crystallinity.[15,18,19]

The hydrolysis process of the PLA depends on various factors such as molecular weight and its distribution, purity, morphology, shape of the specimens and the processing conditions, presence of additives, amount of residual monomer, specific conditions in the degradation environment, etc.[20,21] Interestingly, Höglund et al. have reported crucial differences between the rate of degradation of PLA of industrial origin (with 4.5% D units) compared with that of laboratory scale synthesized poly(L-lactide) of similar molar mass and produced using monomer of higher purity.[22] This study also confirms the effects of the presence of D units in L rich chains known for long.[23]

Hitherto, several works have been reported on the hydrolytic degradation of PLA nanocomposites.[19,24–34] The organoclays are certainly the most studied nanofillers,[19,24,25,31,32] but recent studies also include carbon nanotubes,[27,34] polyhedral oligomeric silsequioxanes,[28] cellulose nanocrystals[29] and titanium dioxide.[30] Here, the review of these results is off topic, but it is important to note that the hydrolytic degradation of aliphatic polyester nanocomposites is under discussion if the influence of nanofillers is considered. However, no comprehensive conclusion can be made by considering that the kinetics of hydrolytic degradation of PLA is dependent only on the ability of nanofillers to modify the water diffusion into PLA matrix.

By considering the multifunctional properties of PLA–ZnO$_s$ products (anti-UV, antibacterial, barrier characteristics) and other expected features (e.g. self-cleaning), these nanocomposites could present a high interest for traditional and special applications as biosourced packaging materials, except for food packaging. On the one hand, it is important to note that depending on the field of application, there are cases where the acceleration of degradation is desirable, while in other cases, it is required to extend the service life of PLA. As reported in the literature, the effective use of (bio)degradable polymers relies on the ability to control the onset and time needed for degradation.[35] We can note that, in practice, industrials exploit the chemical degradation that can be controlled rather than biodegradation essentially controlled by Nature. Thus, not only, the challenge is that PLA properties should be kept at the required level during the specific period of utilization but the material should degrade in a rapid and controlled manner afterward. On the other hand, it is generally assumed that PLA hydrolysis in presence of nanofillers is a complex phenomenon depending on their specific morphology, dispersion, relative hydrophilicity, or in some cases, hydrophobicity, catalytic activity, etc. We agree with the opinion that the factors which are increasing the hydrolysis tendency of neat PLA, ultimately control the degradation of this aliphatic polyester.[36] However, by considering the complex effect of nanofillers, it was reported that they can favor or delay the hydrolytic degradation of PLA.[37]

The hydrolytic stability of PLA based materials can be tailored to obtain predetermined degradation profiles. As aforementioned, PLA is known to mostly degrade through hydrolytic chain cleavage occurring at the level of the aliphatic ester functions and yielding to low molecular weight residues, i.e. lactic acid and related oligomers, able to bio-degrade and ultimately be bioassimilated.[35] Therefore it is of interest to study the effect of nanofillers such as ZnO_s on the hydrolytic degradability of PLA, while understanding the structure/properties relationships in these particular nano-composites. Following the state of the art, a large number of techniques and procedures can be used to study the hy-drolytic degradation of PLA and PLA nanocomposites, ran-ging from the determination of mass loss, and thermal characteristic features [differential scanning calorimetry (DSC), TGA] to molecular parameters via size exclusion chromatography (SEC), FTIR, electrospray ionization mass spectrometry[19,26,35,37] and $_1H$ T_2 NMR relaxometry.[9]

The goal of the present work is to study the effect of the addition of ZnO_s (nanofiller surface treated with triethox-ycaprylylsilane) on the hydrolytic degradation behavior of PLA matrix. To achieve this aim, the controlled hydrolytic degradation behavior of PLA/ZnO_s nanocomposite films in phosphate buffer medium has been studied. The degra-dation process was mainly monitored using SEC and DSC, whereas additional information was obtained via other techniques, e.g. scanning electron microscopy (SEM). As in the study recently published by Qu et al.,[9] the influence of ZnO_s nanofillers surface treated with triethoxycaprylylsilane on the hydrolytic degradation of a hydrolytically degradable matrix such as PLA has been determined, and it is believed that the results of the study have also importance in the perspective of potential application of these new nanocomposites.

Experimental

Materials

Poly(L,L-lactide) was kindly supplied by NatureWorks LLC (USA): PLA98.6/1.4 (L/D-lactide) is a grade for the extrusion of films [$M_{w(PS)}$ = 239 000, molar mass dispersity (M_w/M_n, where M_w is the weight average molar mass and M_n is the number average molar mass): Đ$_M$ = 2.0, according to supplier infor-mation: D-isomer ≈ 1.4%; relative viscosity = 3.94; residual monomer = 0.14%, melt temperature: 210 ± 8°C and heated stabilized]. Commercially available surface treated ZnO nanofiller (hereinafter referred to as ZnO_s) was kindly supplied by Umicore Zinc Chemicals (Belgium) as Zano 20 Plus (surface coated with a silane especially suitable for the treatment of metal oxides, i.e. triethoxycaprylylsilane; ZnO content: 96.2 ± 0.5%, bulk density: 360 g L^{-1}, loss on drying at 105°C (2 h): max. 1%). According to supplier information and our transmission electron microscopy investigations, these nanoparticles are characterized by a rod-like mor-phology, having typically diameters of 15–30 nm and a maximum length of up to 100 nm.[11] Ultranox 626A (bis(2,4-di-t-butylphenyl) pentaerythritol diphosphite) supplied by GE Specialty Chemicals was selected as a thermal stabilizer and used at preferred percentage of 0.3% in PLA.

Preparation of nanocomposites and films

Before processing, PLA was dried overnight at 80°C under vacuum and stored in the presence of humidity absorbent. To minimize the water content for melt blending with PLA, all additives were previously dried. The preparation of PLA

nanocomposites and their processing by extrusion as films was performed in three steps. First, the dried granules of PLA and additives were mixed in a Rondol turbo-mixer (2000 rev min^{-1}, 2 min). Next, PLA nanocomposites (with 1, 2 and 3 wt-%ZnO_s) were prepared by melt compounding using a twin screw extruder Leistritz ZSE 18 HP-40D [screw diameter (D) = 18 mm, L (length)/D ratio = 40] and the fol-lowing conditions of processing: throughput of 1.5 kg h^{-1}, the speed of screws = 100 rev min^{-1}, temperatures of extrusion on the heating zones adapted to the rheological characteristics of PLA, while the temperature of the molten polymer before the extrusion of strands was kept around of 185°C. For the sake of comparison, the neat PLA was pro-cessed in similar conditions of melt-compounding. Then, the previously dried pellets (overnight, at 80°C, under vacuum) were extruded as films of about 70 μm in thickness with a Brabender single screw extruder (D = 19 mm, L/D = 25) adapted with a extrusion die head (ribbon die of 100 mm wide, gap of 0.5 mm) and using for drawing a Univex flat film takeoff unit as downstream equipment. Throughout this contribution, all percentages of nanofiller are given as wt-%.

Hydrolysis tests

Practically, before starting the hydrolysis tests, each film was cut into 1 × 1 cm square specimens (from the middle of film roll) by considering three replicates per sample. Because of the specimen weight (~8 mg), one of the samples will be used for the GPC analysis, another for the DSC analysis and the last one has been kept for a possible replicate of analyses in the case of any problem. Each specimen was then dipped in a laboratory vials containing 10 mL of 0.1M phosphate buffer solution at pH 7.4. The flasks were immersed in a shaking water bath at 37°C. At predetermined periods, the specimens were picked out from the buffered solution and rinsed several times with distilled water. Eventually, the re-sidual water was wiped off from the sample surface before drying (to constant weight) by wrapping it in small paper bag that was placed in a desiccator during up to 3 weeks.

Characterization

Recovery of PLA from selected compositions for molecular weight parameter determination was carried out by firstly dissolving the samples in chloroform and following a similar procedure used in by Murariu et al.[6] The metallic residues were removed by liquid–liquid extraction with a 0.1 N HCl aqueous solution step, followed by intensively washing with demineralized water. Finally, PLA was recovered by precipi-tation in an excess of heptane. After filtration and drying, PLA solutions were prepared in CHCl$_3$ (2 mg polymer/mL solvent). Molecular weight parameters (M_n, M_w and molar mass dispersity Đ$_M$) of pristine PLA and PLA extracted from the studied nanocomposites were determined by SEC using the procedure and relations described elsewhere.[38] SEC was performed in CHCl$_3$ at 30°C using an Agilent liquid chro-matograph equipped with an Agilent degasser, an isocratic HPLC pump (flowrate = 1 mL min^{-1}), an Agilent auto-sampler (loop volume = 100 μL, solution concentration = 2 mg mL^{-1}), an Agilent-DRI refractive index detector and three columns: a PL gel 5 μm guard column and two PL gel Mixed-B 5 μm columns (linear columns for separation of

$M_{W(PS)}$ ranging from 200 to $4 \times 10^5 \, g \, mol^{-1}$). Polystyrene standards were used for calibration. A DSC Q2000 from TA Instruments was used for DSC analysis in nitrogen atmosphere. Samples (weight: about 5–7 mg) were sealed in aluminum DSC pans and placed in the DSC cell. The DSC was calibrated with indium. Samples were heated from 0 to 200°C with a heating rate of 10°C min^{-1}. The events of interest, i.e. the glass transition temperature (T_g), cold crystallization temperature (T_c) and melting temperature (T_m), as well as the enthalpy of cold crystallization (ΔH_c) and melting enthalpy (ΔH_m) were measured on the first heating scan. The degree of crystallization (χ_c) was obtained using the equation (1)

$$\chi_c = \left[\frac{\Delta H_{m(t)} - \Delta H_{c(t)}}{\Delta H_m^0 \times \left(1 - \frac{\%wt_{filler}}{100}\right)} \right] \times 100 \qquad (1)$$

where $\Delta H_{m(t)}$ and $\Delta H_{c(t)}$ are the melting and cold crystallization enthalpies respectively of test sample at the time t of degradation, ΔH_m^0 is the melting enthalpy of the 100% crystalline PLA (93.0 J g^{-1})[39] and $\%wt_{filler}$ is the weight percentage of nanofiller. Sample morphology was observed using a SEM Hitachi SU8020, with field emission gun with landing energy at 3 kV and SE(UL) detector.

Results and discussion

Modification of the aspect and residual mass during hydrolysis

The changes in the general aspect of the samples upon hydrolysis could be first of all visually judged. As revealed in Fig. 1, a significant change in the sample opacity could be noticed for the PLA films with ZnO$_s$, already after 28 days (~1 month) of immersion in buffered solution. For the neat PLA sample, the opacity is only slightly visible after 87 days (~3 months).

Then, the increase of the hydrolysis time led to the rise in relative opacity of all materials. But the most important observation concerns the neat PLA sample, after 147 days (~5 months), it was turned in white colour and became extremely brittle. Furthermore, it was recovered in small pieces after the drying step which followed the sample removal from phosphate buffer solution (Fig. 1). As far as PLA/ZnO$_s$ nanocomposites are concerned, PLA/1%ZnO$_s$ and PLA/2%ZnO$_s$ samples became brittle after 292 days (~10 months and half), while the PLA/3% ZnO$_s$ nanocomposites remained in one piece with rather tough aspect. The modification of the opacity of samples has already been explained in previous papers.[13,14,18,19,40] Authors have noted that the modification of the opacity can be attributed to various phenomena, such as (1) light scattering due to the presence of water; (2) to the degradation products formed during the hydrolytic process, (3) to the formation of holes in the bulk of the specimen during degradation, and (4) to the evolution in crystallinity of the polymer matrix. Indeed, by considering the key role of the degree of crystallinity, it is important to remind that the hydrolytic degradation of the polyester chains is known firstly to take place in the amorphous phase of the matrix.[19,24]

Therefore, the overall opacity of the samples is in fact the translation of the relative PLA crystallinity increase to a higher value (see thermal characterization in following section).

Then, for ~10 months, all ZnO$_s$ filled nanocomposites do not show significant weight changes Since surface degradation usually leads to the loss of degradation products, these results may suggest that the degradation extent at this temperature (37°C) and in this period of time, is less advanced in the case of nanocomposites because no degradation products are released in the hydrolysis medium.[19] By cons, after 5 months, the neat PLA specimens begin to reduce significantly their weight. After 10 months, a weight loss of nearly 50% is measured. It is therefore remarkable to point out that a content in ZnO$_s$ as low as 1 wt-% can trigger such a reduction in weight loss for the entire nanocomposite sample.

An expansion displaying the first month is also proposed into the Fig. 2 for focusing on water uptake. The neat PLA specimen is the sample which absorbs the most water,

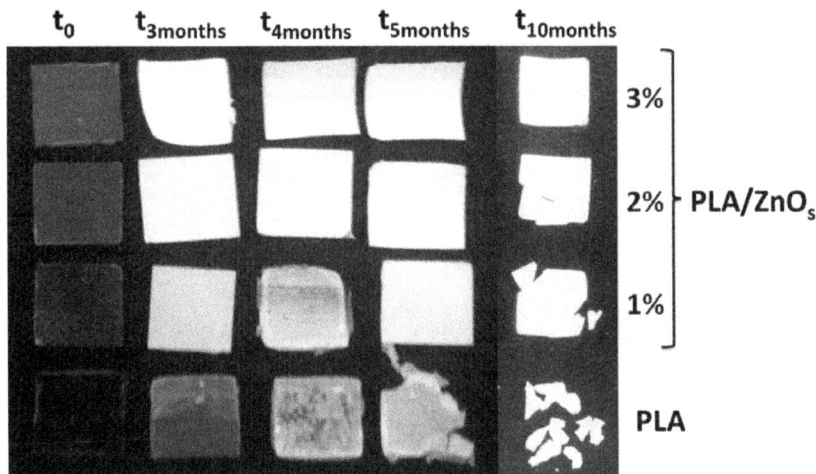

Figure 1   Changes in sample visual aspect at different times of hydrolysis for neat PLA and PLA nanocomposites filled with 1, 2 and 3%ZnO$_s$

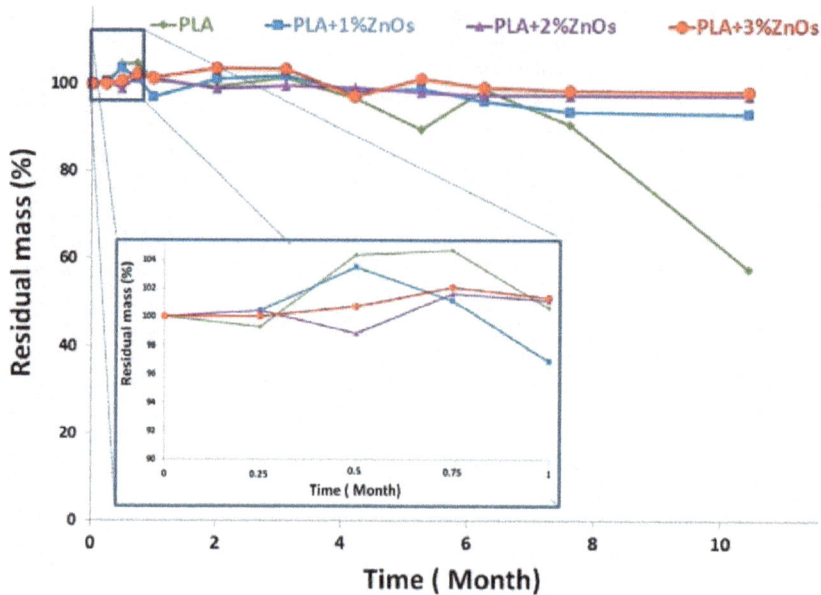

Figure 2   Evolution of residual mass for neat PLA (green) and the nanocomposites filled with 1% (blue), 2% (purple) and 3% ZnO$_s$ (red) as function of degradation time in phosphate buffer solution at 37°C (insert: expansion displaying first month)

whereas the sample which absorbs less water is the one that contains the largest content of ZnO$_s$.

Modification of molecular parameters

Molecular parameter evolution, i.e. as obtained from the analysis of SEC curves, molecular weight evolution (M_w/g mol^{-1}) and molecular mass dispersity (Đ$_M$) are reported in Figs. 3 and 4 respectively. SEC curves before and after hydrolysis at predetermined time (147 days/5 months) are reported in Fig. 3. The traces obtained by SEC analysis of

the different PLA samples reveal lower retention time after a 5 month hydrolytic degradation period for ZnO$_s$ based nanocomposites compared to those of the neat PLA, as consequence of their higher molecular mass. These results suggest that the presence of ZnO$_s$ (significantly) slows down the hydrolytic degradation of PLA. This trend is confirmed by the evolution of M_w versu the degradation time (Fig. 4A). Before degradation and as expected (due to the inherent catalytic effect of ZnO on the degradation of PLA in melt), the initial $M_{w(PLA)}$ of neat PLA films is a little higher than the

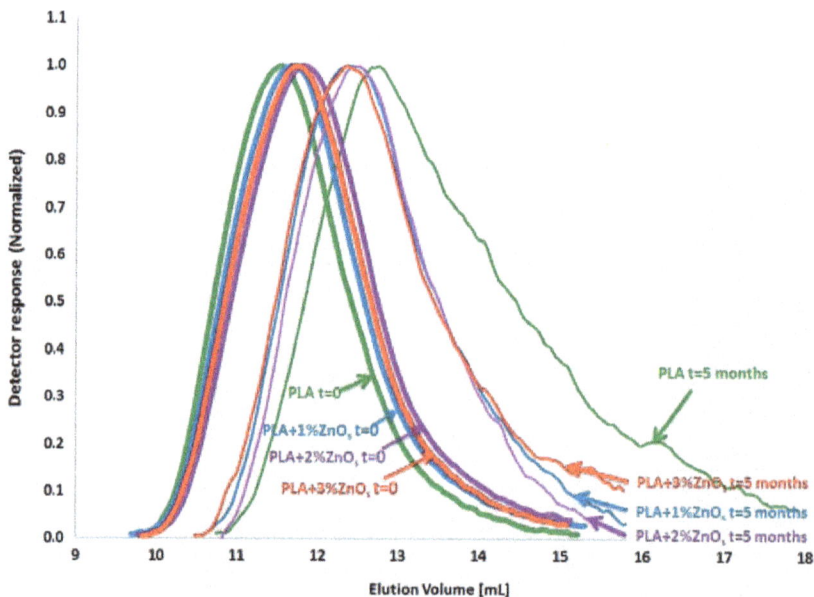

Figure 3   SEC curves of neat PLA (green) and nanocomposites based filled with 1% (blue), 2% (purple) and 3% of ZnO$_s$ (red) before (thick line) and after (thin line) hydrolytic degradation

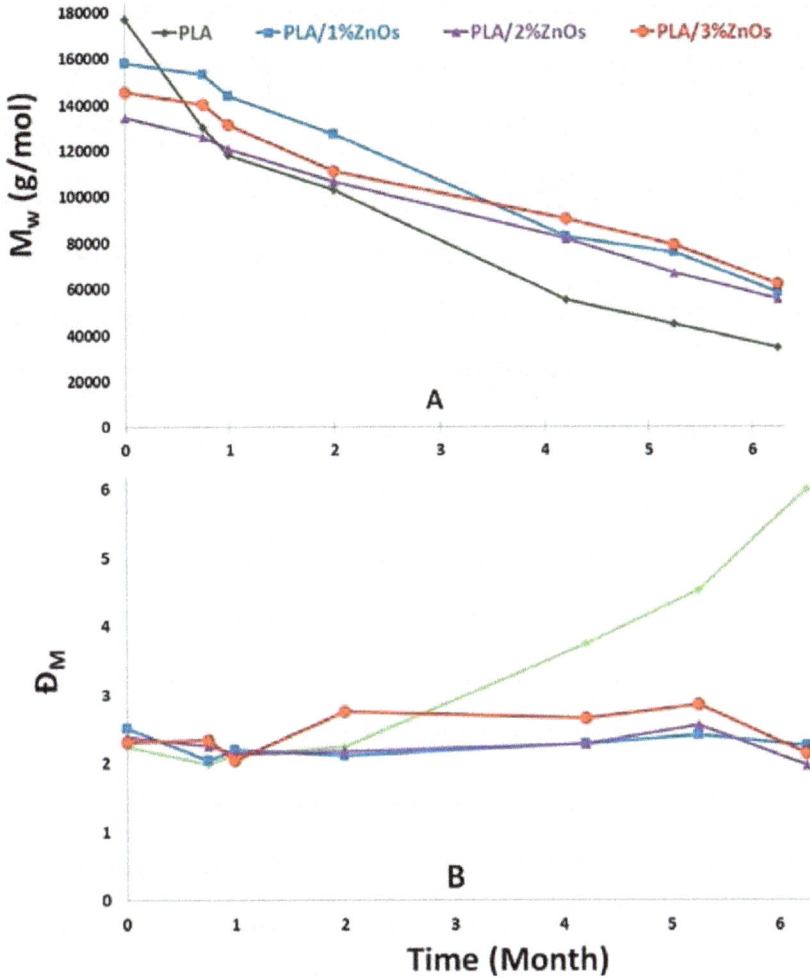

Figure 4 M_w evolution (A) and $Đ_M$ evolution (B) versus hydrolysis time of neat PLA (green) and of nanocomposites based on/filled with 1% (blue), 2% (purple) and 3% ZnO$_s$ (red)

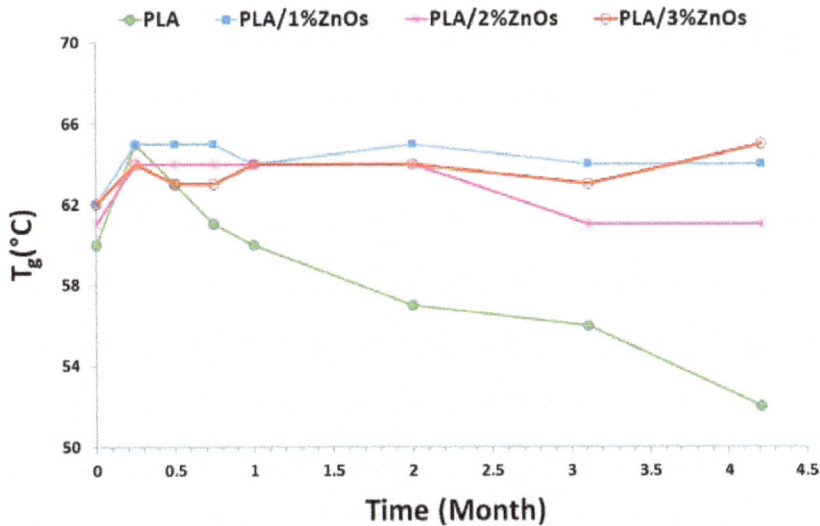

Figure 5 T_g evolution versus hydrolysis time for neat PLA (green) and nanocomposites based filled with 1% (blue), 2% (purple) and 3% ZnO$_s$ (red) (first heating, 10°C min^{-1})

$M_{w(PLA/ZnO_s)}$ of nanocomposites, i.e. $M_{w(PLA)} = 177\,000$ g mol^{-1} and $M_{w(PLA/3\%ZnO_s)} = 145\,700$ g mol^{-1}. However, already after 20 days, the trend is reversed and, then, the M_w of neat PLA gets smaller than those of PLA/1%ZnO$_s$ and PLA/3%ZnO$_s$. During 6 months, M_w of neat PLA decreased regularly up to nearly 80%, while in PLA/ZnO$_s$ nanocomposites, the decrease was about 55% within the same period.

Crucial information about the effect of ZnO$_s$ nanofiller on PLA degradation is also provided by the shape of the SEC curves (Fig. 3). Whereas before the hydrolytic degradation the SEC curves of all PLA samples are relatively narrow, only those of ZnO$_s$ based nanocomposites remain narrow after 5 months of hydrolytic degradation, while the SEC curve of the neat PLA (without ZnO$_s$) is significantly broader as a result of the presence of a significant amount of oligomers. These observations could be related to the evolution of the molecular weight dispersity of the samples (Đ$_M$) (Fig. 4B). More than 2 months of hydrolytic degradation at 37°C, the Đ$_M$ of PLA increases sharply, while for PLA/ZnO$_s$ nanocomposites, it remains relatively stable even up to 6 months.

However, it is important to point out that following the evolution of molecular parameters, as it was reported in previous studies, it comes out that the decrease of molar mass of PLA (from the different samples) is significantly faster than the mass loss.[22] On the other hand, knowing that the initial M_w of PLA of the nanocomposites were somewhat lower, it was surprising to observe that the decrease of molecular weights is rapidly in the case of neat PLA films (even in the first 20 days of degradation, as mentioned before). Accordingly, it was necessary to evidence which factors could influence and explain the better resistance at hydrolysis of PLA–ZnO$_s$ nanocomposites.

Thermal characterization

Regarding the DSC results of the initial films of PLA and PLA nanocomposites, it is noticed that the presence of ZnO$_s$ does not affect significantly the values of T_g ($\sim 61°C$), T_c ($\sim 110°C$) and T_m ($\sim 168°C$).

As far as the thermal transitions and crystallization behavior of the polyester matrix after hydrolytic degradation are concerned, it can be firstly noted that the polyester T_g in unfilled PLA is decreasing steadily from 61 to 52°C during the first 4 months (Fig. 5). It is noteworthy mentioning that for this sample, DSC measurements were not considered after 4 months, because the weight loss was too large. Such a decrease in T_g for the polyester matrix can be explained by both the reduction of the PLA molecular weights and by the known plasticizing effect of lactic acid oligomers formed upon degradation reaction. We will not exclude also the contribution of residual water molecules that can act as plasticizer, thus with potential contribution in the lowering the T_g values.[41] In sharp contrast, in the presence of ZnO$_s$, the T_g of neat PLA and PLA nanocomposites fluctuates around 64°C all along the same period of hydrolytic degradation test (Fig. 5). The evolution of the glass transition temperature observed for PLA/ZnO$_s$ can be attributed to the reduced chain mobility due to a better dispersion of ZnO$_s$ layers in the PLA matrix.[6] No plasticizing effect of lactic acid oligomers or residual water molecules leading to T_g decrease is visible, which might confirm that no PLA degradation occurred. Also the overall PLA crystallinity, actually determined by adding both cooling and heating crystallizations, is different in the presence of ZnO$_s$ (Fig. 6).

The initial crystallinity of all samples is similar before the hydrolytic degradation, i.e. crystallinity degrees of 8.5, 9.1, 8.6 and 7.5% for PLA, PLA/1%ZnO$_s$, PLA/2%ZnO$_s$ and PLA/3%ZnO$_s$ respectively This is further confirmed that following the procedure of extrusion described in the experimental part, once more it is confirmed that ZnO$_s$ is not an effective nucleating agent for PLA.[42]

The degree of crystallinity of neat PLA reaches 10% after 2 weeks of degradation time and remains lower than those of nanocomposites after 4 months. For instance, the

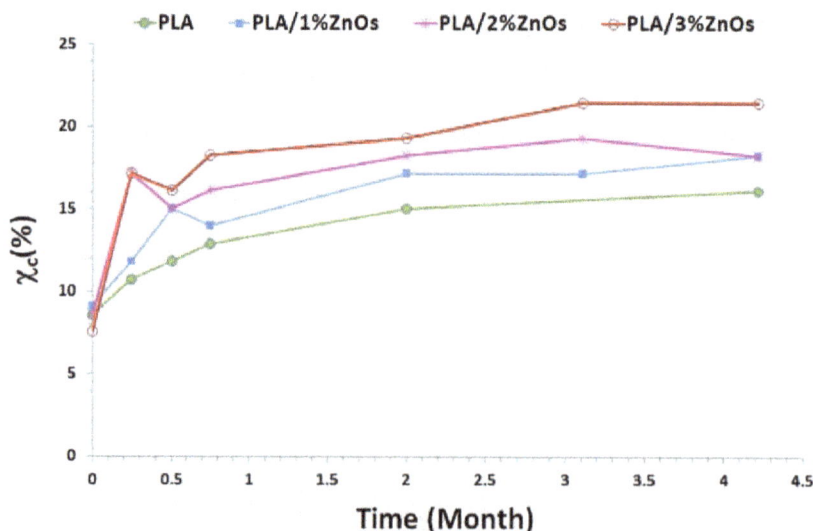

Figure 6 χ_c evolution versus hydrolysis time for neat PLA (green) and nanocomposites based filled with 1% (blue), 2% (purple) and 3% ZnO$_s$ (red) (first heating, 10°C min^{-1})

crystallinity of 3%ZnO$_s$ based nanocomposites increases significantly from 7.5 to 17% after 2 weeks only, and remains constant at ~20% after this interval of time.

Still, the crystallinity of PLA/1%ZnO$_s$ and PLA/2%ZnO$_s$ nanocomposites also increased at the beginning of hydrolytic degradation and remained located in between the values recorded for neat PLA and PLA/3%ZnO$_s$ samples. Interestingly enough, DSC measurements recorded all along the hydrolysis tests assess for a quite good correlation between the level of the degree of crystallinity and the content of ZnO$_s$ added into PLA. These trends in thermal properties have been confirmed with the correlation between the cold crystallization temperature and the degradation time. Considering the time evolution of T_c of PLA (Fig. 7), it can readily be related to the evolution of the molecular weight since it is well known that shorter PLA chains tend to crystallize at lower temperature. During the first 2 weeks, T_c of PLA/ZnO$_s$ nanocomposites decreases slightly before reaching a plateau. This confirms that the presence of ZnO$_s$ promotes initial cold crystallization of PLA matrix.

Clearly, the changes of thermal properties as a function of hydrolytic degradation time identified the ZnO$_s$ as an interesting nanofiller able to alter the PLA crystallization along the degradation process.

It is important to note that, according to a study recently reported by Höglund et al.,[22] the degree of crystallinity can have greater influence on the rate of PLA hydrolysis than the composition of the medium in which PLA is immersed, i.e. phosphate buffer solution or water. Accordingly, it was of interest to verify whether the degree of crystallinity represents the only determining parameter of hydrolytic degradation of PLA/ZnO$_s$ nanocomposites. In accordance with the literature,[25,29,31] the ability of water molecules to diffuse through a polymeric matrix (such as PLA) can be evaluated by water uptake measurements[25] as well as the determination of the diffusion activation energy, i.e. the energy level that a molecule must reach to diffuse inside the polymeric matrix.[11] In the case of the samples investigated in this contribution, the diffusion activation energies have been recently published by Pantani et al.[11] This previous study investigated the water transport properties of PLA/ZnO$_s$ nanocomposites (filled with 1% and 3% nanofiller) at three different temperatures (15, 30 and 45°C) to evaluate the activation energy of the water diffusion phenomenon. The values of energy activation were higher for the films containing 3%ZnO$_s$ (37.63 kJ mol^{-1}) with respect to the neat PLA (29.60 kJ mol^{-1}), confirming the increased difficulty for water molecules to diffuse through the PLA matrix. At the beginning of the degradation, this result could be attributed to the restrictions provoked by nanofiller and most likely to its hydrophobic character and water repellency, as a result of the surface treatment with high amount of triethoxycaprylylsilane. In this context, it is worth pointing out that ZnO showing superhydrophobic properties could be obtained by chemical modification with some selected silanes as highlighted by Li et al.[43] and Ding et al.[44] The increase of the crystallinity after few days of hydrolytic degradation (Fig. 6) also contributes to this slowdown because the diffusion of water is more restricted through PLA crystalline phase. As far as the PLA/1%ZnO$_s$ films are concerned, Pantani et al.[11] calculated a diffusion activation energy of 28.73 kJ mol^{-1} (similar to that of neat PLA film). Additionally, according to our results, only a slight rise of crystallinity is observed for this sample. Therefore, the water diffusion into the PLA is depending on ZnO$_s$ content. PLA hydrolytic degradation has been reported to take place mainly in the bulk material rather than at the surface;[8,16] thus, we can assume that the low water diffusion, attributed to both PLA ability to crystallize and hydrophobic properties

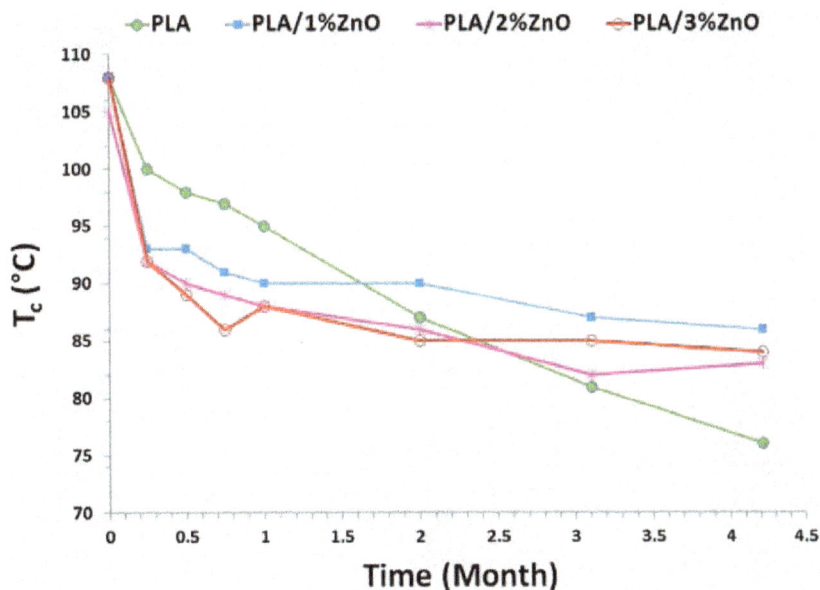

Figure 7 T_c evolution versus hydrolysis time for neat PLA (green) and nanocomposites based filled with 1% (blue), 2% (purple) and 3% ZnO$_s$ (red) (first heating, 10°C min^{-1})

of the nanofiller (PLA/ZnO$_s$ nanocomposites), slows down the process of degradation of nanocomposites. The exciting aspect of this research is thus the possible tuning of the hydrolytic degradation rate of PLA/ZnO$_s$ nanocomposites by varying the nanofiller content.

On the other hand, it is important to point out that the hydrolytic degradation of the polyesters can be influenced by the presence of additives or residual metals, which can decrease or increase the rate of degradation.[45,46] Furthermore, by considering that the presence of ZnO can display some basicity, it can be assumed that its reaction with the acidic end groups formed during autocatalyzed hydrolysis of PLA, has favorable effects in limiting the rate of PLA degradation.

Morphology evolution

The changes in the morphology of neat PLA and PLA/ZnO$_s$ films were studied during up to 10 months of hydrolytic degradation. Field emission SEM images on the surface of materials degraded at 37°C in phosphate buffered solution show after 4 and 10 months a behavior completely different with the presence of ZnO$_s$ nanofillers. Figure 8 allows the comparison of the surfaces of neat PLA and PLA/2%ZnO$_s$ nanocomposites before and after the hydrolytic degradation respectively. The images of the films recovered after pre-determined degradation times are reminded in inserts.

Before degradation, both samples present initially a smooth surface. Some white points within the PLA/2%ZnO$_s$ sample are observed and more likely can be attributed to small clusters of ZnO$_s$ nanoparticles (see the white arrows in Fig. 8A'). As the degradation proceeds, the surface of neat PLA sample becomes rougher and opaque (*vide supra*). After 10 months, the neat PLA sample, which is characterized by

the most advanced decrease of molecular weights, presents large holes ($\sim$ 5 µm of diameter, see the dotted circle in Fig. 8B) confirming the bulk degradation phenomenon as already reported.[17] It is also of interest to note that, by comparison to the PLA/2%ZnO$_s$ nanocomposite, the neat PLA sample is very brittle (see also the insert in Fig. 8B). After 10 months, the PLA/2%ZnO$_s$ sample is seemingly rough. Homogeneous holes over the entire specimen are visible (for instance, see the black arrow in Fig. 8B'). These small pores, i.e. less than 500 nm in diameter, indicate that in presence of ZnO$_s$ the PLA hydrolytic degradation occurs also according to a bulk degradation mechanism with just a delay due to the lower water diffusion into the PLA matrix caused by the hydrophobic nanofiller.

Conclusion

The influence of silanized zinc oxide nanofiller (ZnO$_s$) on the hydrolytic degradation of PLA in a phosphate buffer medium (at 37°C) was studied for more than 10 months. Following the production of PLA–ZnO$_s$ nanocomposites by melt blending, thin nanocomposite films ($\sim$ 70 µm thickness) of similar degree of crystallinity (in the range of 7–9%) were prepared by extrusion.

ZnO$_s$ induced a marked delay in the degradation of commercial PLA since the incorporation of aliphatic silane agents, i.e. triethoxycaprylylsilane, onto the filler surface is not only an efficient way for modifying the polymer/filler interface and for limiting nanofiller catalytic effects during the melt blending with PLA, but also has an effective role in the decreasing the rate of PLA hydrolysis, because the silanization renders the filler and related nanocomposites more hydrophobic. Following the DSC measurements during the hydrolytic degradation, the nanocomposites filled with

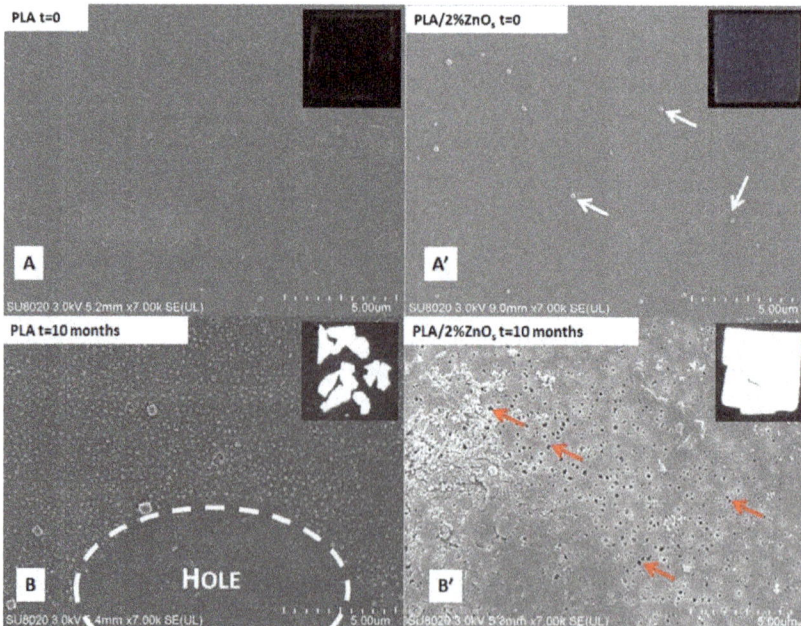

Figure 8 SEM images of neat PLA films (left) and PLA/2%ZnO$_s$ (right) before hydrolytic degradation (A and A' respectively) and after 10 months (B and B' respectively) (× 7000): state of films is reminder in inserts**

(1–3%) ZnO$_s$ have shown a higher increase in the degree of crystallinity with respect to the neat PLA. Both features (crystallinity and presence of hydrophobic nanofiller) lead to hinder the diffusion of water among the polymer matrix and consequently modify the kinetics of the hydrolytic process. The comparative analysis of molecular weights by SEC, the changes in the residual weight of samples, as well as the modification of morphology as evidenced by microscopy (SEM), allowed concluding that the nanocomposites (with a mention for PLA/3%ZnO$_s$) show improved resistance at hydrolysis with respect to the neat PLA. Moreover, it was assumed that PLA hydrolytic degradation is more consistent according to a bulk degradation mechanism with a significant delay due to presence of silanized nanofiller.

Finally, the results presented here are assessing for the possibility to prolong and tune the service life of a commercial PLA products by addition of hydrophobic nanofillers such as ZnO$_s$. Clearly, if applications are targeted in the field of packaging, and specifically food packaging, the potential toxicity of ZnO$_s$ will have to be investigated as well.

Conflicts of Interest

The authors have no conflicts of interest to declare.

Acknowledgements

The authors from University of Mons acknowledge supports by EU (FEDER) and Wallonia Region in the frame of EVE-RWALL research program. This research has also been funded by the Interuniversity Attraction Pole Programme (P7/05) initiated by the Belgian Science Policy Office.

The authors thank also to the Wallonia Region, Nord-Pas de Calais Region and European Community for the financial support in the frame of the INTERREG IV-NANOLAC project. They express their gratitude to all collaborators in the groups of Professor Serge Bourbigot (ENSC Lille, France) and of Professor Eric Devaux (ENSAIT Roubaix, France) for helpful discussions and all mentioned companies for supplying raw materials.

References

1. P. Dubois, O. Coulembier and J. M. Raquez: 'Handbook of ring-opening polymerization'; 2009, New York, Wiley Online Library.
2. J. M. Raquez, Y. Murena, A. L. Goffin, Y. Habibi, B. Ruelle, F. DeBuyl and P. Dubois: *Compos. Sci. Technol.*, 2012, **72**, (5), 544–549.
3. A. Höglund, M. Hakkarainen, U. Edlund and A.-C. Albertsson: *Langmuir*, 2009, **26**, (1), 378–383.
4. M.-A. Paul, C. Delcourt, M. Alexandre, P. Degée, F. Monteverde, A. Rulmont and P. Dubois: *Macromol. Chem. Phys.*, 2005, **206**, (4), 484–498.
5. S. Therias, J.-F. Larche, P.-O. Bussiere, J.-L. Gardette, M. Murariu and P. Dubois: *Biomacromolecules*, 2012, **13**, (10), 3283–3291.
6. M. Murariu, A. Doumbia, L. Bonnaud, A.-L. Dechief, Y. Paint, M. Ferreira, C. Campagne, C. Devaux and P. Dubois: *Biomacromolecules*, 2011, **12**, 1762–1771.
7. G. Schwach, J. Coudane, R. Engel and M. Vert: *Biomaterials*, 2002, **23**, (4), 993–1002.
8. S. Li: *J. Biomed. Mater. Res.*, 1999, **48**, (3), 342–353.
9. M. Qu, H. Tu, M. Amarante, Y.-Q. Song and S. S. Zhu: *J. Appl. Polym. Sci.*, 2014, **131**, (11), 1–7.
10. J. Jayaramudu, K. Das, M. Sonakshi, G. Siva Mohan Reddy, B. Aderibigbe, R. Sadiku and S. Sinha Ray: *Int. J. Biol. Macromol.*, 2014, **64**, 428–434.
11. R. Pantani, G. Gorrasi, G. Vigliotta, M. Murariu and P. Dubois: *Eur. Polym. J.*, 2013, **49**, (11), 3471–3482.
12. G. G. Pitt, M. M. Gratzl, G. L. Kimmel, J. Surles and A. Sohindler: *Biomaterials*, 1981, **2**, (4), 215–220.
13. H. Tsuji, K. Shimizu and Y. Sato: *J. Appl. Polym. Sci.*, 2012, **125**, (3), 2394–2406.
14. S. Li and S. McCarthy: *Biomaterials*, 1999, **20**, (1), 35–44.
15. H. Tsuji and Y. Ikada: *Polym. Degrad. Stabil.*, 2000, **67**, (1), 179–189.
16. I. Grizzi, H. Garreau, S. Li and M. Vert: *Biomaterials*, 1995, **16**, (4), 305–311.
17. S. Li, H. Garreau and M. Vert: *J. Mater. Sci.: Mater. Med.* 1990, **1**, (4), 198–206.
18. M. Hakkarainen: 'Aliphatic polyesters: abiotic and biotic degradation and degradation products', in 'Degradable aliphatic polyesters', 113–138; 2002, Berlin, Springer.
19. K. Fukushima, D. Tabuani, M. Dottori, I. Armentano, J. M. Kenny and G. Camino: *Polym. Degrad. Stabil.*, 2011, **96**, (12), 2120–2129.
20. A. Gupta and V. Kumar: *Eur. Polym. J.*, 2007, **43**, (10), 4053–4074.
21. K. Paakinaho: 'Tampereen teknillinen yliopisto', Julkaisu-Tampere University of Technology, Tampere, Finland, 2013, 1144.
22. A. Höglund, K. Odelius, A.-C. Albertsson and ACS. *Appl. Mater. Interfaces*, 2012, **4**, (5), 2788–2793.
23. M. Vert, F. Chabot, J. Leray and P. Christel: *Makromol. Chem.*, 1981, **5**, (S19811), 30–41.
24. M. A. Paul, C. Delcourt, M. Alexandre, P. Degée, F. Monteverde and P. Dubois: *Polym. Degrad. Stabil.*, 2005, **87**, (3), 535–542.
25. Q. Zhou and M. Xanthos: *Polym. Degrad. Stabil.*, 2008, **93**, (8), 1450–1459.
26. F. Fukushima, D. Tabuani, M. Arena, M. Gennari and G. Camino: *React. Funct. Polym.*, 2013, **73**, (3), 540–549.
27. Y. Zhao, Z. Qiu and W. Yang: *J. Phys. Chem. B*, 2008, **112B**, (51), 16461–16468.
28. Z. Qiu and H. Pan: *Compos. Sci. Technol.*, 2010, **70**, (7), 1089–1094.
29. E. Luiz de Paula, V. Mano and F. V. Pereira: *Polym. Degrad. Stabil.*, 2011, **96**, (9), 1631–1638.
30. Y.-B. Luo, X.-L. Wang and Y.-Z. Wang: *Polym. Degrad. Stabil.*, 2012, **97**, (5), 721–728.
31. Q. Zhou and M. Xanthos: *Polym. Eng. Sci.*, 2010, **50**, (2), 320–330.
32. H. Chen, J. Chen, J. Chen, J. Yang, T. Huang, N. Zhang and Y. Wang: *Polym. Degrad. Stabil.*, 2012, **97**, (11), 2273–2283.
33. S. S. Ray, K. Yamada, M. Okamoto and K. Ueda: *Macromol. Mater. Eng.*, 2003, **288**, (3), 203–208.
34. Y. Zhao, Z. Qiu and W. Yang: *Compos. Sci. Technol.*, 2009, **69**, (5), 627–632.
35. V. Arias, A. Höglund, K. Odelius and A.-C. Albertsson: *Biomacromolecules*, 2013, **15**, (1), 391–402.
36. S. Sinha Ray, K. Yamada, M. Okamoto, Y. Fujimoto, A. Ogami and K. Ueda: *Polymer*, 2003, **44**, (21), 6633–6646.
37. E. Kontou, P. Georgiopoulos and M. Niaounakis: *Polym. Compos.*, 2012, **33**, (2), 282–294.
38. M.-A. Paul, M. Alexandre, P. Degée, C. Calberg, R. Jérôme and P. Dubois: *Macromol. Rapid Commun.*, 2003, **24**, (9), 561–566.
39. E. Fukada: *Biorheology*, 1995, **32**, (6), 593–609.
40. R. Cairncross, J. Becker, S. Ramaswamy and R. O'Connor: *Appl. Biochem. Biotechnol.*, 2006, **131**, (1–3), 774–785.
41. N. Passerini and D. Q. M. Craig: *J. Controlled Release*, 2001, **73**, (1), 111–115.
42. P. O. Bussiere, S. Therias, J.-L. Gardette, M. Murariu, P. Dubois and M. Baba: *Phys. Chem. Chem. Phys.*, 2012, **14**, (35), 12301–12308.
43. M. Li, J. Zhai, H. Liu, Y. Song, L. Jiang and D. Zhu: *J. Phys. Chem. B*, 2003, **107B**, (37), 9954–9957.
44. B. Ding, T. Ogawa, J. Kim, K. Fujimoto and S. Shiratori: *Thin Solid Films*, 2008, **516**, (9), 2495–2501.
45. S. Li and M. Vert: *J. Biomater. Sci., Polym. Ed.*, 1996, **7**, (9), 817–827.
46. J. Mauduit, N. Bukh and M. Vert: *J. Controlled Release*, 1993, **23**, (3), 209–220.

9

Graphene quantum dot functionalized by beta-cyclodextrin: a novel nanocomposite toward amplification of L-cysteine electro-oxidation signals

Nasrin Shadjou[1,2], Mohammad Hasanzadeh*[3,4] ⓘ, Faeze Talebi[1] and Ahmad Poursattar Marjani[5]

[1]Department of Nanochemistry, Nano Technology Research Center, Urmia University, Urmia, Iran
[2]Faculty of Science, Department of Nano Technology, Urmia University, Urmia, Iran
[3]Drug Applied Research Center, Tabriz University of Medical Sciences, Tabriz, Iran
[4]Pharmaceutical Analysis Research Center and Faculty of Pharmacy, Tabriz University of Medical Sciences, Tabriz, Iran
[5]Faculty of Science, Department of Chemistry, Urmia University, Urmia, Iran

Abstract A novel nano-composite of graphene quantum dot-**β**-cyclodextrin was fabricated on the surface of glassy carbon electrode (GCE) using one step and green electrodeposition method. The redox of L-Cys behavior prepared nano-composite-modified of GCE was then characterized by cyclic voltammetry, chronoamperometry, and differential pulse voltammetry. Voltammograms of the modified electrode are recorded in physiological pH (phosphate buffer solution, pH 7.4), the kinetics of charge transfer and mass transport processes across the nano-composite/solution interface were studied. The modified electrode showed an efficient electrocatalytic activity toward the oxidation of L-Cysteine (L-Cys) through a surface-mediated electron transfer. The catalytic rate constant and the L-Cys diffusion coefficient were reported. A sensitive and time-saving method (differential pulse voltammetry) was developed for the analysis of L-Cys. The proposed voltammetric method was also applied to the determination of L-Cys using graphene quantum dot-**β**-cyclodextrin-GCE.

Keywords Graphene quantum dot, Nano-composite, **β**-cyclodextrin, L-cysteine, Fabrication, Electrochemical sensor

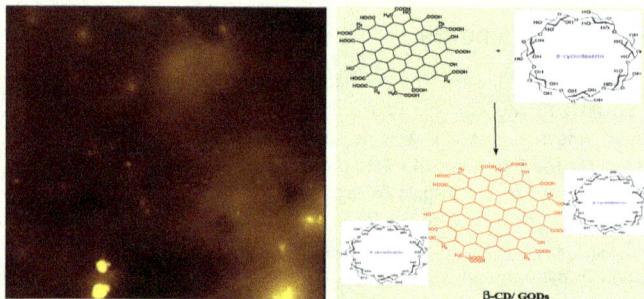

Introduction

L-Cysteine (L-Cys) belongs to sulfur-containing amino acid molecules playing a crucial role in biological systems and food industries.[1-3] L-Cys can be used for the clinical diagnosis of the disease states. For example, hair depigmentation, lethargy, edema, slowed growth, muscle and fat loss, skin lesions and weakness, and liver damage are associated with the deficiency of L-Cys[4] Therefore, it is critically important to develop novel electrode materials for sensitive determination of the trace amounts of different biomolecules, such as L-Cys in biological samples for clinical and pharmaceutical industrial applications.[5-7]

Various methods have been used for the determination of L-Cys, such as high performance liquid chromatography,[8] capillary electrophoresis,[9] chemiluminescent[10] spectrofluorimetry[11,12] and fluorimetry.[13,14] However, some of these methods take the disadvantages of expensive apparatus, time-consuming, and complicated pre-concentration process, multi-solvent extraction and trained technicians. By the comparison with other methods, electrochemical method takes inherent advantages of simple operation, high sensitivity, low cost, good selectivity, easy miniaturization and *in vivo* real time

*Corresponding author, email hasanzadehm@tbzmed.ac.ir

determination.[15–18] Bare carbon electrode showed poor electrocatalytic activity, even no responses for the determination of L-Cys.[19,20] Bulk Au and Pt electrodes usually showed high overpotential owing to the formation of surface oxide resulting in narrow linear range and low selectivity.[21] Various modified electrodes with electrocatalytic properties have been prepared for the determination of L-Cys, such as graphene oxide-Au nanoclusters composites,[22] metal-organic framework,[23] ordered mesoporous carbon-modified glassy carbon electrode (GCE),[24] boron-doped carbon nanotubes (CNTs),[25] copper-based nanoscale materials,[26,27] and vanadate nanoscale materials.[28] Furthermore, the detection limit, linear range, and sensibility are expected to be improved. Therefore, it is necessary to explore novel electrode materials to improve the electrochemical performance for the determination of L-Cys.

On the other hand, the couple L-cystine/L-cysteine is generally used as a model for the role of the disulfide bond and thiol group in proteins in a variety of biological media.[29] Therefore, it is very important to investigate the electrochemical behavior and sensitive detection of CySH.[30–37] Unfortunately, at ordinary electrodes (Pt, Au, graphite), the electrochemical behaviors of CySH are poor; no electrochemical responses could be observed.[38,39] It has been realized that the method to solve this problem may be to utilize new materials as electrodes; at such an electrode surface, the electrochemical responses could be simply obtained directly.[40–48] Over the past decades, several carbon-based materials, including fullerene and boron-doped diamond, have been explored for the electrochemical oxidation and detection of CySH. Recently, due to their unique electronic, chemical, and mechanical properties, CNTs as a new class of carbon nanomaterials have been exploited for the electrochemical oxidation of CySH. The high electrocatalytic activity observed at CNTs is attributed to the presence of the oxygen-containing functional groups on the surface of CNTs and a large number of edge plane graphite sites within the walls and at the ends of CNTs.[49,50] Besides the materials mentioned above, there has been significant interest in the development of one such novel carbon material, i.e. graphene quantum dots (GQDs).

In the recent years, GQDs has become an emerging class of the nano-carbon family, thanks to its low toxicity, it's easy preparation, its high chemical stability, its environmental friendliness, its luminescence or its ability to transfer photo-induced electron.[51,52] It has shown great promise applications in the fields of bioimaging,[53,54] photo-luminescence,[55] catalysts,[56] and fluorescent sensor.[57] In addition, GQDs have been recognized as both excellent electron donors and acceptor, making them interesting candidates for producing electrode materials.[58,59] GQDs can be functionalized[60] especially with oxygen-containing groups such as hydroxyl, carboxyl, and epoxy groups which can greatly enhance their hydrophilia and biocompatibility. Therefore, GQD-based materials have significant potential for biosensor applications. Additionally, various metallic nano-composite materials have been investigated for enzymatic and non-enzymatic amino acid biosensors.[61–69] Traditionally, noble metals are the favored candidates for electrochemical biosensors due to their inherent electrocatalytic activities. For example, previously we used metal nanoparticles and GO as enhanced materials to build amino

acid biosensor.[68] However, due to the high price and scarcity of noble metals, nano-composite materials with great performances and low costs have been intensively explored to develop new functional nanomaterials.

CDs with their largely hydrophobic cavities of variable size and numerous ways of chemical modification are the subject of intensive electrochemical research including both their behavior in homogeneous solutions and in thin films attached to the electrode surfaces.[70–72] Therefore, CDs are employed in electrochemical sensing devices for the determination of selected analytes. On the other hand, literature review show that, integration of CDs to the structure of electroactive materials such as graphene-based materials can be enhancing their electrical conductivity.[73] These improved performances encouraged us to explore the possible leading role played by the presence of β-CD/graphene or its conductive derivatives such as GQDs and functionalized GQDs. GQDs increase the contact area with the analyte, so they could increase the electrochemical active surface to interact with some electroactive analytes.[74] Since the increase in geometric surface area is very important parameter in electrocatalysis, therefore modification of different substrates (such as glass, carbon, graphite, etc.) by GQDs can increase the rate of electrochemical reaction. Therefore, integration of β-CD into GQDs can be provided by the zero-dimensional structure of the deposited films and greatly increases Faradic currents. Thus, in this paper, a nano-composite of GQDs-β-cyclodextrin was fabricated on the surface of GCE using one step electrodeposition method. The redox behavior of prepared nano-composite modified of GCE was then characterized by cyclic voltammetry (CV), chronoamperometry, and differential pulse voltammetry. In the voltammograms of the modified electrode recorded in physiological pH (phosphate buffer solution (PBS), pH 7.4). The kinetics of charge transfer and mass transport processes across the nano-composite/solution interface were studied. The modified electrode showed an efficient electrocatalytic activity toward the oxidation of L-Cysteine (L-Cys) through a surface-mediated electron transfer. The catalytic rate constant and the L-Cys diffusion coefficient were reported. A sensitive and time-saving method (differential pulse voltammetry) was developed for the analysis of L-Cys. The influence factors including scan rate, L-Cys concentration in the electrochemical behaviors have been systematically analyzed. β-CD-GQDs show excellent electrocatalytic activities toward L-Cys and great application potential in electrochemical sensors.

Experimental details

Chemicals and reagents

All chemicals were purchased from Merck (Darmstadt, Germany) and used without further purification. Alumina slurry was purchased from Beuhler (Illinois, USA) and raw material of L-Cys was purchased from Merck (Germany). All solutions were prepared with deionized water. The stock solution of L-Cys (0.001 g per mL) was prepared by dissolving an accurate amount of L-Cys in an appropriate volume of 0.1 M PBS, pH 7.4 (which was also used as supporting electrolyte), and then stored in the dark place at 4 °C. Additional dilute solutions were prepared daily by accurate dilution just before use. Also the other stock solutions were

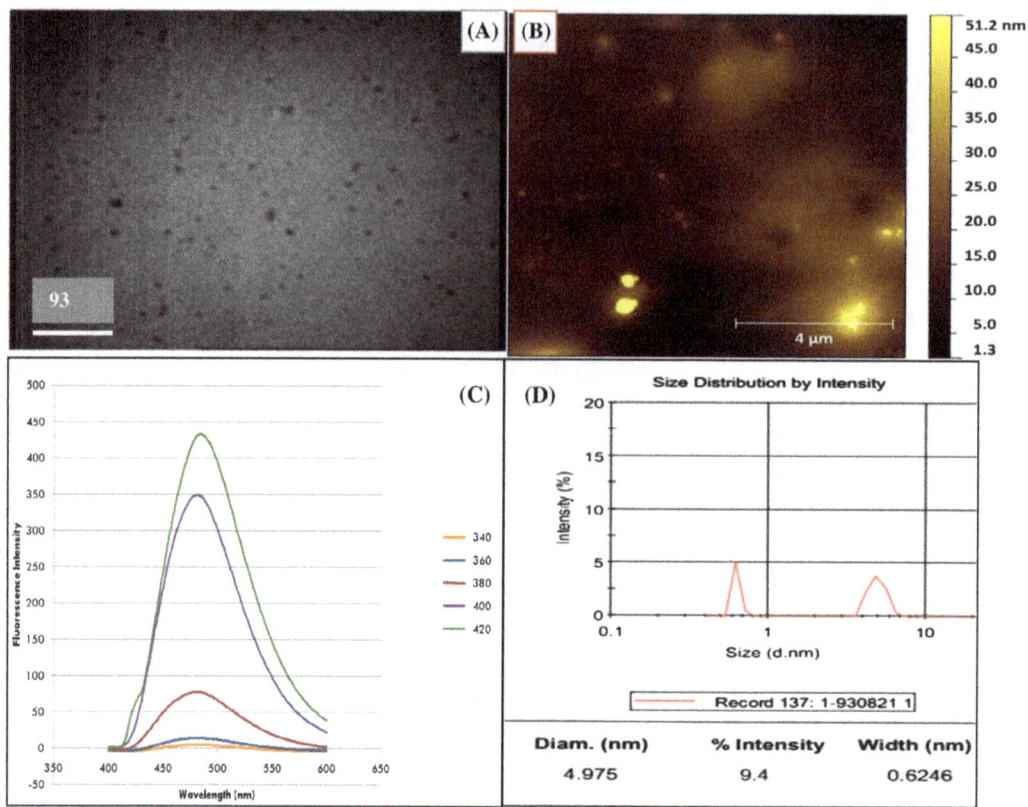

Figure 1   (A, B) TEM and AFM image, (C, D) fluorescence spectra and DLS analysis of GQD

prepared by dissolving an accurate amount equal to molecular weight of each one in an appropriate volume of 1000 mL deionized water and then all stored in the dark place at 4 °C.

Apparatuses and methods

Electrochemical measurements were carried out in a three-electrode cell setup. The system was run on a Personal Computer using NOVA1.7 software. Saturated Ag/AgCl as a reference electrode and the counter electrode (also known as auxiliary electrode), which usually made of an inert material was platinum. All potentials were measured with respect to the Ag/AgCl which was positioned as close to the working electrode as possible by means of a luggin capillary. GCE (from Azar electrode Co., Urmia, Iran) was used as the working electrode. The transmission electron microscope (TEM) images were obtained on Leo 906, Zeiss, (Germany). Atomic force microscopy (AFM) experiments were performed at contact mode by Nanowizard AFM (JPK Instruments AG, Berlin, Germany) mounted on Olympus Invert Microscope IX81 (Olympus Co., Tokyo, Japan). UV–Vis spectroscopy was performed by Cecil, Cambridge, (UK). X-ray powder diffraction (XRD) measurements were performed using Siemens, D500, (Germany). Dynamic light scattering (DLS) were obtained using Malvern 3500 ZS. Spectrofluorimetery test was performed using Jasco, FP-750, (Tokyo, Japan).

Synthesis of GQDs

An easy bottom-up method was used for the preparation of GQDs. At first, GQDs were synthesized by pyrolyzing citric acid and dispersing the carbonized products into alkaline solutions.[75] Briefly, 2 g of citric acid was put into a beaker and heated to 200 °C by a heating mantle until the citric acid changed to an orange liquid. Then, for preparing GQDs, 100 mL of 10 mg/mL NaOH solution was added into the orange homogenous liquid dropwise with continuous stirring. The obtained GQD solution was stable for at least one month at 4 °C.

Characterization of GQDs

Figure 1(A) and (B) presents the AFM and TEM images of synthesized GQDs. The corresponding AFM image shows a single GQD monolayer thin film. Ninety percent of the particles represent dark brown color which assigned to a size range below 10 nm. Furthermore, the DLS study represented hydrodynamic sizes of GQDs with size distribution of 5 ± 4 nm (Figure 1(C)) which findings of DLS analysis confirmed the AFM results.

Figure 1(D) shows fluorescence spectra of the GQD dispersed in water at room temperature. The GQDs have a broad absorption, from 400 to 600 nm. The maximum excitation of 210 nm was obtained with an emission wavelength of ~470 nm. The maximum emission of ~480 nm was obtained with an

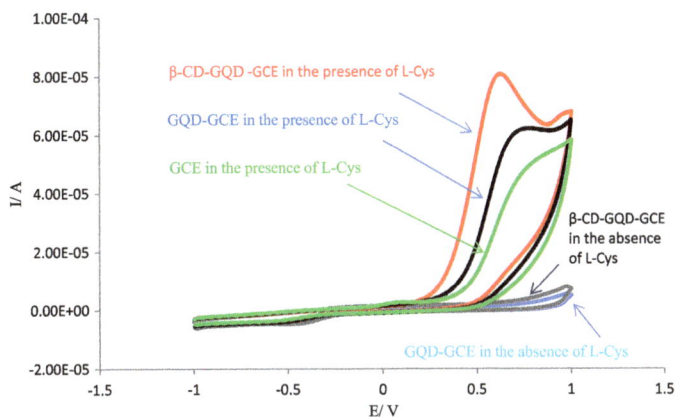

Figure 2   CVs of bare GCE GQDs/GCE, and β-CD/GQDs/GCE in solution without and with 2 mM of L-Cys. Supporting electrolyte: 0.1 M; PBS (pH 7.4); scan rate: 100 mV s^{-1}

excitation wavelength of 400 nm. Also when the excitation wavelength changed from 340 to 420 nm, the maximum peaks were constant. This could be explained by the uniformity both in the size and the surface state of those sp^2 clusters contained in GQD which was responsible for the fluorescence of GQD.

Preparation of GQDs and β-CD-GQDs modified GCE

GCE (2 mm in diameter) was polished to a mirror-like finish with 0.3 and 0.05 μm alumina slurry and then thoroughly rinsed with double distilled water. Then it was successively sonicated in acetone and double distilled water and was allowed to dry at room temperature. Finally, 5 mL of homogenous GQD and β-CD-GQDs films were electrodeposited onto GCE by CV in the potential range from -1.0 to 1.0 V at a scan rate of 200 mV s^{-1} for 50 cycles (Scheme 1). After 50 repeating cycles, the three-electrode system was transferred into 0.1 M PBS (pH 7.4) containing L-Cys and cyclic voltamograms were recorded at a sweep rate of 100 mV s^{-1}.

Results and discussion

The electrochemical behaviors of L-Cys have been analyzed by measuring the CVs at bare GCE and different modified GCEs. Figure 2 shows the CVs at bare GCE with the mixed solution of 2 mM L-Cys and 0.1 M PBS (pH 7.4), GQD and β-CD-GQD modified GCE in PBS (pH 7.4) solution with and without L-Cys. In 0.1 M PBS (pH 7.4) solution, bare GCE exhibits no electrocatalytic activities toward L-Cys. The GQD and β-CD-GQD modified GCE has also no electrochemical response only in the presence of PBS (pH 7.4) solution. Generally, an irreversible CV peaks were observed from CVs of L-Cys at different modified GCEs (GQD-GCE and β-CD-GQD-GCE). Similarly, Gao et al.[76] reported that an irreversible CV peak of 2 mM L-Cys at the graphene-Au modified GCE in 0.1 M phosphate buffered saline (pbs) solution was located at the potential of about +0.6 V. Also, the graphene oxide-Au nanoclusters-modified GCE showed that a pair of reversible CV peaks in the mixed solution of L-Cys and PBS (pH 7.4) were located at about +0.43 V and +0.21 V, respectively at different scan rates.[77] Two pairs of CV peaks of L-Cys in

0.1MPBS (pH 7.4) solution were located at +0.24, +0.07 V, and +0.05, −0.47 V, respectively using polyaniline/CuGeO$_3$ nanowires-modified GCE for the electrochemical determination of L-Cys.[78] Different from the electrochemical responses of L-Cys at the GCEs modified using different nanoscale materials, GQD-GCE and β-CD-GQD-GCE exhibits irreversible oxidation peaks located at 0.75 and 0.62 V, respectively, in the mixed solution of 2 mM L-Cys and 0.1 M PBS (pH 7.4). Therefore, it is reasonably concluded that these CV peaks result from L-Cys because the CV peaks can only be observed in PBS (pH 7.4) solution with L-Cys. The results suggest that the β-CD-GQD-modified GCE exhibits superior performance for L-Cys oxidation.

For comparison of the recognition efficiency, other electrodes including bare GCE, and GQDs-GCE were used for the control experiments. Although the peak currents are remarkably increased at the GQDs-GCE due to the excellent electrical conductivity of GQDs, the I_p is 2 mA which is still too small to distinguish the lower concentrations of L-Cys (see Figure 2). From the above figure, it can be concluded that without active substance, bare GCE or GQDs-GCE cannot result in the successful electrooxidation of L-Cys. When β-CD containing catalyst agent is used to modify GQD-GCE, the differences in both peak currents and potential are obviously enlarged, suggesting that β-CD-GQDs can provide an acceleration center to electron shuttling. These results indicated that β-CD-GQDs film could accelerate the rate of electron transfer of L-Cys and have good electrocatalytical activity for redox L-Cys reaction of L-Cys. Therefore, β-CD-GQDs are a suitable mediator to shuttle electron between L-Cys and working electrode, and facilitate electrochemical regeneration following electron exchange with L-Cys. This observation is also linked to the high conductivity and inherent ability of β-CD-GQDs. That might be related to the excellent properties of β-CD-GQDs such as high-specific surface area and electrical conductivity. Also, the recognition efficiency is further improved at the β-CD-GQDs-GCE with a peak current ratio of 4.1 and ΔE_p of 0.8 V, it is no doubt that β-CD-GQDs play important roles in the improvement of catalytically activity of GQDs toward electrooxidation of L-Cys. As a decisive catalytical component in the composite film, β-CD-MMNPs provides a catalytic platform for selective detection

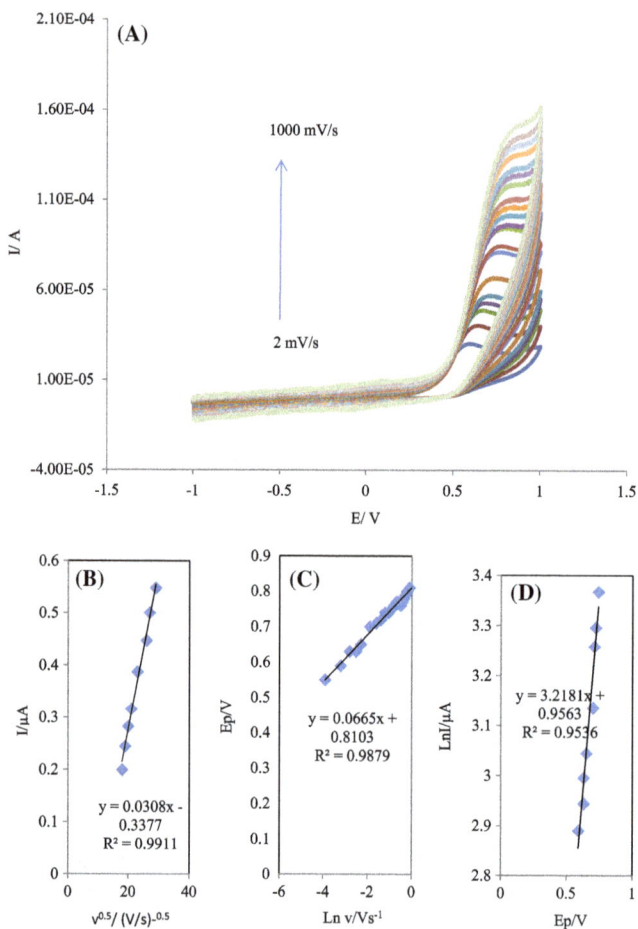

Figure 3   (A) CVs of β-CD/GQDs/GCE in the presence of 2 mM Cys + 0.1 M PBS (pH 7.4) in different scan rates (from inner to outer): 20, 40, 60, 80, 100, 150, 200, 250, 300, 350, 400, 450, 500, 550, 600, 650, 700, 750, 800, 900, and 1000 mV s-1, respectively. (B) Variation of the scan rate-normalized current (Ip/v1/2) with scan rate. (C) Variation of Ep vs. the logarithm of the scan rate. (D) Tafel plot derived from the current-potential curve

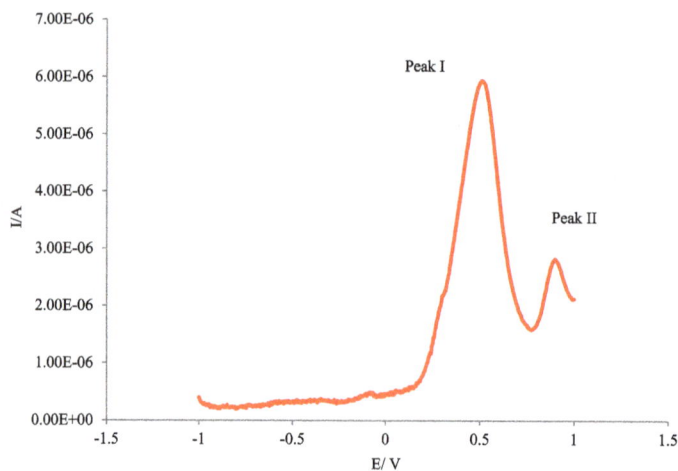

Figure 4   DPV of 2 mM L-Cys at β-CD/GQDs/GCE. Supporting electrolyte: 0.1 M; PBS (pH 7.4); scan rate: 50 mV s⁻¹

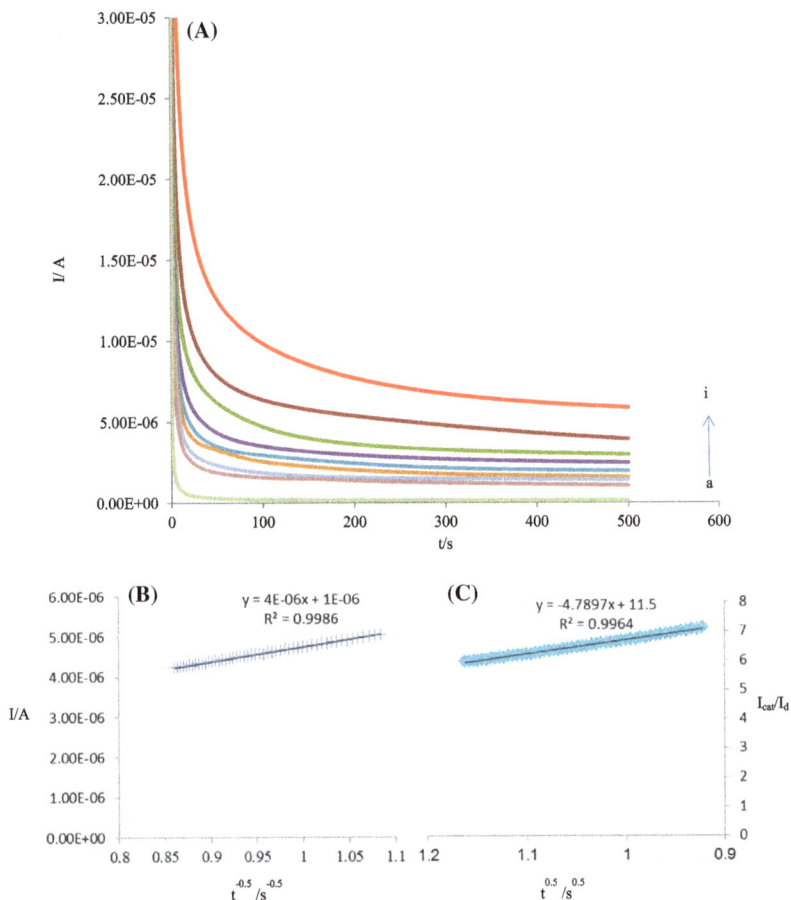

Figure 5 Chronoamperograms of β-CD/GQDs/GCE in 0.1 M PBS in the absence (a) and presence (b–i) of various concentrations of L-Cys: 0, 0.46, 0.15, 0.11, 0.09, 0.076, 0.065, 0.057, and 0.05 M, respectively. Insets: Variation of chronoamperometry currents at vs. time$^{-1/2}$. Anodic currents obtained for a potential step of 0.55 V

of L-Cys meanwhile, GQDs can amplify the electrochemical signals produced during the electrochemical sensing process. In one word, the synergetic effect of GQDs and β-CD leads to a successful and effective detection of L-Cys.

Scan rate is one of parameters significantly affecting electrooxidation of various compounds. Also, useful information involving electrochemical mechanism usually can be acquired from the relationship between peak current and scan rate. Therefore, the electrochemical behaviors of L-Cys at different scan rates were investigated on the surface of β-CD-GQDs-GCE by CV. Thus, the effect of the scan rate on L-Cys electrooxidation was investigated in the range from 20 to 1000 mVs^{-1} using CV method (Figure 3(A)). A linear relationship was obtained between the peak current and the scan rate in the range of 2–300 mVs^{-1}, which revealed that the oxidation of L-Cys was an adsorption-controlled step. Two approaches widely used to study the reversibility of reactions and to determine whether a reaction is adsorption or diffusion controlled consist of the analyses of dependences: I_p vs. $v^{1/2}$ and ln I_p vs. ln v. Figure 3(B) shows these plots for the oxidation peak of L-Cys in 0.1 M PBS. For reversible or irreversible systems without kinetic complications, I_p varies linearly with $v^{1/2}$, intercepting

the origin. Although, the plot of I_p on $v^{1/2}$ presented in Figure 3(B) is linear ($R^2 = 0.9996$), it does not cross the origin of the axes. This is characteristic for the electrodic process preceded by electrochemical reaction and followed by a homogenous chemical reaction. In the scan rate range from 20 to 300 mVs^{-1}, peak current (I_p) of L-Cys electrooxidation depends linearly on square root of the scan rate (v) and is described by the following equation:

$$I_p = 0.0308v(\text{V}^{1/2}\text{s}^{-1})\text{mA} + 0.3377\text{mA} \quad (R = 0.9911)$$

This dependence does not cross the origin (Figure 3(B)). This fact can suggest that the electrode process of L-Cys electrooxidation isn't controlled by diffusion and can be preceded by chemical reaction. On the other hand, a dependence of ln I_p on ln v is linear and described by the following equation:

$$\ln Ip = 0.7540 \ln v(\text{V s}^{-1}) - 2.500\text{mA} \quad (R^2 = 0.9954)$$

Its slope is 0.7540 and indicates adsorption control of the electrode process. A slope close to 0.5 is expected for diffusion-controlled electrode processes and close to 1.0 for adsorption-controlled processes.[79–81]

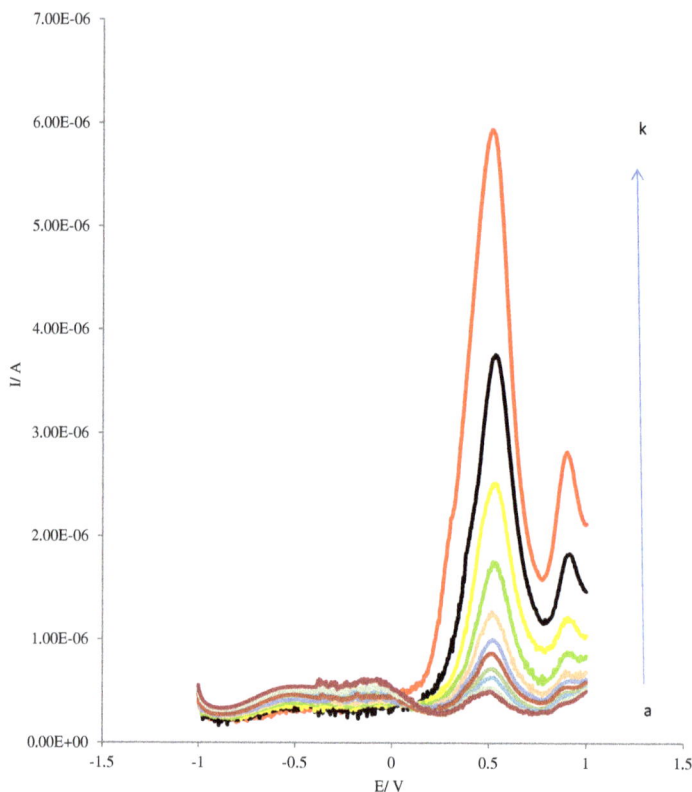

Figure 6   DPVs of β-CD/GQDs/GCE in different concentrations of L-Cys solutions (0.1 (a), 0.2 (b), 0.25 (c), 0.3 (d), 0.33 (e), 0.5 (f), 0.66 (g), 1 (h), 1.33 (i), 2 (j), and 4 (k) mM) Supporting electrolyte: 0.1 M; PBS (pH 7.4); scan rate: 20 mV s^{-1}

Also, a plot of the sweep rate-normalized current ($I_p/v^{1/2}$) vs. sweep rate (Figure 3(B)), exhibits the characteristic shape typical of an EC process.[82]

Figure 3(C) presents a dependence of E_p on scan rate determined from cyclic voltammograms recorded for the L-Cys electrooxidation. If electrochemical reaction is irreversible, then E_p is independent on v. Thus, it can be concluded that heterogeneous electron transfer in L-Cys electrooxidation is irreversible because E_p increases with an increase in the scan rate. In addition, the value of the overall electron transfer coefficient for the reaction can be obtained from the following equation:[83,84]

$$E_p = \frac{RT}{2\beta nF}\ln v + constant$$

where E_p – peak potential (V), R – universal gas constant (8.314 J K^{-1} mol^{-1}), F – Faraday constant (96,487 C mol^{-1}), T – Kelvin temperature (298 K), βn – anodic transfer coefficient, v – scan rate (V s^{-1}).

Using the dependence of anodic peak potential on the logarithm of the potential scan rate (Figure 3(C)), the value of the overall electron transfer coefficient (βn) was obtained as 0.33 for L-Cys electrooxidation.

In order to obtain information on the rate-determining step, Tafel slope ($b = RT/\beta n F$) was determined as 6.0 mV (Figure 3(C)).

According to Bard and Faulkner[81], current-potential characteristic of the oxidation irreversible process is described by the following equation:[81]

$$Ip = 0.227FAC * [\exp(\beta nf)(E_p)]$$

where I_p – peak current (A), $f = F/RT$, A – electrode area (cm^2), C – bulk concentration of L-CYS (mol cm^{-3}). This equation allowed obtaining dependence:

$$\ln I_p = Lna - (\beta nf)E_p$$

A dependence of ln (I_p) vs. (E_p) is linear (Figure 3(D)) and described by the following equation:

$$\ln Ip = 3.2181[(E_p)V]\text{mA} - 0.9563\text{mA}(R^2 = 0.9536)$$

and its slope is 2.30 anodic transfer coefficient (βn) of L-Cys electrooxidation, calculated from the Eq. (3) totals 94.2 mV. This result is close to obtained from previous method (E_p vs. ln v). This slop indicates also the three electron transfer to be rate-limiting, assuming a charge transfer coefficient of $\beta = 0.30$.

We also investigate the electrochemical behavior of L-Cys at GC, GQD/GC, and β-CD-GQDs-GCE electrodes at using DPV technique. Figure 4 shows the electrochemical oxidation of Cys at β-CD-GQDs-GCE electrodes at pH 7.4. As can be seen, there are two anodic peaks at +0.02 (peak I) and +0.38 V (peak II), respectively. This means the oxidation of L-Cys at pH 7.4 undergoes two processes, and the β-CD-GQDs film clearly plays an important role in the observed electrocatalytic behavior. Mao et al. reported the oxidation of L-Cys at CNT-modified electrode at pH 7.0 undergoes two processes: at 0.0 V attributed to the oxygen-containing functional groups of CNTs, and at +0.35 V

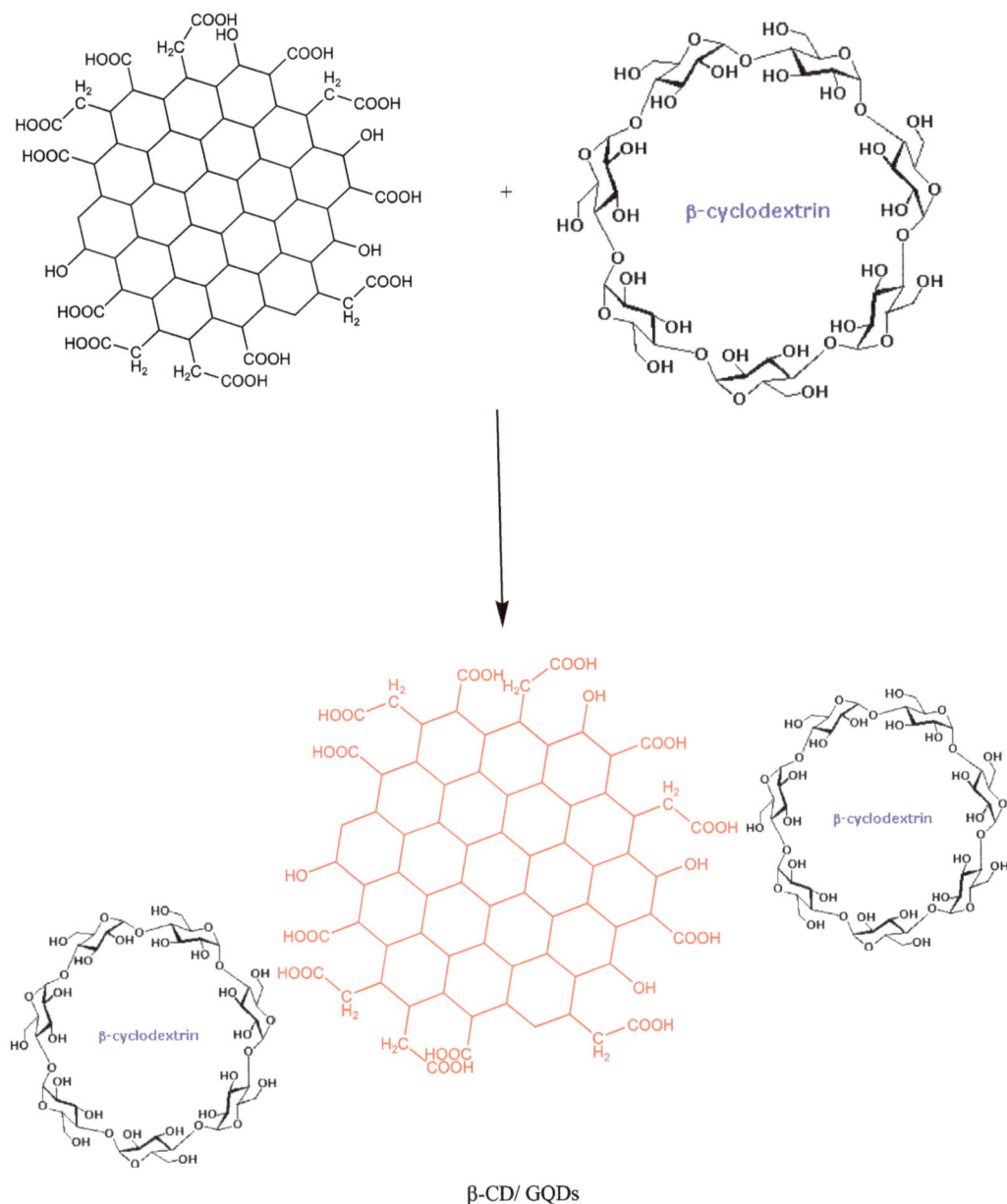

β-CD/ GQDs

Scheme 1   Synthesis procedure of β-CD/GQDs

ascribed to the edge plane-like defective sites of CNTs,[85] which are almost at the same potentials as those at the β-CD-GQD-GCE electrode of this work. This suggests the electrooxidation of Cys at β-CD-GQD-GCE and CNTs modified electrodes at pH 7.4 may have the same processes and similar mechanisms. The redox L-Cys peak pairs observed at β-CD-GQD-GCE electrode in Figure 4 can be also observed for β-CD-GQD and CNT-based materials,[86] which indicates the pairs may originate from pro-tonation/depronation of the oxygen-containing functional groups of β-CD-GQD,[86] So the oxygen-containing functional groups of β-CD-GQDs are thus mainly responsible for peak I. There are also significant edge-plane-like defective sites

existing on the surface of β-CD-GQD,[86] which means peak II at the β-CD-GQD-GCE electrode at pH 7.4 could be logically attributed to the edge-plane-like defective sites of β-CD-GQD.

In order to evaluate the reaction kinetics, the electrocatalyt-ical oxidation of L-Cys on β-CD/GQDs/GCE was investigated by chronoamperometry. Chronoamperometry, as well as CV has been employed for the investigation of the processes occur-ring via an E_rC_i mechanism.[81] Single steps chronoamperograms were recorded by setting the working electrode potentials to desired values and were used to measure the catalytic rate constant on the modified surface. Figure 5(A) shows double steps chronoamperograms for the modified electrode in the absence (a) and presence (b) 0.46, (c) 0.15, (d) 0.11, (e) 0.09, (f) 0.076, (g) 0.065, (h) 0.057, and (i) 0.05 M) of L-Cys over a concentration range of 0.05–0.46 M. The applied potential steps were 0.55 vs. (Ag/AgCl)/V, respectively. The plot of net current vs. $t^{-1/2}$ which has been obtained by removing the background current by the point-by-point subtraction method gives a straight line, Figure 5(B). This indicates that the transient current must be controlled by a diffusion process. The transient current is due to catalytic oxidation of L-Cys which increases as the L-Cys concentration is raised. Using the slopes of these lines, we can obtain the diffusion coefficients of the drugs according to the Cottrell equation:[81], p. 163

$$I = nFAD^{1/2}C^*\pi^{-1/2}t^{-1/2} \tag{2}$$

where D is the diffusion coefficient, and C^* is the bulk concentration. The mean value of the diffusion coefficients of L-Cys was found to be 2.1×10^{-5} cm^2 s^{-1}.

The rate constants of the reactions of L-Cys and the ensuing intermediates with the redox sites of the β-CD/GQDs/GCE electrode can be derived from the chronoamperograms according to Eq. (6):[81], p. 503

$$\frac{I_{catal}}{I_d} = \lambda^{1/2}\left[\pi^{1/2}erf(\lambda^{1/2}) + \frac{\exp(-\lambda)}{\lambda^{1/2}}\right] \tag{3}$$

where I_{catal} is the catalytic current in the presence of L-Cys, I_d the limiting current in the absence of L-Cys and $\lambda = kC_m t$ (k, C_m and t are the catalytic rate constant, bulk concentration of L-Cys and the elapsed time, respectively) is the argument of the error function. For $\lambda > 1.5$, $erf(\lambda^{1/2})$ almost equals unity and Eq. (3) reduces to:

$$\frac{I_{catal}}{I_d} = \lambda^{1/2}\pi^{1/2} = \pi^{1/2}(kC_m t)^{1/2} \tag{4}$$

From the slope of the I_{catal}/I_d plot, the value of k at a given concentration of L-Cys can be derived (Figure 5(C)). The mean value of k in the concentration range of 0.05–0.46 M L-Cys was found to be $8.8 \times 10^{+5}$ cm^3 mol^{-1} s^{-1}.

In order to develop a voltammetric method for determining of L-Cys, we selected the DPV mode, because the peaks are sharper and better defined at lower concentrations of L-Cys than those obtained by CV, with a lower background current, resulting in improved resolution. According to the obtained results, it was possible to apply this technique to the quantitative analysis of L-Cys. As mentioned previously, the PBS of pH 7.4 was selected as the supporting electrolyte for the quantification of L-Cys. The linear range for the determination of L-Cys was investigated under optimal conditions. The analytical parameters for the detection of L-Cys using β-CD-GQD-modified GCE have been investigated by measuring the CVs of L-Cys with different concentrations (Figure 6). A plot between L-Cys concentration and peak current shows a linear relationship (the inset in the upper-right part of Figure 6). The linear range is observed from 0.01 to 2 mM. The results demonstrate that the β-CD-GQD possess good electrocatalytic activity to the oxidation of L-Cys which may due to the large surface area and good catalytic activity providing abundant active sites for L-Cys oxidation.

The stability of the β-CD-GQD-modified GCE electrode under the storage conditions (exposure to air, ambient temperature) was also examined using the same PBS containing 2 mM L-Cys. During the repeating amperometric detection process, a decrease in response current can be observed and 10% of the response current will be lost after 10 repeated detections. This may result from the adsorption of (the intermediate species during L-Cys oxidation). Fortunately, β-CD-GQD-modified GCE electrode can be recovered very well (about 98%) by electrochemical activation in PBS (pH 7.4). The long-term storage stability of the β-CD-GQD-modified GCE electrode was also investigated. The electrocatalytic property of β-CD-GQD-modified GCE electrode to L-Cys oxidation remains satisfactory; only 6% current loss after 6 days and 20% current loss after 1 month are observed.

Conclusion

The β-CD-GQD film was formed electrochemically in a regime of CV on a GCE and checked for electrooxidation of L-Cys at physiological pH. It is concluded that the electrooxidation of formaldehyde that starts around 0.2 V vs. Ag/AgCl occurs through a mediated electron transfer mechanism catalyzed by β-CD-GQD species which form in on the surface of GCE. The results show that the β-CD-GQD-GCE exhibit high electrocatalytic activity toward L-Cys oxidation at around 0.6 V. Therefore, β-CD-GQD can promote L-Cys electron-transfer, thus β-CD-GQD/GCE provides higher electroactive surface area and lower charger transfer resistance compared with GQD-GCE. Using CV and chronoamperometry measurements, the kinetic parameters, such as charge transfer coefficient (a) and the catalytically reaction rate constant (k) for oxidation of L-Cys were determined. It is expected that this work opens new horizons on the design of new nanocatalysts for the electrooxidation studies.

Disclosure statement

No potential conflict of interest was reported by the authors.

Funding

This work was supported by the Nano Technology Research Center, and Faculty of Chemistry, Urmia University, Drug Applied Research Center Research Center, Tabriz University of Medical Sciences [1233]; Iran National Science Foundation (INSF).

References

1. Y. Dong and J. B. Zheng: Probing the effect of temperature and electric field on the low frequency dielectric relaxation in a ferroelectric liquid crystal mesogen.J. Mol. Liq., 2014, **196**, 280–285.

2. P. T. Lee, J. E. Thomson, A. Karina, C. Salter, C. Johnston, S. G. Davies and R. G. Compton: Selective electrochemical determination of cysteine with a cyclotricatechylene modified carbon electrode. Analyst, 2015, **140**, 236–242.

3. L. B. Qu, S. L. Yang, G. Li, R. Yang, J. J. Li and L. Yu: Preparation of yttrium hexacyanoferrate/carbon nanotube/Nafion nanocomposite film-modified electrode: Application to the electrocatalytic oxidation of L-cysteine. Electrochim. Acta, 2011, **56**, 2934–2940.

4. S. Shahrokhian: Lead phthalocyanine as a selective carrier for preparation of a cysteine-selective electrode. *Anal. Chem.*, 2001, **73**, 5972–5978.

5. A. Azadbakht, A. R. Abbasi, Z. Derikvand and Z. Karimi: Synthesis of zinc bismuthate nanorods and electrochemical performance for sensitive determination of L-cysteine. *Nano-Micro Lett.*, 2015, **7**, 152–164.

6. J. C. Claussen, A. D. Franklin, A. ul Haque, D. M. Porterfield and T. S. Fisher: Electrochemical biosensor of nanocube-augmented carbon nanotube networks. *ACS Nano*, 2009, **3**, 37–44.

7. F. Patolsky, G. Zheng and C. M. Lieber: Nanowire sensors for medicine and the life sciences. *Nanomedicine*, 2006, **1**, 51–65.

8. F. Tan, C. Tan, A. Zhao and M. L. Li: Simultaneous determination of free amino acid content in tea infusions by using high-performance liquid chromatography with fluorescence detection coupled with alternating penalty trilinear decomposition algorithm. *J. Agric. Food Chem.*, 2011, **59**, 10839–10847.

9. T. N. Chiesl, W. K. Chu, A. M. Stockton, X. Amashukeli, F. Grunthaner and R. A. Mathies: Enhanced amine and amino acid analysis using pacific blue and the mars organic analyzer microchip capillary electrophoresis system. *Anal. Chem.*, 2009, **81**, 2537–2544.

10. C. J. Lee and J. Yang: α-Cyclodextrin-modified infrared chemical sensor for selective determination of tyrosine in biological fluids. *Anal. Biochem.*, 2006, **359**, 124–131.

11. L. G. Wang, Q. Zhou, B. C. Zhu, L. G. Yan, Z. M. Ma, B. Du and X. L. Zhang: A colorimetric and fluorescent chemodosimeter for discriminative and simultaneous quantification of cysteine and homocysteine. *Dyes Pigments*, 2012, **95**, 275–279.

12. H. L. Liu, Y. H. Wang, A. G. Shen, X. D. Zhou and J. M. Hu: Highly selective and sensitive method for cysteine detection based on fluorescence resonance energy transfer between FAM-tagged ssDNA and graphene oxide. *Talanta*, 2012, **93**, 330–335.

13. H. Wang, W. S. Wang and H. S. Zhang: Spectrofluorimetic determination of cysteine based on the fluorescence inhibition of Cd(II)-8-hydroxyquinoline-5-sulphonic acid complex by cysteine. *Talanta*, 2001, **53**, 1015–1019.

14. P. K. Sudeep, S. B. Joseph, K. G. Thomas and K. Thomas: Selective detection of cysteine and glutathione using gold nanorods. *J. Am. Chem. Soc.*, 2005, **127**, 6516–6517.

15. D. G. Davis and E. Bianco: An electrochemical study of the oxidation of L-cysteine. *J. Electroanal. Chem.*, 1996, **12**, 254–301.

16. Z. N. Liu, H. C. Zhang, S. F. Hou and H. Y. Ma: Highly sensitive and selective electrochemical detection of L-cysteine using nanoporous gold. *Microchim. Acta*, 2012, **177**, 427–433.

17. F. G. Xu, F. Wang, D. G. Yang, Y. Gao and H. M. Li: Electrochemical sensing platform for L-CySH based on nearly uniform Au nanoparticles decorated graphene nanosheets. *Mater. Sci. Eng. C*, 2014, **38**, 292–298.

18. R. Devasenathipathy, C. Karuppiah, S. M. Chen, V. Mani, V. S. Vasantha and S. Ramaraj: Highly selective determination of cysteine using a composite prepared from multiwalled carbon nanotubes and gold nanoparticles stabilized with calcium crosslinked pectin. *Microchim. Acta*, 2015, **182**, 727–735.

19. J. A. Reynaud, B. Malfoy and P. J. Canesson: Electrochemical investigations of amino acids at solid electrodes: Part I. Sulfur components: Cystine, cysteine, methionine. *J. Electroanal. Chem.*, 1980, **114**, 195–211.

20. L. Z. Pei, Y. Q. Pei, Y. K. Xie, C. G. Fan and H. Y. Yu: Synthesis and characterization of manganese vanadate nanorods as glassy carbon electrode modified materials for the determination of L-cysteine. *Cryst. Eng. Comm.*, 2013, **15**, 1729–1738.

21. M. E. Johll, D. G. Williams and D. C. Johnson: Activated pulsed amperometric detection of cysteine at platinum electrodes in acidic media. *Electroanalysis*, 1997, **9**, 1397–1402.

22. S. Ge, M. Yan, J. Lu, M. Zhang, F. Yu, J. Yu, X. Song and S. Yu: Electrochemical biosensor based on graphene oxide-Au nanoclusters composites for L-cysteine analysis. *Biosen. Bioelectron.*, 2012, **31**, 49–54.

23. H. Hosseini, H. Ahmar, A. Dehghani, A. Bagheri, A. Tadjarodi and A. R. Fakhari: A novel electrochemical sensor based on metal-organic framework for electro-catalytic oxidation of L-cysteine. *Biosen. Bioelectron.*, 2013, **42**, 426–429.

24. M. Zhou, J. Ding, L. P. Guo and Q. K. Shang: Electrochemical behavior of L-cysteine and its detection at ordered mesoporous carbon-modified glassy carbon electrode. *Anal. Chem.*, 2007, **79**, 5328–5335.

25. C. Y. Deng, J. H. Chen, X. L. Chen, M. D. Wang, Z. Nie and S. Z. Yao: Electrochemical detection of L-cysteine using a boron-doped

carbon nanotube-modified electrode. *Electrochim. Acta*, 2009, **54**, 3298–3302.

26. H. Razmi, H. Nasiri and R. Mohammad-Rezaei: Amperometric determination of L-tyrosine by an enzymeless sensor based on a carbon ceramic electrode modified with copper oxide nanoparticles. *Microchim. Acta*, 2011, **173**, 59–64.

27. Y. P. Dong, L. Z. Pei, X. F. Chu, W. B. Zhang, and Q. F. Zhang: Electrochemical behavior of cysteine at a CuGeO3 nanowires modified glassy carbon electrode. *Electrochim. Acta*, 2011, **55**, 5135–5141.

28. L. Z. Pei, S. Wang, H. D. Liu and Y. Q. Pei: A review on ternary vanadate one-dimensional nanomaterials. *Recent Pat. Nanotechnol.*, 2014, **8**, 142–155.

29. H. L. Chen and M. S. Li: 'Structure and function of biomacromolecules'; 1999, Shanghai, Shanghai Science Press.

30. J. Zen, A. Kumar and J.-C. Chen: Electrocatalytic oxidation and sensitive detection of cysteine on a lead ruthenate pyrochlore modified electrode. *Anal. Chem.*, 2001, **73**, 1169–1175.

31. N. Spătaru, B. V. Sarada, E. Popa, D. A. Tryk and A. Fujishima: Voltammetric determination of L-cysteine at conductive diamond electrodes. *Anal. Chem.*, 2001, **73**, 514–519.

32. K. Gong, X. Zhu, R. Zhao, S. Xiong, L. Mao and C. Chen: Rational attachment of synthetic triptycene orthoquinone onto carbon nanotubes for electrocatalysis and sensitive detection of thiols. *Anal. Chem.*, 2005, **77**, 8158–8165.

33. M. Teixeira, E. Dockal and E. Cavalheiro: Sensor for cysteine based on oxovanadium(IV) complex of Salen modified carbon paste electrode. *Sens. Actuators B*, 2005, **106**, 619–625.

34. T. R. Ralph, M. L. Hitchman, J. P. Millington and F. C. Walsh: The electrochemistry of L-cystine and L-cysteine: Part 1: Thermodynamic and kinetic studies. *J. Electroanal. Chem.*, 1994, **375**, 1–15.

35. T. R. Ralph, M. L. Hitchman, J. P. Millington and F. C. Walsh: The electrochemistry of L-cystine and L-cysteine part 2: Electrosynthesis of L-cysteine at solid electrodes. *J. Electroanal. Chem.*, 1994, **375**, 17–27.

36. C. Terashima, T. N. Rao, B. V. Sarada, Y. Kubota and A. Fujishima: Direct electrochemical oxidation of disulfides at anodically pretreated boron-doped diamond electrodes. *Anal. Chem.*, 2003, **75**, 1564–1572.

37. S. Fei, J. Chen, S. Yao, G. Deng, D. He and Y. Kuang: Electrochemical behavior of L-cysteine and its detection at carbon nanotube electrode modified with platinum. *Anal. Biochem.*, 2006, **339**, 29–35.

38. M. K. Halbert and R. P. Baldwin: Electrocatalytic and analytical response of cobalt phthalocyanine containing carbon paste electrodes toward sulfhydryl compounds. *Anal. Chem.*, 1985, **57**, 591–595.

39. Z. Wang and D. Pang: Electrocatalysis of metalloporphyrins: Part 9. Catalytic electroreduction of cystine using water-soluble cobalt porphyrins. *J. Electroanal. Chem.*, 1990, **283**, 349–358.

40. W. T. Tan, A. M. Bond, S. W. Ngooi, E. B. Lim and J. K. Goh: Electrochemical oxidation of L-cysteine mediated by a fullerene-C_{60}-modified carbon electrode. *Anal. Chim. Acta*, 2003, **491**, 181–191.

41. S. Fujiwara, C. Pessôa and Y. Gushikem: Hexacyanoferrate ion adsorbed on propylpyridiniumsilsesquioxane polymer film-coated SiO_2/Al_2O_3: use in an electrochemical oxidation study of cysteine. *Electrochim. Acta*, 2003, **48**, 3625–3631.

42. S. Maree and T. Nyokong: Electrocatalytic behavior of substituted cobalt phthalocyanines towards the oxidation of cysteine. *J. Electroanal. Chem.*, 2000, **492**, 120–127.

43. S. Zhang, W. Sun, W. Zhang and W. Qi: Determination of thiocompounds by liquid chromatography with amperometric detection at a Nafion/indium hexacyanoferrate film modified electrode. *Anal. Chim. Acta*, 1999, **386**, 21–30.

44. J. H. Zagal, M. J. Aguirre and C. G. Parodi: Electrocatalytic activity of vitamin B12 adsorbed on graphite electrode for the oxidation of cysteine and glutathione and the reduction of cysteine. *J. Electroanal. Chem.*, 1994, **374**, 215–222.

45. K. Sugawara, S. Hoshi, K. Akatsuka and K. Shimazu: Electrochemical behavior of cysteine at a Nafion° |cobalt(II) modified electrode. *J. Electroanal. Chem.*, 1996, **414**, 253–256.

46. J. A. Reynaud, B. Malfoy and P. Canesson: Electrochemical investigations of amino acids at solid electrodes: Part I. Sulfur components: cystine, cysteine, methionine. *J. Electroanal. Chem.*, 1900, **114**, 195–211.

47. Y. D. Zhao, W. D. Zhang, H. Chen and Q. M. Luo: Electrocatalytic oxidation of cysteine at carbon nanotube powder microelectrode and its detection. *Sens. Actuators B*, 2003, **92**, 279–285.

48. J. Xu, Y. Wang, Y. Xian, L. Jin and K. Tanaka: Preparation of multiwall carbon nanotubes film modified electrode and its application

to simultaneous determination of oxidizable amino acids in ion chromatography. *Talanta*, 2003, **60**, 1123–1130.

49. R. R. Moore, C. E. Banks and R. G. Compton: Basal plane pyrolytic graphite modified electrodes: Comparison of carbon nanotubes and graphite powder as electrocatalysts. *Anal. Chem.*, 2004, **76**, 2677–2682.

50. N. Jia, Z. Wang, G. Yang, H. Shen and L. Zhu: Electrochemical properties of ordered mesoporous carbon and its electroanalytical application for selective determination of dopamine. *Electrochem. Commun.*, 2007, **9**, 233–238.

51. P. Miao, K. Han, Y. Tang, B. Wang, T. Lin and W. Cheng: Recent advances in carbon nanodots: synthesis, properties and biomedical applications. *Nanoscale*, 2015, **7**, 1586–1595.

52. A. Zhao, Z. Chen, C. Zhao, N. Gao, J. Ren and X. Qu: Recent advances in bioapplications of C-dots. *Carbon*, 2015, **85**, 309–327.

53. A. P. Demchenko and M. O. Dekaliuk: Novel fluorescent carbonic nanomaterials for sensing and imaging. *Methods Appl. Fluoresc.*, 2013, **1**, 042001.

54. R. J. Fan, Q. Sun, L. Zhang, Y. Zhang and A. H. Lu: Photoluminescent carbon dots directly derived from polyethylene glycol and their application for cellular imaging. *Carbon*, 2014, **71**, 87–93.

55. B. Shi, L. Zhang, C. Lan, J. Zhao, Y. Su and S. Zhao: One-pot green synthesis of oxygen-rich nitrogen-doped graphene quantum dots and their potential application in pH-sensitive photoluminescence and detection of mercury(II) ions. *Talanta*, 2015, **142**, 131–139.

56. M. Favaro, L. Ferrighi, G. Fazio, L. Colazzo, C. Di Valentin, C. Durante, F. Sedona, A. Gennaro, S. Agnoli and G. Granozzi: Single and multiple doping in graphene quantum dots: Unraveling the origin of selectivity in the oxygen reduction reaction. *ACS Catal.*, 2015, **5**, 129–144.

57. J. Lou, S. Liu, W. Tu and Z. Dai: Graphene quantums dots combined with endonuclease cleavage and bidentate chelation for highly sensitive electrochemiluminescent DNA biosensing. *Anal. Chem.*, 2015, **87**, 1145–1151.

58. C. Yang, L. Hu, H. Y. Zhu, Y. Ling, J. H. Tao and C. X. Xu: rGO quantum dots/ZnO hybrid nanofibers fabricated using electrospun polymer templates and applications in drug screening involving an intracellular H_2O_2 sensor. *J. Mater. Chem. B*, 2015, **3**, 2651–2659.

59. P. Zhang, X. Zhao, Y. Ji, Z. Ouyang, X. Wen, J. Li, Z. Su and G. Wei: Electrospinning graphene quantum dots into a nanofibrous membrane for dual-purpose fluorescent and electrochemical biosensor. *J. Mater. Chem. B*, 2015, **3**, 2487–2496.

60. Y. Li, Y. Zhong, Y. Zhang, W. Weng and S. Li: Carbon quantum dots/octahedral Cu_2O nanocomposites for non-enzymatic glucose and hydrogen peroxide amperometric sensor. *Sens. Actuators B*, 2015, **206**, 735–743.

61. M. Hasanzadeh, N. Shadjou and E. Omidinia: Mesoporous silica (MCM-41)-Fe_2O_3 as a novel magnetic nanosensor for determination of trace amounts of amino acids. *Colloids Surf. B: Biointerfaces*, 2013, **108**, 52–59.

62. M. Hasanzadeh, N. Shadjou, S. T. Chen and P. Sheikhzadeh: MCM-41-NH_2 as an advanced nanocatalyst for electrooxidation and determination of amino acids. *Catal. Commun.*, 2012, **19**, 21–27.

63. L. Saghatforoush, M. Hasanzadeh, N. Shadjou and B. Khalilzadeh: Deposition of new thia-containing Schiff-base iron (III) complexes onto carbon nanotube-modified glassy carbon electrodes as a biosensor for electrooxidation and determination of amino acids. *Electrochim. Acta*, 2011, **56**, 1051–1061.

64. M. Hasanzadeh, G. Karim-Nezhad, N. Shadjou, M. Hajjizadeh, B. Khalilzadeh, L. Saghatforoush, M. H. Abnosi, A. Babaei and S. Ershad: Cobalt hydroxide nanoparticles modified glassy carbon electrode as a biosensor for electrooxidation and determination of some amino acids. *Anal. Biochem.*, 2009, **389**, 130–137.

65. E. Omidinia, N. Shadjou and M. Hasanzadeh: Aptamer-based biosensor for detection of phenylalanine at physiological pH. *Appl. Biochem. Biotechnol.*, 2014, **172**, 2070–2080.

66. E. Omidinia, N. Shadjou and M. Hasanzadeh: Immobilization of phenylalanine-dehydrogenase on nano-sized polytaurine: A new platform for application of nano-polymeric materials on enzymatic biosensing technology. *Mater. Sci. Eng. C*, 2014, **42**, 368–373.

67. E. Omidinia, N. Shadjou and M. Hasanzadeh: Electrochemical nanobiosensing of phenylalanine using phenylalanine dehydrogenase incorporated on amino-functionalized mobile crystalline material-41. *Sens. J. IEEE*, 2014, **14**, 1081–1088.

68. E. Omidinia, N. Shadjou and M. Hasanzadeh: (Fe_3O_4)-graphene oxide as a novel magnetic nanomaterial for non-enzymatic determination of phenylalanine. *Mater. Sci. Eng. C*, 2013, **33**, 4624–4632.

69. E. Omidinia, A. Khanehzar, N. Shadjou, H. Shahbaz Mohamadi, S. Hojati Emami and M. Hasanzadeh: Covalent immobilization of phenylalanine dehydrogenase on glutaraldehyde modified poly (3,4-ethylenedioxy) thiophene: poly(styrenesulfonate)/ polyvinyl alcohol conducting polymer composite films for electrochemical detection of L-phenylalanine. *Anal. Bioanal. Electrochem.*, 2013, **5**, 597–608.

70. A. Ferancová and J. Labuda: Cyclodextrins as electrode modifiers. *Fresenius J. Anal. Chem.*, 2001, **370**, 1–10.

71. A. Ferancová, J. Labuda, J. Barek and J. Zima: Cyclodextrins as supramolecular complex agents in electroanalytical chemistry. *Chem. Listy*, 2002, **96**, 856–862.

72. P. M. Bersier, J. Bersier and B. Klingert: Electrochemistry of cyclodextrins and cyclodextrin inclusion complexes. *Electroanalysis*, 1991, **3**, 443–455.

73. M. Hasanzadeh, M. H. Pournaghi-Azar, N. Shadjou and A. Jouyban: Determination of lisinopril using β-cyclodextrin/graphene oxide-SO_3H modified glassy carbon electrode. *J. Appl. Electrochem.*, 2014, **44**, 821–830.

74. M. Bacon, S. J. Bradley and T. S. Nann: Graphene Quantum Dots. *Part. Part. Syst. Charact.*, 2013, **31**, 415–428.

75. M. Amjadi, J. L. Manzoori and T. Hallaj: Chemiluminescence of graphene quantum dots and its application to the determination of uric acid. *J. Luminescence*, 2014, **153**, 73–78.

76. F. G. Xu, F. Wang, D. G. Yang, Y. Gao and H. M. Li: Electrochemical sensing platform for L-CySH based on nearly uniform Au nanoparticles decorated graphene nanosheets. *Mater. Sci. Eng. C*, 2014, **38**, 292–298.

77. S. Ge, M. Yan, J. Lu, M. Zhang, F. Yu, J. Yu, X. Song and S. Yu: Electrochemical biosensor based on graphene oxide-Au nanoclusters composites for L-cysteine analysis. *Biosen. Bioelectron*, 2012, **31**, 49–54.

78. L. Z. Pei, Z. Y. Cai, Y. Q. Pei, Y. K. Xie, C. G. Fan and D. G. Fu: Electrochemical determination of L-cysteine using polyaniline/$CuGeO_3$ nanowire modified electrode. *Russ. J. Electrochem.*, 2014, **50**, 458–467.

79. J. Xu, F. Shang, J. H. Luong, K. M. Razeeb and J. D. Glennon: Direct electrochemistry of horseradish peroxidase immobilized on a monolayer modified nanowire array electrode. *Biosens. Bioelectron.*, 2010, **25**, 1313–1318.

80. P. Kissinger, and W. R. Heineman: 'Laboratory techniques in electroanalytical chemistry', 2nd edn, 224; 1996, New York, CRC Press.

81. A. J. Bard, and L. R. Faulkner: 'Electrochemical methods: fundamentals and applications', 2nd edn, 236, 503; 2001, New York, Wiley.

82. C. M. Brett, and A. O. Brett: 'Electrochemistry principles, methods and applications', 427; 1993, New York, Oxford University Press.

83. F. Pariente, E. Lorenzo, F. Tobalina and H. Abruna: Aldehyde Biosensor Based on the Determination of NADH Enzymatically Generated by Aldehyde Dehydrogenase. *Anal. Chem.*, 1995, **67**, 3936–3944.

84. J. Harrison and Z. Khan: The oxidation of hydrazine on platinum in acid solution. *J. Electroanal. Chem. Interfacial Electrochem.*, 1970, **28**, 131–138.

85. K. Gong, X. Zhu, R. Zhao, S. Xiong, L. Mao and C. Chen: Rational attachment of synthetic triptycene orthoquinone onto carbon nanotubes for electrocatalysis and sensitive detection of thiols. *Anal. Chem.*, 2005, **77**, 8158–8165.

86. N. Jia, Z. Wang, G. Yang, H. Shen and L. Zhu: Electrochemical properties of ordered mesoporous carbon and its electroanalytical application for selective determination of dopamine. *Electrochem. Commun.*, 2007, **9**, 233–238.

10

Single step functionalization of celluloses with differing degrees of reactivity as a route for *in situ* production of all-cellulose nanocomposites

Koon-Yang Lee[1] ⓘ, and Alexander Bismarck[2,3] ⓘ

[1]The Composites Centre, Department of Aeronautics, Imperial College London, South Kensington Campus, London SW7 2AZ, UK
[2]Polymer and Composite Engineering (PaCE) Group, Faculty of Chemistry, Institute for Materials Chemistry and Research, University of Vienna, Währingerstraße 42, Vienna 1090, Austria
[3]Polymer and Composite Engineering (PaCE) Group, Department of Chemical Engineering, Imperial College London, South Kensington Campus, London SW7 2AZ, UK

Abstract A method of manufacturing all-cellulose nanocomposites using a single-step functionalization of two different celluloses with differing reactivities is presented. All-cellulose nanocomposites are produced by esterification of microcrystalline cellulose (MCC) in pyridine with hexanoic acid in the presence of bacterial cellulose (BC) followed by solvent removal. Neat MCC is more susceptible to esterification, with an accessible amount of hydroxyl groups of 1.79 compared to BC, with an accessible hydroxyl group content of 0.80. As a result, neat MCC undergoes severe bulk modification, turning into a toluene-soluble cellulose hexanoate (C_6-MCC) while BC undergoes surface-only modification. Solution casted C_6-MCC films have a tensile modulus and strength of 0.99 GPa and 23.1 MPa, respectively. The presence of 5 wt.% BC in C_6-MCC leads to an increase in tensile modulus and strength of the resulting nanocomposites to 1.42 GPa and 28.4 MPa, respectively.

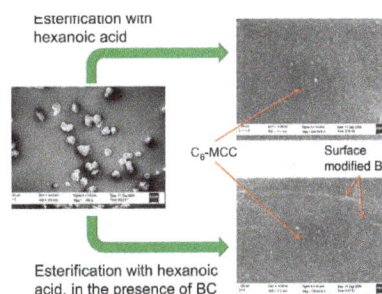

Keywords Bacterial cellulose, Esterification, Mechanical properties, All-cellulose composites, Solution casting, Nanocomposites

Introduction

The concept of 'self-reinforced polymers,' also known as 'all-polymer composites,' was first introduced and explored by Capiati and Porter.[1] They originally used the term 'one-polymer composites.' The authors showed that the interfacial shear strength between polyethylene and polyethylene fibers was much greater than between polyester and glass fibers. All-polyethylene composites were subsequently produced by consolidation of highly oriented high-density polyethylene (HDPE) films/fibers with HDPE or low-density polyethylene in a hot press[2] or co-extrusion of HDPE.[3] Since then, numerous manufacturing methods have been developed to produce all-polymer composites.[4–9] The essence of all-polymer composites lies in the compaction processes, whereby only a fraction of the reinforcing fiber surface is melted under low contact

pressure and subsequently consolidated at high pressure for a short period of time.[10] Once the material is cooled, the reinforcing fibers are bound together, in analogy to conventional fiber-reinforced polymer systems. All-polyethylene[4–7] and all-polypropylene[10] composites have been successfully manufactured and were shown to have significant commercial potential. More recently, all-polylactide composites have also been produced.[11–13] For a comprehensive review on the technology and recent development of all-polymer composites, the readers are referred to a recent publication by Alcock and Peijs.[14]

Cellulose is a linear polymer consisting of two D-anhydroglucose rings linked by β(1 → 4) glycosidic bonds. The concept of all-polymer composites was also applied to cellulose to produce renewable all-cellulose composites.[15] However, different manufacturing approaches had to be used because cellulose cannot be heat processed; currently, two methods are used to produce all-cellulose composites:[16] (i) impregnation of cellulose fibers with a solution containing dissolved

*Corresponding authors, emails koonyang.lee@imperial.ac.uk (K.-Y. Lee), alexander.bismarck@univie.ac.at (A. Bismarck)

cellulose, followed by subsequent regeneration of the dissolved cellulose or (ii) selective dissolution of the surface of cellulose fibers followed by the regeneration of cellulose to bond the (loose) fibers together. The first approach was demonstrated by Nishino et al.[17] They dissolved kraft pulp in 8 wt.% LiCl in dimethylacethylamide (DMAc) solution, which was subsequently infused into a ramie fiber preform. LiCl/DMAc was then removed by a solvent exchange step with methanol, followed by air-drying at room temperature. This essentially creates a composite in which both the reinforcing fibers and the matrix are cellulose. In a separate study by Nishino et al.,[18] filter papers were first activated by immersing them in to distilled water, followed by acetone and DMAc, respectively, and then in 8 wt.% LiCl/DMAc to selectively dissolve only the surface of the fibers. Methanol was once again used to extract the solvent and subsequently air-dried. This manufacturing concept closely resembles that of the manufacturing process of all-thermoplastic polymer composites. The selective dissolution of cellulose on the fiber surface produces the composite matrix after regeneration while the fiber core retains its original structure and acts as reinforcement for the regenerated cellulose matrix. This method has also been used to produce all-cellulose composites from microcrystalline cellulose (MCC)[19] and synthetic cellulose (Cordenka) fibers.[20]

In addition to micrometer-scale cellulosic fibers, nanocellulose fibrils can also be used as reinforcement to produce all-cellulose nanocomposites.[21, 22] In this context, bacterial cellulose (BC) serves as an excellent nanoreinforcement[23, 24] for regenerated cellulose. BC is essentially pure cellulose synthesized by bacteria typically from the *Acetobacter* species.[25] It is inherently nano-sized and highly crystalline in nature, with a fibril diameter of ~50 nm and degree of crystallinity of ~90%, respectively.[26] BC can be produced in static or shaken cultures, as well as in various types of bioreactors.[27] Raman spectroscopy was used to estimate tensile modulus of individual BC nanofibers to be approximately 114 GPa.[28] The idea of using BC (or any nanocellulose) in polymers stems from the possibility of exploiting the high stiffness and strength of cellulose crystals in composite applications.[16] To produce all-cellulose nanocomposites, Soykeabkaew et al.[22] impregnated BC sheets (or nanopapers) with 8 wt.% LiCl/DMAc solution to selectively dissolve the surface of BC nanofibers, followed by subsequent cellulose regeneration in methanol. The authors reported a tensile modulus and strength of up to 20 ± 1.7 GPa and 395 ± 19 MPa.

The manufacturing of all-cellulose (nano)composites requires the dissolution and subsequent regeneration of cellulose. Regeneration of dissolved cellulose is rather laborious as multiple solvent exchange steps are required. While the high vapor pressure of DMAc allows for its simple removal, the gel-like cellulose still needs to be washed with water to fully regenerate the dissolved cellulose.[21] Furthermore, the dissolution time of cellulose in LiCl/DMAc significantly affects the mechanical performance of all-cellulose composites.[17, 18, 22] To avoid these manufacturing issues, Matsumura et al.[29] took a different approach to exploit the properties of cellulose crystals in all-cellulose nanocomposites. This was achieved by the partial derivatization of dissolving wood pulp. The authors managed to partially esterify dissolving wood pulp with hexanonyl groups to produce cellulose crystallite-reinforced cellulose hexanoate, with tensile modulus and strength of up to 1.3 GPa and 25 MPa, respectively. The major advantage of partial derivatization of wood pulp to produce derivatized all-cellulose nanocomposites lies in the fact that these all-cellulose nanocomposites are in principal thermoformable.

In a previous study, we showed[30] that freeze-dried BC will undergo severe bulk esterification, leading to the production of cellulose esters but when never-dried BC was used as starting material and solvent exchanged into the same reaction medium, surface-only esterification occurred. Therefore in this study, we further expand upon the original work by Matsumura et al.[29] to produce derivatized all-cellulose nanocomposites in a single-step containing *surface-modified* BC-reinforced, cellulose crystallite-reinforced cellulose hexanoate. The concept is based on the esterification of MCC and of BC with hexanoic acid in the same reaction medium (into which BC was solvent exchanged).

Experimental

Materials

MCC (Celphere® SCP-100, degree of polymerization = ~100–300, bulk density = ~0.5–0.8 g cm^{-3}) was purchased from Asahi Kasei Chemicals Co, pyridine (analaR NORAMPUR, purity ≥ 99.7%) and ethanol (GPR, purity ≥ 99%) from VWR, hexanoic acid (purity ≥ 99.5%), toluene (analaR NORMAPUR, purity ≥ 99.8%), deuterium oxide (purity ≥ 99.99 atom% D), and p-toluenesulfonyl chloride (purity ≥ 99%) from Sigma-Aldrich and sodium hydroxide (purum grade, pellets) from Acros Organics. All materials were used as received without further purification. BC was extracted from commercially available *nata de coco* (CHAOKOH gel in syrup, Ampol Food Processing Ltd, Nakorn Pathom, Thailand). The extraction method of BC from *nata de coco* is thoroughly described in our earlier work elsewhere.[26, 31]

Esterification of MCC and BC with hexanoic acid

Two grams of MCC was added into a 1-L three-neck round bottom flask containing 400 mL of pyridine and stirred using a magnetic stirrer. To produce derivatized all-cellulose nanocomposites consisting of surface-modified BC-reinforced, cellulose crystallite-reinforced cellulose hexanoate, 100 mg of the extracted and purified neat BC (corresponding to 5 wt.% BC in derivatized all-cellulose nanocomposites) were solvent exchanged from water through methanol (3 × 50 dm^3) into pyridine (2 × 50 dm^3) using a homogenization–centrifugation step as described in our earlier work.[26, 31] The total volume of pyridine was adjusted to 400 mL in the final step prior to the addition of 1.9 g of MCC (to make up a total mass of 2 g cellulose). 92 g (0.48 mol) of p-toluenesulfonyl chloride was added into the reaction vessel followed by an equimolar amount of hexanoic acid. The esterification reaction was carried out at 50 °C for 2 h in nitrogen. Afterward, the reaction medium was subsequently quenched by the addition of 600 mL ethanol. The reaction mixture was washed with ethanol (3 × 400 dm^3) using a homogenization–centrifugation step to remove any unreacted reactants. The reaction product was further washed with water (3×400 dm³) to remove any ethanol residue, flash frozen in liquid nitrogen, and subsequently freeze-dried (Heto PowerDry LL1500 Freeze Dryer, Thermo Scientific, UK). The hexanoic acid-modified MCC and hexanoic acid-modified MCC containing 5 wt.% BC were termed C_6-MCC and 5 wt.% BC-C_6-MCC, respectively.

Manufacturing derivatized all-cellulose nanocomposites

Derivatized all-cellulose nanocomposites were produced by solvent casting, instead of thermal consolidation. Solvent casting was chosen as no first- or second-order transition temperatures (see results and discussion section) were observed. Freeze-dried C_6-MCC and 5 wt.% BC-C_6-MCC were first dissolved/dispersed in toluene at a concentration of 0.05 g mL⁻¹ overnight. The C_6-MCC solution or C_6-MCC solution containing surface-functionalized BC was then poured into a polytetrafluoroethylene mold ($20 \times 20 \times 0.35$ mm) and the toluene evaporated overnight at room temperature.

Characterization of MCC/BC and their derivatized all-cellulose nanocomposites

Attenuated total reflection infrared spectroscopy (ATR-IR) of neat MCC/BC and the derivatized all-cellulose nanocomposites

ATR-IR spectra were recorded using Spectrum One FTIR-spectrometer (Perkin Elmer, MA, USA). The spectra were collected in the range from 4000 to 600 cm⁻¹ at a resolution of 2 cm⁻¹. A total of 16 scans were used for each spectrum. To further quantify the degree of substitution (DS) of our derivatized all-cellulose nanocomposites, the asymmetric vibration of the C–O–C bond at 1158 cm⁻¹ was chosen as an internal standard. A calibration curve to calculate the DS was created using cellulose acetate and cellulose triacetate with known DS. A relationship is then drawn between DS and the ratio of the 1750 and 1158 cm⁻¹ absorption bands. The DS of our derivatized all-cellulose nanocomposites was then calculated based on their respective 1750–1158 cm⁻¹ absorption band ratios.

Differential scanning calorimetric (DSC) characterization of modified MCC and derivatized all-cellulose nanocomposites

Thermal behavior of modified MCC and the derivatized all-cellulose nanocomposites was characterized using DSC (Q2000, TA Instruments, UK). A sample mass of approximately 10 mg was used for each sample. A heat-cool-heat regime was employed; the sample was first heated from room temperature to 200 °C using a heating rate of 10 °C min⁻¹ followed by a cooling step at a rate of 10 °C min⁻¹ to room temperature. The sample was then reheated to 200 °C using a heating rate of 10 °C min⁻¹.

Accessibility of hydroxyl (–OH) groups of neat MCC and BC

In order to quantify the availability of –OH groups of neat MCC and BC for the esterification reaction, we adapted and further developed[26] the hydrogen/deuterium (H/D) exchange method of Frilette et al.[32] The measurement was conducted using dynamic vapor sorption (DVS-Advantage, Surface

Measurement Systems Ltd, Alperton, UK). A sample of 30 mg of neat MCC powder or freeze-dried BC was placed into the sample chamber and pre-conditioned at 0% relative humidity (RH) of deuterium oxide (D_2O) for 10 h at room temperature. The RH of D_2O was then increased to 90% for 48 h to allow for the adsorption of D_2O and hence H/D exchange with accessible –OH groups on neat MCC and BC. The RH was then reduced to 0% for 10 h to remove any adsorbed D_2O molecules. The mass gain due to H/D exchange was recorded *in situ*. As deuterium is one neutron heavier than hydrogen, the mass increase after D_2O desorption and hence the amount of accessible hydroxyl groups available was back calculated from:

$$\Delta m = \frac{[OH] \times m_i \times N_A \times m_n}{162140} \quad (1)$$

where Δm is the mass increase after H/D exchange (mg), [OH] is the concentration of accessible and available hydroxyl groups, m_i is the initial mass of the sample (mg), N_A is the Avogadro number, and m_n is the mass of a neutron (mg). It should be noted that we assumed that cellulose consists only of a single glucose unit with a molecular mass of 162,140 and neglected the slight difference in the molecular structure of the cellulose chain end group containing a reducing group on one end of the molecule and a non-reducing group on the other end.

X-ray diffraction (XRD) of neat MCC/BC and derivatized all-cellulose nanocomposites

XRD patterns were obtained using an X-ray diffractometer (PANalytical X'Pert 1, PANalytical Ltd, Cambridge, UK) equipped with Ni-filtered Cu $K_{\alpha1}$ (1.541 Å) X-ray source. Measurements were taken from $2\theta = 10° – 45°$ using a step size of 0.02°. The crystallinity of the samples was calculated based on the area under the XRD peaks using the following equation:

$$\chi_c[\%] = \frac{A_C}{A_C + A_A} \times 100 \quad (2)$$

where A_C and A_A are the total crystalline and total amorphous areas, respectively, between 10° and 45°. Scherrer's equation was used to determine the lateral dimensions of crystalline domains of the I_α (1 1 0) and I_β (2 0 0) composite reflections (L_4):[33]

$$L_4 = \frac{K\Delta\lambda}{\beta\Delta\cos\theta} \quad (3)$$

where β is the full width at half maximum of the 0 0 2 reflection (in radians), θ is the Bragg angle in degrees and $K = 0.91$.

Thermal stability of neat MCC/BC and derivatized all-cellulose nanocomposites

The thermal degradation behavior in nitrogen atmosphere was investigated using thermogravimetric analysis (Q500, TA Instruments, UK). A sample mass of approximately 5 mg was heated from 30 to 600 °C using a heating rate of 10 °C min⁻¹.

Tensile properties of derivatized all-cellulose nanocomposite films

The derivatized all-cellulose nanocomposite films were first cut into dog-bone shaped specimens using a Zwick cutter. These specimens had a thickness of 75 μm, an overall length of 35 mm and the narrowest part of the specimen was 2 mm.

Figure 1   Solvent resistivity of neat MCC (left), 5 wt.% BC-C$_6$-MCC (middle) and C$_6$-MCC (right). Photographs were taken 10 min after preparation

Prior to tensile testing, the test specimens were secured onto testing cards using a two-part cold curing epoxy resin (Araldite 2011, Huntsman, Advanced Materials, Cambridge, UK). Tensile tests were conducted using a micro-tensile tester (TST350, Linkam Scientific Instruments, Surrey, UK) equipped with a 200N load cell. The crosshead speed used was 1 mm min^{-1}. A total of five specimens were tested. The compliance of the micro-tensile tester was determined to be 6.38×10^{-3} mm N^{-1}.

Results and discussion

As aforementioned, Matsumura et al.[29] introduced hexanonyl groups onto wood pulp. The authors observed that wood pulp could be modified into cellulose hexanoate with DS of up to 2.5 and were able to thermoform the polymer at 150 – 170 °C. C$_6$-MCC and 5 wt.% BC-C$_6$-MCC synthesized in this study, on the other hand, did not show any first- and second-order transition temperatures in DSC, indicating that C$_6$-MCC and 5 wt.% BC-C$_6$-MCC are not thermoformable. This is attributed to the relatively low DS of the C$_6$-MCC and 5 wt.% BC-C$_6$-MCC (see ATR-IR section later). The cellulose is not modified sufficiently to reduce the amount/strength of hydrogen bonds between cellulose molecules to allow for long-range movements of cellulose molecules. Even though the DS of C$_6$-MCC and 5 wt.% BC-C$_6$-MCC are low, it is sufficient to reduce its solvent resistivity. Pure cellulose can only be dissolved in selected solvents, such as concentrated phosphoric acid, N-methylmorpholine N-oxide (NMMO) or LiCl/DMAc. The synthesized C$_6$-MCC dissolved, while 5 wt.% BC-C$_6$-MCC formed a homogenous suspension in toluene at room temperature (Fig. 1). As control, we also dispersed neat MCC in toluene (Fig. 1, left) but as expected, neat MCC did not dissolve in toluene.

ATR-IR spectra of (derivatized) MCC and BC

The ATR-IR spectra of C$_6$-MCC and 5 wt.% BC-C$_6$-MCC, along with neat MCC and BC are shown in Fig. 2. All spectra were normalized against the intensity of the C–O–C absorption band at around 1158 cm^{-1}.[34] The appearance of a new carbonyl (C=O) absorption band at 1750 cm^{-1} can be seen in C$_6$-MCC and

5 wt.% BC-C$_6$-MCC, respectively, which is a direct result of the introduction of hexanonyl groups into cellulose via esterification. A DS of 0.78 was obtained for C$_6$-MCC (see Table 1). When 5 wt.% BC is present in derivatized all-cellulose nanocomposites, the overall DS is 0.70. This implies that the amount of ester bonds relative to the amount of C–O–C bonds is reduced, confirming the difficulty in esterifying BC with hexanoic acid compared to neat MCC. This is consistent with our previous study,[30] in which we showed that never-dried BC underwent surface-only (instead of bulk) modification (see Fig. 3 for the FTIR spectra of hexanoic acid-modified freeze-dried or never-dried solvent exchanged BC). As a result, fewer ester bonds formed within the derivatized all-cellulose nanocomposites when BC is present.

Accessibility of hydroxyl groups

The accessible hydroxyl groups of neat BC and MCC are shown in Table 1. Neat MCC possessed a much higher fraction of accessible –OH groups (1.79 out of 3) compared to neat BC (0.80). As a result, MCC underwent severe bulk esterification, producing toluene-soluble cellulose hexanoate, while solvent exchanged BC underwent surface-only esterification also in the presence of MCC. MCC is typically manufactured by acid hydrolysis of pulp, which is then washed and spray-dried,[35] which might explain the higher reactivity of MCC. This observation is consistent with our previous study,[30] we observed that thorough dehydration processes, such as freeze- or spray-drying, resulted in an increase in the number of accessible –OH groups. Even though neat MCC possesses a high amount of accessible –OH groups, the DS of esterified MCC (C$_6$-MCC) was limited to 0.78 (see Table 1), instead of DS $\geq$ 1.79. This could be attributed to: (i) formation of a 'skin-core' structure as hypothesized by Asai et al.[36] and/or (ii) the relatively short reaction time of 2 h.

XRD of (derivatized) MCC and BC

The XRD patterns of neat MCC and BC, as well as C$_6$-MCC and 5 wt.% BC-C$_6$-MCC are shown in Fig. 4. The diffraction peaks

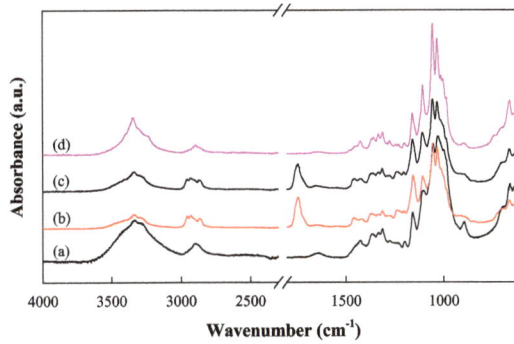

Figure 2 FTIR spectra of *a* neat MCC, *b* C_6-MCC, *c* 5 wt.% BC-C_6-MCC and *d* neat BC, respectively

Table 1 Degree of substitution (DS), accessible –OH groups ([OH]), crystallinity (χ_c), d-spacing of the I_α (1 1 0) and I_β (2 0 0) reflections (d_4), crystallite size of the I_α (1 1 0) and I_β (2 0 0) reflections (L_4) and onset thermal degradation temperature (T_d) of neat MCC, neat BC and the derivatized all-cellulose nanocomposites

Sample	DS[a]	[OH][a]	χ_c (%)	d_4 (Å)	L_4 (Å)	T_d (°C)
Neat MCC		1.79	90 ± 4	4.0 ± 0.5	6.2 ± 0.9	293
C_6-MCC	0.78		49 ± 2	4.0 ± 0.8	3.7 ± 0.3	240
5 wt.% BC-C_6-MCC	0.70		43 ± 12	4.0 ± 0.1	5.0 ± 0.3	251
Neat BC		0.80	85 ± 13	4.0 ± 0.1	6.3 ± 0.2	294

[a]Maximum of 3.

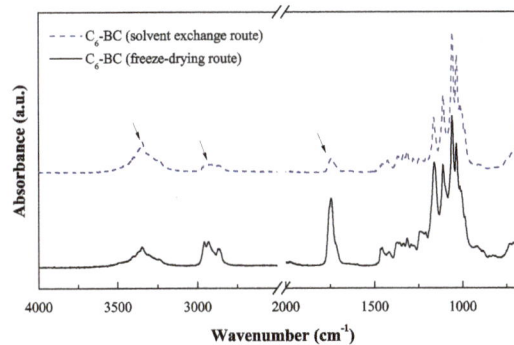

Figure 3 FTIR spectra of neat BC modified with hexanoic acid (C_6-BC). 'Solvent exchange route' refers to neat BC solvent exchanged from water through methanol (3 × 600 cm³), followed by pyridine (2 × 600 cm³) before one final solvent exchange step into pyridine to adjust the concentration to 0.5 (g mL⁻¹)% prior to esterification reaction. 'Freeze-drying route' refers to the direct dispersion of neat BC freeze-dried from water into pyridine directly prior to the esterification reaction

observed for neat BC are centered around 2θ = 14°, 16°, 22.5°, and 34°, while diffraction peaks observed for neat MCC are centered around 2θ = 14°, 16°, 20.5°, 22.5°, and 34°, respectively. Crystallographically, native cellulose is a composite consisting of two different crystalline structures, cellulose I_α and I_β, respectively, co-existing within the same sample.[37] Cellulose I_α has a one-chain triclinic unit cell[38] and cellulose I_β a monoclinic unit cell containing two parallel chains.[39] Based on the two crystalline phase model,[40] peaks 1–5 correspond to the diffraction of I_α (1 1 0) and I_β (1 $\bar{1}$ 0), I_α (0 1 0) and I_β (1 1 0), I_α ($\bar{1}$ $\bar{1}$ 2) and I_β (0 1 2)/(1 0 2), I_α (1 1 0) and I_β (2 0 0), and I_α ($\bar{1}$ $\bar{2}$ 3)/($\bar{1}$ $\bar{1}$ 4) and I_β (0 2 3)/(0 4 0) reflection planes, respectively.

During the esterification of MCC, which resulted in the incomplete formation of cellulose hexanoate (with and without the presence of BC), the crystalline structure of MCC was not retained. Broadening of the XRD peaks was observed for C_6-MCC and 5 wt.% BC-C_6-MCC. The crystallinity of all samples was calculated based on the area under the XRD peaks (equation (2)). The crystallinity, d-spacing, and size of the crystalline domains associated with reflection planes of peak 4 (I_α 1 1 0 and I_β 2 0 0) are tabulated in Table 1. It can be seen from this table that neat MCC and BC are highly crystalline, with a crystallinity of around 85–90%. The esterification of neat MCC (with and without the presence of BC) with hexanoic acid decreased both the crystallinity of our derivatized all-cellulose nanocomposites to about 50%. Combining this with the low DS of the derivatized all-cellulose nanocomposites, it is hypothesized that neat cellulose crystallites are still present within the derivatized all-cellulose nanocomposites.

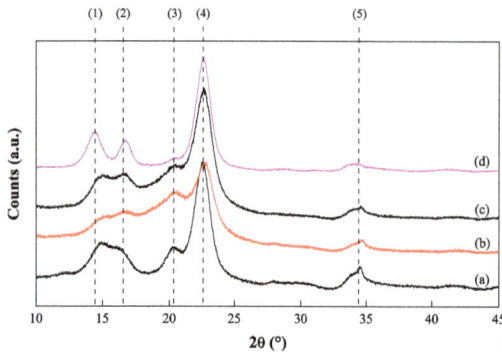

Figure 4   X-ray diffraction pattern of a neat MCC, b C_6-MCC, c 5 wt.% BC-C_6-MCC and d neat BC, respectively

The esterification of cellulose is hypothesized to proceed from the disordered (amorphous) to the ordered (crystalline) regions of cellulose.[41, 42] A different view was presented by Asai et al.,[36] who suggested that the esterification of cellulose should proceed from the surface of cellulose fibers to the core, producing somewhat of 'skin-core structure.' The lack of transition temperatures and the much reduced solvent resistivity would suggest that derivatized cellulose crystals could possess the aforementioned 'skin-core structure.' The d-spacings of the I_α (1 1 0) and I_β (2 0 0) reflection planes stayed constant but the crystallite size of the derivatized all-cellulose nanocomposites decreased compared to neat MCC and BC. This decrease in the crystallite size of the derivatized all-cellulose nanocomposites is in good agreement with the formation of a 'skin-core structure' of the modified cellulose crystallites.

Thermal degradation behavior of (derivatized) MCC and BC

Figure 5 shows the thermal degradation behavior in nitrogen of neat MCC, neat BC, C_6-MCC, and 5 wt.% BC-C_6-MCC. All samples underwent single-step degradation. A small mass loss can be observed at around 100 °C for neat MCC and BC. This is attributed to the loss of water from these samples. Hexanonyl esterified cellulose, on the other hand, is hydrophobic[26] and, therefore, the initial mass loss due to water loss was not observed for C_6-MCC and 5 wt.% BC-C_6-MCC. The onset degradation (defined as the temperature where a sharp change in mass loss occurs) of C_6-MCC and 5 wt.% BC-C_6-MCC shifted to lower temperatures (see Table 1). This is thought to be due to the reduced crystallinity of the modified cellulose. Furthermore, the esterification reaction also the reduced number of effective hydrogen bonds between the cellulose fibers/crystals, which further reduces the thermal stability when compared to neat MCC and BC.

Tensile properties of derivatized all-cellulose nanocomposites

The tensile properties of derivatized all-cellulose nanocomposites are shown in Table 2. For comparison, we have also included the tensile properties of commercially available cellulose esters in Table 2 and the tensile properties of various

all-cellulose (nano)composites in Table 3. C_6-MCC possesses a tensile modulus and strength of 0.99 GPa and 23.1 MPa, respectively. These values are comparable to the esterified wood pulp-based thermoformable all-cellulose nanocomposites manufactured by Matsumura et al.[29] with a similar DS. When C_6-MCC is reinforced with 5 wt.% esterified BC, the tensile modulus and strength increased to 1.42 GPa and 28.4 MPa, respectively. One of the most commonly used micromechanical models for the prediction of tensile moduli of randomly oriented short fiber composites is the Cox-Krenchel model,[43, 44] which was developed based on classical shear-lag theory. It has recently been adopted to analyze the reinforcing efficiency of nanocellulose in various polymer matrices.[24] The micromechanical model is written as:

$$E_{\text{composite}} = \eta_0 \eta_L v_f E_f + \left(1 - v_f\right) E_m \tag{4}$$

where $E_{\text{composite}}$, η_0, η_L, v_f, E_f and E_m represent the predicted tensile modulus of the composite, fiber orientation factor, fiber volume fraction, tensile modulus of the fiber and matrix, respectively. The limited stress transfer efficiency caused by the fact that the reinforcing fibers have a finite length η_L can be obtained from 'shear-lag' model:

$$\eta_L = 1 - \frac{\tanh\left(\frac{\beta L}{2}\right)}{\frac{\beta L}{2}} \tag{5}$$

$$\beta = \frac{2}{d}\left[\frac{2 \times G_m}{E_f \ln\left(\sqrt{\frac{\pi}{X_i \times v_f}}\right)}\right]^{0.5} \tag{6}$$

$$G_m = \frac{E_m}{2 \times (1 + v)} \tag{7}$$

where L, d, G_m, X_i and v denote the fiber length, fiber diameter, shear modulus of the matrix, packing of fibers in the composites and Poisson ratio of the matrix, respectively. In the original Cox publication,[43] $X_i = \sqrt{3}/2$ was used, assuming hexagonal packing of fibers with a mean centre-to-centre fiber spacing of R. Previously we showed that the esterification of BC with hexanoic acid does not lead to changes in the morphology of BC.[31] Therefore, our input parameters for equations (4)–(7) are $E_m = 0.99$ GPa, $d = 50$ nm, for the BC fibril length L we assume 1 μm and $v = 0.34$. A predicted composite tensile modulus of 1.46 GPa was obtained, which is in good agreement with our experimental findings.

While both the tensile modulus and strength of the resulting BC-reinforced derivatized all-cellulose nanocomposites improved over C_6-MCC, the mechanical performance is still lower than that of the highest performing renewable polymer, polylactide ($E = 4$ GPa, $\sigma = 50$ MPa). It should be noted that the tensile properties of our composites are still comparable to commercially available cellulose esters (see Table 2). Nevertheless, the marginal improvements of E and σ upon

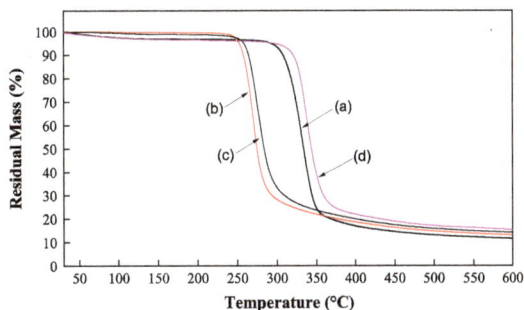

Figure 5 Thermal degradation behavior of *a* neat MCC, *b* C$_6$-MCC, *c* 5 wt.% BC-C$_6$-MCC and *d* neat BC, respectively, in nitrogen atmosphere

Table 2 Tensile properties of derivatized all-cellulose nanocomposites

Sample	*E* (GPa)	σ (MPa)	ε (%)
C$_6$-MCC	0.99 ± 0.06	23.1 ± 2.5	4.3 ± 0.4
5 wt.% BC-C$_6$-MCC	1.42 ± 0.04	28.4 ± 2.7	4.9 ± 0.4
Cellulose acetate[a]	1.90	36.1	
Cellulose acetate propionate[a]	1.40	33.6	
Cellulose acetate butyrate[a]	1.37	36.3	

Note: *E*, σ and ε denote tensile modulus, tensile strength and engineering strain-to-failure, respectively.
[a]Average data obtained from Matweb (http://www.matweb.com) based on molded polymers.
[b]Values estimated from figures published.

Table 3 Tensile properties of all-cellulose nanocomposites manufactured by various authors. E and σ denote tensile modulus and strength, respectively.

Source of cellulose	*E* (GPa)	σ (MPa)	Solvent used	Ref
Ramie/Kraft pulp[a]	20[§]	480 ± 50	LiCl/DMAc	[17]
Filter paper[b]	2[§]	40[§]	LiCl/DMAc	[18]
	2[§]	40[§]		
	6.5[§]	175[§]		
	8[§]	180[§]		
	9[§]	180[§]		
	8[§]	210[§]		
Linen flax	0.8	45	Ionic liquid	[20]
Cordenka fibres	2.5	70	Ionic liquid	
MCC[c]	2.57 ± 0.37	35.8 ± 4.4	LiCl/DMAc	[19]
	3.00 ± 0.37	49.7 ± 3.4		
	1.82 ± 0.32	24.3 ± 7.5		
	1.38 ± 0.74	15.5 ± 3.0		
	1.65 ± 0.39	29.7 ± 6.2		
	2.67 ± 0.30	62.7 ± 5.4		
	2.17 ± 0.24	25.2 ± 4.0		
	2.62 ± 0.49	10.1 ± 3.5		
	2.02 ± 0.70	28.3 ± 9.4		
	1.20 ± 0.17	20.9 ± 5.3		
	3.18 ± 0.12	58.7 ± 3.9		
	2.27 ± 0.43	22.5 ± 1.4		
	1.78 ± 0.39	29.0 ± 4.9		
	2.56 ± 0.43	48.8 ± 8.0		
	2.97 ± 0.59	58.5 ± 7.9		
	3.57 ± 1.07	21.7 ± 15.2		
	7.57 ± 1.56	102.3 ± 21.6		
	5.46 ± 0.82	105.8 ± 9.9		
	5.52 ± 0.80	85.2 ± 9.0		
	6.94 ± 0.74	105.7 ± 29.8		
MCC[d]	0.8[§]	35[§]	LiCl/DMAc	[21]
	0.5[§]	40[§]		
	1.2[§]	55[§]		
	1.6[§]	58[§]		
	1.5[§]	60[§]		
BC[e]	16.0 ± 1.0	392 ± 23	LiCl/DMAc	[22]
	18.0 ± 0.8	411 ± 22		
	16.0 ± 0.6	309 ± 25		
	11.0 ± 0.6	188 ± 19		
	4.4 ± 0.3	97 ± 15		
	2.6 ± 0.4	75 ± 15		
MCC[f]	12.1	218.6	LiCl/DMAc	[45]
	13.6	242.8		
	14.9	215.1		
Beech pulp	12	154	LiCl/DMAc	[45]

the addition of 5 wt.% modified BC to C_6-MCC compared to C_6-MCC are postulated to be due to the low loading fraction of BC within the composites. To produce composites with performance that exceed polylactide, a BC loading of >30 vol.-% should be used.[24] Attempts were made to produce all-cellulose nanocomposites with higher BC loading. However, during the evaporation of toluene, the all-cellulose nanocomposites shrunk and cracked severely, indicating the limitation of this technique to produce high-loading fraction BC-reinforced, cellulose crystallite-reinforced all-cellulose nanocomposites. This is due to the formation of irreversible hydrogen bonds between the remaining accessible –OH groups of (modified) BC within the C_6-MCC matrix. If this (modified) BC network is not activated (i.e. not prevented from shrinking, typically by hot pressing), the BC network will shrink, causing the severe shrinkage of the high BC loading-reinforced derivatized all-cellulose nanocomposites.

Conclusions

A novel method of producing derivatized all-cellulose nanocomposites is presented in this work. By modifying MCC in the presence of never-dried BC by esterification with hexanoic acid in the same reaction medium, BC-reinforced derivatized all-cellulose nanocomposites can be produced in a single step. While such (BC-reinforced) derivatized all-cellulose nanocomposites do not possess glass transition or melting temperatures, these nanocomposites can be readily dissolved/dispersed in toluene and solution casted, thereby avoiding the laborious solvent exchange or dissolution steps to manufacture conventional all-cellulose (nano)composites. ATR-IR spectra showed that neat MCC can be severely esterified with hexanoic acid, even in the presence of 5 wt.% BC. The DS for C_6-MCC and 5 wt.% BC-C_6-MCC was found to be 0.78 and 0.70, respectively. Such low DS explains the lack of glass transition or melting temperatures of the derivatized all-cellulose nanocomposites. XRD data further suggested that most of the accessible hydroxyl groups cannot be accessed as the reaction progresses due to the formation of a 'skin-core' structure. The thermal degradation temperatures of C_6-MCC and 5 wt.% BC-C_6-MCC were found to be lower than that of neat BC and neat MCC. This is attributed to the lower crystallinity of 5 wt.% BC-C_6-MCC and C_6-MCC (~50%) compared to neat MCC and neat BC of around 90%.

Tensile tests showed that the presence of 5 wt.% BC in C_6-MCC led to 43% increase in tensile modulus and 22% increase in tensile strength over C_6-MCC. This showed the reinforcing potential of surface-modified BC in derivatized all-cellulose nanocomposites.

Acknowledgements

The authors would like to thank the UK Engineering and Physical Science Research Council (EPSRC) for funding this work [grant number EP/F028946/1].

References

1. N. J. Capiati and R. S. Porter: *J. Mater. Sci.*, 1975, **10**, (10), 1671–1677.
2. W. T. Mead and R. S. Porter: *J. Appl. Polym. Sci.*, 1978, **22**, (11), 3249–3265.
3. A. E. Zachariades, R. Ball and R. S. Porter: *J. Mater. Sci.*, 1978, **13**, (12), 2671–2675.
4. P. J. Hine, I. M. Ward, R. H. Olley and D. C. Bassett: *J. Mater. Sci.*, 1993, **28**, (2), 316–324.
5. M. A. Kabeel, D. C. Bassett, R. H. Olley, P. J. Hine, I. M. Ward: *J. Mater. Sci.*, 1994, **29**, (18), 4694–4699.
6. M. A. Kabeel, D. C. Bassett, R. H. Olley, P. J. Hine and I. M. Ward: *J. Mater. Sci.*, 1995, **30**, (3), 601–606.
7. R. H. Olley, D. C. Bassett, P. J. Hine and I. M. Ward: *J. Mater. Sci.*, 1993, **28**, (4), 1107–1112.
8. C. Marais and P. Feillard: *Comp. Sci. Tech.*, 1992, **45**, (3), 247–255.
9. A. Teishev, S. Incardona, C. Migliaresi and G. Marom: *J. Appl. Polym. Sci.*, 1993, **50**, (3), 503–512.
10. I. M. Ward and P. J. Hine: *Polym. Eng. Sci.*, 1997, **37**, (11), 1809–1814.
11. J. J. Blaker, K.-Y. Lee, M. Walters, M. Drouet and A. Bismarck: *React. Funct. Polym.*, 2014, **85**, 185–192.
12. K. M. Zakir Hossain, R. M. Felfel, C. D. Rudd, W. Thielemans and I. Ahmed: *React. Funct. Polym.*, 2014, 85, 193–200.
13. F. Mai, W. Tu, E. Bilotti and T. Peijs. *Composites Part A: Appl. Sci. Manuf.*, 2015, **76**, 145–153.
14. B. Alcock and T. Peijs: 'Technology and Development of Self-Reinforced Polymer Composites', in 'Polymer Composites - Polyolefin Fractionation - Polymeric Peptidomimetics - Collagens', (eds. A. Abe, et al.), 1-76; 2013, Springer Berlin Heidelberg.
15. T. Huber, J. Müssig, O. Curnow, S. Pang, S. Bickerton, M. P. Staiger: *J. Mater. Sci.*, 2012, **47**, (3), 1171–1186.
16. S. J. Eichhorn, A. Dufresne, M. Aranguren, N. E. Marcovich, J. R. Capadona, S. J. Rowan, C. Weder, W. Thielemans, M. Roman, S. Renneckar, W. Gindl, S. Veigel, J. Keckes, H. Yano, K. Abe, M. Nogi, A. N. Nakagaito, A. Mangalam, J. Simonsen, A. S. Benight, A. Bismarck, L. A. Berglund and T. Peijs: *J. Mater. Sci.*, 2010, **45**, (1), 1–33.
17. T. Nishino, I. Matsuda and K. Hirao: *Macromolecules*, 2004, **37**, (20), 7683–7687.
18. T. Nishino and N. Arimoto: *Biomacromolecules.* 2007, **8**, (9), 2712–2716.
19. B. J. C. Duchemin, R. H. Newman and M. P. Staiger: *Comp. Sci. Tech.*, 2009, **69**, (7–8), 1225–1230.
20. T. Huber, S. Pang, and M. P. Staiger: *Compos. Part A: Appl. Sci. Manuf.*, 2012, **43**, (10), 1738–1745.
21. A. Abbott and A. Bismarck: *Cellulose*, 2010, **17**, (4), 779–791.
22. N. Soykeabkaew, C. Sian, S. Gea, T. Nishino and T. Peijs: Cellulose, 2009, **16**, (3), 435–444.
23. K.-Y. Lee, T. Tammelin, K. Schulfter, H. Kiiskinen, J. Samela and A. Bismarck: *ACS Appl. Mater. Interfaces*, 2012, **4**, (8), 4078–4086.
24. K.-Y. Lee, Y. Aitomäki, L. A. Berglund, K. Oksman and A. Bismarck: *Comp. Sci. Tech.*, 2014, **105**, 15–27.
25. A. J. Brown: *J. Chem. Soc. Trans.*, 1886, **49**, 172–187.
26. K.-Y. Lee, F. Quero, J. J. Blaker, C. A. S Hill, S. J. Eichhorn and A. Bismarck: *Cellulose*, 2011, **18**, (3), 595–605.
27. K.-Y. Lee, G. Buldum, A. Mantalaris and A. Bismarck: *Macromol. Biosci.*, 2014, **14**, (1), 10–32.
28. Y. C. Hsieh, H. Yano, M. Nogi and S. J. Eichhorn: *Cellulose*, 2008, **15**, (4), 507–513.
29. H. Matsumura, J. Sugiyama and W. G. Glasser: *J. Appl. Polym. Sci.*, 2000, **78**, (13), 2242–2253.
30. K. Y. Lee and A. Bismarck: *Cellulose*, 2012, **19**, (3), 891–900.
31. K.-Y. Lee, J. J. Blaker and A. Bismarck: *Comp. Sci. Tech.*, 2009, **69**, (15–16), 2724–2733.
32. V. J. Frilette, J. Hanle and H. Mark: *J. Am. Chem. Soc.*, 1948, **70**, (3), 1107–1113.
33. A. L. Patterson: *Phys. Rev.*, 1939, **56**, (10), 978.
34. L. M. Ilharco, R. R. Gracia, J. L. daSilva and L. F. V. Ferreira: *Langmuir*, 1997, **13**, (15), 4126–4132.
35. K. Obae, E. Kamada, Y. Honda, S. I. Gomi and N. Yamazaki: 'Cellulose powder', Patent, WO2004106416, 2002.
36. T. Asai, S. Shimamoto, H. Matsumura and T. Shibata: 'Tobacco filter material and a method for producing the same', US Patent 5856006, 1999.
37. R. H. Atalla and DL Vanderhart: *Science*, 1984, **223**, (4633), 283–285.
38. Y. Nishiyama, J. Sugiyama, H. Chanzy and P. Langan: *J. Am. Chem. Soc.*, 2003, **125**, (47), 14300–14306.

39. Y. Nishiyama, P. Langan and H. Chanzy: *J. Am. Chem. Soc.*, 2002, **124**, (31), 9074–9082.

40. M. Wada, J. Sugiyama and T. Okano: *J. Appl. Polym. Sci.*, 1993, **49**, (8), 1491–1496.

41. K. Hess and C. Trogus: *Zeitschrift für Physikalische Chemie-B-Chemie der Elementarprozesse Aufbau der Materie.*, 1931, **15**, (2/3), 157–222.

42. W. A. Sisson: *Ind. Eng. Chem.*, 1938, **30**, 530–537.

43. H. L. Cox: *Br. J. Appl. Phys.*, 1952, **3**, (3), 72–79.

44. H. Krenchel: 'Fibre-reinforcement – theoretical and practical investigation of the elasticity and strength of fibre-reinforced materials', PhD thesis, Technical University of Denmarck, Copenhagen, 1964.

45. Gindl W, Keckes J. Polymer., 2005, **46**, (23), 10221–10225.

46. Gindl W, Schöberl T, Keckes J. Applied Physics A: Materials Science and Processing., 2006, **83**, (1), 19–22.

11

Property enhancement of graphene fiber by adding small loading of cellulose nanofiber

Hong Jiang[1], Weixing Yang[1], Songgang Chai[2], Shuiqin Pu[1], Feng Chen*[1] and Qiang Fu*[1]

[1]College of Polymer Science and Engineering, State Key Laboratory of Polymer Materials Engineering, Sichuan University, Chengdu, China
[2]National Engineering Research Center of Electronic Circuits Base Materials, Guangdong Shengyi Technology Limited Corporation, Dongguan, China

Abstract Herein, a small amount of cellulose nanofiber (CNF) was introduced in graphene oxide (GO) spinning dope to improve the property of graphene fiber. It was found that the one-dimensional CNF could absorb on the GO sheets, and enhance the inter-layer interaction as well as the connection between GO sheets, resulting in an easy formation of lyotropic liquid crystalline structure and a better orientation of the resultant fiber. A large enhancement of tensile strength, modulus, and elongation of GO/CNF hybrid fiber has been achieved. After chemical reduction of GO, the fiber still maintains its original mechanical properties but with high conductivity and improved hydrophilicity. Our work further demonstrates the importance of a stable lyotropic liquid crystalline structure of GO dope in the spinning of graphene fiber and provides a facile way to produce graphene fiber with good performance.

Keywords Graphene fiber, Cellulose nanofiber, Composite fiber, Conductivity

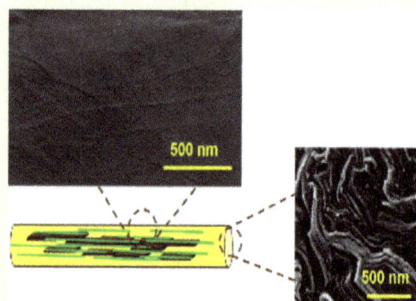

Introduction

Graphene, one of the most attractive materials of twenty-first century, with its unique structure, intrinsic mechanical strength, compelling heat, and electrical conductivity has received persistent attention.[1] Numerous researches have been carried out to translate the intrinsic attributes of individual graphene sheets into macroscopic assemblies for practical applications. Graphene paper, 3D graphene network, and graphene fiber were prepared in succession via various methods and techniques.[2-6] Different from other carbon-based fiber such as traditional carbon fiber and carbon nanotube (CNT) fiber that all needs high temperature in the fiber processing either for post graphitization or in the CNT synthesis, graphene fiber emerges as an extremely attractive and promising carbon-based fiber of the next generation, which can be prepared via the facile and benign wet-spinning at room temperature and was originated from well-sourced graphite powder. This kind of carbon-based fiber could be applied as a supercapacitor, intelligent fiber, and a smart-fiber, via *in situ* or post functionalization. Usually, graphene fiber is obtained via fluidly wet-spinning from the concentrated graphene oxide (GO) aqueous suspensions followed by chemical reduction.[5] Since GO sheets could be well dispersed in many solvents and easy to be functionalized, abundant enhancement method and modification could be proposed to increase the property of graphene fiber.[7,8] Various attempts have been promoted mainly to further enhance the mechanical strength or functionalize the fiber.[9-13] In our previous work we found that the defects, packing density in the fibers, orientation of graphene sheets, and interaction between graphene sheets could significantly affect the properties of the fiber.[14] In a work reported by Gao, the stronger graphene composite fiber has been prepared by grafting PAN on GO sheets to enhance inter-layer adhesion[15], while larger graphene sheets were utilized as building block to construct stronger π-π interaction of adjacent sheets in Tour's group and Gao's group;[16,17] Also diamine cross-linker was introduced in Park's group as coagulation bath to form ion bridges between the adjacent GO sheets.[18] Other polymer such as hyperbranched polyglycerol (HPG), polyvinyl alcohol (PVA)[19] were also introduced in graphene fiber, mainly to enhance the adhesion of the GO sheets. In

*Corresponding authors, emails fengchen@scu.edu.cn (Feng Chen); qiangfu@scu.edu.cn (Qiang Fu)

general, stronger sheet adhesion, better-alignment, less defect benefits a stronger fiber.

Since graphene fiber is prepared via fluidly wet-spinning from the concentrated GO aqueous suspensions and treated with reduction, the original state of GO dope will play an important role in determining the final property of graphene fiber. GO dope with stable lyotropic liquid crystalline structure was of key importance, which usually requires a high concentration or large sized GO sheets.[18,20] Thus, it is highly desired to find an easy way for the formation of stable liquid crystalline structure in GO dope with less concentration or smaller GO sheets. Cellulose nanofiber (CNF) is one-dimensional nano-sized and usually utilized as nanobuilding block for nano-composites. Moreover, it is highly water dispersible and super hydrophilic owing to the abundant hydroxyl groups. Its aqueous suspension shows liquid crystalline behavior and mechanical strong fiber could be wet-spun from the suspension.[21-23] In this work, a small amount of CNF was introduced into the GO dope with a hope to obtain stable liquid crystalline structure, thus graphene fiber with better alignment and mechanical property could be obtained. The interaction between GO sheets and CNF was investigated by Fourier-transform infrared (FTIR). The liquid crystalline behavior of the spinning dope was characterized via Polarized Optical Microscopy, SEM, and rheological property. With the enhancement of liquid crystalline behavior of GO dope via adding CNF, a large enhancement of tensile strength, modulus, and elongation of GO-CNF hybrid fiber has been achieved. After chemical reduction, the reduced GO–CNF (r-GO-CNF) fiber still maintains the original mechanical properties of GO–CNF fibers, but with high conductivity and improved hydrophilicity.

Experimental section

Materials

Graphite powder (200 mesh) was purchased from Qingdao Black Dragon graphite Co., Ltd. Concentrated H_2SO_4 (98%), $KMnO_4$, H_2O_2 (30%), HCl, CH_3COOH, NaBr, NaClO (8% available chorine), and NaOH were purchased from Kermel Chemical reagent plant (Chengdu, China), Chitosan (deacetylation degree of 95%) was purchased from Zhejiang Golden Shell Biological Technology Co., Ltd. Microfibrillated cellulose (Celish MFC KY100-S) was purchased from Daicel Chemical Industries, Ltd., Japan. 2,2,6,6-tetramethylpiperidine-1-oxyl radical (TEMPO) was purchased from Sigma–Aldrich. All reagents were used as received.

Preparation of GO and CNF

GO was prepared from natural graphite by a modified hummer's method. The as-prepared GO was then centrifuged and washed with 1 M hydrochloric acid and deionized water until the supernatant turned to neutral. The clean, washed GO was then concentrated by repeated centrifugation. Nanofibriled cellulose (CNF) was prepared according to TEMPO-mediated oxidation method. The 2wt% of microfibriled cellulose was suspended in water containing TEMPO and NaBr. The pH was maintained at 10.5 by adding 1 M NaOH. The reaction was finished until no more decrease in pH was observed after 3–4 h. The resulting mixtures were then thoroughly washed with deionized water by filtration until the filtrated solution turned neutral and re-dispersed in water at a concentration of around 10 mg/ml with vigorous stirring.

Wet-spinning of GO fibers

A series of spinning solution was prepared by adding quantitative CNF aqueous suspension to concentrated GO gel of 46 mg/ml. The CNF suspension was diluted to a calculated concentration and then ultra-sonic treated for 30 mins using a water bath ultrasonic cleaner before use. The overall concentration of GO was kept at 3 wt%. Different loading of 2/100, 5/100, and 10/100 CNF with respect to GO were prepared and was denoted as GO-CNF-2, GO-CNF-5, and GO-CNF-10, respectively. GO dope without CNF was also prepared as reference sample. These spinning solutions were then vigorously stirred for 20 h to ensure a homogeneous distribution. GO spinning dope with or without CNF was fluidly wet-spun by a capillary with an inner diameter of 450 μm, The spinning dope was injected into the chitosan coagulation (1 wt% chitosan, 0.8 wt% acetic acid) under N_2 pressure. After soaking in the coagulation bath for 1 min, the fiber was then collected on the Teflon drum, and washed with deionized water three times. The fiber was then naturally dried for 24 h, further vacuum dried for 8 h at 60°, finally the GO-CNF fibers were obtained. Reduced GO (r-GO) fiber was obtained by immersing in hydroiodic acid for 15 mins at 80℃, then washed with ethanol and deionized water.

Characterization

Polarized optical microscopy (POM) was carried out on Leica DM EP in transmission mode. Scanning electron microscopy (SEM) was performed on Inspect F (FET) at 5 kV accelerate voltage. Rheology tests were conducted on HAAKE Mars III rheometer equipped with a cone-shaped spindle (1° tapper, 30 mm diameter) and gathered on steady shear mode varied from 0.01 to 10 s^{-1}, logarithmically increased and 24 points in all, and each points was kept for 1 min. Zeta potentials of the GO–CNF suspensions were carried out on a laser electrophoresis zeta-potential analyzer (Zetasizer 3000HSA Malvern Instrument). X-ray diffraction (XRD) tests were carried out in X'Pert Pro XRD instrument to access the graphite intercalating space. Raman spectra were taken at Renishaw invia Raman Microscopy (RM 100) at a Ni-Ne laser of 532 nm. FTIR spectra was collected on Nicolet 6700 (Thermo Electron Corporation) on reflection mode. The Mechanical strength and stress–strain curve were collected from the monofilament tensile testing machine YG-001A with gauge length of 10 mm and a strain rate of 1 mm/min, and the cross-section area was calculated from the SEM images, the final results represented the average value for at least ten samples. Electrical behavior was examined with the standard four-probe method on FIUKE 45 dual display multimeter. Water contact angle tests were performed on OCA20 contact angle goniometer equipped with video capture (Dataphysics, Germany).

Results and discussion

SEM images of GO sheets and CNF are shown in Fig. 1a and b, respectively. The corresponding size distribution images show the diameter of GO sheets ranged from 0.84 to 7.1 μm and the

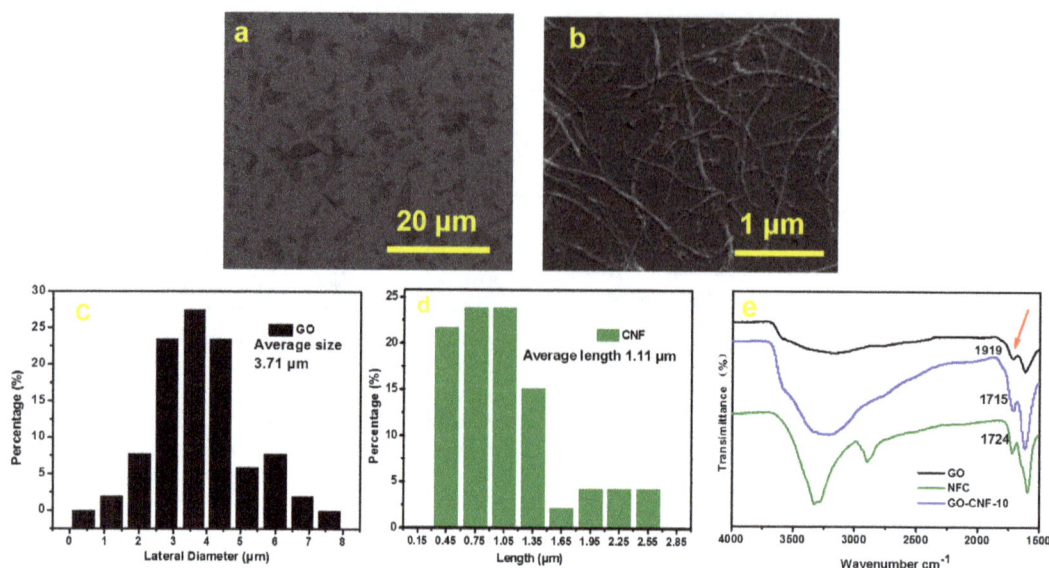

Figure 1 SEM images of *a* as-prepared GO sheets, *b* CNF, *c, d* the corresponding size distribution of GO and CNF, respectively, and *e* FTIR spectra of GO, CNF, and GO-CNF-10 film prepared by vacuum filtration

average diameter is 3.7 μm, which are small-sized GO sheets. And the thickness of GO was 1.0 nm, the corresponding aspect ratio was calculated to be 3700. The relatively small GO sheet size was generally higher with carboxyl and hydroxyl group and is more favored to be functionalized. The length of CNF ranged from 0.38 to 2.64 μm, the average length was 1.11 μm, the average diameter was 32 nm, and the aspect ratio was calculated to be 35. The CNF obtained by TEMPO-mediated oxidation is carboxyl and hydroxyl group rich, while GO sheets are also functionalized with many carboxyl and hydroxyl group after oxidation. Owing to the small size effect, the carboxyl and hydroxyl group of CNF could be fully exposed, thus a strong hydrogen bonding between GO and CNF is expected. FTIR was carried out to confirm the interaction between GO and CNF, this is shown in Fig. 1*e*. For GO, the adsorption peak at 1720 cm^{-1} is assigned to the C=O stretching vibration, similarly for CNF, the adsorption peak at 1723 cm^{-1} originated from the carbonyl group of the carboxyl group.[24,25] However, for mixture of GO-CNF-10 sample, the carbonyl adsorption peak is red shift to 1716 cm^{-1}, on account of the hydrogen bond between GO and CNF, making CNF accessible to attach on GO sheets.

Graphene fiber can be wet-spun from the liquid crystalline behavior of the GO, the pre-ordered GO sheets is the premise for the continuous wet-spinning of the fiber.[5,26] CNF itself could form liquid crystalline for its high long-aspect ratio and high water dispersibility owing to the abundant carboxyl and hydroxyl group on CNF molecular chain.[25,27] First of all, the liquid crystalline phase of hybrid spinning dope at different CNF loadings was examined under polarized optical microscopy observation. The vivid birefringence pattern is the typical characteristic of the liquid crystalline behavior. At the concentration of 10 mg/ml, the GO and GO aqueous suspension display the nematic liquid phase for the schlieren texture, a typical feature for nematic phase could be observed. For comparison, GO aqueous suspension was gradually diluted from 10

to 0.25 mg/ml and also observed under POM as shown in Fig. 2*a*. With the GO concentration decreased, the birefringence pattern became less vivid. At concentration of 0.25 mg/ml, the GO concentration exhibits an isotropic phase. The POM observation of GO–CNF at the CNF loading of 2/100, 5/100, and 10/10 are shown in Fig. 2*b–d*, respectively. As the concentration decreased the birefringence patterns also became faded. To better compare the liquid crystalline structure of GO and GO–CNF building block, the GO and GO–CNF hybrid suspensions at the same concentration were carefully studied, as marked in the rectangular area in Fig. 2. At the same concentration of 0.5 mg/ml, the GO suspension only partially displays birefringence in the tube and the birefringence is hard to observe. As the loading of CNF increased the patterns became more vivid and gradually spreads all over the tube, indicating an increased liquid crystalline behavior of GO suspension by adding CNF. The building block of GO–CNF exhibits liquid crystalline behavior in its aqueous suspension. A small loading of CNF was added in the GO aqueous suspension. The hydrogen bond between CNF and GO makes the CNF available to absorb on GO sheets. The CNF could fix the adjacent GO sheets in the lyotropic liquid crystalline phase, to make it better aligned and reduce the critical concentration needed to achieve a transition from anisotropic to lyotropic phase. To further prove and better observe the pre-ordered structure of GO–CNF building block, the diluted aqueous suspension of GO and GO–CNF suspension at the same concentration were rapidly frozen by liquid nitrogen and vacuum freeze-dried, and then observed in SEM, which is shown in Fig. 2*e–f*. The freeze-dried samples could preserve the original arrangement of the GO sheets and CNF in the liquid state. One observes that the GO foam is less well organized compared with samples of GO-CNF-2 and GO-CNF-5, which displayed more ordered and uniformly organized as CNF loading increased. This phenomenon is in agreement with the POM observation results.

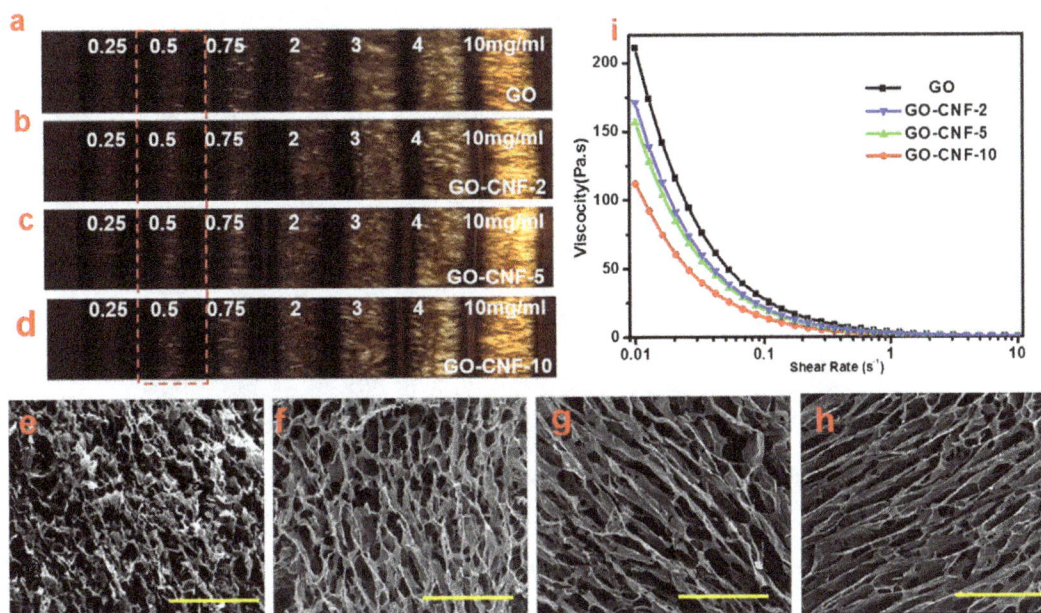

Figure 2 *a–d* Polarized Optical Microscopy of GO, GO-CNF-2, GO-CNF-5, and GO-CNF-10 suspension at different concentrations, *e–h* SEM images of 1 mg/ml spinning dope rapidly cooling by liquid nitrogen and freeze-dried at different CNF loadings, 0, 2,5,and 10, respectively. All the scale bars are 50 μm and *i* Rheological characterization of 10 mg/ml spinning dope at various CNF loadings. All the scale bars are 100 μm

Again it is proved that the nematic liquid crystalline behavior of GO could be facilitated by adding small loading of CNF, this better aligned spinning dope is beneficial for stronger fiber and might benefit for fiber wet-spinning from a relatively low GO concentration and small GO sheet size.

To evaluate the stability of the hybrid spinning dope, the electrostatic repulsion was measured by Zeta potential instrument. The measured zeta potential of GO, GO-CNF-10 is −54.5 mV, −63.5 mV, respectively. The electrostatic repulsion originating from negatively charged GO sheets contributed to the stabilization of the GO sheets in aqueous suspension.[7,15] Such high zeta potential value is favorable for the stability of the GO–CNF building block in aqueous suspensions. Thus, the liquid crystalline dope of GO–CNF could be stable in aqueous surroundings. GO aqueous suspension and CNF aqueous suspension are spinnable separately, the hydrogen bond between GO and CNF makes the CNF accessible to absorb on GO sheets, the GO–CNF building block which also displayed liquid crystalline behavior may provide a possibility for high performance fibers. In practical wet-spinning procedure, GO fiber is wet-spun under shear. Rheological tests were carried out to estimate spinnability and imitate the structure evolution in the shear induced flow during the wet-spinning process, which is shown in Fig. 2*i*. It could be observed that the initial viscosity decreased as the CNF loading increased, which may indict a better pre-alignment of the GO–CNF hybrid suspension.[28] At a low shear rate ranging from 0.1 to 10 s⁻¹, the viscosity of GO–CNF spinning dope is always lower than that of GO dope and the viscosity becomes thinner as the CNF loading is increased. All of the GO–CNF suspensions show shear-thinning behavior, which suggested the non-Newtonian fluids.

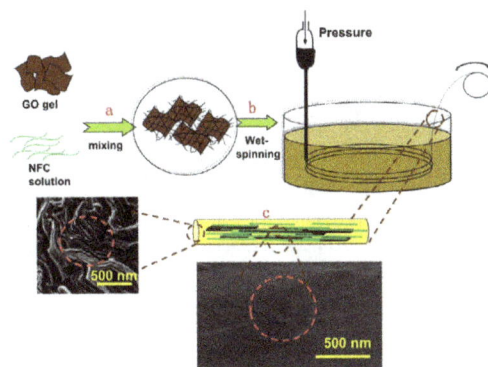

Figure 3 Schematic illustration of wet-spinning procedure of GO–CNF fiber. (a) GO gel was blending with CNF aqueous suspension and (b) pre-aligned spinning dope was wet-spun (c) cartoon images of the structure of GO–CNF composite fiber, the red circle indicts the oriented CNFs. On the surface and cross-section of GO–CNF fiber

The shear-shinning behavior of the samples can be ascribed to the orientation under shear, which may be caused by the fact that GO sheets are better aligned to a more orderly aligned structure from a relatively disordered isotropic phase by adding CNF. This shear-thinning behavior and less viscous state suggested that the GO–CNF suspensions are highly spinnable under shear, which may give birth to a better oriented fiber and could provide instructions for the wet-spinning process.[20]

The GO–CNF composite fiber was wet-spinning as illustrated in Fig. 3. The morphology of the obtained fiber along

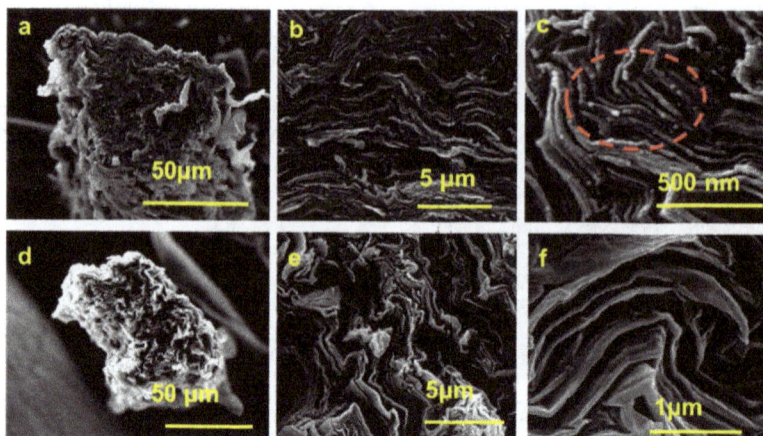

Figure 4   SEM images of *a–c* the cross-section of a GO-CNF-10 fiber, indicting the uniform packing structure and *d–f* cross-section morphology of GO fiber

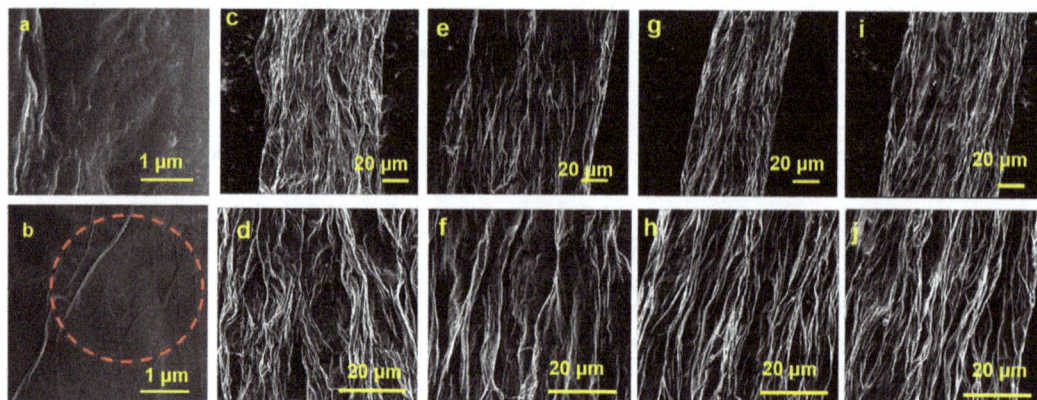

Figure 5   SEM images of fiber surface morphology of *a, b* magnified surface morphology of GO, and GO-CNF-10 fiber, respectively, *c, d* GO fiber at different magnifications, *e, f* GO-CNF-2 composite fiber, *g, h* GO-CNF-5 composite fiber, and *i, j* GO-CNF-10 composite fiber

and vertical to the fiber direction was investigated via SEM (Fig. 3*b* and *c*). The cross-section morphology of the composite fiber shows curled and closely stacked GO sheets (Fig. 4*a–c*). The nanosized one-dimensional CNF could be clearly observed at the magnified cross-section images as shown in Fig. 4*c*, which is small dot-like as circled in red. GO sheets are closely and uniformly piled up on one another inheriting from the pre-aligned original spinning dope. For comparison, the as-spun GO fiber without CNF was also observed (Fig. 4*d–f*), compared with GO–CNF fibers it also displays layered structure similar to GO–CNF fibers, but it is less compactly stacked with more defects and voids as shown in Fig. 4*f*. The voids and defects of GO fiber may originate from the relatively low GO concentration, small size of GO sheets, and the fast solvent change in coagulation procedure. GO sheets are aligned along the flow direction, but on the radical direction it is stacked with several vectors,[24] so there may be some defects and voids during stacking and wrinkling of the GO sheets in the wet-spinning or post-drying procedure. It was reported that a

relatively low concentration GO generally leads to brittle fibers or ribbon-like fibers[6,23,25], especially for small GO sheets. But the GO–CNF composite fiber started from 3 wt.% GO aqueous suspension of small sheets and a small loading of CNF are more densely packed and regular circular shaped; again it is suggested that CNF may play a supporting and bridging role for GO fibre.[29,30]

To better observe the morphology of GO fiber with and without CNF, the surface morphologies of GO fiber and GO–CNF fiber at varied CNF loadings at different magnifications are shown in Fig. 5. As demonstrated in Fig. 5, by comparison, much enhanced orientation of GO sheets by adding CNF can be clearly seen, better alignment of GO–CNF fiber could be observed. For GO-CNF-10 fiber (Fig. 5*i* and *j*), uniformly orientated GO sheets could be observed, at the magnified images (Fig. 5*b*) the CNF is orientated along the fiber axis which is different from the magnified surface section of GO fiber without CNF (Fig. 5*a*). Thus, it can be concluded that the CNF adsorbed on GO sheets well and the GO–CNF building

Table 1   Mechanical Properties of GO and GO–CNF composite fibers

Sample	Tensile modulus (GPa)	Tensile stress (MPa)	Ultimate strain (%)
GO	6.3 ± 0.5	176 ± 14	2.53 ± 0.33
GO-NFC-2	7.1 ± 1.4	199 ± 17	2.89 ± 0.24
GO-NFC-5	8.6 ± 0.9	228 ± 15	3.48 ± 0.32
GO-NFC-10	8.1 ± 0.2	210 ± 5	3.00 ± 0.17

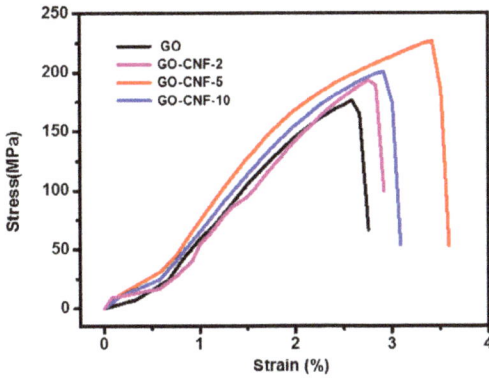

Figure 6   Typical stress–strain curve of GO–CNF composite fibers

blocks are orientated along the flow direction during the wet-spinning procedure, resulting the orientation for both the GO sheets and CNF. The enhanced orientation of GO fiber after adding CNF could be proved by comparing the surface morphologies between GO and GO–CNF fibers as shown in Fig. 5. SEM images clearly reveal that the orientation increased and optimized after the incorporation of CNF. Though CNF cannot be easily observed form the fiber surface, GO sheets indeed show better orientation after adding CNF in GO–CNF fibers. The enhanced orientation degree is ascribed to the better aligned state of the original spinning dope. While at the loading of 10%, the alignment also shows a little drop which may be due to the local aggregation of the long-aspect ratio of CNF. The orientated structure originated from the pre-aligned spinning dope and the alignment was further facilitated by the elongational flow in the spinning procedure. After the solvent

change in the coagulation bath and post-drying process, the GO sheets tend to wrinkle and the adjacent GO–CNF blocks become closely packed accompanying with the diameter shrinking. In the drying process, the fiber contracted both in diameter and length[23] since the fiber was collected under wet state and restrained dried on the drum, the shrinkage in length direction was restricted and the aligned structure could be well preserved.

Mechanical strength is important for the practical application of the fiber. The mechanical properties of GO–CNF composite fibers were measured and the results are summarized in Table 1. Fig. 6 displays the typical stress–strain curves of the fibers at different CNF loadings. The elastic modulus and mechanical strength increased as the CNF loading increased, and for sample at a high loading of 10/100, the mechanical properties show a slight decrease, which may be ascribed to the partial agglomeration of CNF in the highly viscous GO gel. At the CNF loading of 5%, the elastic modulus and fracture strength are promoted up to 8.6 GPa and 228 MPa compared to GO fiber of 6.3GPa and 176 MPa. And the toughness of the fiber is also promoted from 2.53 to 3.48%. The alignment is one important aspect in the mechanical performance enhancement; the better alignment could also reduce the defects and facilitate the load transfer. In addition, the hydrogen bond between GO and CNF can also contribute to the mechanical performance of the GO–CNF fiber. Thus, it can be concluded that a simultaneous enhancement of strength and toughness of GO fiber can be achieved by adding CNF.

It was reported that the alignment of GO sheets, the interaction of the adjacent GO sheets, and the voids and defects in the fiber play a key important role in the mechanical performance of graphene or GO fibers. GO–CNF aqueous suspension shows orientated lyotropic liquid crystalline, the relative orientation degree of the as-prepared

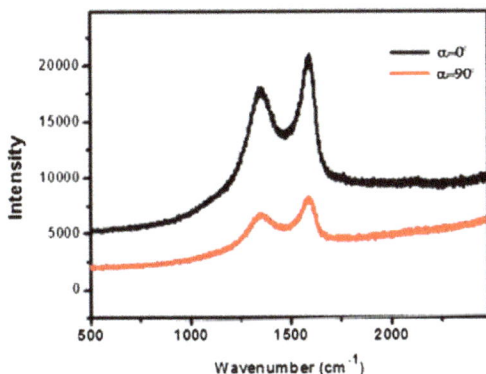

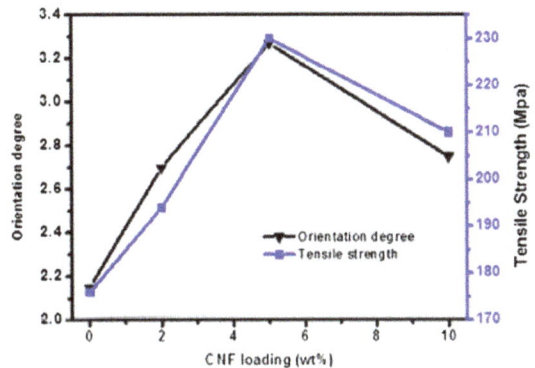

Figure 7   a Polarized Raman spectra of GO-CNF-10 composite fibers in parallel and perpendicular to fiber axis and b orientation degree and tensile strength as a function of CNF loading

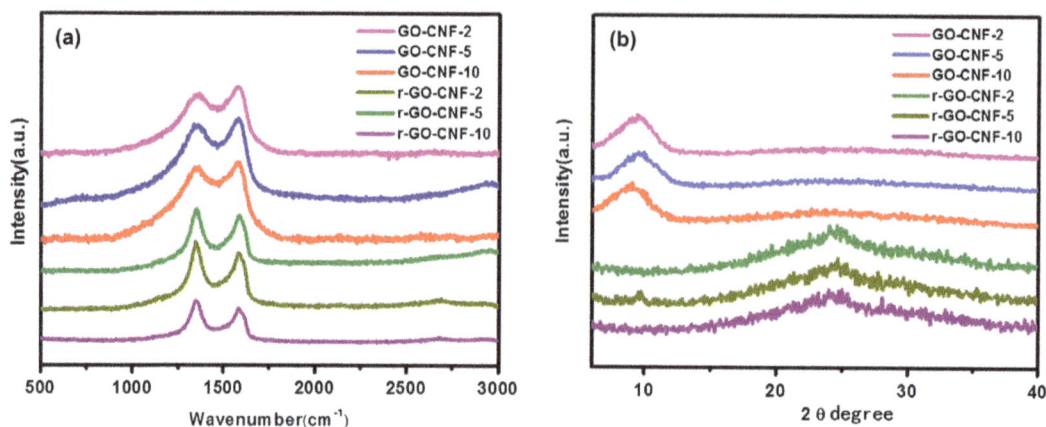

Figure 8   *a* Raman spectra and *b* XRD patterns of composite fibers at different CNF loadings before and after HI reduction

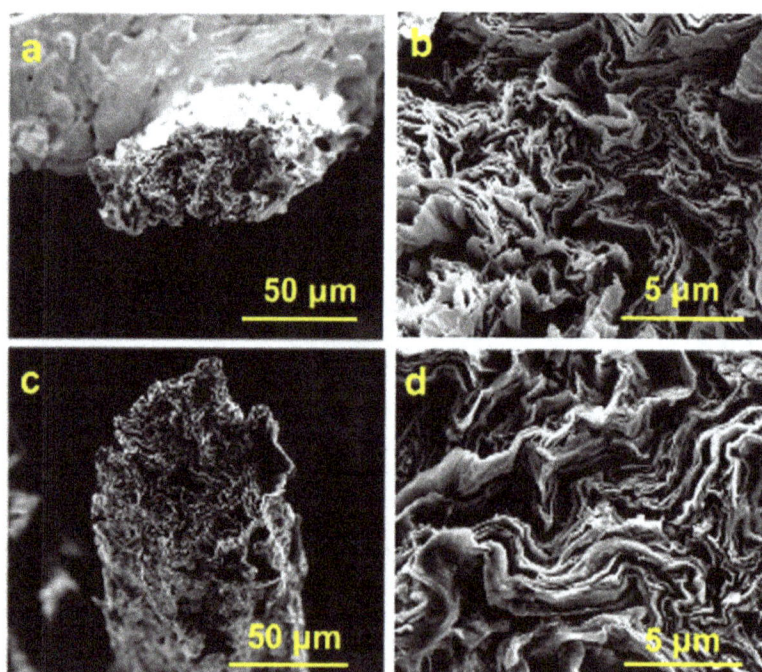

Figure 9   SEM images cross-section of r-GO fiber and r-GO-CNF-10 fiber

fiber from the spinning dope was evaluated by polarized Raman Spectra.[31,32] The orientation degree could be estimated from intensity ratio (I_0/I_{90}) at the parallel direction and vertical direction at ca.1590 cm^{-1} (G band) as shown in Fig. 7a. The relative orientation degree was measured to be 2.16, 2.70, 3.27, and 2.75, respectively for CNF loading at 0, 2–100, 5–100, and 10–100. We correlated the orientation degree with the mechanical strength; the mechanical performance and orientation degree display the synchronous change at varied CNF loadings as shown in Fig. 7b. This result demonstrates the importance of GO sheets orientation in the fiber strength.

The fiber was further reduced to restore the electrical conductivity by immersing in hydroiodic acid aqueous solution (30 wt%) at 80℃ for 15 mins. XRD was carried out to measure the reduction effectiveness. In Fig. 8a, XRD patterns of all the GO–CNF composite fibers show a d-spacing value of 8.4 Å. The r-GO-CNF fiber shows a wide 2θ degree at 25°, and the corresponding d-spacing value is ca. 3.55 Å, which implies that most of the oxygen groups were removed after reduction, and the conjugated network is partially restored. Raman spectra were also utilized to further characterize the reduction effectiveness[33] as demonstrated in Fig. 8b. The intensity ratio of D band (1350 cm^{-1}) to G band (1580 cm^{-1}) is changed from

Table 2 Mechanical and electrical conductivity of r-GO-CNF composite fiber

Sample	Tensile modulus (GPa)	Tensile stress (MPa)	Ultimate strain (%)	Conductivity (S/m)
r-GO	7.05 ± 0.39	193 ± 17	2.53 ± 0.30	13000
r-GO-CNF-2	6.04 ± 0.34	202 ± 8	2.00 ± 0.52	11000
r-GO-CNF-5	7.25 ± 0.21	236 ± 14	3.70 ± 0.31	9300
r-GO-CNF-10	7.51 ± 0.47	201 ± 29	1.99 ± 0.24	8300

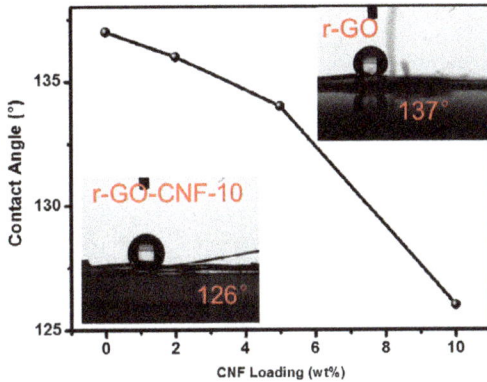

Figure 10 Water contact angle of r-GO-CNF fiber as a function of CNF loading

0.86 to 1.19 after reduction, which also implies an effective reduction.

SEM characterization was carried out to explore the morphology change of GO and GO–CNF after reduction as shown in Fig. 9. After reduction a reduction in diameter could be observed. For r-GO fiber (Fig. 9a and b), no significant change is observed from the cross-section morphology before and after reduction. But for GO–CNF composite fiber (Fig. 9c and d), the packing density of GO sheets decreases after the reduction, which may be correlated to the reduction procedure in which the fiber was immersed in hydroiodic acid (30 wt.%) in the reduction procedure, the super hydrophilic CNF might absorb water and swelling during the reduction thus some little voids were brought. The mechanical properties of the r-GO-CNF fibers are summarized in Table 2. The r-GO-CNF fibre still has a mechanical strength up to 236 MPa and flexibility of 3.7%.

After reduction, the mechanical properties are preserved and the electrical conductivity is simultaneously restored to a large extent as shown in Table 2. The GO–CNF composite fiber displays a high electrical conductivity of 8300 S/m even at a high loading of 10 wt% CNF. Different from the polymer-reinforced GO fiber such as using PAN, HPG, PVA[8] as reinforcing materials, the polymers mainly act as glue and adhere the adjacent GO sheets which displayed nacre mimic structure that severely destructed the electrical conductivity. In our work, the mechanical strength promotion of GO–CNF fiber is mainly resulted from the better alignment of GO sheets as promoted by CNF. The linear shape and small-sized CNF does not entirely cover all the GO sheets. And the conductive path was only partially disrupted. And the GO–CNF composite fiber still shows highly conducted.

CNF is hydrophilic for its abundant carboxyl and hydroxyl groups. In addition, its nanosize, long-aspect ratio, makes

the functional groups fully exposed. The reduced GO sheets were hydrophobic for the removal of oxygen functional group during reduction. After the incorporation of CNF, the water droplets' contact angle decreased from 137° for r-GO fiber to 126°, as the CNF loading increased to 10/100 as shown in Fig. 10. The hydrophilicity of the r-GO fiber is promoted after CNF incorporation owing to the super hydrophilicity of the CNF. The improvement in hydrophilicity is beneficial for its application, for example, in electrochemistry filed. The hydrophobic attributes of graphene sheets will hinder the infiltration of the electrolyte, thus the inner graphene sheets are inaccessible to the electrolyte and the high specific area of graphene material could not be fully used. The highly conductive r-GO-CNF composite fiber with improved hydrophilicity has potential application in fiber-based flexible supercapactior. Furthermore by an appropriate structure design such as porous fiber, r-GO-CNF fiber may lead to supercapactior with much higher capacitance.[34–36]

The resulting mechanical strength for r-GO-CNF fiber was 236 MPa and had an electrical conductivity of 8300–11000 S/m. Compared with other carbon-based fibers such as wet-spun CNT fiber[37] with a mechanical strength of 1.8 GPa, and directly spun twisted CNT yarns[38,39] with a mechanical strength up to 1.0 GPa, traditional carbon fiber with a tensile strength of 6.0 GPa,[40] the GO fiber or graphene fiber emerges as a new family of carbon-based fiber, its mechanical strength could be as high as 500 MPa[17] using large GO sheets as building block and followed with ionic cross-linking up to now. Compared with carbon fiber or CNT fiber which has emerged for a long time, the graphene fibers still have a long distance both in mechanical strength and manufacture procedure, all need to be further promoted. The inferior mechanical strength of graphene fiber may be ascribed to that of the voids and defects in the GO sheets, and the conjugated network could only be partially restored after reduction. Numerous attempts have been proposed to enhance the mechanical properties of graphene or GO fiber,[8,13,15–17] such as polymer blends, post drawing, or wet-drawing; larger size GO flakes have been used. Herein, the incorporation of CNF effectively enhances the mechanical properties of GO fiber from the small GO sheets. Compared with graphene fiber from large GO size or PAN-reinforced graphene fiber, the GO–CNF fiber was weaker. But this method does not require complex polymer grafting, post-drawing process, or large-sized GO sheets , only a small content of CNF is needed. The mechanical strength of the GO–CNF fiber could be further enhanced by post-drawing or ionic cross-linking. More importantly the small-sized CNF does not bring down the electrical conductivity appreciably. After reduction the fiber displays excellent electrical conductivity, good mechanical properties, and improved hydrophilicity.

Conclusion

In conclusion, GO–CNF hybrid fibers have been successfully prepared by directly mixing GO gel with CNF aqueous suspension. The introduction of CNF facilitates the liquid crystalline behavior of the spinning dope; stable, highly spinnable, and well pre-aligned hybrid spinning are formed. GO–CNF fiber with better alignment, improved mechanical performance is obtained from the liquid crystalline of GO–CNF spinning dope. After reduction, the r-GO-CNF fiber was highly conductive and preserved the original mechanical performance of GO-CNF. Besides, the incorporation of CNF also endows r-GO-CNF fiber with improved hydrophilicity compared to the r-GO fiber. This work emphasizes the importance of the original liquid crystalline state in the fabrication of GO fiber, and provides an effective and facile method to reinforce GO fiber by enhancing the pre-alignment of the original spinning dope via adding small loading of CNF. Also multifunctional GO fibre could be obtained using CNF-absorbed GO sheets as new building blocks by further morphology control or functionalization.

Disclosure statement

No potential conflict of interest was reported by the authors.

Funding

This work was supported by the National Natural Science Foundation of China [grant numbers 51421061 and 51210005]. We would like to express our great thanks to Guangdong Shengyi Technology Limited Corporation for financial support.

References

1. A. K. Geim and K. S. Novoselov: The rise of graphene. *Nat. Mater.*, 2007, **6**, (3), 183–191.
2. D. A. Dikin, S. Stankovich, E. J. Zimney, R. D. Piner, G. H. Dommett, G. Evmenenko, S. T. Nguyen and R. S. Ruoff: Preparation and characterization of graphene oxide paper. *Nature*, 2007, **448**, (7152), 457–460.
3. O. C. Compton and S. T. Nguyen: Graphene oxide, highly reduced graphene oxide, and graphene: versatile building blocks for carbon-based materials. *Small*, 2010, **6**, (6), 711–723.
4. H. Hu, Z. Zhao, W. Wan, Y. Gogotsi and J. Qiu: Ultralight and highly compressible graphene aerogels. *Adv. Mater.*, 2013, **25**, (15), 2219–2223.
5. Z. Xu and C. Gao: Graphene chiral liquid crystals and macroscopic assembled fibres. *Nat. Commun.*, 2011, **2**, 571.
6. Z. Chen, W. Ren, L. Gao, B. Liu, S. Pei and H.-M. Cheng: Three-dimensional flexible and conductive interconnected graphene networks grown by chemical vapour deposition. *Nat. Mater.*, 2011, **10**, (6), 424–428.
7. D. R. Dreyer, S. Park, C. W. Bielawski and R. S. Ruoff: The chemistry of graphene oxide. *Chem. Soc. Rev.*, 2010, **39**, (1), 228–240.
8. Z. Xu and C. Gao: Graphene in macroscopic order: liquid crystals and wet-spun fibers. *Acc. Chem. Res.*, 2014, **47**, (4), 1267–1276.
9. D. Yu, Q. Qian, L. Wei, W. Jiang, K. Goh, J. Wei, J. Zhang and Y, Chen: Emergence of fiber supercapacitors. *Chem. Soc. Rev.*, 2015, **44**, (3), 647–662.
10. Y. Hu, H. Cheng, F. Zhao, N. Chen, L. Jiang, Z. Feng and L. Qu: All-in-one graphene fiber supercapacitor. *Nanoscale*, 2014, **6**, (12), 6448–6451.
11. H. Cheng, Z. Dong, C. Hu, Y. Zhao, Y. Hu, L. Qu, N. Chen and L. Dai: Textile electrodes woven by carbon nanotube-graphene hybrid fibers for flexible electrochemical capacitors. *Nanoscale*, 2013, **5**, (8), 3428–3434.
12. Y. Li, K. Sheng, W. Yuan and G. Shi: A high-performance flexible fibre-shaped electrochemical capacitor based on electrochemically reduced graphene oxide. *Chem. Commun.*, 2013, **49**, (3), 291–293.
13. H. Cheng, C. Hu, Y. Zhao and L. Qu: Graphene fiber: a new material platform for unique applications. *NPG Asia Mater.*, 2014, **6**, (7), e113.
14. L. Chen, Y. He, S. Chai, H. Jiang, F. Chen and Q. Fu: Toward high performance graphene fibers. *Nanoscale*, 2013, **5**, (13), 5809–5815.
15. Z. Liu, Z. Xu, X. Hu and C. Gao: Lyotropic liquid crystal of polyacrylonitrile-grafted graphene oxide and its assembled continuous strong nacre-mimetic fibers. *Macromolecules*, 2013, **46**, (17), 6931–6941.
16. C. Xiang, C. C. Young, X. Wang, Z. Yan, C. C. Hwang, G. Cerioti, J. Lin, J. Kono, M. Pasquali and M. Tour: Large flake graphene oxide fibers with unconventional 100% knot efficiency and highly aligned small flake graphene oxide fibers. *Adv. Mater.*, 2013, **25**, (33), 4592–4597.
17. Z. Xu, H. Sun, X. Zhao and C. Gao: Ultrastrong fibers assembled from giant graphene oxide sheets. *Adv. Mater.*, 2013, **25**, (2), 188–193.
18. Y. S. Kim, J. H. Kang, T. Kim, Y. Jung, K. Lee, J. Y. Oh, J. Park and C. R. Park: Easy preparation of readily self-assembled high-performance graphene oxide fibers. *Chem. Mater.*, 2014, **26**, (19), 5549–5555.
19. L. Kou and C. Gao: Bioinspired design and macroscopic assembly of poly (vinyl alcohol)-coated graphen into kilometers-long fibers. *Nanoscale*, 2013, **5**, (10), 4370–4378.
20. R. Jalili, S. H. Aboutalebi, D. Esrafilzadeh, R. L. Shepherd, J. Chen, S. Aminorroaya-Yamini, K. Konstantinov, A. I.Minett, J. M.Razal and G. G. Wallace: Scalable one-step wet-spinning of graphene fibers and yarns from liquid crystalline dispersions of graphene oxide: towards multifunctional textiles. *Adv. Funct Mater.*, 2013, **23**, (43), 5345–5354.
21. J. G. Torres-Rendon, F. H. Schacher, S. Ifuku and A. Walther: Mechanical performance of macrofibers of cellulose and chitin nanofibrils aligned by wet-stretching: a critical comparison. *Biomacromolecules*, 2014, **15**, (7), 2709–2717.
22. A. Isogai, T. Saito and H. Fukuzumi: TEMPO-oxidized cellulose nanofibers. *Nanoscale*, 2011, **3**, (1), 71–85.
23. X. Yao, W. Yu, X. Xu, F. Chen and Q. Fu: Amphiphilic, ultralight, and multifunctional graphene/nanofibrillated cellulose aerogel achieved by cation-induced gelation and chemical reduction. *Nanoscale*, 2015, **7**, (9), 3959–3964.
24. N. D. Luong, N. Pahimanolis, U. Hippi, J. T. Korhonen, J. Ruokolainen, L.-S. Johansson, J.-D. Nam and J. Seppälä: Graphene/cellulose nanocomposite paper with high electrical and mechanical performances. *J. Mater. Chem.*, 2011, **21**, (36, 13991–13998.
25. T. Saito, Y. Nishiyama, J.-L. Putaux, M. Vignon and A. Isogai: Homogeneous suspensions of individualized microfibrils from TEMPO-catalyzed oxidation of native cellulose. *Biomacromolecules*, 2006, **7**, (6), 1687–1691.
26. Z. Xu and C. Gao: Aqueous liquid crystals of graphene oxide. *ACS Nano.*, 2011, **5**, (4), 2908–2915.
27. T. Saito, M. Hirota, N. Tamura, S. Kimura, H. Fukuzumi, L. Heux and A. Isogai: Individualization of nano-sized plant cellulose fibrils by direct surface carboxylation using TEMPO catalyst under neutral conditions. *Biomacromolecules*, 2009, **10**, (7), 1992–1996.
28. J. Lauger, R. Linemann and W. Richtering: Shear orientation of a lamellar lyotropic liquid crystal. *Rheol Acta.*, 1995, **34**, (2), 132–136.
29. Y. Li, H. Zhu, H. Shen, J. Wan, X. Han, J. Dai, H. Dai and L. Hu: Highly conductive microfiber of graphene oxide templated carbonization of nanofibrillated cellulose. *Adv. Funct Mater.*, 2014, **24**, (46), 7366–7372.
30. Y. Li, H. Zhu, S. Zhu, J. Wan, Z. Liu, O. Vaaland, S. Lacey, Z. Fang, H. Dai, T. Li and L. Hu: Hybridizing wood cellulose and graphene oxide toward high-performance fibers. *NPG Asia Mater.*, 2015, **7**, e150.
31. L. M. Ericson, H. Fan, H. Peng, V. A. Davis, W. Zhou, J. Sulpizio, Y. Wang, R. Booker, J. Vavro, C. Guthy, A. N. G. Parra-Vasquez, M. J. Kim, S. Ramesh, R. K. Saini, C. Kittrell, G. Lavin, H. Schmidt, W. W. Adams, W. E. Billups, M. Pasquali, W.-F. Hwang, R. H. Hauge, J. E. Fischer and R. E. Smalley: Macroscopic, neat, single-walled carbon nanotube fibers. *Science*, 2004, **305**, (5689), 1447–1450.
32. V. A. Davis, A. N. G. Parra-Vasquez, M. J. Green, P. K. Rai, N. Behabtu, V. Prieto, R. D. Booker, J. Schmidt, E. Kesselman, W. Zhou, H. Fan, W. W. Adams, R. H. Hauge, J. E. Fischer, Y. Cohen, Y. Talmon, R. E. Smalley and M. Pasquali: True solutions of single-walled carbon nanotubes for assembly into macroscopic materials. *Nat. Nano.*, 2009, **4**, (12), 830–834.
33. K. N. Kudin, B. Ozbas, H. C. Schniepp, R. K. Prud'homme, I. A. Aksay and R. Car: Raman spectra of graphite oxide and functionalized graphene sheets. *Nano Lett.*, 2008, **8**, (1), 36–41.
34. S. H. Aboutalebi, R. Jalili, D. Esrafilzadeh, M. Salari, Z. Gholamvand, S. Aminorroaya Yamini, K. Konstantinov, R. L. Shepherd, J. Chen and S. E. Moulton: High-performance multifunctional graphene yarns: toward wearable all-carbon energy storage textiles. *ACS Nano.*, 2014, **8**, (3), 2456–2466.

35. G. Huang, C. Hou, Y. Shao, B. Zhu, B. Jia, H. Wang, Q. Zhang and Y. Li: High-performance all-solid-state yarn supercapacitors based on porous graphene ribbons. *Nano Energy*, 2015, **12**, 26–32.

36. Z. Weng, Y. Su, D. W. Wang, F. Li, J. Du and H. M. Cheng: Graphene–cellulose paper flexible supercapacitors. *Adv Energy Mater.*, 2011, **1**, (5), 917–922.

37. A. B. Dalton, S. Collins, E. Munoz, J. M. Razal, V. H. Ebron, J. P. Ferraris, J. N. Coleman, B. G. Kim and R. H. Baughman: Super-tough carbon-nanotube fibres. *Nature*, 2003, **423**, (6941), 703–703.

38. Y.-L. Li, I. A. Kinloch and A. H. Windle: Direct spinning of carbon nanotube fibers from chemical vapor deposition synthesis. *Science*, 2004;, **304**, (5668), 276–278.

39. K. Koziol, J. Vilatela, A. Moisala, M. Motta, P. Cunniff, M. Sennett and A. Windle: High-performance carbon nanotube fiber. *Science*, 2007, **318**, (5858), 1892–1895.

40. D. D. Edie: The effect of processing on the structure and properties of carbon fibers. *Carbon*, 1998, **36**, (4), 345–362.

12

Gas barrier efficiency of clay- and graphene-poly(isobutylene-co-isoprene) nanocomposite membranes evidenced by a quantum resistive vapor sensor cell

J. F. Feller*[1], K. K. Sadasivuni[1,2,3], M. Castro[1], H. Bellegou[1], I. Pillin[1], S. Thomas[2] and Y. Grohens[3]

[1]Smart Plastics Group, European University of Brittany (UEB), LIMAT^B-UBS, Lorient, France
[2]Surfaces & Interfaces Group, European University of Brittany (UEB), LIMAT^B-UBS, Lorient, France
[3]School of Chemical Sciences, Mahatma Gandhi University, Kottayam, India

Abstract A quick and newly developed technique has been applied for the determination of gas barrier properties of nanocomposites using conductive polymer nanocomposite (CPC)-based quantum resistive vapor sensors (vQRS). The potential of this device is demonstrated with two kinds of nanocomposites, i.e. poly(isobutylene-co-isoprene) (IIR) filled with either montmorillonite (MMT) or graphene (GR) nanoplatelets. The calibration curves of vQRS were well fitted with the Langmuir–Henry model, whereas the integral time lag method was successfully used to determine the diffusion coefficient D of nanocomposite membranes. The values obtained allowed to derive fillers' shape factor from the Cussler's model and the results obtained with the vQRS cell show that the diffusion coefficient D of volatile organic compounds (VOC such as toluene at 25°C), through IIR membranes can be significantly decreased by the dispersion of nanoplatelets in the polymer matrix. Quantitatively, the gas barrier efficiency of pristine IIR can be increased of 67% and 80% by the incorporation of only 4.76% mm^{-1} of Cloisite10A, and 4.76% mm^{-1} of reduced GR oxide, respectively. Inversely, partially exfoliated expanded graphite and organo-modified clay dispersed without maleic anhydride functions present on poly(isobutylene-co-isoprene) (IIR) result in lower gas barrier efficiency, i.e. 17% and 57%, respectively.

Keywords Diffusion coefficient, Gas barrier, Quantum resistive vapor sensor, Reduced graphene oxide, Clay

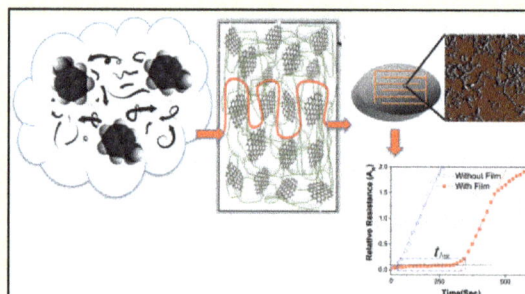

Introduction

Many applications covering biomedical, laboratory, and industrial fields rely on the diffusion properties of polymer membranes. These include the controlled drug delivery and transfusion of body fluids through polymer membranes in medicine,[1] fabrication of impermeable tires, balls and inflatable boats in industries, etc. Moreover, the information on barrier properties of polymers helps in the selective removal of toxic solvent vapors. Owing to these reasons, extensive works were done on the solvent barrier properties of various polymer–solvent systems.[2–12] On the other hand,

classical techniques to measure diffusion properties require demanding preparation and are often time consuming. In this context, the use of quantum resistive sensors (QRS) made of conductive polymer nanocomposites (CPCs) developed in our group[13–15] appeared to be a good alternative to existing methods of determination of the diffusion coefficient, as it was found to combine rapidity, convenience in use, and reliability.[16] The low amount of solvent required for the analysis (only a few cubic centimeters) is really remarkable. To demonstrate the effectiveness of the QRS-based technique, it has been applied to the evaluation of the gas barrier properties of two kinds of polymer nanocomposites, i.e. poly(isobutylene-co-isoprene) IIR filled with nanoplatelets of either montmorillonite

(MMT)[17,18] or graphene (GR),[19,20] known for their efficiency to modify the diffusion of gases through polymers.

Indeed, most polymers are permeable to small molecules, which has motivated researchers to improve their gas barrier performances by the addition of nanofillers in order to decrease their diffusivity. The superior features of nanometric fillers offer great possibilities to develop polymer nano-composites. Compared to conventional microcomposites, they exhibit both enhanced mechanical properties[16,21–23] and gas impermeability[24–26] even at low filler content. The nano-composites containing flakelike or platelike inorganic fillers of high aspect ratio have particular interest as barrier materials. For two-dimensional nanosheets, the individual layers have a much higher aspect ratio than typical microscopic aggregates, which have to be exfoliated to yield high-performance nano-composites.[27] The large interfacial area and attractive inter-actions per unit volume with the matrix as a result of exfoliation help nanoclays in fabricating mechanically strong impermeable elastomer membranes.[28,29]

Poly(isobutylene-co-isoprene) (IIR) is a special purpose elastomer because of its excellent properties such as impermeability/air retention, weathering resistance, ozone resistance, vibration dampening, sealing efficiency, etc. This synthetic rubber is a copolymer of isobutylene (97–98%) and a small amount of isoprene (2–3%). The low levels of unsa-turation between long poly(isobutylene) segments impart good flexing properties as well to this polymer.[30] Various nanofillers and other additives can be added to IIR to improve its performances further. However, IIR-clay nano-composites are difficult to prepare because of the hydro-phobicity of IIR resulting in poor interactions with clay silicate layers.[31] Therefore, a grafting with maleic anhydride (MA) is often performed to introduce polar groups on the non-polar IIR backbone and to improve IIR/organoclay compatibility.[32,33] Among the various methods, the solvent assisted grafting is quite useful.[31] Apart from clay, GR, which exhibits a maximum theoretical specific surface of $2630\ m^2\ g^{-1}$,[34] has focused the attention of many scientists interested in the improvement of rubber-based nano-composites' properties. In fact, this atomically thick two-dimensional sheet of sp^2-bonded carbon atoms has an ultra-high specific area expected to improve simultaneously mechanical, electrical, and permeability properties.[35–41]

Both GR and layered silicate nanocomposites can be used as barrier materials in packaging applications for instance. The process of permeation through the composite films depends on the chemical nature, size, and shape of penetrants, the type and crosslinking density of polymers, the nature of fillers used, and some environmental factors such as temperature, con-centration, etc. To correlate solvent diffusivity in polymers, many models such as solution–diffusion model, dual mode model, Eyring's hole theory, percolation theory, fractal theory, random walk theory, and others are used.[42] The simple sol-ution–diffusion model[43] proposes a three-step process for sol-vent diffusivity through the matrix: occupation of solvent molecules in the free volume present in polymers, swelling caused by the binding of solvent to the network sites and the solvent entry at the crosslinked regions. The possibility of sol-vent molecules to cluster inside the polymer network has been addressed by Barrie et al.[43] by explaining the observed non-linear isotherm for high activity solvents. Aris[44] proposed another three mode model of sorption including the bulk dis-solution of solvent in polymer networks, solvent absorption onto the surfaces of vacuoles and hydrogen bonding between polymer hydrophilic groups and solvents. The diffusion beha-vior of small molecule penetrants in dense polymer membranes is also reported by Chen et al.[45] based on the percolation theory, the fractal theory, and random walk theory. Finally, Fredrickson and Bicerano,[46] Cussler et al.,[47] and Bharadwaj[48] models explain theoretically the gas barrier properties of nanocomposites.

This paper focuses on the effectiveness of quantum resistive vapor sensors (vQRS) made of CPCs to quickly characterize gas barrier efficiency of nanocomposite mem-branes. In order to quantify QRS performances their results were compared to those of the more traditional Fourier Transform InfraRed Spectroscopy (FT-IR) technique.[49] The choice of pertinent nanocomposites to illustrate the capa-bilities of the technique went on poly(isobutylene-co-iso-prene) IIR filled with GR and MMT nanoplatelets, which are promising materials for the design of membranes impermeable to gases. This work gave also the opportunity to study the influence of the chemical nature, interactions, and shape factor of nanoplatelets on their barrier properties. In addition, the investigation of sorption and diffusion mechanisms of molecules expected to have good diffusion ability, i.e. vapors of toluene allowed the determination of transport parameters through nanocomposite films with the `time lag' approach.[13] In contrast, by using the diffusion coefficient values, the effective average aspect ratio of fillers could also be calculated.

Materials & Methods

Materials used for nanocomposites synthesis

IIR (Butyl 065) was purchased from Exxon Company (Notre Dame de Gravenchon, France) (density $0.92\ g\ cm^{-3}$). Multiwall carbon nanotubes (CNTs) N-7000 were kindly provided by Nanocyl (Sambreville, Belgium). They are produced via cata-lytic carbon vapor deposition (CCVD) process and have a carbon purity of 90% (10% catalyst). This grade corresponds to CNT with an average diameter of 10 nm and a mean length between 100 and 1000 nm.[14] The nanoclay (Cloisite 10A), an organo-modified MMT has a density $1.5\ g\ cm^{-3}$ and an inter-layer distance of 1.9 nm; it was obtained from Southern Clay Products (Gonzales, USA). The materials were dried under vacuum at 60°C temperature for 24 h before the composite preparation. Poly(caprolactone) (PCL) ($M_w = 14\,000\ g\ mol^{-1}$) and poly(carbonate) (PC) Lexan141R were obtained from General Electric Plastics (Evry, France). The graphite particles used as precursor of GR platelets (average size $> 45\ \mu m$, density $1.82\ g.cm^{-3}$) and all the other chemicals such as $KMNO_4$, H_2SO_4 etc. as well as the solvents like toluene, ethanol, THF, HCl, etc. were procured from Aldrich, (Saint-Quentin Fallavier, France) and were used as received. Expanded graphite (EG) eGR TIMREX BNB90 was kindly provided by Timcal (Willebroek, Belgium).

Fabrication of vQRS

The CPCs vQRS were fabricated in two steps. First, PC and PCL-based nanocomposites were prepared by separately dissolving the polymers in chloroform under constant stirring and by adding a constant amount of CNT (1% w/w) into each polymer solution. Carbon nanotubes are well dispersed under sonication for 1 h at 50°C and kept degassing for 30 min. Then the nanocomposite solution mixtures were deposition on interdigitated electrodes (IDEs) by the spray layer-by-layer (sLbL) developed in our group, which allows the control of the sensor's hierarchical conducting architecture.[50–56] The spray conditions were set as follows: a nozzle scanning speed of $V_s = 10 \, cm \, s^{-1}$, a stream pressure of ps = 0.20 MPa, a solution flowrate adjusted to index 2 and an IDE to nozzle distance of $d_{tn} = 8 \, cm$. During solvent evaporation, CPC microdrops of 50–100 nm size can weld to form a three-dimensional network by percolation that leads to a film of final thickness between hundreds of nanometers and several micrometers, depending on the number of sprayed layers.[51] After fabrication, vQRS were conditioned overnight at 30°C in a controlled atmosphere. The type of sensor, a number of layers composing the transducer and average initial resistances (R_0) were pure CNT (4 layers)/$R_0 = 58\pm5 \, k\Omega$, PC-CNT (8 layers)/$R_0 = 5.2\pm1.2 \, k\Omega$, and PC-CNT (12 layers)/$R_0 = 65\pm10 \, k\Omega$, respectively. All vQRS were exposed three consequent times to the studied organic volatile compound (VOC), i.e. toluene, to study the diffusion properties through the membrane (polymer or nanocomposite film). The high sensitivity of vQRS results from the progressive conversion of simple ohmic conduction (when CNT are in close contact) into tunnel conduction (when a gap appears between CNT) on the adsorption of analytes at the CNT/CNT nanojunctions, which results in an exponential variation of the resistance.[14] Temperature variations can also influence the conduction mechanism; this is why this parameter must be controlled during the experiment.[57,58] Other factors influencing the sensitivity of QRS sensor are the nature of the polymer matrix, the conducting nanofiller content, the initial resistance, the chemical nature of analytes, particularly the Flory–Huggins intermolecular interaction parameter (χ_{12}) based on the difference in solubility parameters δ of the VOC and the polymer.[54] Finally, the selectivity of vQRS to analytes can be tuned by changing the

chemical nature of the polymer matrix of the CPC[51] or by grafting organic functions on the surface of the conducting nanofiller.[55,56]

Preparation of MA-modified poly(isobutylene-co-isoprene) (MA-g-IIR)

Maleic anhydride-modified poly(isobutylene-co-isoprene) MA-g-IIR was prepared by Makoto et al.[31] in the following method. The desired amount of MA was dissolved in xylene and mixed with IIR in inert medium initiated by benzoyl peroxide at 110°C for 3 h by maintaining the same temperature. When the reaction was completed, the solution was poured into acetone to remove the unreacted MA. The obtained MA-g-IIR precipitate was dried at 80°C in a vacuum oven till it attained a constant weight. FT-IR spectroscopy was used to evidence the grafting reaction. The appearance of peaks at 1728 cm^{-1} and 1860 cm^{-1} in the FT-IR spectrum of MA-g-IIR indicates the grafting of MA groups on the IIR backbone (Fig. 1). The peaks at 1780 and 1785 cm^{-1} are, respectively, assigned to the symmetric and asymmetric stretched vibration absorbance of the carbonyl groups (Fig. 1).

Preparation of reduced GR oxide (RGO) from GR oxide (GO)

Graphene oxide was prepared by improved GO synthesis method[59] by adding concentrated H_2SO_4/H_3PO_4 and $KMnO_4$ mixture to graphite flakes (3.0 g, 1 wt eq.). The mixture was heated to 50°C for 12 h under stirring and cooled by pouring onto ice. The excess amount of $KMnO_4$ was removed by using H_2O_2 and the suspension was simultaneously filtered through a metallic US standard testing sieve (300 µm; W. S. Tyler, Mentor, Ohio, USA) for removing the unreacted graphite flakes. After an extended multiple-wash process with water, HCl, and ethanol, the material remaining was coagulated with ether and filtered through a PTFE membrane. The obtained GO was subjected to high temperature to prepare thermally reduced GO in the second step. For this the nitrogen gas is passed through GO for 30 min in sealed container to create an inert atmosphere on its surface and thereafter placed in a pre-heated muffle furnace at 150°C for 30 min.

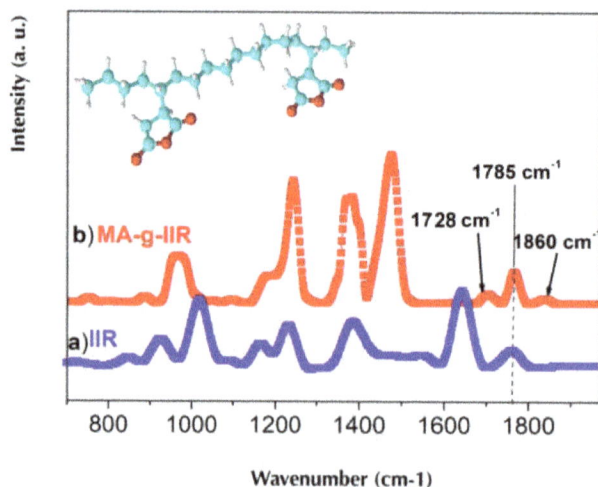

Figure 1   FT-IR spectra of a IIR and b MA-g-IIR

Preparation of the composite membranes

Poly(isobutylene-co-isoprene) IIR composites containing micro- as well as nanofillers are prepared by solution compounding using THF as solvent. Initially the filler was homogeneously dispersed in THF by sonication for 1 h. The THF-solubilized IIR suspension was mixed with this filler solution using mechanical stirring for 3 h. The obtained rubber composite solution was dried at 80°C in an oven under vacuum till it attained constant weight. Specifically, cured composites were obtained by mixing the dried composites with stearic acid at 1.7 wt-%, zinc oxide (activator) at 4.4 wt-%, MBTS (2-bisbenzothiazole-2,2'-disulfide) at 0.4 wt-%, TMTD (tetra methyl thiuram disulfide) at 0.9 wt-% and sulfur (elemental sulfur S8) at 1.3 wt-%, in an internal Brabender mixer at 20 rpm and at 50°C for 30 min, Aldrich (Saint-Quentin Fallavier, France). It was then compression molded at 160°C for the respective curing times determined from the moving die rheometer. The composition of the prepared samples is given in Table 1.

Experimental setup for the analysis of diffusion with vQRS

A schematic representation of the experimental set-up integrating the vQRS to determine the time lag is illustrated in Fig. 2. The vQRS provides the relative variation of resistance (A_R) with time that can be easily recorded using a Keithley 6517A electrometer Its chemo-resistive behavior will depend on the molecules to which it is exposed and its composition must be selected accordingly. Here the polymer chosen for the QRS matrixes are poly(carbonate) (PC) and poly(caprolactone) (PLC) because they were found very sensitive to toluene vapors resp.[51,56] In this experimental set-up, the polymer or nanocomposite membrane to be tested separates the solvent chamber (filled with a saturated vapor of the analyte) from the QRS chamber. Both the pressure difference between both sides of the membrane and the temperature in the cell are controlled.

The QRS relative resistance (A_R) changes with time can be recorded from the introduction of vapors into the chamber until the end of diffusion through the membrane.[16] The QRS chemo-resistive response (A_R) is defined by equation (1)

$$A_R = \frac{R - R_0}{R_0} \tag{1}$$

where R and R_0 are, respectively, the QRS resistance in the presence of vapor and its initial resistance (in the dried state) in inert atmosphere.

Typical experimental results such as those presented in Fig. 3 show that the area under the curve is increasing with time as a function of the arrival of analyte molecules on the second face of the membrane in the chamber containing the QRS.

During the experiment, the pressure inside the solvent chamber is nearly equal to the saturating vapor pressure and that in the QRS chamber is equal to atmospheric pressure. The vapor diffusion will proceed through the membrane up to a state of equilibrium provided that the concentration balance constitutes the driving force.[13,60] Classically the diffusion coefficient value D can be calculated by following the integral `time lag' method on the basis of the mathematical description made by Fick's law, which explains an analogy between the heat and mass flows in a polymer, based on the free chain mobility.[61] The transfer of a substance through a polymer matrix is thus proportional to the concentration

Table 1   Details of samples used

No.	Sample code	Composition
1	IIR	Poly(isobutylene-co-isoprene)
2	IIR-Cloisite 10A	Poly(isobutylene-co-isoprene) (95.24 wt-%) + Cloisite 10A (4.76 wt-%)
3	MA-g-IIR-Cloisite 10A	Maleic anhydride grafted poly(isobutylene-co-isoprene) (95.24 wt-%) + Cloisite 10A (4.76 wt-%)
4	IIR-RGO	Poly(isobutylene-co-isoprene)(95.24 wt-%) + reduced graphene oxide (4.76 wt-%)
5	IIR-EG	Poly(isobutylene-co-isoprene)(95.24 wt-%) + expanded graphite (4.76 wt-%)

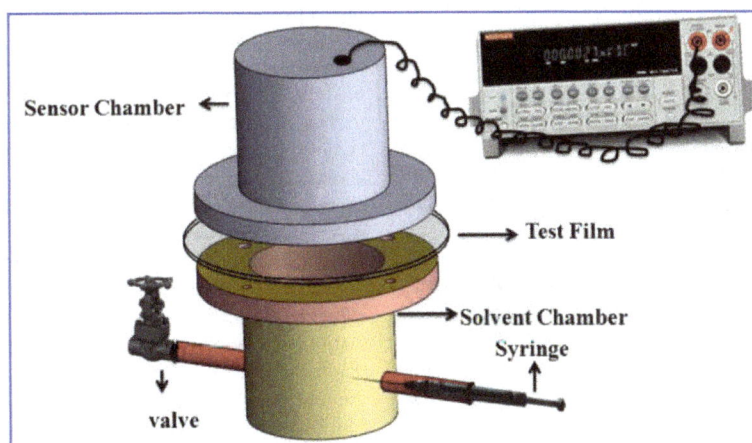

Figure 2   Schematic representation the measurement cell in which QRS are implemented

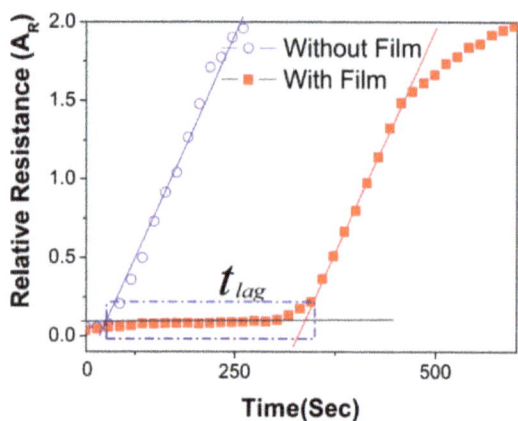

Figure 3   Typical chemo-resistive response of QRS when exposed to toluene molecules with and without IIR film[16]

gradient within a volume unit, which can be represented by equation (2)

$$F = -D\frac{\partial c}{\partial x} \qquad (2)$$

where F is the transfer rate (mass flow) per unit volume and unit cross-sectional area, c is the concentration of the substance being diffused (diffusant), x is the distance of diffusion perpendicular to the cross-section, and D is the diffusion coefficient.

The diffusion coefficient is a substance specific variable, which is regarded as a constant at low concentrations, whereas it varies at higher concentrations. With respect to equation (2), Fick derived a second law for one-dimensional diffusion in isotropic substances, i.e. in the x direction,[13,60] which is given by equation (3)

$$\frac{\partial c}{\partial t} = D\frac{\partial^2 c}{\partial^2 x} \qquad (3)$$

where t is time and x is vertical distance from the membrane surface using the boundary conditions,

$$c = c_1 \text{ at } x = 0 \text{ and } c = c_2 \text{ at } x = 1.$$

All expressions can be simplified into the final form of equation (4) for explaining the total solute transferred related to time.[62]

$$Q(t) = \frac{AD\Delta c}{l}\left[t - \frac{l^2}{6D}\right] \qquad (4)$$

where $Q(t)$ is the total amount of solute transferred through a membrane of area A at time t and D is the effective diffusion coefficient. A graph of Q versus t is a straight line intercepting the y axis at $t_{lag} = l^2/6D$.

The determination of the intercept of the linear part of the curve with the x axis (Fig. 3) gives the time lag (t_{lag}), which allows to calculate the diffusion coefficients D using equation (5) provided that the membrane thickness (l) is known

$$D = \frac{l^2}{6t_{lag}} \qquad (5)$$

In the experimental device of Fig. 2, a multimeter records the QRS resistance changes as soon as they there are exposed to

the VOC, in this case toluene. At the beginning of the test, a 0.3-cm^3 of VOC was injected in the bottom half of the vessel and vapor gets distributed with equal pressure throughout the chamber sealed with the test membrane.[13] The diffusion time of analytes through the QRS transducer is evaluated to a fraction of second, which is assumed to be negligible compared with the diffusion time through the membrane of several seconds. All the experiments were performed at 25°C and were repeated thrice at each operating condition for reproducibility.

Techniques for structural & morphological characterization

The nanoscale characterization of GO was done using atomic force microscopy (AFM) in ambient conditions using normal tapping mode (TM-AFM) on a caliber multimode scanning probe microscope from Bruker-Veeco (Dourdan, France). The set point amplitude of antimony-doped silicon tapping mode cantilever (LTESP model, Veeco, New York, USA) was ~4.5 V. The cantilever with tip radius between 5 and 20 nm had typical resonance frequency about 270 kHz, and a cantilever spring constant (k) was 20–80 N m^{-1}. The substrate used for filler deposition was oxidized smooth silica wafer.

The ultra-thin transmission electron microscopy (TEM) samples of 70-nm thickness were cut using a cryogenic ultramicrotome Leica ultracut UCT at −90°C and images were taken using a JEOL JEM-1400 electron microscope at 100 kV.

The scanning electronic microscope (SEM) images were obtained with a Jeol JSM-6460LV SEM.

An expert model of Philips diffractometer with CuKα radiation generator (α = 0.15404 nm, 40 kV, 40 mA) was used for XRD measurements of the composite samples.

Results & Discussion

Structure & morphology of nanocomposites

Before the characterization of gas barrier properties of nanocomposites, it is important to check the level of individualization of nanoplatelets and their organization in the polymer matrix. The thickness of GO observed by AFM was evaluated to be >1.4 nm, which corresponds to two to four atomic layers, clearly indicating the successful exfoliation driven by oxidation (Fig. 4). The average surface area of the GO platelets ranged from 1 to 4 μm^2.

The XRD analysis allows to investigate the effectiveness of individualization of platelets in a polymer matrix characterized by an increase of interlayer distance (intercalation) and complete separation (exfoliation). Figure 5a shows that IIR-EG still presents a weak peak at $2\theta = 26.5°$, that corresponds to the (002) basal plane, with an interlayer spacing of $d_{002} = 0.345$ nm characteristic of pure EG. This suggests that IIR-EG still contains graphite flakes not completely exfoliated. Conversely, IIR-RGO shows no such peak at $2\theta = 26.5°$ that supports the formation of well individualized RGO nanoplatelets. In the case of MMT-based nanocomposites, the diffraction peak at $2\theta(4.8°$ visible in Fig. 5b corresponds to the (001) basal plane of clay. The increase in the interlayer spacing from $d_{001} = 1.83$ nm (corresponding peak at $2\theta(4.8°$ to pure Cloisite 10A platelets) to $d_{001} = 4.61$ nm for IIR–Cloisite composite (peak at $2\theta(1.92°$). The MA-g-IIR-Cloisite

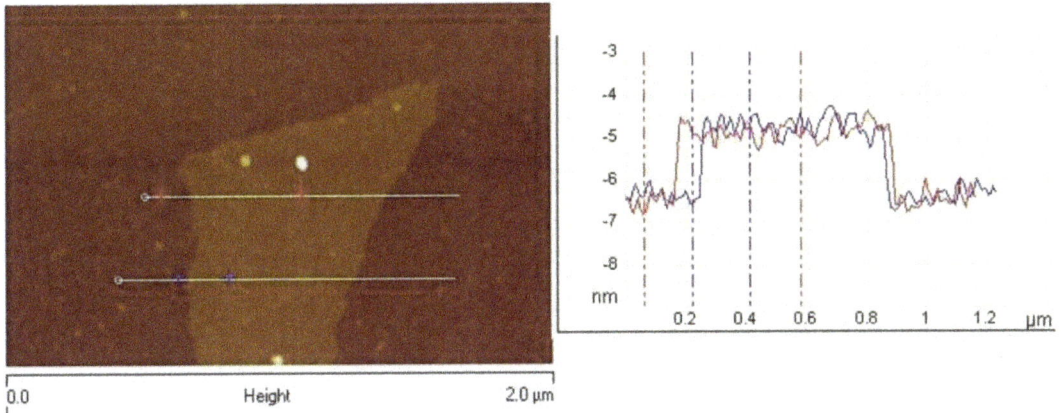

Figure 4   Atomic force microscopy (AFM) images of GO platelets deposited on an oxidized smooth silica surface

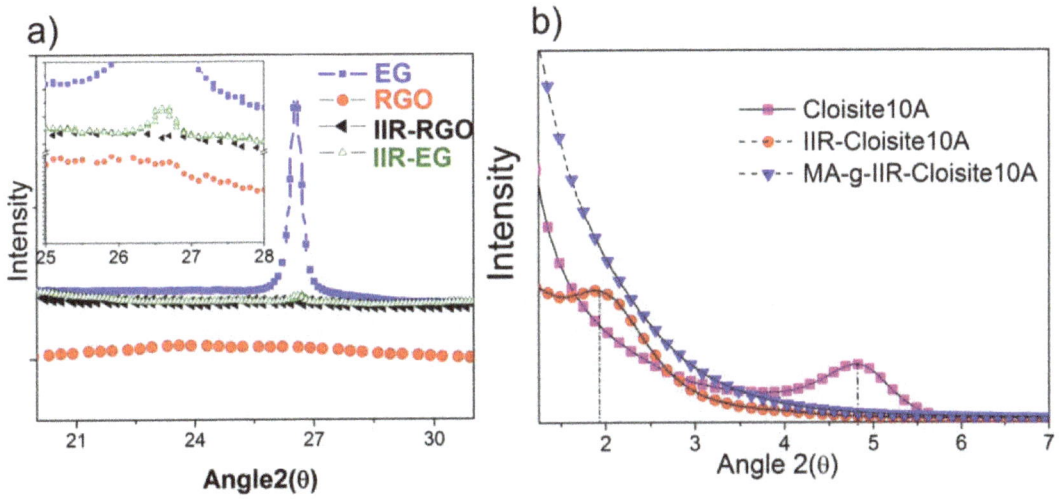

Figure 5   WAXD patterns of a EG, RGO, IIR-EG, and IIR-RGO (insight shows closed view) and b Cloisite10A, IIR-Cloisite10A, and MA-g-IIR-Cloisite10A

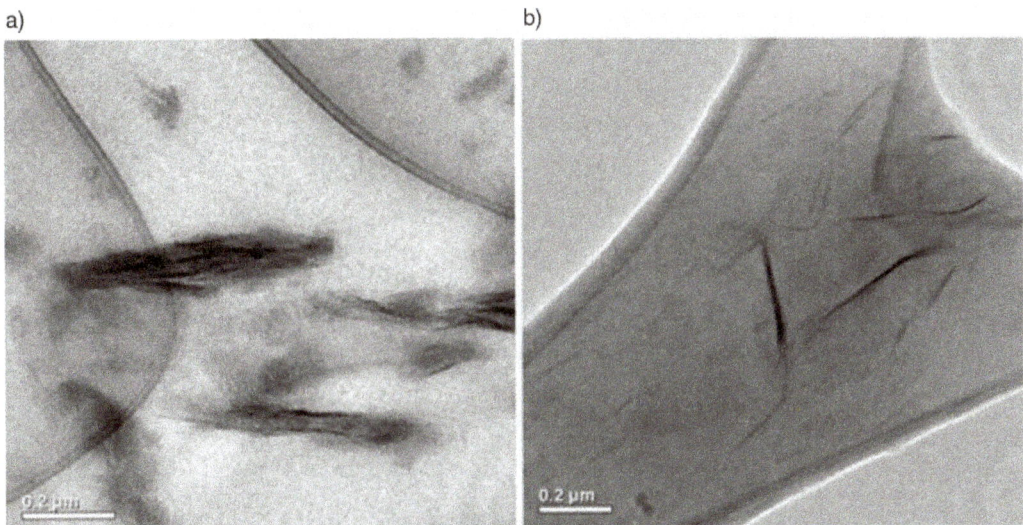

Figure 6   Transmission electron microscopy (TEM) photographs of a IIR and b MA-g-IIR filled with 1.12 v/v% of cloisite 10A

shows no characteristic peak, substantiating the complete exfoliation of clay in MA-g-IIR (Fig. 5b).

The TEM images give information about the dispersion of clay in IIR and MA-g-IIR matrixes (Fig. 6). It is clear from Fig. 6a that in pure IIR some stacks of 20–100 nm thickness remain present, whereas in Fig. 6b the clay sheets appear well dispersed in MA-g-IIR. Since MMT has a much higher electronic density than neat polymers, it appears dark in TEM images.

The morphology of RGO, IIR-RGO, and IIR-EG can be visualized by SEM in Fig. 7. The homogeneous dispersion of RGO in IIR is clear from the SEM micrograph for IIR-RGO nanocomposites (Fig. 7b). Reduced GR oxide particles have a platelike shape, with average sizes of 1–10 μm (Fig. 7a). Each flake consists of multiple layers of GR nanoplatelets with aspect ratios of ~20–50. The surface of IIR-RGO composite is very homogeneous and smooth, but IIR-EG contains rough surface because of the large particle size of EG as evidenced from Fig. 7c.

Characterization of vQRS chemo-resistive responses

It is expected from a good gas sensor that on exposure to vapors, it will have a short response time (within the second), a high sensitivity [in the ppm (parts per million) range], and a

response amplitude proportional to the amount of analytes to detect (quantitativity of signals). As it is clearly visible from Fig. 3 (curve without nanocomposite film), vQRS give a quick response (within the second) of large amplitude (200% after 250 s). In order to determine the relationship between the chemo-resistive response Ar and the toluene vapor concentration in the cell, this parameter has been varied from 25 and 500 ppm for the three kinds of QRS: PC-CNT, PCL-CNT, and CNT. To make the comparison easier, the data obtained with the different systems have been collected in Fig. 8 and fitted with two different models that will be presented as in what follows.

In previous studies on the quantification of chemo-resistive responses of vQRS in the ppt (parts per thousand) range,[51–56] it had been shown that such behavior could be well fitted with the complete Langmuir–Henry Clustering (LHC) model derived from a dual mode sorption law as expressed in equation (6)

$$A_R = \frac{b_L(f'' - f)}{1 + b_L f} + k_H f + (f - f'')f^{n'} \qquad (6)$$

where b_L is the Langmuir affinity constant, f'' is the vapor fraction over which Langmuir's diffusion is replaced by

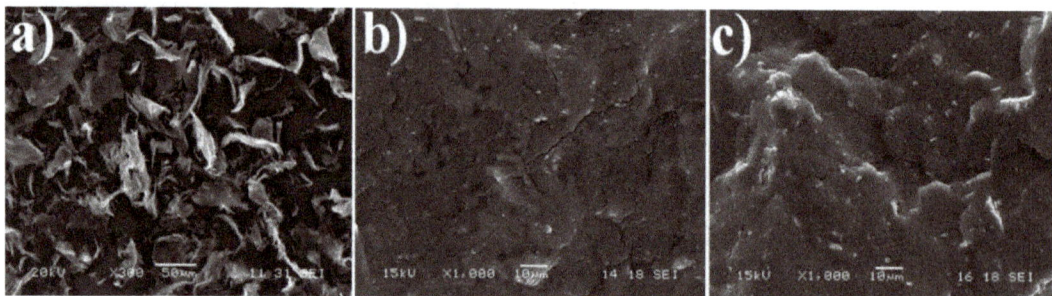

Figure 7   SEM of a RGO, b IIR-RGO, and c IIR-EG

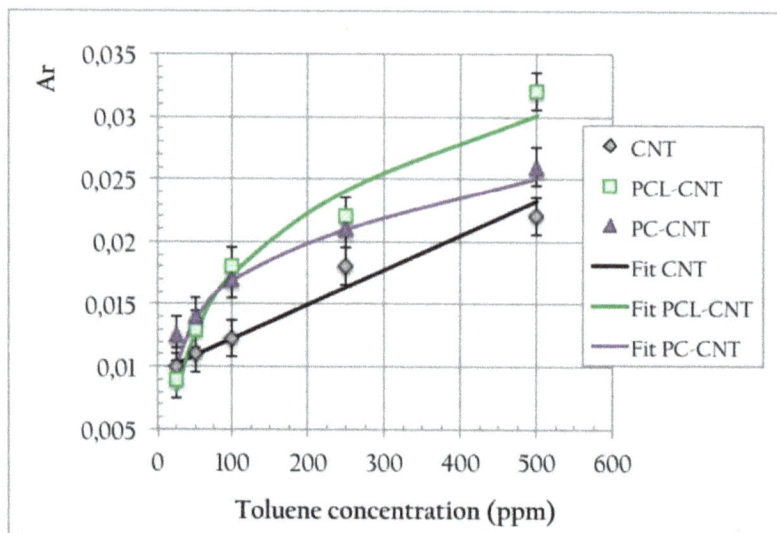

Figure 8   Evolution of PC-1%CNT (w/w), PCL-1%CNT (w/w), and neat CNT (1%w/w) QRS response with toluene vapor concentration in ppm

Henry's diffusion, f is the solvent fraction, k_H is Henry's solubility coefficient, and n' is the number of vapor molecules associated in clusters.

Here obviously, for PC-CNT and PCL-CNT the third term of this equation, describing the clustering of VOC molecules when they aggregate in large amount on macromolecules and nanofillers, is not relevant for parts per million concentrations and the model can be reduced to the first two terms LH. Even for the sensor made of CNT alone it seems that the chemo-resistive behavior can be well described by a simple linear law such as that proposed in equation (7). The fitting parameters are collected in Table 2

$$A_r = k_H f + C \qquad (7)$$

k_H is Henry's solubility coefficient, f is the solvent fraction, and C is a constant.

These laws can be used to predict the concentration of a known vapor from the value of the QRS response, but also make possible to determine their limit of detection (LOD) by extrapolation, here LOD # 10 ppm. Moreover, each sensor has been individually calibrated at 25°C with VOC flowrates varying from $Q_v = 0\text{-}100\ cm^3\ min^{-1}$ as explained in Ref. 51.

Thanks to the performances demonstrated above, it is expected that all the vQRS selected in this study will allow to determine with accuracy when only some parts per million of vapor molecules will pass through the polymer membrane from the vapor chamber to sensor chamber (Fig. 2).

Effect of nanofillers on diffusion in nano-composites

The incorporation of nanofillers with a layered structure into polymer matrixes has been shown to be highly effective in reducing the diffusion of toluene vapor. The relative diffusion coefficient in such nanocomposites has been defined by Cussler et al.[47] using equation (8). This parameter

is expressing the gas barrier efficiency resulting from the proper dispersion of nanofillers in a polymer matrix

$$\frac{D_c}{D_0} = \left[\frac{1 + \alpha^2 \phi^2}{1 - \phi} \right]^{-1} \qquad (8)$$

where (ϕ is the volume fraction, α is the particle's aspect ratio, D_c and D_0 are, respectively, the diffusion coefficients with and without filler.

The relative diffusivity of toluene molecules in IIR, IIR-Cloisite, MA-g-IIR-Cloisite, and IIR-RGO nanocomposites and the values of filler aspect ratio derived from equation (8) have been collected in Table 3.

It can be noticed that the most effective nanocomposite to slow down the diffusion of toluene vapor molecules is IIR-RGO, which exhibits the lowest diffusion coefficient and the highest nanofiller shape factor $L/e = 130$. This suggest a good dispersion/exfoliation of RGO in the IIR matrix, which could nevertheless be improved as the value of GR shape factor calculated from equation (8) is still inferior to the value of $L/e = 1000$ calculated from the geometric parameters extracted from Fig. 4. Moreover, Fig. 9 gives a clear overview of the evolution of both the diffusion coefficients and the gas barrier efficiency (through the diffusion coefficient reduction calculated in equation (9)) of the different samples studied

$$\Delta D = \frac{D_m - D_c}{D_m} \qquad (9)$$

where D_m and D_c are, respectively, the diffusion coefficients of the polymer matrix and the nanocomposite.

Although it was observed from the XRD spectra (Fig. 5) that both intercalated and exfoliated structures are present in the nanocomposites, when the interface between nanofillers and macromolecules is of good quality, this leads to a better exfoliation and finally results in nanocomposites with higher gas barrier efficiency as clearly visible in Fig. 9. For instance, only 2.92%v/v of clay can decrease the diffusion coefficient of toluene of 57-67%, respectively, without and with MA. This effect is even more remarkable for GR platelets dispersed in IIR, for which only 2.43%v/v leads to a decrease in diffusion coefficient of toluene of 17-80% for respective IIR-EG and IIR-RGO. The very high barrier effect obtained with this later system is because of the specific properties of the interfacial area, which are determined not only by the physico-chemical characteristics of the polymer matrix and the nanofillers, but also by the minimization of side effects that can lead to a less compact material. This can happen when penetrant molecules are clustering, which generates the crazing of the matrix, and/or the partial dissolution of

Table 2   Parameter obtained after fitting with the Langmuir–Henry Clustering (LHC) model expressed by equations (6) and (7)

Toluene	CNT	PC-CNT	PCL-CNT
k_H	0.275	1.13	1.16
C	0.0095	–	–
f'	–	170	190
b_L	–	400	180

CNT: carbon nanotubes; PC: poly(carbonate); PCL: poly(caprolactone).

Table 3   Diffusion coefficients of toluene vapor through IIR-based nanocomposites

Sample codes	Filler (m m^{-1}%)	Filler (v/v%)	D dif. coeff. (cm^2 s^{-1})	Rel. dif. coeff.	Aspect ratio
IIR	0	0	8.11E-08 ± 6.1E-09	1	–
IIR-EG	4.76	2.43	6.71E-08 ± 4.1E-09	0.75 ± 0.03	33 ± 4
IIR-Cloisite	4.76	2.92	3.51E-08 ± 2.1E-09	0.40 ± 0.03	108 ± 8
MA-g-IIR-Cloisite	4.76	2.92	2.66E-08 ± 1.2E-09	0.33 ± 0.03	123 ± 7
IIR-RGO	4.76	2.43	1.64E-08 ± 2.3E-09	0.20 ± 0.04	130 ± 2

EG: expanded graphite; RGO: reduced GR oxide; IIR: poly(isobutylene-co-isoprene);

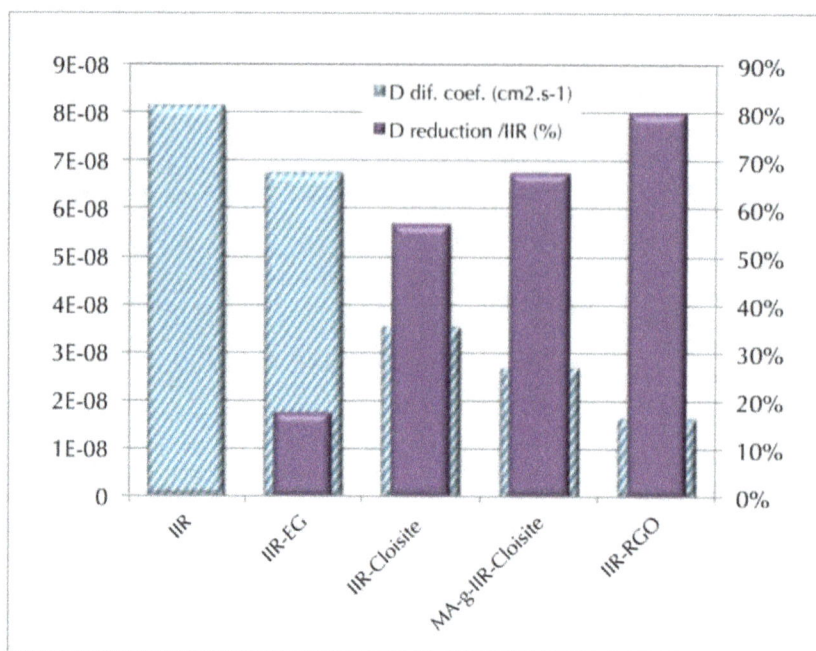

Figure 9 Decrease in diffusion coefficient ($cm^2 s^{-1}$) and increase in diffusion coefficient reduction of toluene in IIR, IIR-2.92%v/v Cloisite, MA-g-IIR-2.92%v/v Cloisite, and IIR-2.43%v/vRGO composites (from left to right)

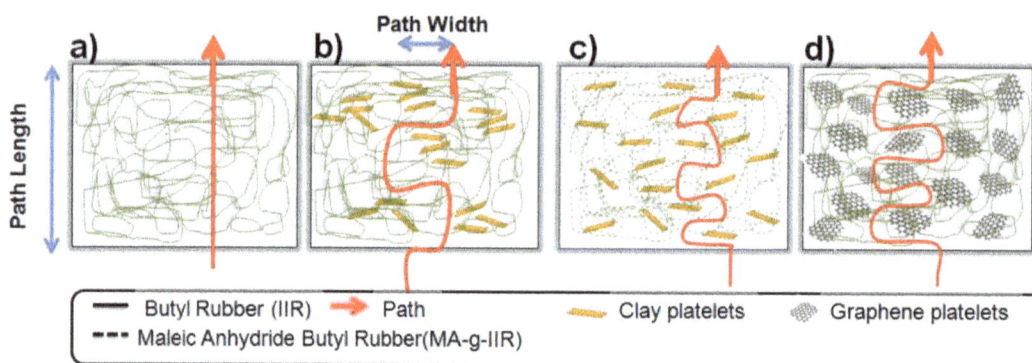

Figure 10 Schematic representation of tortuous path and dispersion in *a* IIR, *b* IIR-Cloisite10A, *c* MA-g-IIR-Cloisite10A, and *d* IIR-RGO nanocomposites

some components; in this case, the diffusion becomes less dependent or even independent of the amount of sorbed molecules.[49] Moreover, the fact that MA-g-IIR-Cloisite10A exhibits a lower permeability than IIR-Cloisite10A nanocomposite is another illustration of the interest of enhancing the interactions between MMT and IIR to promote the dispersion of clay at the nanometric level in IIR and produce more tortuous pathways that will slow down the solvent molecules diffusion. An attempt to conceptualize the vapor molecules' transport through the different composites is made in Fig. 10. Direct flow through a polymer matrix without any obstacle (Fig. 10*a*), polymer matrix in which stacked and exfoliated platelets of MMT have been dispersed respectively (Fig. 10*b* and *c*), nanocomposites in which GR platelets are well exfoliated leading to long tortuous paths and high surface of interface as evidenced for IIR-RGO composites (Fig. 10*d*). These facts are comforted by theoretical calculations showing that GR platelets have a much larger surface area (# $2630\,m^2 g^{-1}$) than clay (# $750\,m^2 g^{-1}$), which justifies that solvent molecules need more time to cross the network made of GR than that of clay.

Conclusions

The new experimental set-up designed by integrating highly sensitivity vQRS made of CPCs in a diffusion cell appeared to be a low cost technique, easy to operate, and effective to determine the barrier properties of nanocomposite films with the conventional time lag method through the determination of the diffusion coefficients. The diffusion of toluene, a volatile organic compound among the most common laboratory reagents, was chosen to demonstrate the efficiency of the vQRS cell to discriminate between fillers' dispersion at the micro- and nanoscales in composite membranes. Through the determination of gas barrier properties, this technique can be used indirectly to evaluate the efficiency of fillers or macromolecules' functionalization to enhance interatomic/intermolecular interactions, the dispersion state and aspect ratio the filler's once dispersed in the matrix. This study confirms the interest of grafting poly(isobutylene-co-isoprene) with MA to allow a better dispersion of organo-modified MMT in this matrix. Consequently, the gas barrier efficiency of this polymer already known to be very high, could be improved by 67%. Moreover, RGO, because of its larger specific surface than MMT, has demonstrated even better gas barrier performances by reaching a reduction in diffusion coefficient of 80% which, compared to the 17% obtained for EG, illustrates again the necessity of reaching a high level of exfoliation. Finally, it is expected that the simplicity of this technique will favor the development of gas barrier materials of interest in many technological fields such as packaging, fire-retardancy, batteries, biomedical applications, etc.

Conflicts of interest

The authors have no conflicts of interest to declare.

Acknowledgements

The authors acknowledge the Department of Science & Technology, Delhi, India and EU FEDER, French Ministry for Research and the Brittany Region for the financial support. The authors are grateful to Anthony MAGUERESSE and Françoise PERESSE for their contribution to this work.

References

1. B. Krause, M. Storr, T. Ertl, R. Buck, H. Hildwein, R. Deppisch and H. Göhl: *Chem. Ing. Tech.*, 2003, **75**, 1725–1732.
2. P. E. Cassidy and T. M. Aminabhavi: *Rubber Chem. Technol.*, 1983, **56**, 594–618.
3. S. B. Harogoppad and T. M. Aminabhavi: *Macromolecules*, 1991, **24**, 2598–2605.
4. R. D. Khinnavar and T. M. Aminabhavi: *J. Appl. Polym. Sci.*, 1991, **42**, 2321–2328.
5. C. S. Proikakis, N. J. Mamouzelos, P. A. Tarantili and A. G. Andreopoulos: *Polym. Degrad. Stab.*, 2006, **91**, 614–619.
6. S. Gopakumar and M. R. Gopinathan Nair: *Polymer*, 2005, **46**, 10419–10430.
7. K. Azaar, I. D. Rosca and J. M. Vergnaud: *Polymer*, 2002, **43**, 4261–4267.
8. M. A. Kader and A. K. Bhowmick: *Polym. Degrad. Stab.*, 2003, **79**, 283–295.
9. E. Mathai, R. P. Singh and S. Thomas: *J. Membr. Sci.*, 2002, **202**, 35–54.
10. F. El-Tantawy: *Polym. Degrad. Stab.*, 2001, **73**, 289–299.
11. S. Erdal, I. Bahar and B. Erman: *Polymer*, 1998, **39**, (10), 2035–2041.
12. P. Sen Majumder, A. B. Majali, V. K. Tikku and A. K. Bhowmick: *J. Appl. Polym. Sci.*, 2000, **75**, 784–795.
13. T. T. Tung, M. Castro, T. Y. Kim, K. S. Suh and J. F. Feller: *J. Mater. Chem.*, 2012, **22**, 21754–21766.
14. B. Kumar, M. Castro and J. F. Feller: *Chem. Sens.*, 2013, **3**, (20), 1–7.
15. S. Chatterjee, M. Castro and J. F. Feller: *J. Mater. Chem. B*, 2013, **1**, (36), 4563–4575.
16. S. K. Kumar, M. Castro, I. Pillin, J. F. Feller, S. Thomas and Y. Grohens: *Polym. Adv. Technol.*, 2013, **24**, (5), 487–494.
17. G. Gorrasi, M. Tortora, V. Vittoria, E. Pollet, B. Lepoittevin, M. Alexandre and P. Dubois: *Polymer*, 2003, **44**, 2271–2279.
18. O. Gain, E. Espuche, E. Pollet, M. Alexandre and P. Dubois: *J. Polym. Sci. B*, 2005, **43**, 205–214.
19. Y. H. Yang, L. Bolling, M. A. Priolo and J. C. Grunlan: *Adv. Mater.*, 2013, **25**, 503–508.
20. B. Stevens, E. Dessiatova, D. Austin Hagen, A. D. Todd, C. W. Bielawski and J. C. Grunlan: *ACS Appl. Mater. Interf.*, 2014, **5**, (13), 9942–9945.
21. A. Usuki, Y. Kojima, A. Okada and O. Kamigaito: *J. Mater. Res.*, 1993, **8**, 1185–1189.
22. T. Lan and T. J. Pinnavaia: *Chem. Mater.*, 1994, **6**, 2216–2219.
23. S. Sadhu and A. K. Bhowmick: *J. Polym. Sci. B*, 2004, **42**, 1573–1585.
24. X. Tong, H. Zhaw, T. Tang, Z. Feng and B. Huang: *J. Polym. Sci. A*, 2002, **40**, 1706–1711.
25. P. B. Messermith and E. P. Giannelis: *J. Polym. Sci. A*, 1995, **33**, 1047–1057.
26. G. Gorrasia, M. Tortoraa, V. Vittoriaa, E. Polletb, B. Lepoittevinb, M. Alexandreb and P. Dubois: *Polymer*, 2003, **44**, 2271–2279.
27. H. Van Olphen: 'An introduction to clay colloid chemistry', 2nd edn; 1977, New York, J. Wiley & Sons.
28. P. Le Baron, Z. Wang and T. Pinnavaia: *Appl. Clay Sci.*, 1999, **15**, 11–29.
29. R. A. Vaia, G. Price, P. N. Ruth, H. T. Nguyen and J. Lichtenhan: *Appl. Clay Sci.*, 1999, **15**, 67–92.
30. A. Saritha, K. Joseph, S. Thomas and R. Muraleekrishnan: *Compos. A*, 2012, **43**, 864–870.
31. K. Makoto, T. Azusa, T. Hiromitsu, U. Amritsu and I. Isamu: *J. Polym. Sci. A*, 2006, **44**, 1182–1188.
32. N. Salahuddin and A. Akelah: *Polym. Adv. Technol.*, 2002, **13**, 339–345.
33. X. C. Li and C. S. Ha: *J. Appl. Polym. Sci.*, 2003, **87**, (12), 1901–1909.
34. D. Chen, L. Tang and J. Li: *J. Chem. Soc. Rev.*, 2010, **39**, 3157–3180.
35. L. Huiqin, L. Shuxin, L. Kelong, X. Liangrui, W. Kuisheng and G. Wenli: *Polym. Eng. Sci.*, 2011, **51**, (11), 2254–2260.
36. S. H. Song, H. K. Jeong and Y. G. Kang: *J. Ind. Eng. Chem.*, 2010, **16**, 1059–1065.
37. R. Prud'Homme, B. Ozbas, I. Aksay, R. Register and A. Douglas: US patent, 7,745,528 B2; 2010
38. B. Xin, W. Chaoying, Z. Yong and Z. Yinghao: *Carbon*, 2011, **49**, 1608–1613.
39. H. Kim, A. A. Abdala and C. W. Macosko: *Macromolecules*, 2010, **43**, 6515–6530.
40. A. K. Geim and K. S. Novoselov: *Nat. Mater.*, 2007, **6**, 183–191.
41. M. A. Rafiee, J. Rafiee, Z. Wang, H. Song, Z. Z. Yu and N. Koratkar: *ACS Nano*, 2009, **3**, 3884–3890.
42. C. E. Rogers: Permeation of gases and vapours in polymers, In: 'Polymer permeability', (ed. J. Comyn); 1986, London, Elsevier applied science, pp. 11–74.
43. A. S. Michaels, W. R. Vieth and J. A. Barrie: *J. Appl. Phys.*, 1963, **34**, (1), 13–20.
44. R. Aris: *Arch. Ration Mech. Anal.*, 1986, **95**, 83–91.
45. C. Chen, B. Han, J. Li, T. Shang, J. Zou and W. Jiang: *J. Membr. Sci.*, 2001, **187**, 109–118.
46. G. H. Fredrickson and J. Bicerano: *J. Chem. Phys.*, 1999, **110**, 2181–2188.
47. E. L. Cussler, W. J. Hughes, W. J. Ward III and R. Aris: *J. Membr. Sci.*, 1988, **38**, (2), 161–174.
48. R. K. Bharadwaj: *Macromolecule*, 2001, **34**, 9189–9192.
49. I. Linossier, F. Gaillard, M. Romand and J. F. Feller: *J. Appl. Polym. Sci.*, 1997, **66**, 2465–2473.
50. J. F. Feller, H. Guézénoc, H. Bellégou and Y. Grohens: 'Smart poly(styrene)/carbon black conductive polymer composites films for styrene vapour sensing', *Macromol. Symp.*, 2005, **222**, 273–280.
51. J. Lu, B. Kumar, M. Castro and J. F. Feller: *Sens. Actua. B*, 2009, **140**, 451–460.
52. B. Kumar, J. F. Feller, M. Castro and J. Lu: *Talanta*, 2010, **81**, 908–915.
53. J. Lu, J. F. Feller, B. Kumar, M. Castro, Y. S. Kim, Y. T. Park and J. C. Grunlan: *Sens. Actua. B*, 2011, **155**, 28–36.
54. S. Nag, L. Duarte, E. Bertrand, V. Celton, M. Castro, V. Choudhary, P. Guegan and J. F. Feller: *J. Mater. Chem. B*, 2014, **2**, 6571–6579.
55. J. F. Feller, N. Gatt, B. Kumar and M. Castro: *Chemosensors.*, 2014, **2**, 26–40.
56. M. Castro, J. Lu, S. Bruzaud, B. Kumar and J. F. Feller: *Carbon*, 2009, **47**, 1930–1942.
57. P. Edwards, H. Gray, M. Lodge and R. Williams: *Angew. Chem. Int. Ed.*, 2008, **47**, 6758–6765.
58. T. A. Ezquerra, M. Kulescza, C. S. Cruz and F. J. Balt-Calleja: *Adv. Mater.*, 1990, **2**, 597–600.
59. C. M. Daniela, V. K. Dmitry, M. B. Jacob, S. Alexander, S. Zhengzong, S. Alexander, B. A. Lawrence, L. Wei and M. T. James: *ACS Nano*, 2010, **4**, (8), 4806–4814.
60. J. Crank: 'The mathematics of diffusion', 2nd edn; 1975, Oxford, Clarendon Press.
61. A. Fick: *Ann. Phys. Lpz.*, 1855, **170**, 59–86.
62. H. S. Carslaw: 1st edn; 1906, London, Macmillan.

13

Influence of drying procedure on glass transition temperature of PMMA based nanocomposites

Maria Eriksson[1], Han Goossens*[1] and Ton Peijs[1,2]

[1]Laboratory of Polymer Materials, Department of Chemical Engineering and Chemistry, Eindhoven University of Technology, 5600 MB Eindhoven, The Netherlands
[2]Centre for Materials Research & School of Engineering and Materials Science, Queen Mary University of London, Mile End Road, London E1 4NS, UK

Abstract A literature review of poly(methyl methacrylate) (PMMA) nanocomposites shows inconsistent results regarding the influence of the addition of nanofillers on the mobility of the system. In academic research, solvent based preparation methods are often used to prepare nanocomposites with the aim to obtain a good and controlled dispersion. However, little attention is paid to the influence of the used solvent on the properties of the nanocomposites. We show that in PMMA nanocomposites prepared via solution casting from different solvents, the apparent decrease in glass transition temperature is caused by insufficient drying and that when an adequate drying procedure is used, an increase in glass transition temperature is always observed.

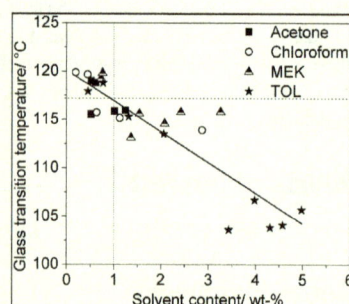

Keywords Nanocomposites, Polymer, Silica, Glass transition temperature, Solution casting, Processing

Introduction

It is now generally accepted that the introduction of nanofillers alters the mobility of the polymer matrix and that, as a consequence, the dynamics of polymer nanocomposites are significantly different from the neat polymer. In summary, systems with strong polymer filler interactions show a decreased mobility while in systems where the interactions are weak or non-existing an increased mobility is observed. The glass transition temperature (T_g) is a relatively easy material property to measure and it is often used as a measure of changes of the dynamics in a particular system. For a polymer such as poly(2-vinylpyridine) with a pronounced polarity, large increases in the glass transition temperature were found in composites with silica as well as alumina nanoparticles.[1,2] For apolar polymers such as poly(styrene) (PS) the absence of polar (side)groups implies a weak interaction with polar fillers and in these systems a decrease in glass transition temperature is normally observed with the introduction of nanofillers.[1-3] Following the reasoning presented above, a polymer such as poly(methyl methacrylate) (PMMA) is also expected to show an increase in glass transition temperature with the addition of a polar filler such as silica. However, contradictory results can be found in literature, where both increased, unchanged and decreased values of the glass transition temperature were reported. At this point, it is relevant to point out that solvent based methods are commonly employed to overcome dispersion issues in polymer nanocomposites. The polymer and the filler are dissolved/dispersed in a common solvent and the solvent is subsequently removed by solvent casting, spin coating or precipitation in a non-solvent. It is known that, similar to non-interacting fillers, also small diluents, molecules such as water or other solvents can significantly reduce the glass transition temperature of a polymer. However, even though solvent based methods are commonly applied to prepare nanocomposites, the literature is scarce on the potential effects of solvent retention on the properties of the obtained nanocomposites. The objective of this article is to discuss the influence of solvent retention on the glass transition temperature of PMMA based nanocomposites. First, relevant results from literature are presented. This is followed by a discussion on results from recent experiments from our group.

---

*Corresponding author, email t.peijs@qmul.ac.uk

Literature review

Glass transition temperature in PMMA/silica nanocomposites

Many groups reported an increase of the glass transition temperature in PMMA/silica nanocomposites[4–11] and the general explanation given is that strong interactions between the hydroxyl groups at the silica surface and oxygen of the acrylate groups reduce the mobility of the polymer chains, thereby increasing the T_g. Some groups, however, reported an unchanged value, or even a decrease in T_g with increased filler loading.[2,12,13] The explanations are varying. Ash et al.[14] attributed the decrease to residual solvent after the in situ polymerization, Fu et al.[13] stated that the polymer was in a non-equilibrium state after drying and that entrapped air increased the free volume in the polymer, thereby increasing the mobility and lowering the T_g. They showed that after annealing at elevated temperatures, a T_g similar to the one of the pure polymer was found. Rittigstein and Torkelson[2] observed a decrease in T_g when the samples were prepared by solvent casting in acetic acid and attributed this decrease to the screening of the interaction between the silica and the polymer by the acid molecules. When the samples were prepared in methyl ethyl ketone (MEK), no such screening took place, and an increase in T_g was found.

Glass transition temperature in other PMMA based nanocomposites

Several other nanosized particles have also been used in PMMA based nanocomposites. Ash et al.[14,15] prepared PMMA/alumina composites via in situ polymerization of MMA in the presence of alumina particles. At low filler loadings, the glass transition temperature remained unchanged, but upon increasing the loading, a decrease in T_g was found. The onset of the decrease was found to be dependent on the size of the filler, but when T_g was plotted as a function of the surface to volume ratio the curves overlapped. The same group also showed that if the particles are coated, the reduction disappears. A decrease of the glass transition temperature in alumina based PMMA nanocomposites was also confirmed by Rittigstein and Torkelson.[2]

Also gold particles have been used as filler material in PMMA composites. Srivastava and Basu[16] prepared different sizes of gold particles by in situ preparation of gold particles in the presence of PMMA. These PMMA capped particles were then mixed with PMMA homopolymer by using both good and bad solvents. Both increase and decrease in T_g were observed and were found to depend on the width of the interface region between the polymer matrix and the filler as calculated from SAXS data. This behavior was explained in terms of sharpness of the interface. A thin interfacial region gives a very sharp interface and the mean chain segment relaxation is dominated by the relaxation in the surface layer, which in this case is increased due to the non-interacting filler, and an overall decrease in T_g is observed. If, on the other hand, the interface is very diffuse, the relaxation is dominated by chain segments further away from the filler, which due to the presence of the filler have a lower mobility and an increase in T_g compared to an unfilled system is therefore observed.

In general, the changes in glass transition temperature in composites based on spherical or near spherical fillers are not so large, even at high filler loadings. In the case of anisotropic fillers, the situation is quite different. In nanocomposites made from PMMA and graphite prepared by Ramanathan et al.,[17] increases in the glass transition temperature of up to almost 40°C were observed. Also in PMMA/clay nanocomposites, large increases were found.[18,19] In Table 1, an overview of PMMA based nanocomposites and their corresponding glass transition temperatures can be found. As stated before, an increase in the glass transition temperature is observed in most cases. There are, however, some systems where a decrease was observed. Most notable are the samples spin coated in acidic acid which showed a decrease of almost 20°C.[2] In the table, the drying procedures applied before the analysis are also listed. It is observed that for samples showing large T_g reductions, no or very short drying procedures were applied and it can therefore be hypothesized that the observed decreases could also be due to remaining solvent. Therefore, also the influence of solvent retention in polymer films is included in the literature review.

Related research on glass transition temperature in thin films

Another research area where the effect of confinement on the glass transition temperature has been discussed thoroughly is the area of thin and ultrathin films. For more details, the reviews of the topic by Forrest and Dalnoki-Veress[25] and Alcoutlabi and McKenna[26] are recommended as only the main findings are cited here below. It was shown by several groups that the glass transition temperature in a free standing thin film decreases with decreasing film thickness.[27–31] Most of the research focused on thin films of PS, but a decrease in glass transition temperature was also observed in free standing PMMA films, even though the reduction is small compared to PS films.[29,30] The decrease in the glass transition temperature was attributed to an enhanced mobility at the surface of the film, and as the thickness of the film is decreased, the effect on the mobility of the whole sample is naturally increased. For supported films, the situation is different. It was shown that the glass transition temperature can increase,[32–36] decrease[33–42] or remain unaltered[29,32] depending on the interaction between the film and the substrate upon which the film is cast.[43] The magnitude of the shift was shown to depend not only on the thickness of the film, but also on the type of polymer and substrate. These results are in analogy with findings in nanocomposite research that show that the magnitude of the shift in glass transition temperature depends on the strength of the polymer–filler interaction. Supported thin films have therefore been successfully used as a simplified model for nanocomposites where the thickness of the film between supports can be assumed to represent the interparticle distance in the composite.[1,3] By varying the thickness of the film, different filler loadings can be simulated. For PMMA films, an increase in glass transition temperature with decreased film thickness is generally observed, even though a decrease was occasionally observed.

Table 1 Overview of PMMA based nanocomposites and their changes in glass transition temperature relative to PMMA bulk polymer

Group	Composite	Filler content (wt-%)	ΔT_g (°C)	Preparation method	Solvent	Drying time and temperature	Analysis method
Garcia et al.[4]	PMMA/silica	0–35	10 (max.)	Solution mixing	Toluene	12 h at 140°C	DSC
Moll and Kumar[11]	PMMA/silica	0–62.5	7 (max.)	Solution mixing	MEK	48 h at 150°C	DSC
Jouault et al.[20]	PMMA/silica	10 vol.-%	1 (at 5 vol.-%)	Solution mixing	DMAc	8 days at 130°C	DSC
Hub et al.[12]	PMMA/silica (14 nm)	4 / 8 / 12 / 10	−7* / 1* / −2* / 0* and 2** / 0** / 1** / 2**	Solution mixing	MEK	*Dried for 72 h at 140°C **Annealed for 72 h more at 140°C	DSC
Chinthamani-peta et al.[9]	PMMA/silica	0–23	~15 (max.)	In situ polymerization / In situ polymerization	... / THF	12 h at 80°C	DSC/DMTA
Zhang et al.[10]	PMMA/silica (vinyl functionalized)	0–10	13 (max.)		Water	24 h at 60°C	DSC
Hu et al.[6]	PMMA/fumed silica	0–4	~15 (max.)		...	1 h at 120°C	DMA
Sargsyan et al.[8]	PMMA/silica	0–65	6 (max.)	In situ preparation of silica	Water	8 h at 150°C	DSC
Li et al.[5]	PMMA/silica	0–15, 37	DSC: no change DMA: increase		THF	12 h at 60°C +12 h at 110°C	DSC/DMTA
Ferguson[61]	PMMA/silica	0–28	18 (max.)		THF/chloroform	5 days at 65°C	DMA
Fu et al.[13]	PMMA/silica	0–2	±2		THF	1 day at RT	DSC
Kyriakos et al.[21]	PMMA/silica	0–21.9	DSC: 6 (max.) TSDC: 10 (max.)		THF	6 h at 40°C +8 h at 140°C +2 h at 120°C	DSC/TSDC
Parker et al.[22]	PMMA/silica	0–5	DSC: ~14 (max.) Fluorescence: ~10 (average)	Solution mixing of grafted silica particles followed by precipitation	DMAc	Drying at 80°C	DSC/fluorescence
Rittigstein and Torkelsson[2]	PMMA/silica (10–15 nm)	0–0.6 vol.-%	−18 (max.)	Spin coating in acetic acid	Acetic acid	>21 days in fumehood	Fluorescence
	PMMA/silica (10–15 nm)	0–0.6 vol.-%	+6 (max.)	Spin coated in MEK	MEK		
	PMMA/alumina (47 nm)	1–10 vol.-%	−5 (max)	Spin coating in MEK	MEK		
Ash et al.[14,15]	PMMA/alumina (39 nm)	1–5	−24 (max.)	In situ polymerization	...	2 h at 115°C	DSC/DMTA
	PMMA/coated alumina (39 nm)	1–5	0				
Pandis et al.[23]	PMMA/silver	0–0.5	−9 (max.) +6 (42 Å grafts) −22 (1.4 Å grafts)	In situ polymerization	...	n.a.	DSC/DMA
Srivastava and Basu[16]	PMMA/gold	0–8		In situ polymerization of the particles in presence of PMMA and redistribution of grafted particles in PMMA matrix via solvent mixing	Acetone, toluene, acetone/water 1:10	n.a.	DSC
Meneghetti and Qutubuddin[18]	PMMA/MMT	10 / 10	18/16 / 8/8	Emulsion polymerization (exfoliated) / In situ polymerization (intercalated)	Deionized water	n.a.	DSC/DMTA / DSC/DMTA
Yeh et al.[19]	PMMA/MMT	0–5 / 0–5	22 (max.) / 16 (max.)	In situ emulsion polymerization / Solution mixing	Water / NMP	24 h at 50°C	DSC/DMTA / DSC

Table 1 Continued.

Group	Composite	Filler content (wt-%)	ΔT_g (°C)	Preparation method	Solvent	Drying time and temperature	Analysis method
Huang and Brittain[24]	PMMA/MMT	5 5 5	14 (compared to MM) 15 (compared to MM)	Mechanical mixing (MM) Suspension polymerization Emulsion polymerization	Water/MMA 5:1	Dried in vacuum at 60°C	DSC
Ramanathan et al.[17]	PMMA/graphite PMMA/expanded graphite PMMA/graphite nanoplatelets	1 5 1 5 1 5	43 30 20 0 10 30	Solution mixing and precipitation	THF+methanol	10 h at 80°C	DMA

Influence of solvent on glass transition temperature in thin films

In thin film research, the films are almost exclusively prepared by spin casting where centrifugal forces contribute to a fast solvent evaporation and enables formation of homogeneous thin films down to the thickness of a few nanometres. However, since vitrification of the film generally occurs before solvent removal, complete solvent removal is not obtained in the spin casting process. For an amorphous polymer such as PMMA, with a high glass transition temperature, it was shown that a glassy material is obtained once the solvent fraction in the film is less than 0.19, and the formation of the glass will considerably decrease the evaporation rate of the solvent.[44,45] The amount of retained solvent in different cast films was studied by several groups.[44,45,47–49] García-Turiel and Jérôme[47] used gas chromatography to study solvent retention in PS films spin cast from toluene. They concluded that while in thick films, there are only a few percent of solvents left after annealing at a temperature above the glass transition temperature, the amount of retained solvent can be much higher in thin films. Moreover, they found that the solvent molecules are mainly located at the interface between polymer and substrate. Their findings were supported by the observations of Perlich et al.[47] who used neutron reflectometry to assess the amount of retained toluene in spin cast PS films annealed at different temperatures, both above and below the glass transition temperature. They found that solvent retention was molecular weight dependent and increased with increasing molecular weight. Even though annealing at temperatures above the glass transition temperature reduced the amount of solvent in the samples, there was always residual solvent present in the samples, also after annealing.

Zhang et al.[48] also used neutron reflectometry to study the amount of toluene in spin coated films of both PS and PMMA. They observed that, while the PS films appeared solvent free already after drying at room temperature, PMMA films of a thickness of 121 nm contained 0.8 vol.-% of toluene even after two annealing steps. They also found that the toluene content in the PMMA films was thickness dependent, with a larger amount of toluene retained in the thinner films. This was attributed to strong interactions between the polymer, the solvent and the substrate upon which the polymer was cast. Strong interactions decrease the mobility of the PMMA and thus, the diffusivity of the solvent out of the polymer.

While many groups focused on the quantification of the amount of solvent in the films, less researchers reported on the influence of the solvent retention on the properties of cast films. In two different papers, Patra et al. studied the thermal and mechanical properties of PMMA films prepared from different solvents.[49,50] Their samples were dried for 4 or 15 days respectively at room temperature before tested with DSC. When the films were cast from toluene, THF or chloroform the glass transition temperatures were 20–35°C lower than bulk values, while it was 8°C higher in films cast from DMF. They explained the increase in the latter case with the ability of the solvent to form strong bonds with the carbonyl groups of the polymer thereby strengthening it. Bistac and Schultz studied the mobility in solution cast films of PMMA using DSC and dielectric spectroscopy.[51] They also used different solvents and concluded that the type of solvent largely influences the α and β relaxation temperatures of samples dried at room temperature for 48 h. In contradiction to Patra *et al.* who found similar values of the glass transition temperature for samples cast from toluene or chloroform, Bistac and Schultz found values differing by almost 20°C. Glass transition temperatures were 12 and 33°C lower than in bulk PMMA respectively. The difference was explained by the difference in strength of the acid base interaction between the PMMA and the different solvent molecules. These results were confirmed by Serghei and Kremer, who found a 37°C decrease in the dynamic glass transition temperature in films solvent cast from chloroform when no annealing was performed. However, after annealing at 127°C for 12 h under nitrogen atmosphere, the bulk PMMA dynamics were recovered.[52]

Another study of interest was made by Ellison *et al.*[53] and concentrated on the effect of small molecular diluents on the thickness dependence of the glass transition temperature in different films. They showed that the confinement effect in spin coated thin films could be reduced. This was shown both for non-interacting PS films which would normally show a decrease in glass transition temperature and for poly(2-vinylpyridine), a polymer that is very strongly interacting with the substrate, which displayed an increase in glass transition temperature. In the former, a thickness independent glass transition temperature equal to that of the bulk polymer was found and in the latter only a very modest increase was found compared to an undiluted system. Yet another study on the water sorption of poly(vinyl acetate) thin films shows that the water uptake suppresses the confinement effect noted in dry films (decrease in glass transition temperature with decreasing film thickness), and that the wet samples have a glass transition temperature that is actually higher than for dry samples.[54] It was also shown that annealing at elevated temperatures for more than 300 min was necessary to obtain dry samples.

In addition to the solvent retention effects, it is well known that the fast solvent evaporation used in the spin casting process can result in non-equilibrium conformations of the polymer chains and cause residual stresses which might contribute to altered dynamics.[55,56] The amount of residual stresses and the annealing time required to remove them depends on the solvent quality and the relaxation time towards an fully unperturbed state is very long.[57]

Influence of solvent on glass transition temperature in nanocomposites

Solvent based methods are also commonly applied in the preparation of nanocomposites. In order to improve the dispersion, the filler and the polymer are dispersed in a common solvent and thereafter the solvent is evaporated. Even though literature on nanocomposites is abundant, there is little information on how sample preparation affects the properties of the obtained nanocomposites. Sen *et al.* prepared PS samples by solvent casting and a fast evaporating method, and they concluded that the traces of solvent (around 0.2 wt-%) left in the samples after annealing were not responsible for the decrease in the

glass transition temperature found in the solvent cast samples. The explanation was that the samples prepared via their fast evaporation route contained similar amounts of residual solvent, but showed no glass transition temperature reduction.[58] For a very different system, i.e. a fluoroelastomer filled with carbon black and swollen in methyl ethyl ketone, it was shown that even at very high filler loadings (up to 35 wt-%), the filler did not influence the glass transition temperature of the system, which was solely dependent on the solvent fraction.[59] Ferguson worked on a more relevant system for our research.[60] He prepared PMMA films by solvent casting from different solvents where, in order to account for ethanol formation during the sol–gel process used to form silica from TEOS, also ethanol was added to the mixture. The influence of the drying procedure on properties such as the glass transition temperature was thereafter studied. Samples were dried in an oven at 65°C for several days and he concluded that it was impossible to remove all the solvent at this temperature. After 6 days of drying, about 4 wt-% of solvent was still present. The obtained nanocomposites showed an increase in glass transition temperature with increasing filler content, in accordance with most trends in literature, but it has to be mentioned that all the obtained values were lower than those found for reference PMMA films prepared from a solution without added ethanol.

In conclusion, it has been shown for both nanocomposites and thin films, that when an interacting filler or surface is present, a moderate increase in glass transition temperature in PMMA based systems can be expected. If the polarity of the matrix is increased, a larger increase in glass transition temperature is observed and if the polarity is decreased, the opposite is observed. There are, however, some examples where a decrease in glass transition temperature was found in PMMA composites. With the reasoning provided above, these findings are rather surprising. On the other hand, it was shown that solvents used in sample preparation are generally difficult to remove and a lower glass transition temperature as compared to the bulk polymer is often observed. It is therefore hypothesized that solvent retention is responsible for many of the observed reductions in glass transition temperature. In order to investigate the influence of the fillers on the properties of the prepared composite, a proper drying procedure is therefore needed. In the following section, experimental evidence for this hypothesis is put forward and it is shown that extensive drying is needed in order to observe the true influence of filler addition.

Experimental

Materials

PMMA v825 from Arkema, France, and the colloidal silica suspension from Nissan Chemicals, Japan in different solvents (MEK, toluene) were used as received. MEK, toluene, chloroform and acetone, all from Biosolve, Valkenswaard, the Netherlands, were used as received.

Sample preparation

The polymer was dissolved in the chosen solvent and for the nanocomposite samples colloidal silica in the same solvent was added to the dissolution. The samples were left on a shaker for two nights in order to fully dissolve and mix. Samples were subsequently poured into Petri dishes and left to evaporate at room temperature. Subsequently, they were dried in a vacuum oven under nitrogen flow at different temperatures and for varying periods of time.

Characterization techniques

A Q500 TGA from TA Instruments was used to confirm the filler loading in the samples and to assess how much solvent was present in the films. 10–15 mg of sample was heated to 600°C at a rate of 10°C min^{-1} under nitrogen atmosphere. For the determination of the solvent content after drying, the samples were heated to 150°C at a rate of 10°C min^{-1} and thereafter kept isothermally for 90 min. Thereafter, the samples were heated up to 450°C. The solvent content was set to be equal to the weight loss at 300°C. The molecular weight distributions both before and after drying experiments were determined using a size exclusion chromatography set-up from Waters with a Waters 510 pump and a Water 712 WISP chromatograph with an injection volume of 50 µL. The state of dispersion of the silica particles in the polymer matrix was studied with transmission electron microscopy. The measurements were performed using a FEI Tecnai 20 microscope, operated at 200 kV. Ultrathin sections were microtomed at room temperature and put on a copper grid with a carbon support layer. For the determination of the glass transition temperature via DSC measurements, a Q1000 DSC from TA Instruments was used. The samples were heated from room temperature to 200°C at a rate of 10°C min^{-1} and held at that temperature for 5 min. Thereafter, they were cooled to room temperature, also at a rate of 10°C min^{-1}. The cycle was repeated once. The glass transition temperature was determined from the second heating run. In addition, a Q800 DMA, also from TA Instruments, was used to determine the glass transition temperature. The samples were undergoing a sinusoidal deformation of an amplitude of 10 µm at a frequency of 1 Hz, and simultaneously heated from room temperature to 180°C at a rate of 3°C min^{-1}. The glass transition temperature was set to be the maximum of the storage modulus (E'') peak.

Results and discussion

First, the influence of residual solvent on PMMA films will be discussed and in a second step, the corresponding nanocomposites will be considered. Some characteristics for the different solvents used in the sample preparation are listed in Table 2.

The samples were solvent cast into Petri dishes, left to evaporate in a fume hood for 1 week and thereafter dried in a vacuum oven at 120 or 140°C for one to five nights. Directly after solvent evaporation in the open air, DMA measurements were performed on the as cast films. The temperature at the maximum of tan δ and E'' are reported in Fig. 1 together with the solvent content as measured from TGA experiments. Even though the maximum of tan δ often is used as an equivalence to the glass transition temperature, E'' is corresponding better to the values obtained in DSC

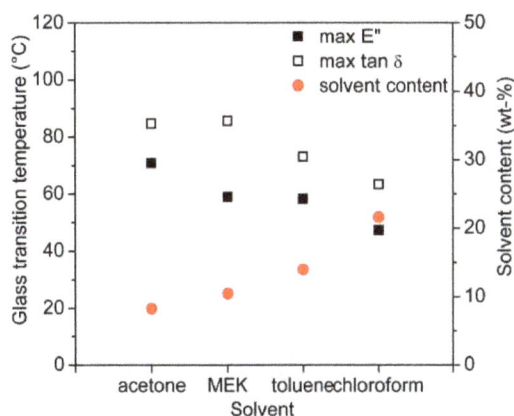

Figure 1 Glass transition temperature for solvent cast films prepared from different solvents: solvent content as measured with TGA is also plotted in graph

measurements and will hereafter be used as a measure of the glass transition temperature.

In Fig. 1, it is observed that all the samples still contain more than 10 wt-% solvent after 1 week evaporation at room temperature, and that in chloroform, there is as much as 20 wt-% solvent left after the evaporation step. Moreover, the glass transition temperatures in the as cast films are far below the temperature measured for neat PMMA ($\sim$117°C) and it decreases below 50°C in samples prepared from chloroform. Bistac and Shultz attributed the higher solvent retention in PMMA films prepared in chloroform compared to films prepared in toluene or acetone to the strong acid base interactions between the basic PMMA and the acidic chloroform.[51] Acetone and toluene belong to a group of lightly basic solvents and do therefore only present weak interactions with PMMA. Similar results were obtained by Patra et al., who attributed the difference between toluene and chloroform to a weaker interaction with the PMMA molecules in the former.[50] From these results, it is evident that the drying of samples at elevated temperatures before measurements is important, and that the drying procedure needs to be carefully reviewed when the solvent is changed.

Figure 2a shows the TGA traces measured on films prepared using MEK as solvent. Directly after solvent evaporation in air the samples contain a visible amount of solvent, but drying at 120°C for one night reduces this amount significantly and further drying does bring modest improvement. When the drying is performed at 140°C, the sample already appears dry after one night in the oven. In

Table 2 Presentation of different materials used in drying experiment

Solvent	Boiling point (°C)	Density (g cm^{-3})
Acetone	56	0.79
Chloroform	61	1.48
MEK	80	0.81
Toluene	110	0.87

Polymer	Glass transition temperature (°C)	Density (g cm^{-3})
PMMA	118	1.18

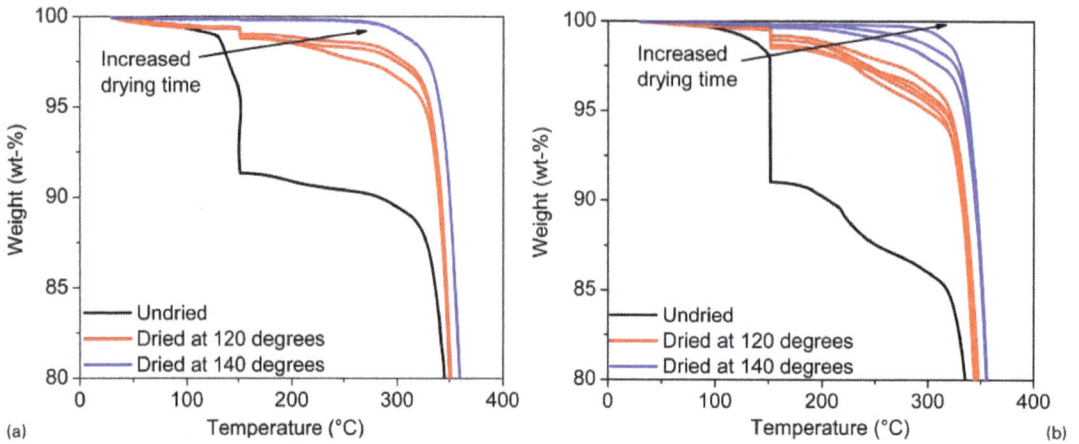

Figure 2 TGA results for samples dried with different drying procedures: weight loss as function of temperature for *a* samples prepared with MEK and *b* samples prepared with toluene

Fig. 2*b*, TGA results for the samples prepared in toluene are presented. After solvent evaporation, there is still a considerable amount of solvent present in the samples. Drying at 120°C for one night removes some of the solvent and further drying at 120°C does decrease the solvent content further but this is still not to an acceptable level. Drying overnight at 140°C is more efficient than drying for five nights at 120°C and further drying will reduce the solvent content, but only after three nights at 140°C the sample contains less than 1 wt-% solvent.

In Fig. 3*a*, the solvent content after various times of drying and in Fig. 3*b*, the corresponding glass transition temperatures for samples prepared with different solvents are presented. When drying is carried out at elevated temperatures, the mobility of the solvent becomes important. Toluene has a high boiling point and is therefore less mobile than the other solvents at the drying temperature and the complete solvent removal is only possible when the drying temperature is increased to 140°C. Solvents with low boiling points such as acetone and chloroform are easily removed at both drying temperatures. In Fig. 4, the glass

transition temperature is plotted as a function of solvent content for the different solvents. The type of solvent determines how easily the residual solvent can be removed from the sample, but all the data can be collapsed onto a curve showing a linear dependence of the glass transition temperature on the solvent content. Only when the samples contain less than 1 wt-% of solvent, the glass transition temperature of the neat polymer is recovered.

Influence of silica on glass transition temperature of polymer nanocomposite films

Samples with 5 wt-% of silica was prepared using either MEK or toluene as a solvent. In Fig. 5, the transmission electron microscopy pictures of the as prepared systems can be found. The use of these two solvents results in a very different state of dispersion. The samples prepared in MEK show a good dispersion with individually dispersed particles, while the silica particles in the samples prepared using toluene are highly agglomerated. In the same figure, samples dried for 5 days at 140°C are also presented. The state of dispersion is similar to the samples measured before drying, indicating that

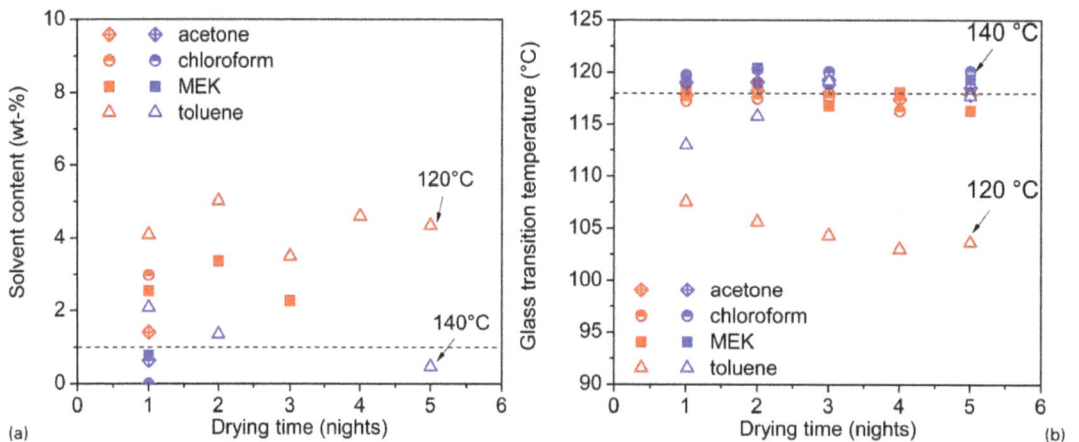

Figure 3 *a* solvent content as function of drying time in nights for films prepared with different solvents and *b* corresponding glass transition temperature: red symbols correspond to drying temperature of 120°C and blue symbols correspond to drying temperature of 140°C

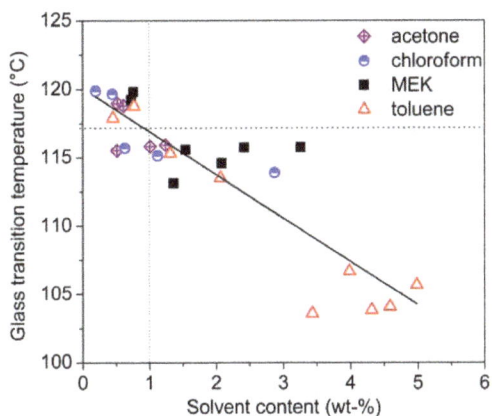

Figure 4 **Glass transition temperature as function of solvent content as function of for films prepared with different solvents and dried at different temperatures: dotted lines are guides for eye showing glass transition temperature for neat PMMA and 1 wt-% solvent limit; solid line corresponds to linear fit of data**

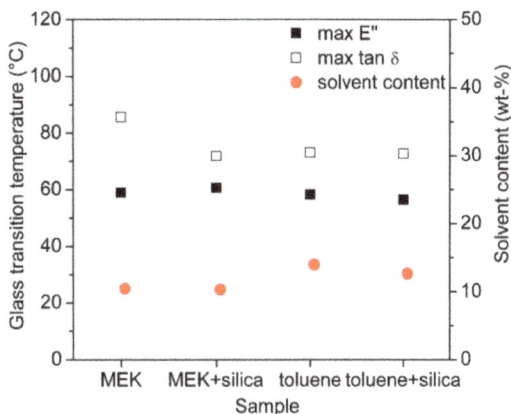

Figure 6 **Glass transition temperature and solvent content of air dried samples: results for both filled and unfilled samples prepared using either MEK or toluene are presented**

the drying procedure has no influence on the state of the dispersion to the samples and that changes in glass transition temperature upon drying have a different origin.

The glass transition temperature as measured by DMTA and the solvent content in the as cast samples of samples containing 5 wt-% of silica can be seen in Fig. 6. The data of the unfilled samples are added as reference. The filled and unfilled samples contain similar amounts of solvent, and the samples prepared in toluene contain more solvent than the samples cast in MEK. All the samples do again show a lower glass transition temperature than the neat PMMA (117°C) and there is no significant effect of filler addition on the glass transition temperature, as measured from the maximum of the E''. The glass transition as determined from the maximum of the tan δ is also added as a reference. It is noted that the tan δ of the unfilled MEK sample is much higher than the tan δ of the filled samples, and that a large decrease upon filler addition would be measured if this peak would be used to quantify the glass transition temperature. This result shows that not only the sample preparation but also the analysis method and the interpretation of the data are important issues to consider in order assessing the real effect of filler addition.

In Fig. 7a, the difference in glass transition temperature between the neat film and the nanocomposite prepared in

the same solvent is presented as a function of the drying time. For samples prepared in MEK, a modest increase in glass transition temperature with the increase in drying time is observed, which is similar for both drying temperatures. However, for the sample dried in toluene, a decrease in glass transition temperature upon filler addition is initially observed, whereas a large increase is observed in samples dried at the higher temperature. After 3 days of drying at either temperature, a similar trend as in the MEK samples is observed. This is in line with the results presented by Fu *et al.* who showed that an initial decrease in the glass transition temperature disappeared after the samples was annealed for an additional time at 140°C.[13] The glass transition temperature as a function of solvent fraction is presented in Fig. 7b. The difference in glass transition temperature is increased with decreasing amount of residual solvent and that for both the solvents used, an increase in glass transition temperature upon filler addition is found when the samples are dry, consistent with the majority of the results in literature (Table 1).

Finally, in Fig. 8, the differences in glass transition temperature between PMMA/silica nanocomposites with different filler contents and neat PMMA are presented. The samples prepared in MEK (dried for two nights at 120°C) show a good dispersion and a modest increase in glass transition temperature with increasing filler content. The

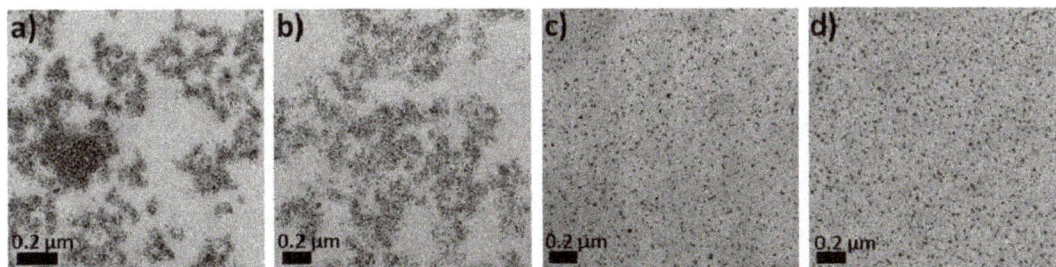

Figure 5 *a* PMMA+5 wt-% silica in toluene, undried, *b* PMMA+5 wt-% silica in toluene, dried, *c* PMMA+5 wt-% silica in MEK, undried and *d* PMMA+5 wt-% silica in MEK, dried

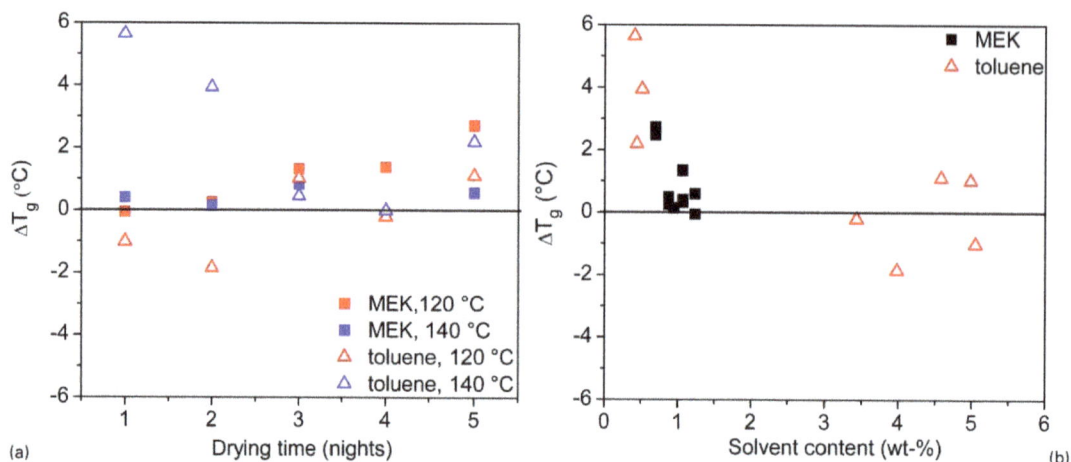

(a)

(b)

Figure 7 Difference in glass transition temperature between samples with 5 wt-% silica and films of neat PMMA expressed as function of *a* drying time and *b* solvent content for films prepared with different solvents and dried at different temperatures

samples prepared in toluene (dried for two nights at 120°C and an additional two nights at 140°C) do also show a modest increase in glass transition temperature as compared to the neat polymer. However, as shown in Fig. 5, samples prepared from a toluene solution are prone to agglomeration and, as a consequence, the glass transition temperature is decreasing again upon increased filler content.

Conclusions

In this paper, we show that solvent retention in nanocomposites containing PMMA and silica nanoparticles can lead to a decrease in the glass transition temperature. This effect counteracts the effect of the filler addition and might lead to misinterpretations of the real influence of the filler addition. The drying step in a process method including the use of solvents has a major influence on the properties of the prepared nanocomposites and has to be carefully adapted to the employed polymer–filler–solvent system. It is concluded

that improper matching of the drying time and temperature with the investigated polymer–filler system can partly explain the scattered results of the influence of filler addition on the glass transition temperature reported in the literature.

Conflicts of interest

The authors have no conflicts of interest to declare.

Acknowledgements

The authors wish to acknowledge Anne Spoelstra for the help with the TEM images.

References

1. P. Rittigstein, R. D. Priestley, L. J. Broadbelt and J. M. Torkelson: *Nat. Mater.*, 2007, **6**, 278–282.
2. P. Rittigstein and J. M. Torkelson: *J. Polym. Sci. Part B: Polym. Phys.*, 2006, **44B**, 2935–2943.
3. A. Bansal, H. Yang, C. Li, K. Cho, B. C. Benicewicz, S. K. Kumar and L. S. Schadler: *Nat. Mater.*, 2005, **4**, 693–698.
4. N. García, T. Corrales, J. Guzmán and P. Tiemblo: *Polym. Degrad. Stabil.*, 2007, **92**, 635–643.
5. C. Li, J. Wu, J. Zhao, D. Zhao and Q. Fan: *Eur. Polym. J.*, 2004, **40**, 1807–1814.
6. Y.-H. Hu, C.-Y. Chen and C.-C. Wang: *Polym. Degrad. Stabil.*, 2004, **84**, 545–553.
7. R. Avolio, G. Gentile, M. Avella, D. Capitani and M. E. Errico: *J. Polym. Sci. Part A: Polym. Chem.*, 2010, **48A**, 5618–5629.
8. A. Sargsyan, A. Tonoyan, S. Davtyan and C. Schick: *Eur. Polym. J.*, 2007, **43**, 3113–3127.
9. P. S. Chinthamanipeta, S. Kobukata, H. Nakata and D. A. Shipp: *Polymer*, 2008, **49**, 5636–5642.
10. F.-A. Zhang, D.-K. Lee and T. J. Pinnavaia: *Polymer*, 2009, **50**, 4768–4774.
11. J. Moll and S. R. Kumar: *Macromolecules*, 2012, **45**, 1131–1135.
12. C. Hub, S. E. Harton, M. A. Hunt, R. Fink and H. Ade: *J. Polym. Sci. Part B: Polym. Phys.*, 2007, **45B**, 2270–2276.
13. H. P. Fu, R. Y. Hong, Y. J. Zhang, H. Z. Li, B. Xu and Y. Zheng: *Polym. Adv. Technol.*, 2009, **20**, 84–91.
14. B. J. Ash, R. W. Siegel and L. S. Schadler: *J. Polym. Sci. Part B: Polym. Phys.*, 2004, **42B**, 4371–4383.
15. B. J. Ash, R. W. Siegel and L. S. Schadler: *Macromolecules*, 2004, **37**, 1358–1369.
16. S. Srivastava and J. K. Basu: *Phys. Rev. Lett.*, 2007, **98**, 165701.
17. T. Ramanathan, S. Stankovich, D. A. Dikin, H. Liu, S. T. Shen and L. C. Brinson: *J. Polym. Sci. Part B: Polym. Phys.*, 2007, **45B**, 2097–2112.
18. P. Meneghetti and S. Qutubuddin: *Langmuir*, 2004, **20**, 3424–3430.

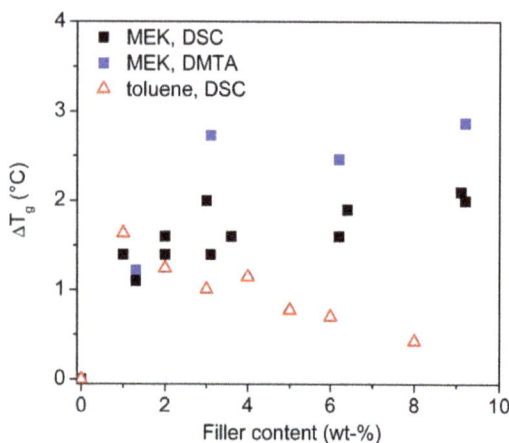

Figure 8 Glass transition temperature as function of filler content for nanocomposites of PMMA and silica prepared using either MEK or toluene as solvent

19. J. M. Yeh, S.-J. Liou, M.-C. Lai, Y.-W. Chang, C.-Y. Huang, C.-P. Chen, J.-H. Jaw, T.-Y. Tsai and Y-H. Yu: *J. Appl. Polym. Sci.*, 2004, **94**, 1936–1946.

20. N. Jouault, F. Dalmas, F. Boué and J. Jestin: *Polymer*, 2012, **53**, 761–775.

21. K. Kyriakos, K. R. Raftopoulos, P. Pissis, A. Kyritsis, F. Näther, L. Häußler, D. Fischer, A. Vyalikh, U. Scheler, U. Reuter and D. Pospiech: *J. Appl. Polym. Sci.*, 2013, **128**, 3771–3781.

22. K. Parker, R. T. Schneider, R. W. Siegel, R. Ozisik, J. Cabanelas, B. Serrano, C. Antonelli and J. Baselga: *Polymer*, 2010, **51**, 4891–4898.

23. Ch. Pandis, E. Logakis, A. Kyritsis, P. Pissis, V. V. Vodnik, E. Džunuzović, J. M. Nedeljković, V. Djoković, J. C. Rodriguez Hernez and J. L. Gomez Ribelles: *Eur. Polym. J.*, 2011, **47**, 1514–1525.

24. X. Huang and W. J. Brittain: *Macromolecules*, 2001, **34**, 3255–3260.

25. J. A. Forrest and K. Dalnoki-Veress: *Adv. Colloid. Interface Sci.*, 2001, **94**, 167–195.

26. M. Alcoutlabi and G. B. McKenna: *J. Phys.: Condens. Matter*, 2005, **17**, R461–R524.

27. J. S. Sharp and J. A. Forrest: *Phys. Rev. Lett.*, 2003, **91**, 235701.

28. C. B. Roth and J. R. Dutcher: *Eur. Phys. J. E*, 2003, **12E**, S103–S107.

29. C. B. Roth and J. R. Dutcher: *J. Electroanal. Chem.*, 2005, **584**, 13–22.

30. V. M. Boucher, D. Cangialosi, H. Yin, A. Schönhals, A. Alegría and J. Colmenero: *Soft Matter*, 2012, **8**, 5119–5122.

31. Y. Grohens, L. Hamon, G. Reiter, A. Soldera and Y. Holl: *Eur. Phys. J. E*, 2002, **8E**, 217–224.

32. J. L. Keddie, R. A. L. Jones and R. A. Cory: *Faraday Discuss.*, 1994, **98**, 219–230.

33. D. S. Fryer, P. F. Nealey and J. J. de Pablo: *Macromolecules*, 2000, **33**, 6439–6447.

34. D. S. Fryer, R. D. Peters, E. J. Kim, J. E. Tomaszewski, J. J. de Pablo, P. F. Nealey, C. C. White and W. Wu: *Macromolecules*, 2001, **34**, 5627–5634.

35. D. Labahn, R. Mix and A. Schönhals: *Phys. Rev. Lett. E*, 2009, **79E**, 011801:1–011801:9.

36. R. N. Li, F. Chen, C.-H. Lam and O. K. C. Tsui: *Macromolecules*, 2013, **46**, 7889–7893.

37. O. Prucker: *Macromol. Chem. Phys.*, 1998, **199**, 1435–1444.

38. J. H. Kim, J. Jang and W.-C. Zin: *Langmuir*, 2000, **16**, 4064–4067.

39. J. H. Kim, J. Jang and W.-C. Zin: *Langmuir*, 2001, **17**, 2703–2710.

40. L. Singh, P. J. Ludovice and C. L. Henderson: *Thin Solid Films*, 2004, **449**, 231–241.

41. J. S. Sharp, J. H. Teichroeb and J. A. Forrest: *Eur. Phys. J. E*, 2004, **15E**, 473–487.

42. S. Kawana and R. A. L. Jones: *Phys. Rev. E*, 2001, **63E**, 021501.

43. G. B. de Maggio, W. E. Frieze, D. W. Gidley, M. Zhu, H. A. Hristov and A. F. Yee: *Phys. Rev. Lett.*, 1997, **78**, 1524–1527.

44. M. O. Ngui and S. K. Mallapragada: *J. Appl. Polym. Sci.*, 1999, **72**, 1932–1920.

45. H. Richardson, I. Lopez-Garcia, M. Sferrazza and J. L. Keddie: *Phys. Rev. E*, 2004, **70**, 051805:1–051805:2.

46. J. García-Turiel and B. Jérôme: *Colloid. Polym. Sci.*, 2007, **285**, 1617–1623.

47. J. Perlich, V. Körstgens, E. Metwalli, L. Schulz, R. Georgii and P. Müller-Buschbaum: *Macromolecules*, 2009, **42**, 337–344.

48. X. Zhang, K. G. Yager, S. Kang, N. J. Fredin, B. Akgun, S. Satija, J. F. Douglas, A. L. Karim and R. L. Jones: *Macromolecules*, 2009, **43**, 1117–1123.

49. N. Patra, A. C. Barone and M. Salerno: *Adv. Polym. Technol.*, 2011, **30**, 12–20.

50. N. Patra, M. Salerno, A. Diaspro and A. Athanassiou: *J. Mater. Sci.*, 2011, **46**, 5044–5049.

51. S. Bistac and J. Schultz: *Int. J. Adhes. Adhes.*, 1997, **17**, 197–201.

52. A. Serghei and F. Kremer: *Macromol. Chem. Phys.*, 2008, **209**, 810–817.

53. C. J. Ellison, R. L. Ruszkowski, N. J. Fredin and J. M. Torkelson: *Phys. Rev. Lett.* 2004, **92**, 095702:1–095702:4.

54. S. Kim, M. K. Mundra, C. B. Roth and J. M. Torkelson: *Macromolecules*, 2010, **43**, 5158–5161.

55. K. R. Thomas, A. Chenneviere, G. Reiter and U. Steiner: *Phys. Rev. E*, 2011, **83E**, 021804:1–021804:8.

56. J. Y. Chung, T. Q. Chastek, M. J. Fasolka, H. W. Ro and C. M. Stafford: *ACN Nano*, 2009, **4**, 844–852.

57. A. Raegen, M. Chowdhury, C. Calers, A. Schmatulla, U. Steiner and G. Reiter: *Phys. Rev. Lett.*, 2010, **105**, 227801:1–227801:4.

58. S. Sen, Y. Xie, A. Bansal, H. Yang, K. Cho, L. S. Schadler and S. K. Kumar: *Eur. Phys. J.: Spec. Top.*, 2007, **141**, 161–165.

59. M. C. Righetti, M. Arjoldi, M. Vitali and G. Pezzin: *J. Appl. Polym. Sci.*, 1999, **73**, 377–384.

60. M. L. Ferguson: 'Preparation and characterization of polym.–SiO₂ hybrids', Oregon State University, Corvallis, OR, USA, 2011.

14

The loading rate effect on Mode II fracture toughness of composites interleaved with CNT

T. Lyashenko-Miller*[1], J. Fitoussi[2] and G. Marom[1]

[1]Casali Center of Applied Chemistry, The Institute of Chemistry and the Center for Nanoscience and Nanotechnology, The Hebrew University of Jerusalem, Jerusalem, Israel
[2]Arts et Métiers ParisTech, PIMM – UMR CNRS 8006, 151 Boulevard de l'Hôpital, 75013 Paris, France

Abstract The loading rate effect on Mode II interlaminar fracture toughness (ILFT) is examined in this study with interleaved epoxy/carbon fabric laminates tested under dynamic conditions. Specifically, an SP1 protein-treated carbon nanotube-reinforced epoxy leaf is inserted at the midplane of the laminates, and the fracture properties are measured by the crack lap shear method at two different loading rates. Whereas our preliminary study performed under quasi-static conditions showed that this specific interleaving generated an ~85% improvement in the Mode II ILFT, the current work shows that the occurrence and magnitude of the improvement depend on the loading rate and crack velocity, with different effects on the initiation and the instable/stable propagation stages. Improvements in Mode II ILFT for both the crack initiation and propagation phases can reach up to ~145% for certain dynamic loading conditions.

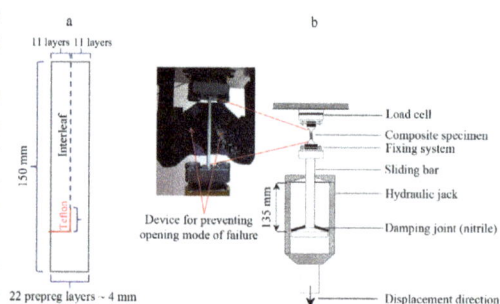

Keywords Nanocomposites, Carbon nanotubes, Fracture toughness, High-rate loadings, Delamination, Interleaving

Introduction

One of the common failures in laminated composite materials is delamination caused by microcracks in the laminated structure that grow under a sufficiently high shear stress. Improvement of the mechanical properties in general and particularly of the interlaminar fracture toughness (ILFT) is one of the most investigated issues in the field of composite materials. Delamination of composite structures under Mode I (opening mode of failure where the stress is perpendicular to the crack) or Mode II (shear mode of failure where the stress is in parallel to the crack) loadings is frequent and has been studied extensively, wherein different configurations of the edge notch fracture (ENF) test are utilized in opening and shear modes, respectively. A limitation of this test is that it is often performed under slow loading in quasi-static conditions that do not simulate adequate practical scenarios. In fact, the service conditions of many composite structures, such as those that are used in aircraft, marine, automobile, and sports industries, generate fatigue-related behavior and dynamic crack growth.

Over the years, different solutions have been utilized for the delamination problem, working to increase the fracture toughness of laminated composites under a shear stress field. In many of them, the increased delamination resistance has been accompanied by a deterioration of other mechanical properties, such as strength and stiffness.[1,2] More recently, with the remarkable progress in nanotechnologies, minute quantities of different types of nanoreinforcement, e.g., CNT, have been mixed into the polymer matrices to produce improved ILFT without a penalty of reduction of other essential mechanical properties and with no weight gain.[3,4]

Accordingly, in our preceding study, we have applied a modified interleaving technology, based on the original concept of placing a thermoplastic polymer leaf at the midplane of the laminated composite.[5] The guiding concept of our study was that nanoparticles for improved fracture toughness should be placed specifically at critical zones of high stress concentrations, e.g., regions of potential delamination, instead of dispersing them wastefully throughout the matrix. Thus, a thin leaf of epoxy/protein-treated carbon nanotubes (t-CNT) was inserted in the midplane of a carbon fabric–epoxy composite laminate (see Fig. 1), generating a remarkable improvement in the ILFT without deteriorating the static properties.[6] Essential details of the SP1 protein can be found in the literature.[7]

*Corresponding author, email tatiana.liashenko@mail.huji.ac.il

Because the ability to resist crack propagation in composite materials depends on the interfacial properties of the matrix and fibers, the environmental conditions and the loading rate and mode, an effective test method is required, which is more sensitive than the classic ENF test to the broad range of such factors that affect fracture toughness.[8,9] Regarding the rate factor, there are many studies in the literature of the behavior of polymer composites under dynamic conditions, which cover different materials, test specimens, and modes of loading.[9–15] The diversity of those studies and of their results indicates that such studies ought to be case specific; hence, every material and testing configurations must be studied particularly.

In view of that, the present research was undertaken to broaden the scope of our previous work[6] – on Mode II fracture toughness of a CNT interleaved laminate (described in Fig. 1) – to study the effect of loading rate. However, unlike the studies on rate effects cited above, where conventional test methods (such as Mode II center-notch-flexure under impact

loading, or end-notch-flexure under different loading rates or under drop-weight impact), here the crack lap shear (CLS) test under extremely controlled rate and failure mode conditions was employed.

Experimental section

Materials and preparation of samples

Prepregs of J.D. Lincoln, INC, product name L-930 were used. The prepregs are based on a carbon fiber fabric (50–70 wt.%)/bisphenol A/epichlorohydrin-based epoxy resin with a ply thickness of approximately 0.25 mm. Epoxy LY1556 with anhydride hardener HY917 and accelerator DY070 was chosen as the matrix for the interleaf.

Multi-walled carbon nanotubes C-100 (CNT) (Arkema), functionalized in a non-covalent mode by SP1 protein with a maximal CNT-to-protein weight-to-weight ratio of 20:1 (e.g., t-CNT), were supplied by SP Nano Ltd. Additional information on the SP1 protein and the CNT treatment process can be found in[7]. The wt.% of t-CNT in the epoxy matrix was 0.5 wt.%, wt.% of CNT pristine in the epoxy matrix was 0.5 wt.%, and of SP1 0.025 wt.%.

One gram of t-CNT or CNT powder was added to 100 g of LY1556 epoxy. Dispersion by manual mixing was followed by probe sonication using Hielscher Ultrasonics processor UIP1000 (1000 W) for 10 min.

The specimens were prepared by stacking of twenty-two layers of prepreg fabric in a [0/90] configuration into $150 \times 150 \times 4$ mm laminates as presented in Fig. 1. In order to form an artificial initiation crack, a Teflon film of 15 mm width and 20 μm thickness was inserted in the laminate between the 11th and 12th plies (at the midplane) as shown (Fig. 2a). The non-cured interleaves, which were prepared by mixing of all the compounds (LY1556 pure or with the t-CNT/CNT pristine/SP1 protein addition/HY917/DY070) in a separate flask, were introduced utilizing small brush at the midplane of the

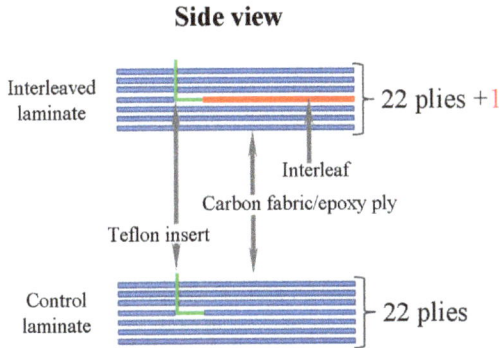

Side view

Figure 1 Schematic presentation of interleaved and non-interleaved laminates

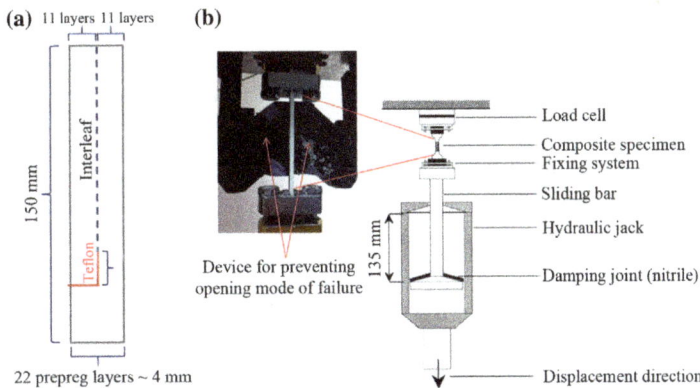

Figure 2 *a* Side view of CLS specimen and *b* tensile instrument and close-up of specimen used in performing CLS tests

Table 1 Composite laminates and their nomenclature

Sample	Interleaf	% Nanoreinforcement in the interleaf	Dispersion method
Control	N/A	N/A	N/A
Epoxy	LY1556/HY917	N/A	N/A
CNT	LY1556/HY917	0.5 wt.% CNT	Ultrasonic energy
SP1	LY1556/HY917	0.025 wt.% SP1	Ultrasonic energy
t-CNT	LY1556/HY917	0.5 wt.% CNT/SP1	Ultrasonic energy

laminate from edge to edge (Fig. 1). Curing was performed under a pressure of 1.7 MPa at 130 °C for 30 min. The laminate was cut to the specimens of dimensions $150 \times 10 \times 4$ mm. Five different types of laminates were prepared. The laminates were labeled according to the interlayer type as summarized in Table 1.

CLS test

CLS test was performed in order to examine Mode II (shear) fracture behavior of the composite materials, although usually this kind of test is used in order to calculate mixed-mode (I and II) fracture toughness. The specimen for CLS test is presented in Fig. 2a and consists of 2 parts: lap and strap. During the loading (tension), the crack propagates between lap and strap, while strap is fixed at the lower extremity, as shown in Fig. 2b.

High strain-rate tensile tests

High strain-rate tensile tests were conducted using servo-hydraulic machine at different strain rates until the total failure of composite specimen. The test machine is equipped with a launching system. The composite specimen is positioned between the load cell (upper extremity) and the moving device (lower extremity) as presented in Fig. 2b. Prior to the contact between the sliding bar and the hydraulic jack, the latter one is accelerated over a straight displacement of 135 mm in order to reach the nominal crosshead velocity before loading begins. Once the contact occurs, the specimen is then subjected to a tension at a constant load rate. The damping joint placed between the slide and the hydraulic jack may attenuate partially the wave effects caused by the dynamic shock.

In order to prevent the mixed mode of failure of the composites and ensure the maximum shear failure, special 'arms' were used (Fig. 2b) which suppressed the Mode I component during the loading.

Ultra high-speed camera was used during each test to record all specimen failures and calculate crack extension and speed.

The CLS specimens were loaded using servo-hydraulic machine at load rate of 4 and 8 m/s. Load–displacement records were obtained during the test, and the crack front location was recorded during crack propagation. The energy release rate (G_{II}) was calculated according to[16–18]:

$$G_{II} = \frac{P_c^2}{2b^2 Et} \tag{1}$$

where P_c is the critical load for crack initiation/propagation, b is the specimen width, E is the laminate modulus, and t is the total thickness.

The G_{II} values for initiation were calculated from the critical P_c value at the onset of pre-crack propagation.

Morphology characterization

The fracture morphology of the specimens and t-CNT exfoliation in the epoxy matrix was characterized utilizing a high-resolution scanning electron microscope Sirion, and the operating voltage was 5.00 kV. The samples were coated with an Au–Pd nano-layer using a SC7640 Sputter.

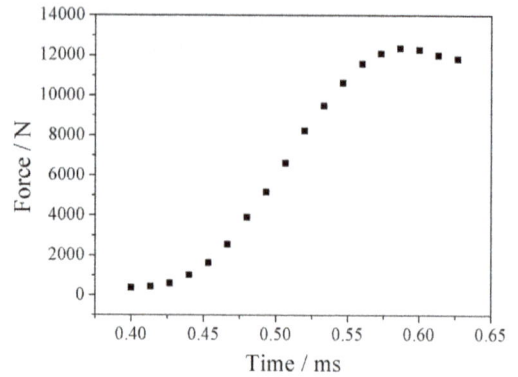

Figure 3 Typical force–time curve for the Mode II fracture toughness dynamic test

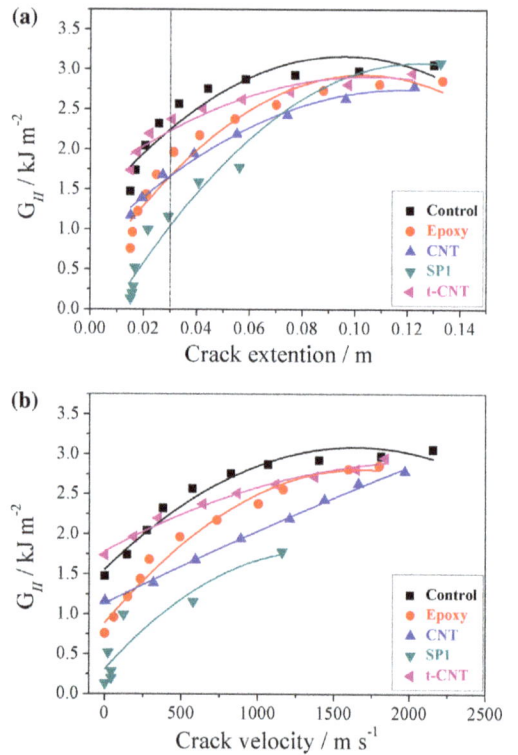

Figure 4 Comparison of G_{II} values as a function of crack extension a and crack velocity b of five types of specimens at a loading rate of 4 m/s

The quantitative measurements of surface roughness were performed utilizing Scanning Probe Microscope (SPM) Dimension 3100 Nanoscope V.

Results and discussion

We have recently shown that under quasi-static conditions by interleaving of a protein-treated CNT/epoxy layer at the midplane of a carbon fabric-reinforced epoxy laminate, the Mode II ILFT of the composite material is increased by 85% ($\sim$3.48 kJ/m^2) compared to the control laminate without an interleaf ($\sim$1.88 kJ/m^2).[6] In view of the results under quasi-static conditions, it became appropriate to broaden the study to include dynamic conditions that are more relevant to typical service conditions of such structures. Also, as pointed out above, dynamic measurements might have an additional benefit by uncovering new characteristics of the composite material – not seen under quasi-static loadings. As previously stated,[6] we focus on the t-CNT sample in comparison with four reference samples that consist of interleaves based on the separate constituents of the t-CNT leaf.

Figure 3 presents a typical force–time curve as received during the dynamic Mode II loading (the specific curve is for control specimen loaded at crosshead speed of 8 m/s). Figs. 4 and 5 present the plots of $G_{||}$ as a function of the crack extension and velocity at two different loading rates – 4 m/s (Fig. 4a and b) and 8 m/s (Fig. 5a and b), respectively – of the five samples that have been examined (Table 1). The results represent testing of two similar sets. The plots constitute the crack growth resistant curves (R-curves) that characterize the points of fracture initiation and the regions of stable (linear) and unstable (non-linear) crack growth prior and beyond the dotted lines, respectively, as shown in Fig. 4a and 5a. The points where the vertical line intercept the traces (the instability points) are related to the fracture toughness of each composite laminate, which is an intrinsic feature of the material.[19]

First, considering the onset of fracture, it is seen that the t-CNT interleaved laminate exhibits higher $G_{||}$ of ILFT of initiation values: at a loading rate of 4 m/s, the fracture resistance increases from $\sim$1.5 kJ/m^2 (control) to $\sim$1.75 kJ/m^2 (t-CNT), which is $\sim$17% improvement (Fig. 4a); and at a loading rate of 8 m/s, the initiation $G_{||}$ increases from $\sim$0.22 kJ/m^2 (control) to $\sim$0.54 kJ/m^2 (t-CNT), which is $\sim$145% improvement (Fig. 5a).

Generally, as can be seen in the graphs of Figs. 4 and 5, the fracture resistance increases as the crack extension/velocity increase for all the samples. There is, however, a number of noticeable differences – related mostly to the loading rates. Under the lower loading rate (Fig. 4), the crack instability point is reached at around a crack extension of 0.03 m, beyond which $G_{||}$ approaches a plateau. Below and at the instability point, a small improvement ($\sim$17%) of the ILFT $G_{||}$ from $\sim$1.5 kJ/m^2 (control) to $\sim$1.75 kJ/m^2 (t-CNT) is observed; thereafter, the control sample exhibits the highest fracture toughness values throughout the crack extension/velocity range. Under the higher loading rate (Fig. 5), the crack instability point is reached at a significantly higher crack extension of about 0.10 m, beyond which $G_{||}$ continues to increase monotonically. Here, the t-CNT interleaf contributes to a dramatic increase in the fracture toughness of about 70% compared to the control along the whole crack extension/velocity range. Obviously, the loading rate affects the fracture resistance through its effect on the crack stability and velocity.

The straightforward expression of crack stability is its velocity. Looking specifically at the $G_{||}$-crack velocity relationship (Figs. 4b and 5b), we see that under the two loading rates (4 and 8 m/s), the crack velocity increases – and so do the $G_{||}$ values – as the initial crack propagates, until a steady-state plateau (that is more distinct at 4 m/s loading) is attained. The crack extension starts at a relatively low velocity and as the fracture process proceeds the crack velocity grows. Under dynamic loading, the crack initiation occurs before the critical

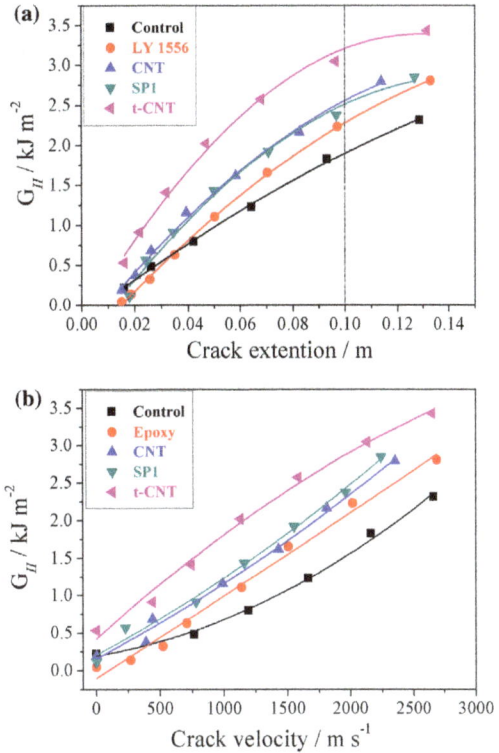

Figure 5 Comparison of $G_{||}$ values as a function of crack extension *a* and crack velocity *b* of five types of specimens at a loading rate of 8 m/s

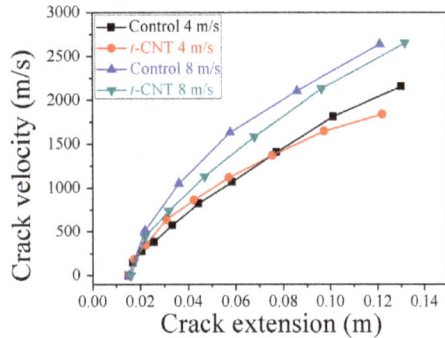

Figure 6 Comparison of crack velocity and crack extension of the control and *t*-CNT specimens at loading rates of 4 and 8 m/s

fracture load is reached – unlike the case of quasi-static loading, where the crack initiates at maximum loading, as in[6]. At a loading rate of 4 m/s, the maximum fracture toughness value of $G_{||}$ $\sim$3 kJ/m^2 is reached at a crack velocity of 1000–1700 m/s, while at a loading rate of 8 m/s, the maximum $G_{||}$ value is higher – $\sim$3.5 kJ/m2 (for the t-CNT sample), and is attained for a crack velocity over 2500 m/s, which is below the expected steady-state plateau. Based on the quasi-static $G_{||}$ values[6], it is estimated that under quasi-static loading, the initial crack velocity at fracture onset is around 3000 m/s.

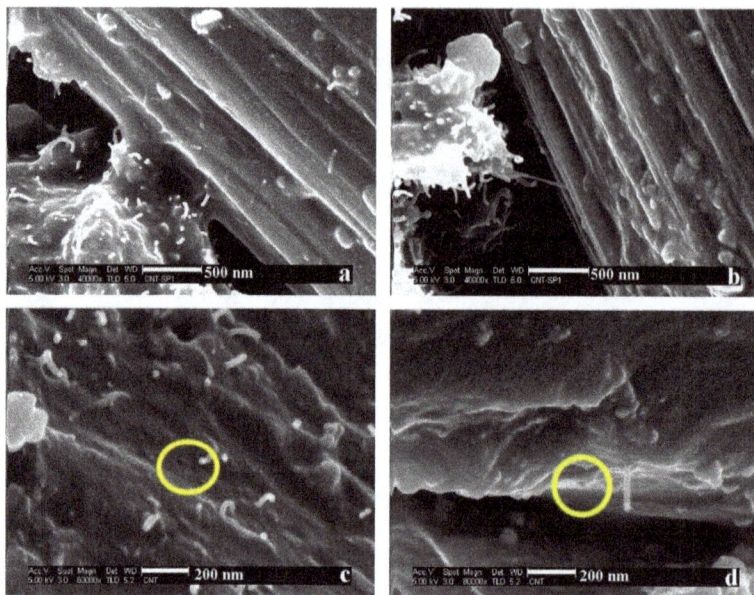

Figure 7   SEM images of fracture surface of t-CNT *a, b* and CNT *c, d* specimens at a loading rate of 8 m/s

The main question that stems of these results pertains to the loading rate effect: For example, why the *t*-CNT sample in the crack propagation stage does not exhibit an improved ILFT at a loading rate of 4 m/s, while a significant improvement is observed under 8 m/s. Because the fracture toughness is sensitive to crack velocity, which in turn is determined by the loading rate, the answer is given in part by comparing the respective crack velocities at a given crack extension. Whereas under a loading rate of 4 m/s, the maximum $G_{\parallel}$ value of above 3.0 kJ/m² is attained already at a crack velocity of 1500 m/s, at a loading rate of 8 m/s, for the same crack velocity, the corresponding $G_{\parallel}$ value is less than 2.5 kJ/m². To demonstrate the effect of crack velocity, Fig. 6 presents plots of crack velocity against crack extension for the *t*-CNT and the control samples under the two loading rates, where it is clearly seen that at the same crack extensions, the crack velocity under the 8 m/s loading rate is notably higher. Based on all these observations, it is concluded that fracture toughness is significantly higher at high crack velocities, where it becomes more sensitive to energy dissipation mechanisms of crack front interactions.

To identify the energy dissipation mechanisms and as fracture toughness correlates with fracture surface features, a SEM comparative fractography was performed.

First, we refer to the improvement of the ILFT in laminates with incorporated *t*-CNT leaf attributed to SP1 protein treatment of the CNT, shown to have created strong interactions between the *t*-CNT and epoxy and *t*-CNT and carbon fibers. These interactions result from the presence of graphite-specific binding peptides on the SP1 N-terminus originally engineered to react with the CNT. So, the SP1 protein of the SP1-CNT assembly – in which on the one hand the protein binds to the CNT to form a complex and on the other it bonds covalently to the epoxy matrix – also binds to the carbon fiber. In fact, the SP1 protein plays a role of a coupling agent

between the CNT and the carbon fibers of the fabric; thus, the adhesion between the epoxy matrix and carbon fibers increases. As the adhesion between the matrix and the fibers increases, the cohesive failure is promoted which ensures the increase in the $G_{\parallel}$ for crack initiation. In addition to the cohesive failure, there is an interfacial failure that is not less important because it is required for the formation of bridging CNT that constitute the main energy absorption mechanism in composites. Actually, the optimal combination of both cohesive and interfacial failure provides the best value of ILFT of the composite material.[20] The CNT bridging mechanism can be clearly seen in Fig. 7 that reveals the significance of *t*-CNT addition to the midplane of the laminate, with reference to the other interleaves (epoxy, epoxy-CNT, and epoxy-SP1), where no significant improvements or no improvement at all has resulted. Fig. 7 presents the fracture surfaces of two different specimen types: the *t*-CNT (Fig. 7*a* and *b*) and the CNT (Fig. 7*c* and *d*) specimens at a loading rate of 8 m/s. As was already mentioned, the bridging mechanism, where the *t*-CNT binds to the carbon fiber on the one hand and to the epoxy matrix on the other, is seen in Fig. 7*b*, implying that the bonding between the three components – epoxy matrix, *t*-CNT, and carbon fiber – is effective. For the untreated CNT specimens, however, Fig. 7*c* and *d* shows that the dominant fracture mechanism is pullout, as seen by both the protruding CNT and the corresponding cylindrical holes in the matrix (both marked with circles). This emphasizes the crucial binding capability of the SP1 protein.

Considering the loading rate effect, Fig. 8*a* and *b* presents typical fracture surfaces of the *t*-CNT samples tested at 4 and 8 m/s, respectively. It can be seen that the two-dimensional images display much the same features of the fracture morphology of the *t*-CNT interlayer suggesting that a quantitative surface roughness analysis is essential to identify differences, if any. Accordingly, SPM measurements were performed and Fig. 9 presents typical SPM images of fracture surfaces of *t*-CNT

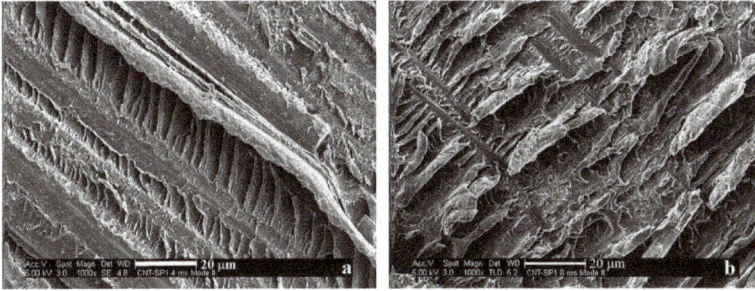

Figure 8   Comparison of fracture surfaces of *t*-CNT specimens after loading at different loading rates: 4 m/s *a*, 8 m/s *b*

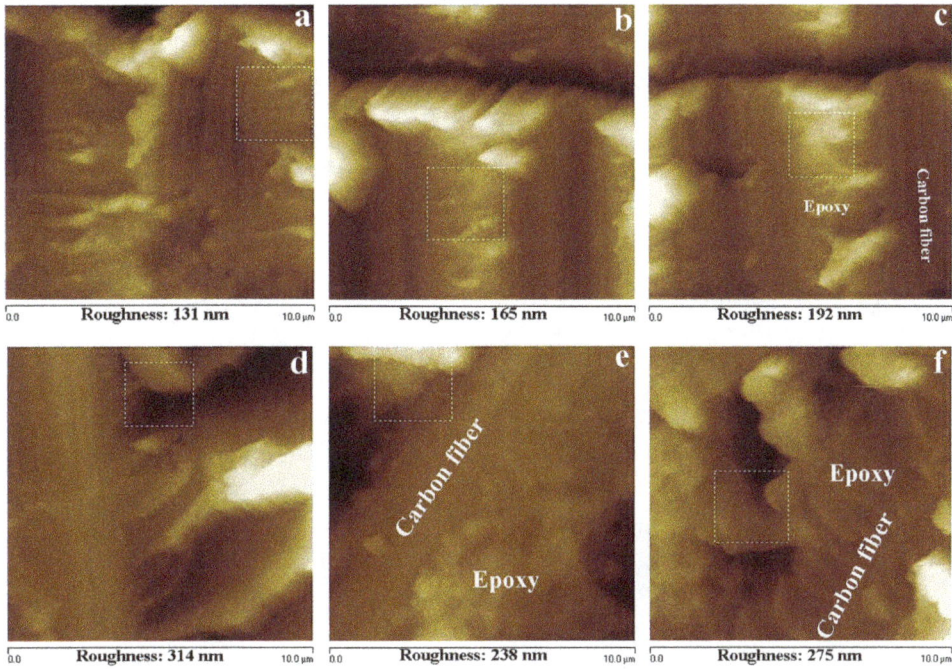

Figure 9   SPM images of fracture surfaces of *t*-CNT specimens loaded at different rates: 4*a*–*c* and 8*d*–*f* m/s

specimens loaded at different rates: 4 m/s (*a*–*c*), 8 m/s (*d*–*f*). For every sample, the roughness measurements were performed at different spots of the epoxy matrix in close proximity to a carbon fiber or a carbon fiber imprint (marked by the squares in the images), and the surface roughness was recorded at 3–4 sites at each spot.

A qualitative visual scrutiny of the SPM micrographs suggests a priori that *t*-CNT samples loaded at 8 m/s (Fig. 9*d*–*f*) generate much higher surface areas during fracture propagation than the same samples loaded at 4 m/s (Fig. 9*a*–*c*). Furthermore, the coverage of the carbon fibers by the epoxy matrix is significantly more extensive in the first rather than in the latter. The quantitative roughness measurements support the qualitative observations wherein the average roughness of the *t*-CNT samples tested at 8 m/s is approximately 300 nm (Fig. 9*d*–*e*) compared with 160 nm for the 4 m/s samples. This implies that excessive energy

is dissipated at higher loading rates through cohesive fracture and roughness of the epoxy matrix.

Conclusions

Two main conclusions can be drawn from this research pertaining to the specific composite system studied here and – more generally – to loading rate effects on delamination fracture toughness.

In this study, we expand the scope of the quasi-static study[6] to a higher range of loading rates, showing that a very small content of protein-treated CNT, interleaved into the composite structure, produces a significant improvement in the Mode II interlaminar fracture resistance of carbon fabric/epoxy laminate. The improvement is achieved by interleaving the protein-treated CNT layer specifically into the highest stress concentration region – the midplane of the laminate. In addition, it is demonstrated that such improvements dependent strongly on the loading rate.

The loading rate effect is expressed through the crack velocity, wherein at higher crack velocities during the crack expansion process, higher energies are dissipated – particularly in the protein-treated CNT laminates. High magnitude crack velocities are implicit under high loading rates, e.g., 8 m/s. In the latter scenario, the stored elastic energy prior to the onset of delamination results in crack initiation velocities at burst rates similar to those recorded under quasi-static conditions. Hence, the observed improvements in the ILFT, particularly in the protein-treated CNT laminates, occur only at high crack velocities.

Acknowledgments

We would like to thank to Zeev Cang from Tango Engineering, Ltd. for supplying the prepregs for this research.

Disclosure statement

No potential conflict of interest was reported by the authors.

References

1. J. L. Tsai, B. H. Huang and Y. L. Cheng: *J. Compos. Mater.*, 2009, **43**, (25), 3107–3123.
2. L.-C. Tang, X. Wang, Y.-J. Wan, L.-B. Wu, J.-X. Jiang and G.-Q. Lai: *Mater. Chem. Phys.*, 2013;**141**, (1), 333–342.
3. F. H. Gojny, M. H. G. Wichmann, U. Köpke, B. Fiedler and K. Schulte: *Compos. Sci. Technol.*, 2004, **64**, (15), 2363–2371.
4. Y. Li, N. Hori, M. Arai, N. Hu, Y. Liu and H. Fukunaga: *Composites Part A*, 2009, **40**, (12), 2004–2012.
5. N. Sela and O. Ishai: *Composites*, 1989, **20**, (5), 423–435.
6. T. Lyashenko, N. Lerman, A. Wolf, H. Harel and G. Marom: *Compos. Sci. Technol.*, 2013, **85**, 29–35.
7. A. Wolf, A. Buchman, A. Eitan, T. Fine, Y. Nevo, A. Heyman and O. Shoseyov: *J. Adhes.*, 2012, **88**, (4–6), 435–451.
8. J. Fitoussi, M. Bocquet and F. Meraghni: *Composites Part B*, 2013, **45**, (1), 1181–1191.
9. A. J. Smiley and R. B. Pipes: *Compos. Sci. Technol.*, 1987, **29**, (1), 1–15.
10. W. J. Cantwell. *J. Mater. Sci. Lett.*, 1996, **15**, (7), 639–641.
11. P. Compston, P. Y. B. Jar and P. Davies: *Composites Part B*, 1998, **29**, (4), 505–516.
12. K. Friedrich, R. Walter, L. A. Carlsson, A. J. Smiley and J. W. Gillespie Jr. *J. Mater. Sci.*, 1989, **24**, (9), 3387–3398.
13. H. Maikuma, J. W. Gillespie and D. J. Wilkins: *J. Compos. Mater.*, 1990, **24**, (2), 124–149.
14. B. R. K. Blackman, J. P. Dear, A. J. Kinloch, H. MacGillivray, Y. Wang, J. G. Williams and P. Yayla: *J. Mater. Sci.*, 1996, 31, (17), 4467–4477.
15. P. Compston, P. Y. B. Jar, P. J. Burchill and K. Takahashi: *Compos. Sci. Technol.*, 2001, **61**, (2), 321–333.
16. S. M. Lee: *Compos. Sci. Technol.*, 1992, **43**, (4), 317–327.
17. P. Mangalgiri and W. Johnson: *J. Compos. Tech. Res.*, 1986, **8**, (2), 58–60.
18. A. J. Russel and K. N. Street: 'Moisture and temperature effects on the mixed-mode delamination fracture of unidirectional graphite/epoxy', in 'Delamination and debonding of materials ASTM STP 876', (ed. W. S. Johnson), 349–370; 1985, ASTM International, Philadelphia, PA.
19. B. D. Agarwal and L. J. Broutman: 'Analysis and performance of fiber composites', 1990, New York, John Wiley & Sons.
20. H. Albertsen, J. Ivens, P. Peters, M. Wevers and I. Verpoest: *Compos. Sci. Technol.*, 1995, **54**, (2), 133–145.

15

Recent advances in production of poly(lactic acid) (PLA) nanocomposites: a versatile method to tune crystallization properties of PLA

Marius Murariu*, Anne-Laure Dechief, Rindra Ramy-Ratiarison, Yoann Paint, Jean-Marie Raquez and Philippe Dubois*

Center of Innovation and Research in Materials and Polymers (CIRMAP), Laboratory of Polymeric and Composite Materials (LPCM), University of Mons & Materia Nova Research Center, Place du Parc 20, 7000 Mons, Belgium

Abstract A new approach leading to poly(lactic acid) (PLA) nanocomposites designed with improved nucleating/crystallization ability has been developed. As proof of concept, nanofillers of different morphology (organo-modified layered silicates, halloysite nanotubes and silica) were surface-treated with ethylene bis-stearamide (EBS), a selected fatty amide able to promote chain mobility during PLA crystallization from the melt and nucleation. The fine dispersion of the nucleating additive via nanoparticles (NPs) as 'nano-template' is leading to nanocomposites

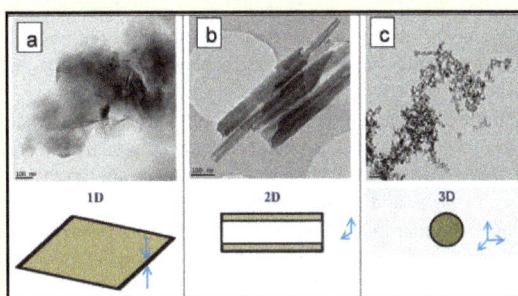

showing unexpected improvements in PLA crystallization rate. This was evidenced by differential scanning calorimetry (DSC) from the high values of the degree of crystallinity (20–40%) with respect to neat PLA (4.3%) and the sharp decrease in crystallization half-time under isothermal conditions (at 110°C), even below one minute. Furthermore, after injection molding the outstanding crystallization properties of PLA were again confirmed. Accordingly, the PLA-nanofiller/EBS nanocomposites revealed remarkable degree of crystallinity (in the range of 30–40%). Surprisingly, the presence of EBS can significantly increase the impact resistance of PLA and PLA based nanocomposites. By considering the remarkable increasing in crystallinity, a key parameter to allow PLA utilization in durable applications, the development of the new approach is expected to lead to significant improvements in the processing and performances of PLA products.

Keywords Poly(lactic acid), Nanofillers, Ethylene bis-stearamide (EBS), Crystallisation, Impact resistance

Introduction

The extraordinary interest for biopolymers encompasses various factors including consumer demand for more environmentally sustainable products, the development of bio-based feed stocks, the increasing of restrictions for the use of petro-polymers with high `carbon footprint' and others.

Poly(lactic acid) or polylactide (PLA), is industrially obtained respectively, through the polymerization of lactic acid or by ring opening polymerization (ROP) of lactide (the

cyclic dimer of lactic acid, as an intermediate). PLA is an environmentally friendly and commercially available aliphatic polyester produced from renewable resources and that has a key position on the market of biopolymers. Characterized by very interesting properties, it is presently considered between the most promising biomaterials with the brightest development prospect.[1] Currently, PLA is receiving a considerable attention not only for traditional utilization in packaging and textile products, but more recently it has shown an increasing interest for technical applications.[1–7] The last tendencies show clearly that the growth of PLA production indeed comes from the demand

*Corresponding authors, emails marius.murariu@materianova.be; philippe.dubois@umh.ac.be

of long lasting bioplastics in industry sectors such as electronics and automotive, end-user markets requiring similar performances and processing characteristics that match those of existing polymers, traditionally derived from petroleum or other fossil resources.[8,9]

However, although PLA shows interesting physical and mechanical properties (high tensile strength and stiffness, transparency, biodegradability...), the poor impact resistance and low elongation at break, sensitivity to hydrolysis, permeability to gas, as well as the low rate of crystallization, impede its utilization in technical applications.[1,10–12] In this respect, the profile of PLA properties [rigidity, dimensional stability, heat deflection temperature (HDT), etc.] was improved by combining this polyester matrix with micro- and nano-fillers, impact modifiers, flame retardants, plasticizers, other polymers, etc.[1,13–21] The increase in HDT (i.e. the temperature at which the polymer deforms under a specified load) is a main goal and an important property required for the design, engineering and manufacture of PLA products considered for durable applications. However, significant improvements regarding this parameter are reported by production of PLA nanocomposites and by increasing PLA crystallinity.[22,23]

Unfortunately, the use of PLA in technical applications is actually not so high because PLA has a slow crystallization rate when compared with many other thermoplastics. In fact, this is a kind of `Achilles' heel', limiting PLA use in a high performance application. Therefore, it is believed that new PLA grades with improved properties are needed. The control and increase in crystallinity will extend the use of PLA to electrical appliances and automotive parts (as substitution of petroleum based polymers). This parameter is particularly essential to control PLA's degradation rate, thermal resistance, as well as optical, mechanical and barrier properties.[23,24] It is important to remind that under the real injection molding conditions, because of the high cooling rate, only almost amorphous items from PLA can be obtained, which results in a lower HDT and mechanical properties.[25] Therefore, to produce PLA of high crystallinity many efforts are made in this direction by both academia and industry. Furthermore, this subject has generated a wide interest and an impressive number of studies and reviews are focusing specifically on the current understanding of PLA crystallization properties.[19,24–28]

To increase the crystallization rate of PLA, different methods[24] have been considered such as: addition of organic nucleating agents,[29,30] microfillers,[31–33] nanofillers,[34–37] stereocomplexes,[27,38] derivatives of carboxylic and fatty acids (low molecular weight aliphatic amides),[23,39] layered metal phosphonates,[40,41] etc. The nucleating additives can increase the number of primary nucleation sites reducing the nucleation induction period, and therefore initiate PLA crystallization at higher temperature on cooling. An alternative is to add plasticizers, which increase the polymer chain mobility, and therefore enhance the crystallization rate by reducing the energy required during crystallization for the chain folding process.[32] For this reason, various nucleating agents and plasticizers (in combination or not) have been used in order to increase the crystallization rate of PLA.[13,24,42]

However, a decrease in thermal stability and mechanical properties (tensile and flexural strength, rigidity, etc.) after the addition of plasticizers is noticed.

A challenge to improve PLA nucleation and crystallization kinetics is appealing, and accordingly tremendous efforts are made in this direction. Following this objective, the main goal of this study is to propose a new approach allowing to tune-up the crystallization properties of PLA, whereas the performances and specific end-use properties of PLA nanocomposites are maintained or even improved. In this regard, the experimental study was directed to the production (as proof of concept, PoC) of PLA nanocomposites characterized by improved features, going from crystallization to mechanical properties. Commercially available nanofillers (OMLS, halloysite, silica) of different morphology and high surface area, which potentially could act as nucleation sites for PLA, were used as `nano-template' for a selected fatty amide, i.e. ethylene bis-stearamide (EBS), an organic additive claimed to promote lubrication, chain mobility and nucleating ability of PLA during the crystallization from the melt. Compared to the anterior studies[23,24] the nanofillers will play a key role allowing not only obtaining a PLA with specific properties, but also the finer dispersion of EBS.

For the illustration of the concept, the nanofillers were previously surface-treated with EBS and used for melt-compounding with PLA. Using adapted techniques, the resulting polymer nanocomposites (PNC) were characterized to highlight the properties of nucleation and overall crystallization rate of PLA. Furthermore, the tuning of loading and nanofiller/EBS ratio was presented as additional tool for tailoring the mechanical properties of PLA nanocomposites.

Experimental

Materials

Poly(L,L-lactide) – hereafter called PLA, was the 4032D grade (supplier NatureWorks LLC) with $M_n = 133\,000$, dispersity, $M_w/M_n = 1.94$ (M_w and M_n, being respectively, weight- and number-average molar mass, determined by size exclusion chromatography (SEC) using polystyrene standards for column calibration), whereas upon the supplier the other characteristics are as follows: D isomer = 1.4%; relative viscosity = 3.94; residual monomer = 0.14%.

Nanofillers having respectively, one, two and three dimensions of the particulates in the order of few nanometers (according to ISO/TS 27687 (2008), i.e. Terminology and definitions for nano-objects – nanoparticle, nanofibre and nanoplate) were considered: organo-modified layered silicates (with one dimension in nanometer scale, thus will be hereinafter mentioned as `1D'), halloysite naotubes (2D) and equi-axed (isodimensional) silica (3D).

As 1D nanofiller, the selected organo-modified layered silicate (OMLS) used in this study was Cloisite 25A (C25A) as supplied by Southern Clay Products. Upon the supplier, C25A consists of nanometric platelets of layered magnesium aluminum silicate organically modified with dimethyl 2-ethylhexyl (hydrogenated tallow alkyl) quaternary ammonium at a modifier concentration of 95 meq/100 g clay. Based on thermogravimetric analysis (TGA), C25A contains about 70 wt-% inorganic silicate.

As 2D-nanofiller, halloysite nanotubes (HNT) were supplied by Aldrich with the following characteristics: 30–70 nm in diameter, 1–3 µm the length of nanotubes, surface area of $64\,m^2\,g^{-1}$ and 1.26–$1.34\,mL\,g^{-1}$ pore volume.

A high surface fumed silica (SiO_2) supplied by Cabot as CAB-O-SIL H5 was used as 3D nanofiller. The principal characteristics are as follows: BET surface area = $300\,m^2\,g^{-1}$; specific gravity = $2.2\,g\,cm^{-3}$; purity: $>99.8\%\,SiO_2$; average particle (aggregate) length = 0.2–$0.3\,\mu m$; whereas the primary particle size is given to be about 8 nm. N,N'- Ethylenebis(stearamide), a fatty amide with linear formula $[CH_3(CH_2)_{16}CONHCH_2-]_2$ and traditionally known as `EBS', was supplied by Sigma–Aldrich.

Treatment of nanofillers by EBS and realization of PLA blends using internal kneaders

After the dry-mixing of nanofillers and EBS, at a weight ratio of 80/20 (for this study, *vide infra*) using a laboratory Rondol turbo-mixer, this first stage was followed by the `dry-coating' at temperature (160°C) in an internal mixer (5 min at $30\,rev\,min^{-1}$ for adequate feeding, then 15 min at $100\,rev\,min^{-1}$) to allow the melting of EBS on the surface of nanofiller used as nano-template for the nucleating agent.

To minimize the water content before processing, PLA, nanofillers and EBS have been dried overnight at 80°C under vacuum. To produce various PLA formulations (Table 1), 3 wt-% of nanofiller were firstly dry mixed with PLA pellets (Rondol turbo-mixer, 2 min, $2000\,rev\,min^{-1}$), step followed by the moderate mixing (cam blades) at 200°C by using a Brabender bench scale kneader (model 50 EHT) following a specific procedure: 3 min premixing at $30\,rev\,min^{-1}$, speed in order to avoid an excessive increase in the torque during melting of PLA, followed by 7 min mixing at $70\,rev\,min^{-1}$. For the sake of comparison, the neat PLA matrix was processed by melt compounding under similar conditions.

In fact, the experimental program was more exhaustive, involving different nanofiller/EBS ratios and various loadings of nanofiller. For simplicity and to focus on the PoC, only the compositions from Table 1 are mainly discussed hereafter.

To get first information about the mechanical properties of the nanocomposites produced at laboratory scale, plates (~ 3.1 mm thickness) were produced by compression molding at 190°C by using an Agila PE20 hydraulic press. More specifically, the material was first pressed at low pressure for 200 s (three degassing cycles), followed by a high pressure cycle at 150 bars for 150 s. The resulting samples were then cooled under pressure (50 bars) for 300 s using water as cooling agent (temperature slightly $>10°C$, allowing a rapid cooling) and used in the next step to produce by gridding (Ray-Ran CNC Test Sample Profile Cutter) the specimens for mechanical characterization. Specimens (discs of 25 mm diameter and 1.5 mm thickness) required for both, WAXS and additional DSC analyses, were performed with a DSM micro injection molding machine using the following conditions: temperature of injection = 200°C, temperature of the mold = 110°C, residence time in the mold of 30 s.

Characterization

Differential scanning calorimetry (DSC)

DSC measurements were performed by using a DSC Q200 (TA Instruments) under nitrogen flow. The procedure was as follows: first heating scan at 10°C/min from 0°C up to 200°C, isotherm at this temperature for 2 min, then scan at different cooling rates (typically at $10°C\,min^{-1}$) down to $-20°C$ and finally, second heating scan from -20 to 200°C at $10°C\,min^{-1}$. Additional cooling rates (from $2.5°C\,min^{-1}$ up to $40°C\,min^{-1}$) have been used to evidence the PLA crystallization from the melt.

The first scan was realized to erase the prior thermal history of the samples. For all samples only the amount of PLA was considered. In order to evidence and quantify the crystallization of PLA during non-isothermal crystallization experiments, the peak of crystallization temperature upon cooling (T_{cc}) and the crystallization enthalpy upon cooling (ΔH_{cc}) were measured and the correct integration was attested systematically in the subsequently heating run. Then, the events of interest upon heating (h), i.e. the glass transition temperature (T_{gh}), cold crystallization temperature (T_{ch}), enthalpy of cold crystallization (ΔH_{ch}), temperature and enthalpy of polymer chain rearrangement known also as pre-melt crystallization (T_{rh} and ΔH_{rh} respectively), melting temperature (T_{mh}) and melting enthalpy (ΔH_{mh}) were determined from the second scan. The degree of crystallinity was determined by subtracting ΔH_{ch} and ΔH_{rh} (if available) from ΔH_m and by considering a melting enthalpy of $93\,J\,g^{-1}$ for 100% crystalline PLA.[43,44] It is also important to mention that the specimens performed by injection molding for WAXS investigations were used for additional DSC analyses to assess the values of the degree of crystallinity following the first scanning.

Screening tests to evaluate the crystallization kinetics of PLA were performed through the determination of crystallization half-time ($t_{1/2}$) during isothermal crystallization. Using DSC technique, the PLA samples were heated to 200°C at a rate of $10°C\,min^{-1}$, held 2 min at this temperature to erase their thermal history, step followed by high speed cooling ($40°C\,min^{-1}$) to the iso-crystallization temperature of interest, typically 110°C, and maintained under isothermal conditions for up to 120 min. The relative crystallinity was calculated by integrating the total area under the curve for each crystallization exotherm. The $t_{1/2}$ was taken to be the time at which the relative crystallinity (area) was equal to 50%.

Mechanical testing measurements

Tensile testing measurements were performed using a Lloyd LR 10K tensile bench in accordance to the ASTM D 638-02a

Table 1 Composition and codification of main PLA samples selected for this study

| Entry | Sample code* | Sample composition/wt-% | | | | |
		PLA	C25A	HNT	SiO_2	EBS
1	PLA	100	...	...	...	...
2	PLA- EBS	99.25	...	...	...	0.75
3	PLA- 1D	97	3	...	...	...
4	PLA- 1D/EBS	96.25	3	...	...	0.75
5	PLA- 2D	97	...	3	...	...
6	PLA- 2D/EBS	96.25	...	3	...	0.75
7	PLA- 3D	97	...	...	3	...
8	PLA- 3D/EBS	96.25	...	...	3	0.75

* 1D, 2D and 3D are symbolizing the number of dimensions in nanometer scale according to ISO/TS 27687 (2008).

norm at a tensile rate of 1 mm min^{-1} using specimens type V and a distance of 25.4 mm between grips. Notched impact resistance (Izod) measurements were performed using a Ray-Ran 2500 pendulum impact tester and a Ray-Ran 1900 notching apparatus, according to the ASTM D 256 norm (method A, 3.46 m s^{-1} impact speed, 0.668 kg hammer). All mechanical tests were carried out using specimens previously conditioned for at least 48 h at 20($\pm$2)°C under a relative humidity of 50($\pm$3)% and the values were averaged out over five measurements.

Scanning electron microscopy (SEM)

SEM was performed using a scanning electronic microscope Philips XL at an accelerated voltage up to 30 kV and various magnitudes. SEM was equipped for both secondary electron (SE) and back scattered electron (BSE) imaging.

Transmission electron microscopy (TEM)

Transmission electron micrographs were obtained with a Philips CM200 apparatus using an accelerator voltage of 120 kV. The nanocomposite samples (70–80 nm thick) were prepared with a Leica UCT ultracryomicrotome by cutting at −100°C. Reported microphotographs represent typical morphologies as observed at, at least, three various places.

Wide angle X-ray scattering (WAXS) characterizations

The specimens for WAXS characterization (discs of 25 mm diameter and 1.5 mm thickness) were produced by injection molding. The WAXS analysis was performed on a Siemens D5000 diffractometer using Cu K_a radiation (wavelength, 1.5406 Å) at room temperature with a scanning rate of 2° min^{-1}.

Results and discussion

Preliminary considerations in relation to nano-filler treatment

Following the interest in the utilization of bionanocomposites, nanofillers of different morphologies and with at least one dimension (<100 nm) within the nanoscale range (three-dimensional `isotropic' nanofillers, two-dimensional such as nanotubes or nanofibers, or one-dimensional sheet-like geometry of nanoparticles[45]) were used with satisfactory achievements in the design of PLA nanocomposites.[46] It has been also reported that the effectiveness of nanofiller addition on PLA properties can be evidenced even at very low loadings, i.e. 1–3%.[47,48,49] To qualitatively assess the versatility of our new experimental approach, nanofillers of different morphology were selected for melt-mixing with PLA, knowing that previously they have been tested with positive results in the production of PLA nanocomposites.[44,50–53]

Fig. 1a–c shows the typical TEM images at high magnification of the NPs considered in the work, attesting for their nanoscale morphological features (1D, 2D or 3D). As mentioned before (see experimental part) these NPs have one (C25A), two (HNT) or three dimensions of nanometer scale (silica). Silica (SiO$_2$) is also classified in the category of isodimensional nanofillers.[45] As illustrated by the representative SEM images shown in Fig. 2a–c, it is noteworthy that the raw NPs behave as aggregates and agglomerates before dispersion into PLA. The specific surface of NPs is different, going from the highest specific surface for C25A (1D) (about 725 m^2 g^{-1} for exfoliated montmorillonite), the intermediate one for silica (3D) (about 300 m^2 g^{-1}), to the lowest one for HNT (2D) (above 60 m^2 g^{-1}).

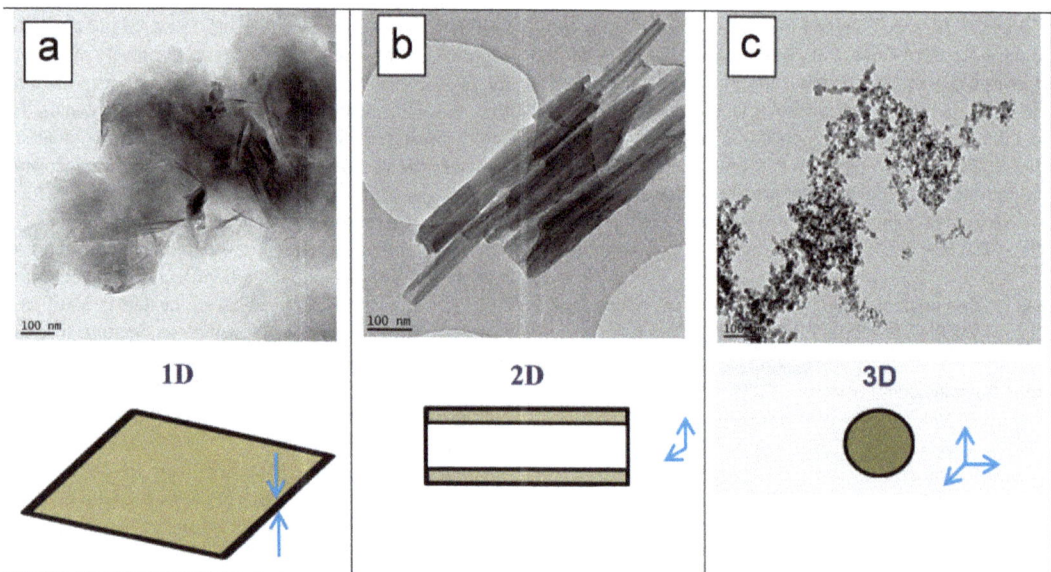

Figure 1 a–c TEM images to illustrate morphology at nano-scale of designated fillers: 1D = OMLS (C25A), 2D = HNT, 3D = SiO$_2$ (scale bar is of 100 nm)

Figure 2 *a–c* scanning electron micrographs (SEM) to illustrate initial aggregate structure of as received fillers: 1D = OMLS, 2D = HNT, 3D = SiO$_2$ (NB: for better evidencing of NPs different morphology, dissimilar magnifications were used)

In the first experimental step, the fillers of different morphology and surface area were mixed at 160°C with EBS to enable its complete melting (the melting point as determined by DSC is at 145°C) following a `dry-coating' process. EBS is currently used as wax, dispersing agent or as lubricant additive to facilitate and to improve the dispersion of solid materials, to enhance the processability and to decrease the friction in polymer applications (PLA is included).[54] Furthermore, Harris *et al.*, have found that this additive can act as nucleating additive for PLA, leading to the increase of crystallinity and crystallization rate through an optimized injection molding process. Accordingly, they have reported significant improvements in PLA mechanical performances (HDT, flexural strength, flexural modulus, etc.).[23]

Regarding the treatment with EBS, as illustrated in Fig. 3*a–d* for HNT and silica in the presence of water, it is very clear that the dry coating with this additive modifies the surface properties of the fillers, therefore these nanofillers turn from hydrophilic to show hydrophobic behaviour. The treatment with EBS yields a hydrocarbon-like surface for the nanofillers, making them much less polar than the starting nanofillers. This process also shows some similarities with the `dry-coating' of fillers with fatty acids, such as stearic acid (SA), or stearate salts.[55,56]

C25A is already hydrophobic due to the intercalation of hydrogenated tallow alkyl quaternary ammonium ions between montmorillonite (MMT) layers. In this respect no comparative images are shown here. It is important also to mention that the molecular characterizations by SEC (size exclusion chromatography) of the samples considered in this study attest that the addition of NP/EBS is leading to the same molecular parameters for the PLA matrix (for sake of clarity, these results are not shown here). Therefore, these nanofillers are considered to be less sensitive to moisture, while it is reasonable to ascribe the outstanding increase of crystallinity of PLA nanocomposites to the effective role of EBS as nucleating agent (see sections on `Non-isothermal crystallization' and `Isothermal crystallization and injection molding experiments').

Regarding the `coating' of the considered nanofillers, it is important to mention that other techniques (not reported here) have also been tested with promising results such as the `wet' method, i.e. the surface treatment and intensive pre-mixing of NPs with solutions of EBS in selected solvents (ethanol, isopropanol, etc.). Nevertheless, it is believed that the `dry-coating' is more interesting for extrapolation at larger scale than the techniques based on the utilization of organic solvents. Furthermore, it is noteworthy that TGA can be successfully used as rapid method to quantify and assess the presence of EBS in the desired nanofiller/EBS ratio, whereas the SEM and TEM techniques were considered as additional tools, giving information about the morphology of the nanofillers. Unfortunately, as revealed in the case of treated or untreated HNT (Fig. 4*a* and *b*), TEM observations cannot give a clear conclusion about the main differences related with the presence of layers of EBS onto the surface of nanofiller. However, it is assumed that these fillers of different morphology and specific surface area can be considered as templates acting as `nano-support' of high surface for EBS. As a result of this treatment, the amount of hydrophobic molecules (EBS) might overload the nanofiller surface, whereas the slight disaggregation of particle agglomerates can be also assumed. An overloading with EBS can be deliberately aimed by considering that the additive will be finally found at the polyester/nanofiller interface and finely dispersed throughout the PLA matrix. On the other hand, here was preferred the nanofiller/EBS ratio of 80/20 based on a previous experimental investigation, especially to induce the crystallization from the melt at higher temperature. In fact, by decreasing the content of EBS the crystallization peak was recorded somewhat at lower temperatures. Therefore, it was preferred to remain in the optimum temperature range traditionally indicated for PLA crystallization,[27] but additional parameters (e.g. mechanical properties) were also considered (*vide infra*).

Finally, it is assumed that various factors can affect, more or less, the crystallization kinetics of PLA, going from the nature of PLA (molecular weight, dispersity, isomeric purity, etc.), nanofiller loading, morphology and treatment, quality of distribution and dispersion through the polyester matrix, etc.

For simplicity and to focus on the PoC, only selected results are discussed hereafter at nanofiller-to-EBS weight ratio of 80/20. Furthermore, as it will be disclosed in the section on `Morphology of nanocomposites and mechanical properties', it is possible to modulate the performances of PLA nanocomposites following different objectives: medium to high crystallinity, tensile strength and stiffness, toughness, etc., optimizing the loading of nanofiller and nanofiller/EBS ratio.

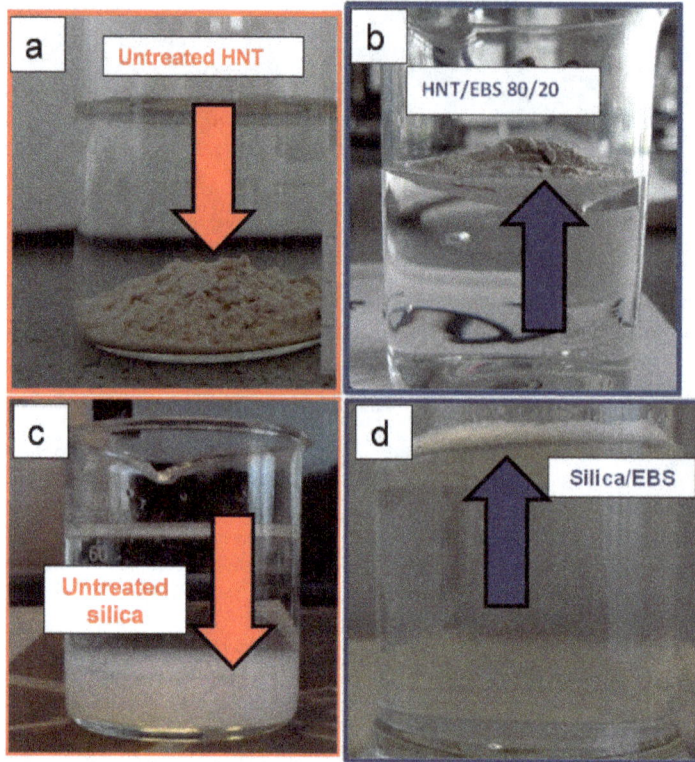

Figure 3   *a–d* illustration of 'hydrophobicity' in the presence of water of HNT and SiO_2 treated with *b*, *d* EBS with respect to *a*, *c* respectively untreated nanofillers

Non-isothermal crystallization

To evidence the crystallization of polymers such as PLA, various techniques can be successfully used, going from the differential scanning calorimetry (DSC), wide-angle X-ray scattering (WAXS), infrared spectroscopy (IR), nuclear magnetic resonance (NMR), dynamic mechanical analysis (DMA), to atomic force microscopy (AFM) and polarized optical microscopy (POM).[28]

Primarily, we have considered that DSC is probably the simplest technique which could give relevant and fast information about the crystallization properties of PLA nanocomposites. For an easier understanding of the key-points, hereafter are commented mainly the results obtained by studying the non-isothermal crystallization from the molten state.

Fig. 5*a* and *b* shows the representative DSC curves of different PLA samples as they were recorded respectively, during cooling from the melt (rate of $10°C \, min^{-1}$) after erasing the prior thermal history, followed by the second DSC heating (the quantification of data is reported in

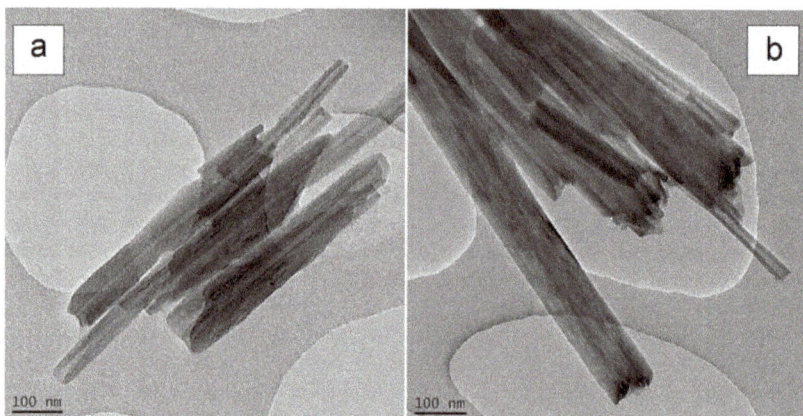

Figure 4   *a*, *b* TEM images of *a* untreated HNT and *b* HNT/EBS

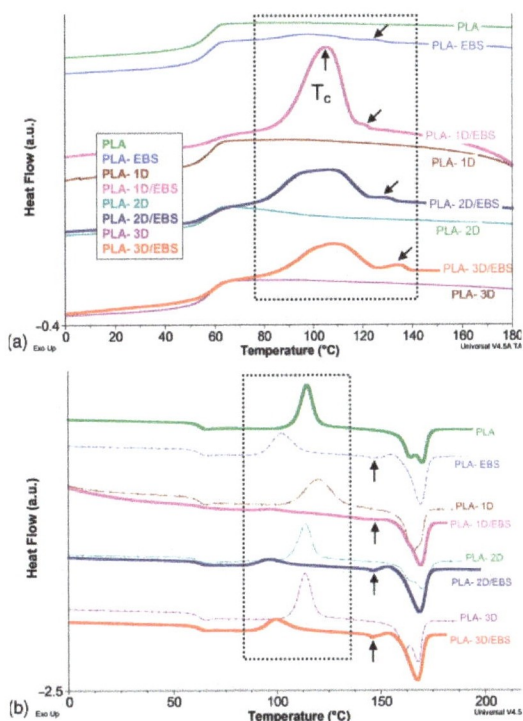

Figure 5 *a, b* comparative DSC traces of PLA (with and without EBS) and those of PLA nanocomposites containing nanofillers (with/without EBS) as they were recorded *a* during cooling and *b* during second heating (rate of $10°C\,min^{-1}$ was used in both DSC scans)

Table 2). By considering the crystallization from the molten state of PLA samples (Fig. 5*a*) it is seen that, in all cases addition of surface treated NPs (1D/EBS, 2D/EBS, 3D/EBS) leads to the clear PLA crystallization during cooling (i.e. the presence of well evidenced exothermic crystallization) in contrast to the pristine PLA and to PLA nanocomposites without any EBS-based treatment. For instance, the nanocomposites containing EBS display a crystallization process with a maximum peak (T_{cc}) recorded at 105–107°C. From the quantification of crystallization enthalpy recorded during cooling (ΔH_{cc}), addition of 1D/EBS filler shows the highest effectiveness (a ΔH_{cc} of $24\,J\,g^{-1}$, assigned to a gain of crystallinity higher than 26%), compared to the addition of 2D/EBS and 3D/EBS, that are giving lower values (i.e. ΔH_{cc} of

about $17\,J\,g^{-1}$, a gain in crystallinity higher than 18%). Furthermore, neat PLA does not show any detectible crystallization, whereas the PLA-EBS sample shows only poor and broad exotherms of crystallization (ΔH_{cc} of about $4\,J\,g^{-1}$) with a T_{cc} at 99°C. Similar results were reported by Harris *et al.*, following the evaluation of EBS and talc as nucleating agents for PLA.[23] Additionally, from the DSC traces recorded during cooling of PLA containing EBS, it can be seen very small exothermic peaks within the temperature range 125–135°C (indicated by arrows in Fig. 5*a*) that can be reasonably ascribed to the crystallization of EBS itself. Furthermore, it is believed that the additive (EBS) is either largely localized at the interface between PLA-nanofiller or finely dispersed through PLA matrix, especially in the case of an overloading with EBS of nanofillers of lower specific surface. The additive plays a complex role in the crystallization of PLA (from the melt) promoting nucleation as well as chains mobility. Besides, dramatic improvements in crystallization can be obtained by combining as nucleating agents EBS and NPs, both having ability to endorse the crystallization of PLA, first of all, due to formation of large fraction of nuclei. Concerning the nanocomposites without EBS, it is obvious from the traces recorded during cooling that they do not reveal any noticeable crystallization peak. Nevertheless, the data reported in the literature on this subject are very different, showing that the nanofillers could promote, more or less, the PLA crystallization.

The data derived from DSC curves shown in Fig. 5*b* (recorded during second DSC heating) are reported in Table 2. Because the addition of nanofillers does not modify significantly the T_g values, we will focus our comments mainly on the differences of crystallinity for different PLA based materials as obtained during second heating step. The multiple melting peaks (T_m) or the presence of shoulders on DSC curves, are usually ascribed to the melting of crystalline regions of various size and perfection formed during cooling and crystallization processes. It is worth noting that the influence of additional factors may further affect the melting of PLA (different crystalline structures, presence of fractions of low molecular weight, the effect of thermo-mechanical processing, etc.). Furthermore, the lower temperature peak is generally associated to the melting of the small crystals produced by secondary crystallization, whereas the peak recorded at higher temperature corresponds to the melting of the major crystals formed in the primary crystallization process.[57]

Table 2　Comparative DSC data of PLA and PLA nanocomposites containing nanofillers with/without EBS (second DSC heating, $10°C\,min^{-1}$)

Sample	$T_{ch}/°C$	$\Delta H_{ch}/J\,g^{-1}$	$T_{rh}/°C$	$\Delta H_{rh}/J\,g^{-1}$	$T_{mh}/°C$	$\Delta H_{mh}/J\,g^{-1}$	$\chi_c*/\%$
PLA	115	30.3	...	...	164; 170	34.3	4.3
PLA-EBS	102	21.3	155	1.4	169	31.7	9.7
PLA-1D	120	31.5	...	...	164; 167	36.4	5.3
PLA-1D/EBS	98	1.3	...	...	169	36.6	38.0
PLA-2D	114	25.1	...	...	163; 170	28.8	4.0
PLA-2D/EBS	97	9.9	153	2.3	169	31.3	20.5
PLA-3D	113	31.6	...	...	161; 168	36.2	4.9
PLA-3D/EBS	99	11.4	154	0.6	167	31.8	21.3

*χ_c, crystallinity as calculated by substracting ΔH_{ch} and ΔH_{rh} from ΔH_{mh} and by considering an enthalpy of $93\,J\,g^{-1}$ for 100% crystalline PLA.

Interestingly in Table 2, the addition of EBS and NPs/EBS into PLA leads to a single T_m (not multiple melting peaks or additional shoulders), assuming a higher uniformity for the crystals formed during cooling from the melt. Some statements for this hypothesis were obtained via POM analyses (see Supplementary Material, Fig. S1: www.maneyonline.com/doi/suppl/10.1179/2055033214Y.0000000008), that have proved for PLA-(1D-3D)/EBS nanocomposites compared to the neat PLA not only higher kinetics of crystallization, but the formation of a microcrystalline/microspherulitic structure as well. This appears to be another main characteristic supporting the improvements in the mechanical properties of these nanocomposites. The presence of EBS is attested from the DSC traces obtained during second heating by the small melting peaks at above 145°C (evidenced by arrows on Fig. 5b). It is also important to highlight that some nanocomposites have shown pre-melting crystallization of relatively low enthalpy, which has been considered in the calculation of the degree of crystallization. The second DSC scanning also reveals that neat PLA and its nanocomposites in absence of EBS undergo evident cold crystallization as it is attested by the high cold crystallization enthalpies (ΔH_{ch}). Moreover, the values of T_{ch} are in the range 110–120°C, temperatures significantly higher than those attributed to PLA samples containing EBS (T_{ch} from 97 to 102°C). Therefore, the presence of EBS onto nanofiller surface can facilitate the cold crystallization of the polymer matrix during heating as well.

Furthermore, linked to the behaviour recorded during cooling scan, the degree of crystallinity (χ_c) of PLA in the nanocomposites without EBS remains very low (i.e. in the range 4–5.3%), meaning that the considered nanofillers in absence of EBS do not behave as efficient nucleating agent for PLA under the specific experimental conditions. However, the χ_c of PLA in the nanocomposites (without EBS) is comparable to those recorded for the pristine PLA (4.3%), proving the low crystallization ability of PLA. In contrast, the nanocomposites containing NP/EBS show remarkable values of χ_c (going from 20 to 38%) and the following order of effectiveness for the treated nanofillers can be suggested: 1D/EBS > 3D/EBS ≥ 2D/EBS. This could be explained by the surface area of these nanofillers. For instance, the 1D-nanofiller has the highest surface area among them and leads to the better results in terms of PLA crystallization rate. However we cannot exclude that other factors could influence PLA crystallization such as the nature of nanofiller, existence or not of any previous treatment, etc.

Last, in relation to the non-isothermal crystallization from the melt, it is necessary to mention that comparative investigations using the conventional DSC technique were also realized using different rates of cooling to attest the versatility of the new approach (please consider the Supplementary Material, Fig. S2 www.maneyonline.com/doi/suppl/10.1179/2055033214Y.0000000008).

Isothermal crystallization and injection molding experiments

Another confirmation of the effectiveness of NPs/EBS as nucleating agents comes from the determination of the crystallization half-time ($t_{1/2}$) using the isothermal crystallization experiments. Commonly, the data reported in the

literature attest that the crystallization rate of PLA is the highest at temperatures between 100 and 120°C.[26] It is worth mentioning that various nanofillers, talc, multi-amide products, etc. were already tested to nucleate PLA matrices, determining the remarkable decreasing of $t_{1/2}$ and spectacular changes in the crystallization rate, polymer morphology and properties of the resultant PLA compounds.[23,58]

The effects of NP/EBS addition on the crystallization kinetics of PLA were here compared through $t_{1/2}$ obtained during isothermal crystallization experiments. Crystallization exotherms were measured as a function of time as reported elsewhere.[23,35] Selected data obtained from the isothermal crystallization at 110°C of neat PLA and those of nanocomposites with or without EBS are reported in Table 3. These results fully confirm that NP/EBS combinations behave as remarkable nucleating agents for PLA, their addition into the polyester matrix leading to lower $t_{1/2}$ with respect to the nanocomposites containing untreated NPs and neat PLA. All nanocomposites containing EBS show intense and narrow exotherms on the recorded DSC diagrams, while as received or processed PLA displayed much broader traces. PLA before processing show a huge $t_{1/2}$ (about 52 min), which attests for the low crystallization ability of this polyester. These results are in good agreement with the values currently reported in literature using a similar method.[23] Again, it is worth reminding that the $t_{1/2}$ of the PLA depends on crystallization temperature, optical purity, molecular weight, processing parameters, etc. For instance Battegazzore et al. have reported a $t_{1/2}$ of 82 min for a PLA containing 4% D-isomer following its isothermal crystallization at 110°C.[32] However, Pantani et al.[28] have assumed that following the mechanical degradation induced by processing, a limited number of smaller chains can induce a significant rise in the nucleation rate, leading to a significant increase of PLA crystallization kinetics.

From Table 3, the most interesting results are obtained by addition of NP/EBS that shows a very high effectiveness in the reduction of $t_{1/2}$, a parameter of real interest in the frame of applications related with injection-molding processes. It ($t_{1/2}$) was found to be less than 1.5 min using 3D/EBS treated nanofiller, and equal or even less than 1 min (using 1D/EBS filler) respectively, in the case of PLA- 2D/EBS and PLA- 1D/EBS samples.

It has also been reported in the literature that very interesting and useful information regarding the crystallization properties can be obtained via injection molding experiments.[23,27] Thus, the injection molding technique was used to

Table 3 Half-time of crystallization of neat PLA and PLA nanocomposites containing nanofillers with/without EBS

Sample	Time/min
PLA (as received)	51.6
PLA (processed)*	7.1
PLA-1D	6.2
PLA-1D/EBS	0.7
PLA-2D	6.3
PLA-2D/EBS	1.0
PLA-3D	4.7
PLA-3D/EBS	1.4

*Processed following procedure described in experimental section.

Table 4 Comparative DSC data on specimens performed by injection molding of PLA and PLA nanocomposites containing nanofillers with/without EBS (first DSC heating, 10°C min^{-1})

Sample	$T_c/°C$	$\Delta H_c/J\,g^{-1}$	$T_r/°C$	$\Delta H_r/J\,g^{-1}$	$T_m/°C$	$\Delta H_m/J\,g^{-1}$	$\chi_c^*/\%$
PLA	109	27.1	...	...	163; 169	33.0	6.3
PLA-1D	92	20.1	152	2.2	168	39.0	17.9
PLA-1D/EBS	90	12.0	152	1.4	167	39.8	28.4
PLA-2D	96	20.5	154	1.7	168	36.8	15.7
PLA-2D/EBS	...	...	...	...	167	38.2	41.1
PLA-3D	100	22.2	152	1.3	168	34.1	11.4
PLA-3D/EBS	76	3.8	...	...	167	36.5	35.2

* χ_c, crystallinity as calculated by subtracting ΔH_c and ΔH_r from ΔH_m and by considering an enthalpy of 93 J g^{-1} for 100% crystalline PLA.

practically prove the increase of crystallinity, results confirmed by DSC analyses and comparative WAXS measurements (see hereafter). Following DSC data (Table 4), neat PLA is again characterized by a low degree of crystallinity (χ_c above 6%) using this particular technique. The degree of crystallinity is higher in the case of nanocomposites containing untreated NPs and was found to be in the range 11–18%. Interestingly, all nanocomposites containing both, nanofiller (as template) and EBS, show unconventional crystallization ability, and consequently an increased χ_c (in the range 28–41%) is evidenced under similar conditions of processing (i.e. 30 s at 110°C). In other terms, under our experimental procedure, the considered NPs can also develop significant crystallization ability, but not as high as surface treated NPs. It may be assumed that this parameter (χ_c) can be also influenced by the composition of the nanocomposites (e.g. loading in EBS and nanofiller) and additional other factors (e.g. processing conditions).[42]

Fig. 6 shows the WAXS patterns of the specimens performed by injection molding using an annealing time in the mold (at 110°C) of only 30 s. By comparing the effects of untreated and treated NPs, it comes out that the nanocomposites containing NPs/EBS show in all cases stronger diffraction peaks at 2θ of about 16.5° with respect to those of nanocomposites obtained using untreated NPs. These peaks are ascribed to the presence of the crystalline phase of PLA attributed to (200) and/or (110) planes crystal facets, which are characteristic to the α-crystal form.[59] Additional small peaks, e.g. at 2θ of about 19° that are traditionally attributed to (203) planes, are also observed for nanocomposites.[59,60]

Furthermore, in good agreement with the DSC results, the neat PLA shows only a small crystallization peak with a maximum at 2θ = 16.5°, indicating that the sample is almost amorphous.

Additionally, by analyzing the comparative WAXS spectra of PLA-1D and PLA-1D/EBS samples (shown in Fig. 6), it can be concluded that the treatment of nanofiller (C25A) by EBS may lead to some improvement in the morphology of nanocomposites, mainly in the quality of intercalation of PLA chains between the silicate layers. As shown in Fig. 6, for instance the d_{001} value slightly increased (above 2 Å) in the nanocomposites containing 1D/EBS filler with respect to the nanocomposite containing non-treated fillers (d_{001} of about 30 Å). It is also interesting to underline that the intensities of the peaks (at 16.5°) were found almost correlated to the degree of crystallinity, as determined by DSC on the samples obtained by injection molding (results shown in Table 4).

As preliminary conclusion coming from these characterizations, it is clear that the PLA- (1D-3D)/EBS nanocomposites lead to a noteworthy degree of crystallinity (e.g. 30–40%), at shorter injection molding cycle time. The increase in the crystallinity and substantial reduction of $t_{1/2}$ through co-addition of NPs and EBS as (co)nucleating agents is seen as a versatile way that allows obtaining direct and significant improvements in the processing of commercial PLA grades.

Morphology of nanocomposites and mechanical properties

It is well known that the NPs have a strong tendency to form agglomerates and aggregates (see also the section on 'Preliminary considerations in relation to nanofiller treatment'). Therefore, even non-reactive surface treatments can lead to important changes not only at the interface with the polymer matrix, as well as in the interactions between particles, by reducing their surface energy and/or allowing a finer dispersion or overcoming their re-agglomeration.[55] It is worth pointing out that PLA nanocomposites containing the considered nanofillers (OMLS, HNT, SiO$_2$) already did the subject of previous studies published in the literature.[46,52,53,61–63] We will discuss succinctly hereinafter some aspects related only to the morphology of PLA-(1D- 3D)/EBS nanocomposites.

Fig. 7a–c and d–f shows selected TEM pictures of the nanocomposites produced by addition into PLA of 3% nanofiller (previously treated with EBS), at low and high magnification respectively. It is obvious from the TEM

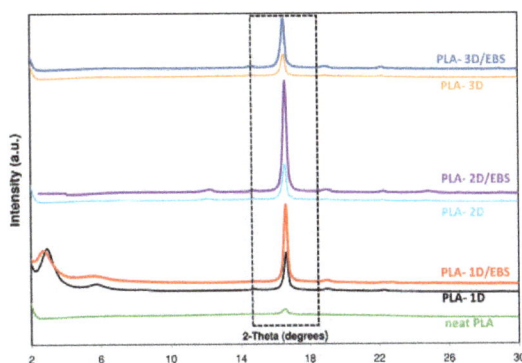

Figure 6 Comparative WAXS patterns on specimens performed by injection molding of PLA and PLA nanocomposites containing nanofillers with/without EBS

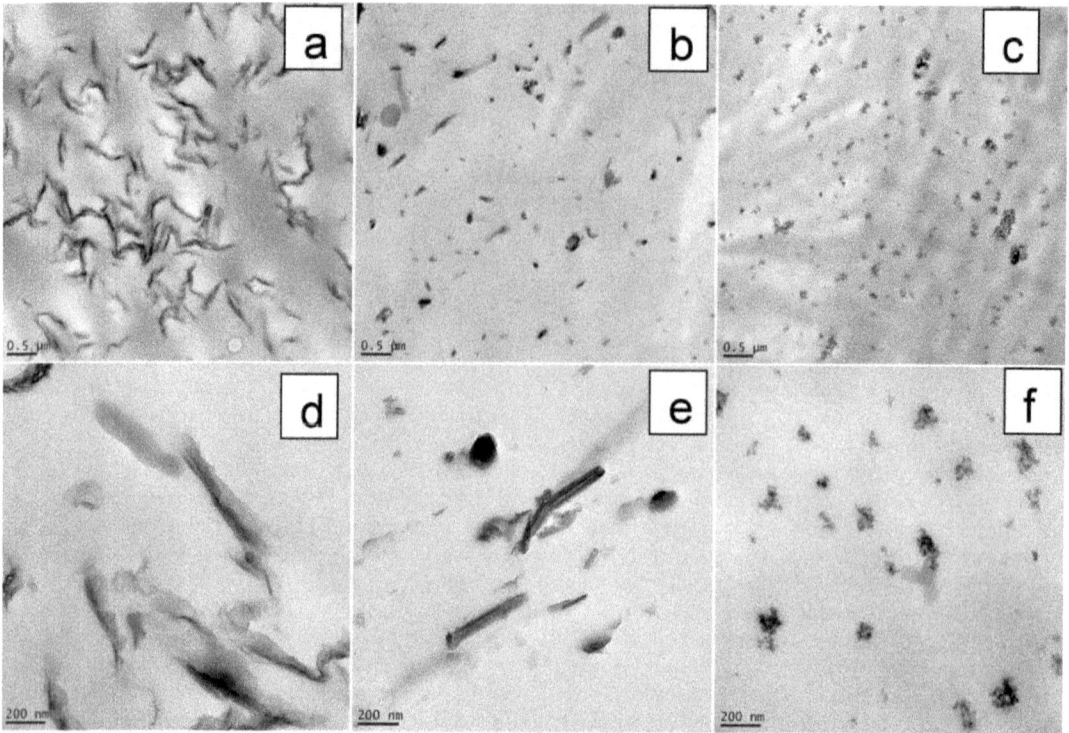

Figure 7 *a–f* selected TEM images at low and high magnification of nanocomposites loaded with 3% NPs (coated by EBS): *a, d* PLA-1D/EBS; *b, e* PLA-2D/EBS; *c, f* PLA-3D/EBS

images that a quite good NPs distribution/dispersion was reached within the polyester matrix, whereas the melt mixing was performed under moderate shear using internal kneaders. Besides, the TEM images of PLA-3D/EBS samples containing amorphous silica (Figs. 7c and f) attest for the presence of small associations of NPs of nanometric dimension (lower than 100 nm), that are not totally separated. Following the results reported elsewhere,[53] the dimension of aggregates can be even of micron size, being quite problematic to separate the individual silica nanoparticles. This may be explained by the very low dimension of silica NPs (about 8 nm) and the presence of strong interactions endorsed by the high concentration of silanol groups. The NPs thereby show strong self-networking ability and therefore it is more difficult to break the big aggregates during the melt-blending process.[64]

By considering also the results obtained in the frame of other experimental programs, it is assumed that the presence of EBS onto nanofiller surface (used as 'nano-template' for the nucleating additive) has no negative effects in relation to the morphology of nanocomposites.

Table 5 summarizes the results of the mechanical characterization of PLA nanocomposites loaded with 3% nanofiller to evidence the main effects of EBS addition. The stress-strain and impact tests were carried out at room temperature after the previous conditioning of specimens produced from plates obtained by compression molding (see the experimental part), whereas comparison to neat PLA and PLA-EBS samples is also given. First, due to non-reactive interaction with the polymer matrix, it comes out that the presence of EBS into PLA or onto the surface of NPs slightly affects the values of tensile strength (values in range

Table 5 Comparative mechanical properties of PLA and PLA nanocomposites (with/without EBS)

Entry	Sample code	Max. tensile strength/MPa	Young's modulus/MPa	Nominal strain at break/%	Impact resistance (Izod)/kJ m^{-2}
1	PLA	62 ($\pm$1)	2000 ($\pm$100)	5.9 ($\pm$0.8)	2.7 ($\pm$0.2)
2	PLA-EBS	58 ($\pm$2)	2400 ($\pm$170)	10.4 ($\pm$5.4)	5.5 ($\pm$1.1)
3	PLA-1D	64 ($\pm$2)	2850 ($\pm$50)	4.6 ($\pm$0.5)	2.8 ($\pm$0.2)
4	PLA-1D/EBS	59 ($\pm$2)	2950 ($\pm$50)	3.6 ($\pm$0.6)	3.2 ($\pm$0.4)
5	PLA-2D	67 ($\pm$2)	2490 ($\pm$130)	4.9 ($\pm$0.5)	2.8 ($\pm$0.2)
6	PLA-2D/EBS	64 ($\pm$3)	2650 ($\pm$120)	4.5 ($\pm$0.6)	3.6 ($\pm$0.3)
7	PLA-3D	66 ($\pm$1)	2400 ($\pm$180)	4.3 ($\pm$0.2)	2.7 ($\pm$0.1)
8	PLA-3D/EBS	60 ($\pm$3)	2670 ($\pm$170)	3.5 ($\pm$0.3)	5.8 ($\pm$1.1)

58–64 MPa compared to 64–67 MPa respectively, for nanocomposites with and without EBS).

Surprisingly, the rigidity (Young's modulus) of PLA nanocomposites increases in all cases in presence of EBS, while preserving the low nominal strain at break. Following the data reported in Table 5, addition of EBS has a key role in the improvement of the impact resistance of PLA, i.e., two-fold ($5.5\,kJ\,m^{-2}$) in the case of PLA/EBS sample with respect to the neat PLA ($2.7\,kJ\,m^{-2}$). To the best of our knowledge, this is one of the first papers dealing with the possibility to tune PLA toughness by the addition of tiny amount of EBS as additive. The nanocomposites containing NPs/EBS show in all cases improved impact resistance, with a special mention for PLA-3D/EBS sample. These results shall need rigorous and comprehensive interpretation connected to the morphology of nanofiller and nanocomposites.[64] However, different hypotheses are concerned: the importance of microcrystalline structure knowing that formation of spherulites of lower dimension could positively affect the impact resistance, the nanofillers or fine dispersed EBS (via nanofiller-template) may impart enhanced impact resistance, the specific effect of nanofiller morphology and stiffness, and so on.[44,65]

On the other hand, it is assumed that additional improvements in properties can be expected following the industrial processing by injection molding. The level of tensile strength, rigidity or impact resistance can be tuned up by modifying the loading of nanofiller and by optimizing the NPs/EBS ratio, while maintaining advanced crystallization properties. In the frame of additional studies was experimentally demonstrated that by finely varying the NPs/EBS ratio it is possible to obtain, respectively, a higher tensile strength or to induce both, the crystallization from the melt at higher temperature, while improving simultaneously the impact resistance. As exemplified in Fig. 8 in the case of HNT (`2D'), the nanofiller/EBS ratio can play a key role in relation to the mechanical properties of the nanocomposites, e.g. impact resistance. Thus, following the requirements of the application, different parameters can be optimized, including the nanofiller/EBS ratio and nanofiller loading.

Finally, the improvements of mechanical properties and increases in crystallinity (other characteristics can be also concerned: HDT, flexural strength and modulus, water vapour barrier, etc.), suggest that the nanocomposites loaded with NP/EBS are potentially interesting for engineering applications.

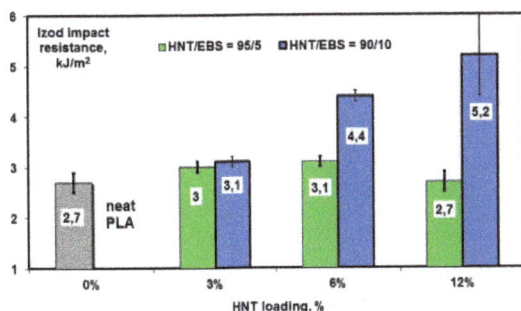

Figure 8 Effects of HNT/EBS ratio and nanofiller loading on impact resistance (Izod) of PLA–(3–12)%HNT/EBS nanocomposites: HNT/EBS ratio of 95/5 and 90/10

Conclusions

The use of PLA in technical applications is currently limited because PLA shows a slow crystallization rate under typical processing conditions. A new approach leading to PLA nanocomposites designed with improved nucleating/crystallization ability and accordingly, better processing and enhanced properties has been developed. As PoC and to attest the flexibility of the method, various nanocomposites were produced by melt blending PLA with nanofillers of different morphology (one to three nano-dimensions). The nanofillers used as nano-templates for the nucleating agent, were surface treated with ethylene bis-stearamide (EBS), a selected fatty amide having the role to promote both chain mobility in the melt and nucleating ability. This treatment leads to changes of interfacial properties in nanocomposites as revealed by mechanical testing, whereas the fillers become hydrophobic and less sensitive to moisture.

The properties of crystallization were evidenced using DSC as a main tool of investigation. In all cases addition of treated NPs into PLA leads to surprising properties of crystallization evidenced during cooling (crystallization from the melt), a higher effectiveness being associated to the use of 1D/EBS as nanofiller. The degree of crystallinity of the nanocomposites containing EBS was found in the range 20–40%, values much higher than those of pristine PLA and of PNCs without EBS (crystallinity limited to ~5%). Another confirmation of the effectiveness of the NPs/EBS as nucleating agents was obtained from the determination of the crystallization half-time ($t_{1/2}$) in isothermal crystallization experiments. These results fully confirmed that NP/EBS combinations show very important increases in PLA crystallization rate leading to much lower $t_{1/2}$ (even in the range 0.5–1 min). Furthermore, it was evaluated the crystallinity development in molding conditions. WAXS in full agreement with DSC measurements confirmed the unconventional crystallization ability (χ_c of 30–40%) of PLA containing NPs/EBS. By analyzing the morphology of PLA–(1-3D)/EBS nanocomposites, good nanofiller distribution/dispersion was observed (via TEM images) whereas by considering their mechanical properties, some increase in rigidity and impact resistance was noticed. Interestingly, it is assumed that the new approach, due to its versatility, can be regarded as a remarkable possibility to design new PLA products with enhanced processing and properties.

Conflicts of Interest

The authors have no conflicts of interest to declare.

Acknowledgements

Authors thank the Wallonia Region, Nord-Pas de Calais Region and European Commission for financial supports in the frame of INTERREG IV: NANOLAC project. They also thank their partners in the frame of NANOLAC project for helpful discussions. Authors also thank the Wallonia Region and their partners in the frame of BEETPACK project. We are grateful to Professor Mirosław Pluta (Center of Molecular and Macromolecular Studies Lodz, Poland), Alice Belfiore and Lisa

Dangreau (Materia Nova, Belgium) for assistance in analyses. LPCM thanks the Belgian Federal Government Office of Science Policy (SSTC- PAI 6/27) for general support and is much indebted to both `Région Wallonne' and the European Commission `FSE and FEDER' for financial support in the frame of Phasing-out Hainaut: Materia Nova. J.-M. Raquez is FRS-FNRS research associate.

References

1. K. Madhavan Nampoothiri, N. R. Nair and R. P. John: *Bioresour. Technol.*, 2010, **101**, (22), 8493–8501.
2. R. Babu, K. O'Connor and R. Seeram: *Prog. Biomater.*, 2013, **2**, (1), 1–16.
3. R. Auras, B. Harte and S. Selke: *Macromol. Biosci.*, 2004, **4**, (9), 835–864.
4. B. Gupta, N. Revagade and J. Hilborn: *Progr. Polym. Sci.*, 2007, **32**, (4), 455–482.
5. R. E. Drumright, P. R. Gruber and D. E. Henton: *Adv. Mater.*, 2000, **12**, (23), 1841–1846.
6. M. Jamshidian, E. A. Tehrany, M. Imran, M. Jacquot and S. Desobry: *Comprehens. Rev. Food Sci. Food Safety*, 2010, **9**, (5), 552–571.
7. P. Dubois and M. Murariu: *JEC Compos. Magaz.*, 2008, **45**, 66–69.
8. C. Smith: *Compound. World*, Jun. 2012, 45–51.
9. C. Smith: *Compound. World*, Sep. 2013, 31–36.
10. N. Bitinis, R. Verdejo, E. M. Maya, E. Espuche, P. Cassagnau and M. A. Lopez-Manchado: *Compos. Sci. Technol.*, 2012, **72**, (2), 305–313.
11. R. M. Rasal, A. V. Janorkar and D. E. Hirt: *Progr. Polym. Sci.*, 2010, **35**, (3), 338–356.
12. P. Bordes, E. Pollet and L. Avérous: *Progr. Polym. Sci.*, 2009, **34**, (2), 125–155.
13. M. Murariu, A. Da Silva Ferreira, M. Alexandre and P. Dubois: *Polym. Adv. Technol.*, 2008, **19**, (6), 636–646.
14. M. Murariu, L. Bonnaud, P. Yoann, G. Fontaine, S. Bourbigot and P. Dubois: *Polym. Degrad. Stabil.*, 2010, **95**, (3), 374–381.
15. M. Murariu, A. D. S. Ferreira, E. Duquesne, L. Bonnaud and P. Dubois: *Macromol. Sympos.*, 2008, **272**, (1), 1–12.
16. K. Fukushima, M. Murariu, G. Camino and P. Dubois: *Polym. Degrad. Stabil.*, 2010, **95**, (6), 1063–1076.
17. G. Kfoury, J.-M. Raquez, F. Hassouna, J. Odent, V. Toniazzo, D. Ruch and P. Dubois: *Front. Chem.*, 2013, **1**, (32), 1–46 (Published online 2013 December 17, doi:10.3389/fchem.2013.00032).
18. S. Bourbigot and G. Fontaine: *Polym. Chem.*, 2010, **1**, (9), 1413–1422.
19. K. S. Anderson, K. M. Schreck and M. A. Hillmyer: *Polym. Rev.*, 2008, **48**, (1), 85–108.
20. Y. Wang, S. M. Chiao, T. F. Hung and S. Y. Yang: *J. Appl. Polym. Sci.*, 2012, **125**, (5), E402–E412.
21. M. Murariu, A. Da Silva Ferreira, P. Degée, M. Alexandre and P. Dubois: *Polymer*, 2007, **48**, (9), 2613–2618.
22. S. Sinha Ray, K. Yamada, M. Okamoto and K. Ueda: *Polymer*, 2003, **44**, (3), 857–866.
23. A. M. Harris and E. C. Lee: *J. Appl. Polym. Sci.*, 2008, **107**, (4), 2246–2255.
24. S. Saeidlou, M. A. Huneault, H. Li and C. B. Park: *Progr. Polym. Sci.*, 2012, **37**, (12), 1657–1677.
25. Z. Gui, C. Lu and S. Cheng: *Polym. Test.*, 2013, **32**, (1), 15–21.
26. M. L. Di Lorenzo: *Eur. Polym. J.*, 2005, **41**, (3), 569–575.
27. K. S. Anderson and M. A. Hillmyer: *Polymer*, 2006, **47**, (6), 2030–2035.
28. R. Pantani, F. De Santis, A. Sorrentino, F. De Maio and G. Titomanlio: *Polym. Degrad. Stabil.*, 2010, **95**, (7), 1148–1159.
29. Y. Cai, S. Yan, Y. Fan, Z. Yu, X. Chen and J. Yin: *Iran Polym. J.*, 2012, **21**, (7), 435–444.
30. Y. Cai, S. Yan, J. Yin, Y. Fan and X. Chen: *J. Appl. Polym. Sci.*, 2011, **121**, (3), 1408–1416.
31. F. Yu, T. Liu, X. Zhao, X. Yu, A. Lu and J. Wang: *J. Appl. Polym. Sci.*, 2012, **125**, (S2), E99–E109.
32. D. Battegazzore, S. Bocchini and A. Frache: *Expr. Polym. Lett.*, 2011, **5**, (10), 849–858.
33. H. W. Xiao, P. Li, X. Ren, T. Jiang and J.-T. Yeh: *J. Appl. Polym. Sci.*, 2010, **118**, (6), 3558–3569.
34. S. H. Park, S. G. Lee and S. H. Kim: *Compos. A: Appl. Sci. Manuf.*, 2013, **46**, (0), 11–18.
35. M. Murariu, A. L. Dechief, L. Bonnaud, Y. Paint, A. Gallos, G. Fontaine, S. Bourbigot and P. Dubois: *Polym. Degrad. Stabil.*, 2010, **95**, (5), 889–900.
36. V. Krikorian and D. J. Pochan: *Macromolecules*, 2004, **37**, (17), 6480–6491.
37. J. Lee and Y. Jeong: *Fibers Polym.*, 2011, **12**, (2), 180–189.
38. H. Tsuji, H. Takai and S. K. Saha: *Polymer*, 2006, **47**, (11), 3826–3837.
39. J. Y. Nam, M. Okamoto, H. Okamoto, M. Nakano, A. Usuki and M. Matsuda: *Polymer*, 2006, **47**, (4), 1340–1347.
40. Q. Han, Y. Wang, C. Shao, G. Zheng, Q. Li and C. Shen: *J. Compos. Mater.*, 2014, **48**, 2737–2746.
41. L. Suryanegara, H. Okumura, A. Nakagaito and H. Yano: *Cellulose*, 2011, **18**, (3), 689–698.
42. H. Li and M. A. Huneault: *Polymer*, 2007, **48**, (23), 6855–6866.
43. E. W. Fischer, H. J. Sterzel and G. Wegner: *Kolloid-Z.u.Z. Polym.*, 1973, **251**, (11), 980–990.
44. M. Murariu, A.-L. Dechief, Y. Paint, S. Peeterbroeck, L. Bonnaud and P. Dubois: *J. Polym. Environ.*, 2012, **20**, (4), 932–943.
45. A. P. Kumar and D. Depan: *Progr. Polym. Sci.*, 2009, **34**, (6), 479–515.
46. J.-M. Raquez, Y. Habibi, M. Murariu and P. Dubois: *Progr. Polym. Sci.*, 2013, **38**, (10–11), 1504–1542.
47. L. Zaidi, S. Bruzaud, A. Bourmaud, P. Médéric, M. Kaci and Y. Grohens: *J. Appl. Polym. Sci.*, 2013, **130**, (3), 1357–1365.
48. Y.-T. Shieh, Y.-K. Twu, C.-C. Su, R.-H. Lin and G.-L. Liu: *J. Polym. Sci. B: Polym. Phys.*, 2010, **48**, (9), 983–989.
49. M. Murariu, A. Doumbia, L. Bonnaud, A. L. Dechief, Y. Paint, M. Ferreira, C. Campagne, E. Devaux and P. Dubois: *Biomacromolecules*, 2011, **12**, (5), 1762–1771.
50. M. Pluta, M.-A. Paul, M. Alexandre and P. Dubois: *J. Polym. Sci. Part B: Polym. Phys.*, 2006, **44**, (2), 299–311.
51. G. Gorrasi, R. Pantani, M. Murariu and P. Dubois: *Macromol. Mater. Eng.*, 2014, **299**, (1), 104–115.
52. V. Krikorian and D. J. Pochan: *Chem. Mater.*, 2003, **15**, (22), 4317–4324.
53. K. Fukushima, D. Tabuani, C. Abbate, M. Arena and P. Rizzarelli: *Eur. Polym. J.*, 2011, **47**, (2), 139–152.
54. J. B. McDaniel: `Polylactic acid shrink films and methods of manufacturing same', *US7846517 B2*, published 7 December 2010, Google Patents: 2010
55. R. N. Rothon: `Functional polymers and other modifiers', in `Functional fillers for plastics', 105–128; 2005, New York, Wiley-VCH.
56. M. A. Osman, A. Atallah and U. W. Suter: *Polymer*, 2004, **45**, (4), 1177–1183.
57. Z. Su, Q. Li, Y. Liu, G.-H. Hu and C. Wu: *J. Polym. Sci. B: Polym. Phys.*, 2009, **47**, (20), 1971–1980.
58. P. Song, Z. Wei, J. Liang, G. Chen and W. Zhang: *Polym. Eng. Sci.*, 2012, **52**, (5), 1058–1068.
59. J. Y. Nam, S. Sinha Ray and M. Okamoto: *Macromolecules*, 2003, **36**, (19), 7126–7131.
60. H. Wang and Z. Qiu: *Thermochim. Acta*, 2012, **527**, 40–46.
61. Y. Li, C. Han, J. Bian, L. Han, L. Dong and G. Gao: *Polym. Compos.*, 2012, **33**, (10), 1719–1727.
62. V. Ojijo and S. Sinha Ray: *Progr. Polym. Sci.*, 2013, **38**, (10–11), 1543–1589.
63. S. Solarski, M. Ferreira, E. Devaux, G. Fontaine, P. Bachelet, S. Bourbigot, R. Delobel, P. Coszach, M. Murariu, A. Da Silva Ferreira, M. Alexandre, P. Degee and P. Dubois: *J. Appl. Polym. Sci.*, 2008, **109**, (2), 841–851.
64. H. Xiu, C. Huang, H. Bai, J. Jiang, F. Chen, H. Deng, K. Wang, Q. Zhang and Q. Fu: *Polymer*, 2014, **55**, (6), 1593–1600.
65. M. Du, B. Guo and D. Jia: *Polym. Int.*, 2010, **59**, (5), 574–582.

16

Mechanistic insights on nanosilica self-networking inducing ultra-toughness of rubber-modified polylactide-based materials

Jérémy Odent[1], Jean-Marie Raquez[1]*, Jean-Michel Thomassin[2], Jean-Michel Gloaguen[3], Franck Lauro[4], Christine Jérôme[2], Jean-Marc Lefebvre[3] and Philippe Dubois[1]

[1]Laboratory of Polymeric and Composite Materials (LPCM), Center of Innovation and Research in Materials and Polymers (CIRMAP), University of Mons (UMONS), Place du Parc 20, B-7000 Mons, Belgium
[2]Center for Education and Research on Macromolecules (CERM), University of Liege, Sart-Tilman, Allée de la Chimie 3 B6, B-4000, Liege 1, Belgium
[3]Unité Matériaux et Transformations (UMET), UMR CNRS 8207, Université Lille1, Sciences et Technologies/CNRS, Cité Scientifique C6, 59655 Villeneuve d'Ascq, France
[4]Industrial and Human Automatic Control and Mechanical Engineering Laboratory (LAMIH), UMR CNRS 8201, University of Valenciennes and Hainaut-Cambresis, Le Mont Houy, BP 311, 59304 Valenciennes Cedex, France

Abstract Developing novel strategies to improve the impact strength of PLA-based materials is gaining a significant importance in order to enlarge the range of applications for this renewable polymer. Recently, the authors have designed ultra-tough polylactide (PLA)-based materials through co-addition of rubber-like poly(ε-caprolactone-co-D,L-lactide) (P[CL-co-LA]) impact modifier and silica nanoparticles (SiO$_2$) using extrusion techniques. The addition of silica nanoparticles into these immiscible PLA/P[CL-co-LA] blends altered their final morphology, changing it from rubbery spherical inclusions to almost oblong structures. A synergistic toughening effect of the combination of P[CL-co-LA] copolymer and silica nanoparticles on the resulting PLA-based materials therefore occurred. To explain this particular behavior, the present work hence aims at establishing the mechanistic features about the nanoparticle-induced impact enhancement in these immiscible PLA/impact modifier blends. Incorporation of silica nanoparticles of different surface treatments and sizes was thereby investigated by means of rheological, mechanical and morphological methods in order to highlight the key parameters responsible for the final impact performances of the as-produced PLA-based materials. Relying on video-controlled tensile testing experiments, a toughening mechanism was finally proposed to account for the impact behavior of resulting nanocomposites.

Keywords Polylactide, Impact modifier, Morphology, Compatibilization, Silica nanoparticles, Toughening mechanism

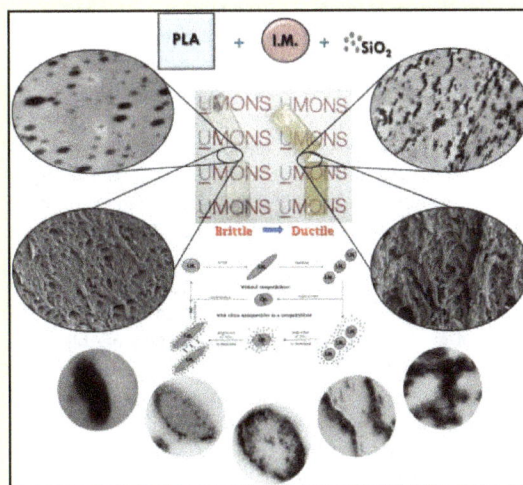

Introduction

To generate rubber-toughened thermoplastics with optimal mechanical properties, the rubber phase morphology

*Corresponding author, email: jean-marie.raquez@umons.ac.be

(domain size and related size distribution) and polymeric matrix–rubbery domain interfacial properties must be properly controlled.[1–5] The morphology of a two-phase system usually results to a balance between the breakup and coalescence of the dispersed domains in the flow field.[6] For uncompatibilized blends, the final particle size increases with the dispersed phase concentration because of increased coalescence. Adding a compatibilizing agent to immiscible polymer blends enables to control the final morphology of these blends by lowering their interfacial tension,[7] resulting in a suppression, or at least, a decrease of coalescence extent in the dispersed phase.[8–10] Apart from reducing the interfacial tension, the role of a compatibilizer is to significantly enhance the adhesion between the continuous and dispersed phase within blend. Many authors reported different compatibilization routes through the utilization of organic compatibilizers like block or graft copolymers either preformed or generated *in situ* during flow-induced processes.[7,11–14] Recently, another compatibilization method has been reported through the addition of solid nanoparticles into immiscible polymer blends.[15–18] The role of the nanoparticles is to get specifically localized at the interface of both polymeric partners, to strengthen the interfacial adhesion between the partners and thus to enhance the overall material performances.[19–21] The use of nanoadditives (e.g. clay or silica nanoparticles) provides further advantages for polymeric blends such as enhanced material properties, ease of processing, and lower costs in comparison to block and graft copolymer compatibilizers.[22]

Ternary polymer composites containing glassy polymer, soft elastomer and rigid filler are also being the subject of an increasing number of studies.[23,24] The mechanical properties of such composites depend not only on the components but also on the phase structure (i.e. the relative arrangement of the components) and the phase size.[25] Concerning the phase morphology, separated microstructures where the elastomer and the filler are dispersed in the polymer matrix separately[26,27] and core–shell microstructures with the filler encapsulating the elastomer,[28,29] are the two current morphologies for these ternary composites. These distinct morphologies gave different mechanical performances with higher modulus and/or toughness in function of the dispersion extent.[30,31] Lipatov *et al.*[32–34] reported that the addition of solid particles into an immiscible polymer blend increases the thermodynamic stability of the mixture and changes the compositions of the separated phases. The authors proposed a simultaneous action of two mechanisms to explain the effect of fillers on the phase behavior: (i) the thermodynamical alterations at the interface because of the selective adsorption of one of both components and (ii) the redistribution of components at the interface and in the bulk that may diminish the phase-separation temperature.

The present contribution concerns ultra-tough immiscible polymer blends made of polylactide (PLA) rubber-toughened with poly(ε-caprolactone-*co*-D,L-lactide) (P[CL-*co*-LA]) copolyester and silica nanoparticles (SiO₂).PLA represents the most extensively investigated renewable polymer for substituting conventional petroleum-based polymers in different types of applications such as in automotive and electronic industries,[35,36] but suffers from its brittleness. This impedes its applications in some areas particularly like packaging materials.[37,38] Recently, the authors have reported the use of as-synthesized hydrolytically degradable P[CL-*co*-LA] copolymers as effective biodegradable impact modifiers for PLA.[39] Directly related to a control over the rubbery character of the dispersed phase and the morphological features of resulting blends, an optimum toughening effect was reached

using a P[CL-*co*-LA] copolymer at a relative molar content of 28 mol%LA. The authors also reported that adding silica nanoparticles into these blends could alter their final morphology and then enhance the impact strength of these blends in such a way to achieve ultra-tough PLA-based materials.[40] As far as a toughening effect was evidenced, more understanding about the role of silica nanoparticles on the morphological features, the compatibility degree and the toughening mechanism of immiscible PLA/P[CL-*co*-LA] blends are however required to control the ultra-toughness of these blends. The authors here emphasize all issues aiming to clarify the morphology alteration for these immiscible PLA/P[CL-*co*-LA] blends through the use of silica nanoparticles of different surface treatments and particle size, leading to the construction of those silica nanoparticle-based self-networks. To achieve the best control over the blend morphology, it is well known that the final phase structure of ternary composites is governed by both thermodynamic (e.g. interfacial tension) and kinetic factors (e.g. shear stress and processing conditions).[19,41] In this respect, the authors have attempted to control the phase structure and the related-toughness of PLA/P[CL-*co*-LA]/SiO₂ composites upon coalescence and coarsening processes, i.e. by adjusting the nanoparticles dispersion and related processing methods used, and tuning the nanoparticle surface chemistry, mean size and relative content. Three types of silica nanoparticles, i.e. having different ability of self-agglomeration regarding surface treatment and surface area, and two processing methods (one-step or two-step) were mainly used to prepare the ternary composites for morphological, rheological and thermo-mechanical investigations. In addition, a special emphasis is made to highlight the toughening mechanisms related to the final mechanical performances of the as-produced PLA-based materials.

Experimental section

Materials

ε-caprolactone (99%, Acros) was dried for 48 h over calcium hydride and distilled under reduced pressure. D,L-lactide (> 99.5%, Purac) was conserved in a glove box. *n*-heptanol (98%, Aldrich) was dried over molecular sieve (4 Å) and tin(II) octoate (Sn(Oct)₂) (95%, Aldrich) was used as received without any purification, and diluted in dry toluene (0.01M). A commercially available extrusion-grade PLA (NatureWorks 4032D) designed especially for production of biaxially oriented films was used as received ($\overline{M_n}$ = 133 500 ± 5,000 g mol⁻¹, Đ = 1.94 ± 0.06 as determined by size-exclusion chromatography (SEC), 1.4 ± 0.2% D-isomer content as determined by the supplier). CAB-O-SIL TS-530 (TS-530, 225 m² g⁻¹) is of high surface area fumed silica, which has been surface modified with hexamethyldisilazane while CAB-O-SIL M-5 (M-5, 200 m² g⁻¹) and CAB-O-SIL H-5 (H-5, 300 m² g⁻¹) are untreated fumed silica and were supplied by Cabot.

Synthesis of poly(ε-caprolactone-*co*-D,L-lactide) copolymer

The copolymerization was carried out by bulk ring-opening polymerization (ROP) of ε-caprolactone and D,L-lactide promoted by *n*-heptanol and tin(II) octoate at an initial molar

[alcohol]/[tin(II) octoate] ratio of 100. The reaction was carried out for 24 h in an oil bath at 160°C, and quenched it in an ice bath. The crude product was dissolved in a minimum volume of chloroform, followed by precipitation into a 10-fold excess of heptane. They were recovered after filtration and drying under vacuum, until reaching a constant weight (yield: 99%).The LA molar content of the resulting copolymer was of 28 mol% (as determined by ^{1}H NMR analyses), while a number-average molecular weight of 35 400 g mol^{-1} (equivalent polystyrene) and a dispersity index (Đ) of 2.0 (as determined by GPC analyses) were achieved. This as-synthesized copolymer is amorphous with a glass transition temperature of about −36°C [as determined by differential scanning calorimetry (DSC) analyses].

Sample preparation and compounding

Before extrusion, PLA pellets were dried for at least 12 h at 80°C in an oven under reduced pressure of 10^{-1} mbar. Both binary and ternary PLA-based systems were prepared in a DSM twin-screw micro-compounder (15 cc) operated at 200°C and 60 rpm for 3 min. Two processing methods were employed to prepare PLA/P[CL-co-LA]/SiO$_2$ternary systems upon how these silica nanoparticles were dispersed, i.e. either directly into the blend (so-called one-step process) or first into one of the partners before blending (so-called two-step process) into PLA/P[CL-co-LA]-based systems. In the one-step process, both P[CL-co-LA] and silica were directly compounded with PLA, while in the two-step process silica nanoparticles were mixed first in one of the two partners by dissolving them in a minimal of chloroform. It was then followed by a slow evaporation of solvent under stirring and subsequently blended upon the same procedure as the one-step process. Standard samples of resulting PLA-based materials were then prepared by compression molding at 200°C for 10 min.

Characterization techniques

Proton nuclear magnetic resonance (^{1}H NMR) spectra were recorded in CDCl$_3$ using a Bruker AMX-500 apparatus at a frequency of 500 MHz. Size-exclusion chromatography was performed in THF (containing 2 wt-% NEt$_3$) at 35°C using a Agilent liquid chromatograph equipped with a Agilent G322A degasser, an isocratic HPLC pump G1310A (flowrate: 1 mL min^{-1}), a Agilent autosamplerG1329A (loop volume: 100 μL, solution concentration: 1 mg mL^{-1}),a Agilent-DRI refractive index detector G1362A and three columns: a guard column PLgel 10 μm and two columns PLgel mixed-B5 μm. Molecular weight and molecular weight distribution were calculated by reference to a relative calibration curve made of polystyrene standards. Differential scanning calorimetry was performed using a DSC Q2000 from TA Instruments at heating and cooling rates of 10°C min^{-1} under nitrogen flow (second scan). Notched Izod impact tests were performed according to ASTM D256 using a Ray-Ran 2500 pendulum impact tester ($E = 4$ J, mass = 0.668kg and speed = 0.46 m s^{-1}). Tensile tests were performed according to ASTM D638 using a Zwick universal tensile testing machine (speed = 1 mm min^{-1} and preload = 5 N). Dynamic mechanical thermal analyses (DMTAs) were performed under ambient atmosphere using a DMTA Q800

apparatus from TA Instruments in a dual cantilever mode. The measurements were carried out at a constant frequency of 1 Hz, a temperature range from − 100°C to 150°C at a heating rate of 2°C min^{-1}. Data acquisition and analysis of the storage modulus (E'), loss modulus (E''), and loss tangent (tan δ) were recorded automatically. Rheological measurements were performed using a ARES Rheometer from Rheometrics. Frequency sweep measurements were recorded in linear regime conditions at a temperature of 200°C at a strain of 1%. Video-controlled mechanical testing was performed using an electro-mechanical machine Instron 5800 series. This equipment provides access to the true axial stress–strain curve, as well as to the local evolution of volume strain during the test. To ensure the localization of the plastic deformation at the center of the sample, a slight geometrical defect is machined out in the middle of the gage length of the dumbbell-shaped specimen (reduction about of 5% of the specimen width). Seven dot markers, made of dark ink, are printed on the front flat face of the sample (see Supplementary Material Fig. 1). Five of these dots are aligned and equally spaced in the tensile direction, x_3, while the two others are aligned with the central dot along the transverse direction, x_1.The axial true strain in the analysis zone, ε_3, is obtained by a polynomial interpolation of partial strains measured from the displacement of axial markers using Lagrange Transform and following Hencky's definition. For uniaxial tensile testing, the transverse strains, ε_1 and ε_2, are equal if the strain field is transversally isotropic in the center of the neck.[42–45] The volume strain, is simply computed from the trace of the true strain tensor: $\varepsilon_v = \varepsilon_1 + \varepsilon_2 + \varepsilon_3$. The volume strain can be positive, zero or negative. In particular, a volume increase is related to the occurrence of either crazing or cavitation (as in rubber-toughened blends), and in the opposite case, the authors consider that a compaction mechanism is active. Isovolumetric deformation mechanism refers to the activation of

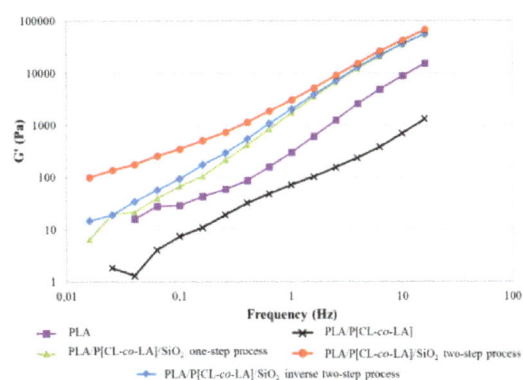

Figure 1 Rheological curves (storage modulus G' in function of frequency v) of neat PLA (square, purple curve), PLA-based materials containing 10 wt-% P[CL-co-LA] (cross, dark curve) and corresponding blends containing 5 wt-% of silica nanoparticles (TS-530) produced via: one-step process (triangle, green curve), two-step process in which SiO$_2$ were first added into the P[CL-co-LA] partner (lozenge, blue curve) and two-step process in which SiO$_2$ were first added into the PLA partner (circle, red curve)

shear banding mechanisms (see Supplementary Material Appendix for more details). Room-temperature fractured surfaces of specimens were sputter-coated with gold and then examined through scanning electron microscopy (SEM) to highlight plastic deformation and possible toughening mechanisms. Accordingly, the room-temperature fracture was performed with a single column tensile test machine type Hounsfield H5KT with 5 kN cell force at a speed of 1 mm min^{-1} while SEM analyses were carried out using a Hitachi SU8020 (100 V–30 kV). Transmission electron microscopy (TEM) was carried out using a Philips CM20 microscope operated at 200 kV to investigate the morphological structure of resulting materials. To record TEM images, the samples were cryomicrotomed at $-100°C$ by a Leica UCT microtome. Analysis software ImageJ was used for the analysis of TEM images to estimate the particle size and their distribution within the matrix.

Results

When P[CL-co-LA] copolymer and silica nanoparticles are used altogether, a simultaneous increase in toughness and stiffness in ternary PLA-based composites can be achieved in function of the processing method and the final blend morphology.[40] In this respect, a close relationship between mechanical performance and microstructure was first investigated on ternary PLA-based materials containing 10 wt-% of P[CL-co-LA] copolymer and 5 wt-% of TS-530 (225 m^2 g^{-1}, hexamethyldisilazane-modified SiO$_2$) upon how these silica nanoparticles were dispersed: either directly into the blend (so-called one-step process) or first into one of the polymeric partners (within PLA or P[CL-co-LA] partner) before blending with the second partner (so-called two-step process). Rheological investigations can be useful to monitor the formation of peculiar morphologies. Indeed, the rheological properties of the materials are strongly influenced by the extent of silica dispersion. It results to a significant increase of the storage modulus at low frequency when a physical network of (nano)particles is formed in the polymeric matrix (Fig. 1). It can be here observed that the storage modulus of PLA/P[CL-co-LA] blends containing 5 wt-% of TS-530 is only altered upon the aforementioned two-stage melt-process in which silica nanoparticles were first added into PLA. As a result, the two-stage melt-process gave access to the formation of a filler-network structure compared to the other methods.

To further explain this network, morphological investigations were carried out by TEM in order to highlight the blend morphology in function of the processing methods used (Fig. 2). In contrast to the one-step process and the two-step process in which SiO$_2$ were first added into the P[CL-co-LA] partner, it can be observed that almost oblong morphologies are achieved in the two-step process in which silica nanoparticles are first dispersed into the PLA partner (see see Fig. 2d). The achievement of such peculiar structures in which a physical network of silica nanoparticles is reached upon this selected processing method might ensure great overall material performances, implying the self-networking of silica nanoparticles at the PLA/P[CL-co-LA] interface.

As a result, toughness of resulting blends is definitely impacted by the way in which these nanoparticles were added (Table 1). Both one-step process (entry A) and reverse two-step process in which silica nanoparticles were first added into P[CL-co-LA] partner (entry B) show decreased toughness. In contrast, in the case of the two-stage melt-process, i.e. by first adding the silica nanoparticles into PLA matrix (entry C), a significant 10-times improvement of impact strength (as high as 27.3 kJ m^{-2} with 5 wt-% TS-530) and an increase of the materials elongation at break (closed to 42%) are recorded.

The effect of the surface treatment and surface area of silica nanoparticles at different contents was subsequently investigated on the morphology and properties of the PLA-based blends containing 10 wt-% of P[CL-co-LA] (Fig. 3). The aforementioned two-stage melt-process was in this regard considered, i.e. by adding first silica nanoparticles into PLA. Accordingly, TS-530 (225 m^2 g^{-1}, hexamethyldisilazane-modified SiO$_2$) and M-5 (200 m^2 g^{-1}, unmodified SiO$_2$) were selected as hydrophobic and hydrophilic silica nanoparticles of similar surface area respectively, while H-5 (300 m^2 g^{-1}, unmodified SiO$_2$) was selected as hydrophilic silica of higher surface area. From our TEM investigations, the major fraction of silica nanoparticles is aggregated around the P[CL-co-LA] dispersed phase as well as at the PLA/P[CL-co-LA] interface (e.g. *Figure 3a and b*). This so-implies the alteration of their final morphology, i.e. the conversion of regularly obtained spherical inclusions (as observed within PLA/P[CL-co-LA] binary blend, see Fig. 2a) into almost oblong structures in the presence of silica nanoparticles (e.g. *Figure 3c*). Surprisingly, whichever the silica nanoparticles considered, the as-formed oblong microdomains are interconnected at higher nanosilica content, leading to the formation of almost co-continuous morphologies (e.g. *Figure 3d*). Indeed, the junction of such peculiar oblong microstructures gradually appears at increasing nanoparticle content to reach an optimum at 10 wt-% TS-530, 5 wt-% M-5 and 3 wt-% H-5. However, it is worth noting that further increasing the nanoparticle content did not promote a larger extent of interconnection, but led to the main aggregation of silica nanoparticles within the blends more likely because of saturation of the PLA/P[CL-co-LA]interface.

Rheological measurements supports this statement, confirming the formation of interconnected structures (i.e. self-networked structures resulting from the migration of nanosilica into the PLA/P[CL-co-LA] interface) within PLA-based materials at these optimum contents of silica nanoparticles (Fig. 4). The large increase of the storage modulus in the lower frequency region again asserts the formation of filler-network structures, as indicated by the formation of a plateau in the low frequency regime.

As a result, the toughness of PLA-based materials is enhanced, up to 18 times, in the presence of silica nanoparticles within the PLA-based blends (Fig. 5) (see Supplementary Material Table 1 for tensile data). Indeed, synergistic toughening effects are obtained after the co-addition of P[CL-co-LA] copolymer and silica nanoparticles into PLA matrix, imparting enhanced toughness compared to PLA/SiO$_2$ (i.e. $\approx$3.5 kJ m^{-2}, 5 wt-%) and PLA/P[CL-co-LA] (i.e. 11.4 kJ m^{-2}, 10 wt-%) binary blends. For instance, the impact strength increases from

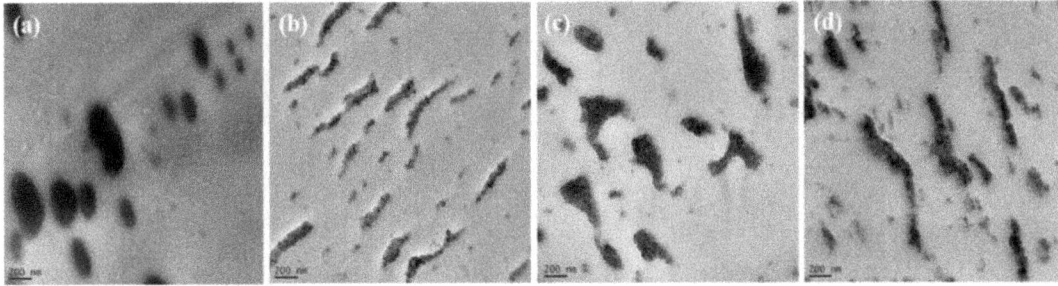

Figure 2 Transmission electron microscopy (TEM) micrographs (scale of 200 nm) of sample sections of PLA-based materials containing 10 wt-% of P[CL-*co*-LA] *a* and corresponding blends containing 5 wt-% of silica nanoparticles (TS-530) depending on the processing methods. Samples produced via: one-step process *b*, two-step process in which SiO$_2$ were first added into P[CL-*co*-LA] partner *c* and two-step process in which SiO$_2$ were first added into PLA partner *d*

Table 1 Influence of the processing method on the mechanical properties of PLA-based materials containing 10 wt-% of P[CL-*co*-LA] and 5 wt-% of silica nanoparticles (TS-530)

Entry	Processing methods	E^{a}/MPa	σ_y^{a}/MPa	ε_y^{a}/%	σ_b^{a}/MPa	ε_b^{a}/%	I.S.^{b}/kJ m^{-2}
Ref. 1	PLA	1860 ± 40	–	–	58 ± 3	4 ± 0.5	2.7 ± 0.1
Ref. 2	PLA/P[CL-*co*-LA]	1680 ± 50	48 ± 3	4.0 ± 0.1	10 ± 0.5	21 ± 5	11.4 ± 0.3
A	One-step PLA/P[CL-*co*-LA]/SiO$_2$	1920 ± 54	52 ± 4	4.0 ± 0.2	11 ± 2	14 ± 4	7.9 ± 0.5
B	Two-step Pre-mix P[CL-*co*-LA]/SiO$_2$ + PLA	1920 ± 48	50 ± 1	4.0 ± 0.2	10 ± 0.2	16 ± 3	8.0 ± 0.4
C	Two-step Pre-mix PLA/SiO$_2$ + P[CL-*co*-LA]	1850 ± 53	47 ± 1	3.6 ± 0.1	13 ± 5	42 ± 16	27.3 ± 1.2

As determined by tensile tests at a drawing speed of 1 mm min^{-1} (ASTM D638). Young's modulus (E), yield stress (σ_y), yield strain (ε_y), fracture stress (σ_b) and strain at break (ε_b).
As determined by notched Izod impact tests (ASTM D256). Impact strength (I.S.).

2.7 kJ m^{-2} for the neat PLA up to 47.2 kJ m^{-2} upon the co-addition of 10 wt-% of P[CL-*co*-LA] and 5 wt-% of M-5. Interestingly enough, a maximum on the ultra-toughness implemented by the peculiar phase-morphologies was even recorded at 10 wt-% of TS-530 and 3 wt-% of H-5 (see Fig. 5). However, once higher loading of silica nanoparticles was used, actually higher than 10 wt-% of TS-530, 5 wt-% of M-5 or 3 wt-% of H-5, a drop in the toughness of resulting PLA-based blends occurs in relation to the presence of silica nanoparticle aggregates within the PLA-based blends (see Fig. 5).

As far as the extent of PLA crystallinity could largely improve impact toughness through a change in the deformation mechanism as crystallization occurs,[46] more investigations about thermo-mechanical features are also required in this work. In this respect and even if the crystallinity degree of PLA-based blends increased from 3 to 13% after adding 10 wt-% of P[CL-*co*-LA] copolymer within PLA, the DSC results did not show any major change on the thermal properties for the PLA/P[CL-*co*-LA] blends in the presence of silica nanoparticles (not reported here). In other terms, whichever the nanoparticles used and the silica content, the crystallinity of the blends remains in the same range of ca. 13–15%, highlighting no real relationship between thermal and impact properties of resulting blends. Dynamic mechanical thermal analysis was then used to explain this statement, measuring the thermodynamic response in flexion mode as a function of temperature for the neat PLA and immiscible PLA/P[CL-*co*-LA] blends upon the relative silica nanoparticles content (Fig. 6). While neat PLA is characterized by an apparent alpha transition (T_α) of about 61°C, two distinct peaks corresponding to both rubbery P[CL-*co*-LA] copolymer and PLA matrix are visible in the case of the binary blend at temperatures of $-37°C$ and 59°C, respectively. Loading silica nanoparticles into these blends led to a shift of the characteristic T_α peak of P[CL-*co*-LA] to higher temperatures and the T_α peak of PLA toward lower temperatures (the difference of $\Delta T_{\tan\delta}$ was, respectively, about 5°C and 7°C at 5 wt-% and 10 wt-% TS-530), both in adequacy with an improved compatibility between the blend components (see Supplementary Material Fig. 2, for additional details).

To give access to the plastic response (i.e. elastic and non-elastic deformations) of resulting PLA-based materials in the presence or not of silica nanoparticles, a video-controlled mechanical testing method was used under uniaxial tension in this work (Figs. 7 and 8). Typical stress–strain curves are then noticed here, displaying very limited plastic deformation in neat PLA in contrast to large plasticity with a significant increase in ductility for the resulting rubber-toughened PLA-based blends (Fig. 7, top). This brittle-to-ductile transition is further supported by volume variation data analysis in which a dual response in terms of non-elastic deformation behavior was clearly recognized, i.e. the coexistence of dilatational (mechanisms inducing volume change such as cavitation, fibrillation and mainly crazing) (Fig. 7, middle) and non-dilatational (mechanisms inducing shearing without any volume change such as shear banding) (Fig. 7, bottom) mechanisms under sub-T_g drawing conditions.

In this respect, Fig. 8 reported the different contributions to the true strain, i.e. elastic strain, dilatational strain and shear strain within these PLA-based materials. As a result, the involved toughening mechanisms are mostly dilatational

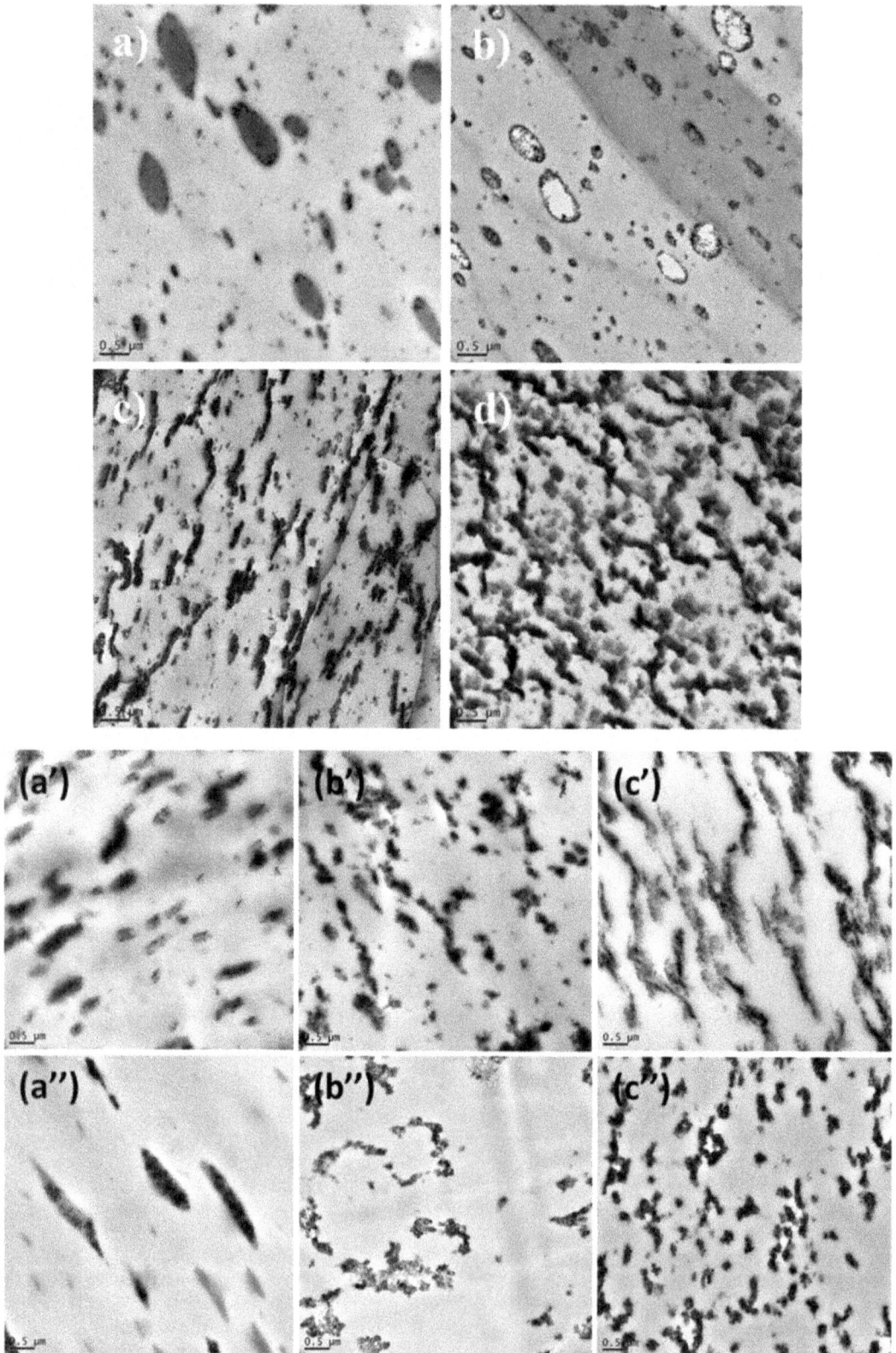

Figure 3   Transmission electron microscopy (TEM) micrographs (scale of 0.5 μm) of sample sections of PLA/P[CL-*co*-LA] blends containing TS-530 *a*, *b*, *c*, *d*, M-5 *a′*, *b′*, *c′* or H-5 *a″*, *b″*, *c″*: 1 wt-% *a*, *a′* and *a″*, 3 wt-% *b*, *b′* and *b″*, 5 wt-% *c*, *c′* and *c″* and 10 wt-% *d* of silica nanoparticles

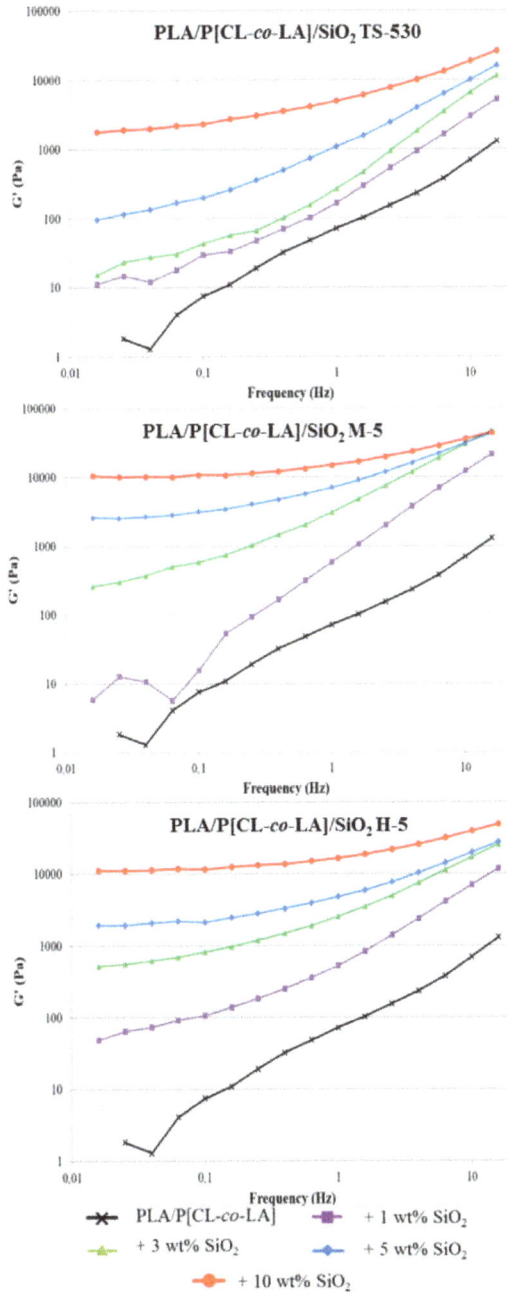

Figure 4 **Rheological curves (storage modulus *G*′ in function of frequency *ν*) of PLA-based materials containing 10 wt-% P[CL-*co*-LA] and TS-530 (top), M-5 (middle) or H-5 (bottom): 1 wt-% (square, purple curve), 3 wt-% (triangle, green curve), 5 wt-% (lozenge, blue curve) and 10 wt-% (circle, red curve) of silica nanoparticles in comparison with PLA/P[CL-*co*-LA] binary blend (cross, black curve)**

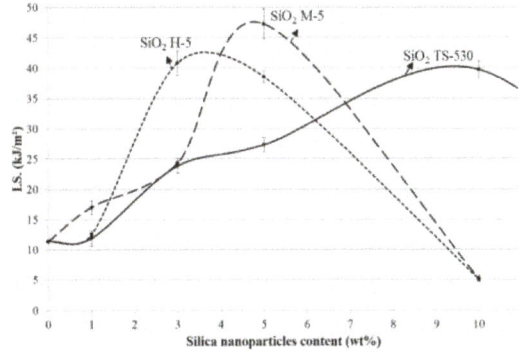

Figure 5 **Dynamic mechanical thermal analysis (DMTA) thermograms (loss modulus in function of temperature) of unfilled PLA (green, solid line), PLA-based materials containing 10 wt-% of P[CL-co-LA] (dark, broken dash line) and corresponding blends filled with 5 wt-% (blue, long dash line) and 10 wt-% (red, short dash line) of silica nanoparticles (TS-530)**

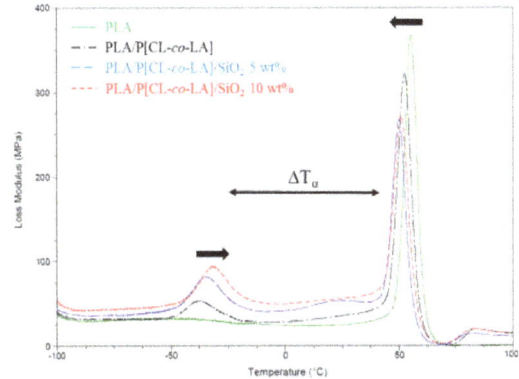

Figure 6 **Notched Izod impact strength of PLA-based materials containing 10 wt-% of P[CL-*co*-LA] copolymer upon addition of silica nanoparticles: TS-530 (solid line), M-5 (long dash line) and H-5 (dash line)**

in nature, i.e. the occurrence of crazing and/or cavitation and fibrillation appeared as the prevailing toughening mechanism for e.g. rubber-toughened PLA/P[CL-co-LA] binary blends (Fig. 8, top). As far as silica nanoparticles are loaded, greater energy dissipative micromechanisms are reached, i.e. enhanced cavitational modes (Fig. 7, middle) and the occurrence of shear banding (Fig. 7, bottom). In this respect, volume variation data analysis on PLA-based blends containing 5 wt-% of nanosilica clearly evidences that the early stage of plasticity is dominated by the occurrence of shear banding (Fig. 8, middle). Surprisingly, reaching the saturation of the PLA/P[CL-co-LA] interface in order to promote the formation of a filler-network and an interconnected structure obviously favored cavitational behavior at an early stage, which became more prominent than shear banding mechanism (Fig. 8, bottom).

High deformations and plasticity are furthermore evidenced on room-temperature fractured surfaces of these PLA-based materials (Fig. 9). As attested by SEM, macroscopic localized zones of micro-voids and micro-fibrils are found after addition of a rubbery P[CL-co-LA] impact modifier within the PLA matrix (Fig. 9, top). Surprisingly, the emergence of greater energy dissipative micromechanisms are further highlighted by loading silica nanoparticles within

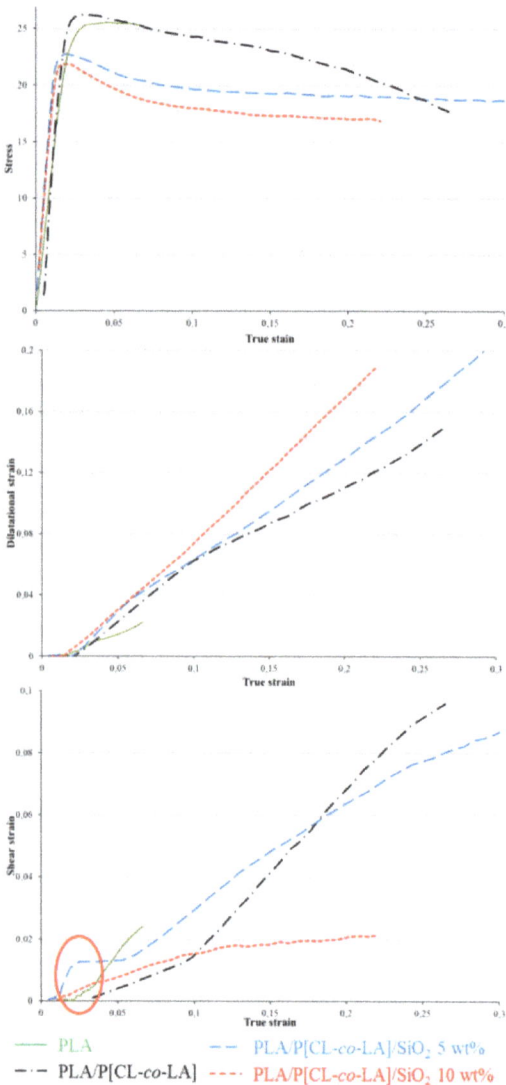

Figure 7 Typical stress-strain curves (top) and respective contributions of dilatational strain (middle) and shear strain (bottom) in unfilled PLA (green, solid line), PLA-based materials containing 10 wt-% of P[CL-co-LA] (dark, broken dash line) and corresponding blends filled with 5 wt-% (blue, long dash line) and 10 wt-% (red, short dash line) of silica nanoparticles (TS-530)

Figure 8 Contributions of elastic strain (ε_{el}, green, solid line), dilatational strain (ε_{cav}, red, short dash line) and shear strain (ε_{sh}, blue, long dash line) as a function of true strain in PLA-based materials containing 10 wt-% of P[CL-co-LA] (top) and corresponding blends filled with 5 wt-% (middle) and 10 wt-% (bottom) of silica nanoparticles (TS-530) as determined by video-controlled mechanical tests

these PLA/P[CL-co-LA]-based blends (Fig. 9, bottom). Indeed, SEM investigations reveal large-scale deformation and interpenetrating micro-fibrils within PLA/P[CL-co-LA]/SiO₂ ternary blends.

Discussions

For polymer systems, both thermodynamic (e.g. interfacial tension) and kinetic factors (e.g. shear stress and processing conditions) govern the final phase structure. This can explain how loading silica nanoparticles into the immiscible PLA/P[CL-co-LA] impact modifier blends may affect the morphological characteristic features. Indeed, thermodynamic

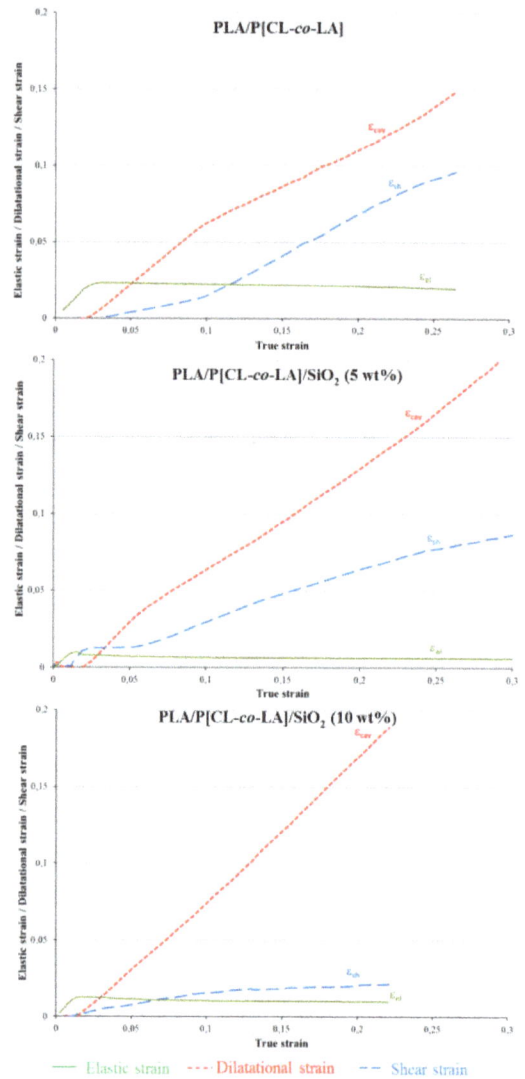

factors are in favor of an embedded structure reducing the interfacial tension, while kinetic factors lead to a two-phase system independently of shear forces and viscosity of each polymeric partner. In this work, the kinetic contribution could be a prevailing parameter to explain why the silica nanoparticles can lead to these peculiar morphologies (i.e. oblong and even co-continuous structures), and therefore affect the overall performances of resulting nanocomposites. For instance, the reverse two-process strategy (where silica nanoparticles were first dispersed into the P[CL-co-LA] copolymer or the one-process strategy, rheological measurements indicate the absence of filler-network structure when 5 wt-% of TS-530 was used. In contrast and in the case of the

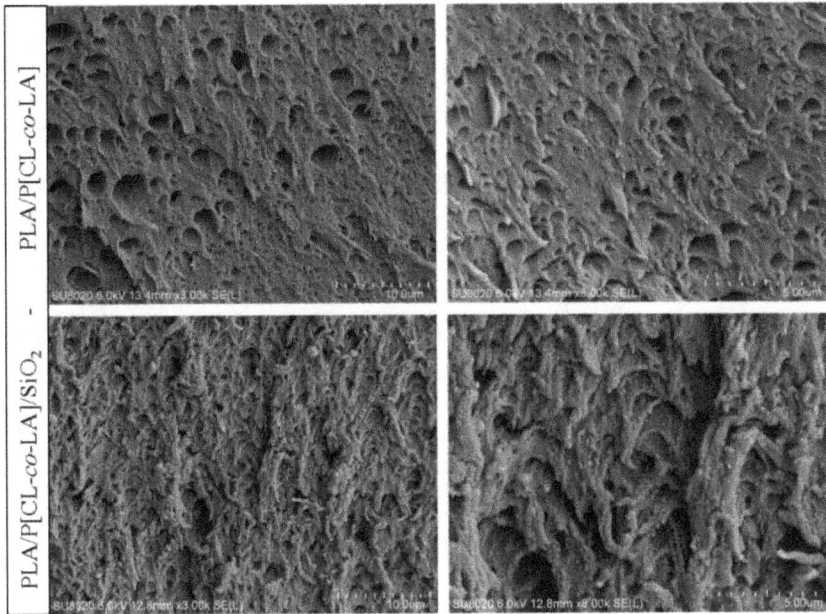

Figure 9 Scanning electron microscopy (SEM) micrographs (scale of 10 μm on left and zoom to 5 μm on right) of notched surfaces of PLA-based materials containing 10 wt-% of P[CL-*co*-LA] (top) and corresponding blend filled with 5 wt-% of silica nanoparticles (TS-530) (bottom)

two-process strategy in which silica nanoparticles were first dispersed into the PLA partner, the formation of filler-network structures (as previously evidenced by rheology, Figs. 1 and 4) can be noticed and be explained by the fact that the silica nanoparticles were mainly located in the PLA phase and could readily migrate toward the interface of the blends on processing. As a result, oblong structures are here obtained as evidenced by TEM measurements (Figs. 2 and 3). Similar trend was already reported by Yang et al.[19] highlighting the dependence of the phase morphology and the formation of the filler-network structure upon interfacial interactions in relation to the migration of silica to the interface. However, it could not be excluded that these silica nanoparticles can affect the overall viscosity of these blends and therefore can control the migration extent of silica nanoparticles within these blends as an additional feature.

From a thermodynamic viewpoint, wetting coefficient and interfacial tension are among the main parameters that can also affect the preferential affinity between the particles and polymeric phases. Such a behavior has been reported after addition of nanoparticles into immiscible polymer blends as an effective method to compatibilize the resulting blend, owing to their migration and localization at the blend interface used in these works.[15–21] However, Lipatov et al.[32–34] highlighted that the effect of filler introduction is mostly based on the thermodynamics of interaction near the surface. In this work, the silica nanoparticles do not have any specific affinity with respect to both polymeric partners, and tend to readily migrate at the PLA/P[CL-*co*-LA] interface where the interfacial adhesion is the lowest. If considered, the selective location of nanoparticles could be merely affected upon the hydrophilic–hydrophobic balance (and then the nanoparticles surface treatment). As a result,

different optimum contents for these silica nanoparticles are reached upon their ability to self-agglomerate them at the interface upon their non-affinity towards both polymeric partners (PLA and P[CL-*co*-LA]). Accordingly, these different types of silica nanoparticles can be classified as follows: TS-530 < M-5 < H-5. A possible explanation to support this classification may arise from the preferential wetting between silica nanoparticles and P[CL-*co*-LA] phase upon both surface treatment and size of silica nanoparticles, resulting in the migration of the major fraction of silica nanoparticles at the PLA/P[CL-*co*-LA] interface. For instance, related to the facilitated motion of these silica nanoparticles to migrate to the interface as a function of their particle size, hydrophilic (without any surface treatment) silica nanoparticles of smaller particle size (higher surface area) can easily migrate from PLA matrix phase to the P[CL-*co*-LA] dispersed phase and their interface. In other terms, the lowest optimum content is achieved when hydrophilic silica nanoparticles (H-5) are used by decreasing their interfacial adhesion more significantly than others.

The morphological characteristic features, a key-parameter of this study, may explain how the nanoparticles in function of their surface treatment and surface area can strongly control ultra-toughness of these blends (Fig. 5 as a reminder). It is well known that the overall shape, average dimensions and related size distribution of the dispersed rubbery phase must be properly controlled to reach the compositions with optimal mechanical properties. From this work, the highest alteration in the blend morphology is guided by the extent of migration of silica nanoparticles into the PLA/P[CL-*co*-LA] interface. Zhang et al.[47] reported similar trends within immiscible polypropylene/polystyrene blends, showing a drastic reduction of polystyrene phase-size and a very homogeneous size distribution through the addition of silica nanoparticles. This owes to the role of silica nanoparticles acting as interfacial compatibilizer and allowing enhanced compatibility degree and lower interfacial tension between both immiscible partners (as previously evidenced by DMTA, Fig. 6).This therefore induces the self-networking of silica nanoparticles at the interface and leads to the formation of these peculiar structures (see Fig. 3). As a result, the as-modified interfacial adhesion between polymeric partners upon the addition of silica nanoparticles led to a marked improvement of fracture-resistance for the resulting blends through these peculiar energy dissipative structures. At this stage, it can be claimed out that the resulting `infinite' network made of elongated rubbery domains affords a more efficient toughening scheme, as compared to the individually dispersed spherical microdomains.

Regarding the mechanism involved, the addition of silica nanoparticles to the immiscible PLA/P[CL-co-LA] blend influences the coalescence and breakup of the rubbery droplets and causes the shape and average size of the dispersed microdomains depending on the location of nanosilica (Scheme 1). For uncompatibilized blends such as PLA/P[CL-co-LA] blends, the morphology evolution is derived from the formation of large droplets of the dispersed rubbery P[CL-co-LA] phase within the PLA matrix in the molten state and the deformation of P[CL-co-LA] droplets under the shear force. During processing, the breakup of the dispersed/deformed P[CL-co-LA] phase into smaller microdomains is followed by a coalescent process. However, the system tends to minimize its total free energy and loading silica nanoparticles within this

PLA/P[CL-co-LA] blend results in the migration of nanosilica into the phase for which they exhibit the stronger affinity and the lower interfacial tension, i.e. mainly at the PLA/P[CL-co-LA] interface. The nanoparticle selective location at the PLA/P[CL-co-LA] interface stabilizes the polymer phase morphology by decreasing the interfacial interaction, while the nanosilica migration and their self-networking among the P[CL-co-LA] domains take part in the interconnected or even co-continuous organizations. As reported, loading silica nanoparticles into PLA/P[CL-co-LA] blends leads to the increase in viscosity of the resulting compositions, retarding the coalescence of P[CL-co-LA] droplets and therefore affecting the overall shape and average size of dispersed microdomains (as proposed in Scheme 1). Similar results were reported by e.g. Ahn and Paul,[48] showing that the addition of clay into rubber-toughened Nylon-6 affected the dispersion of the rubber phase resulting in larger and elongated rubber micro-domains. Co-continuous structures were even reported by several authors as a result of enhanced compatibility upon the addition of nanoadditives within immiscible polymer blends.[49-51] Accordingly, thermodynamic and kinetic driving forces are involved by enhancing the compatibility of PLA/P[CL-co-LA] blends depending on the specific localization of silica nanoparticles at their interface and by playing on the breakup/coalescence equilibrium, respectively. Importantly, such peculiar morphologies and the as-induced high toughness were fully preserved after reshaping/remolding the blends, giving similar values of the impact strength and confirming the stability of the observed structures.

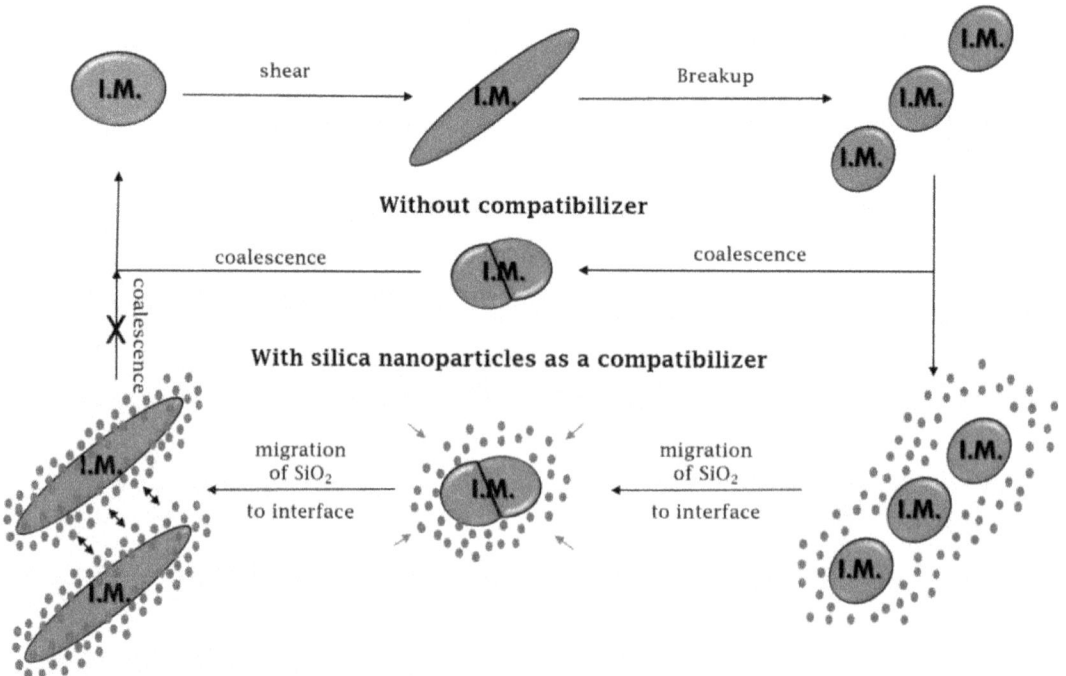

Scheme 1 Representation of the morphology evolution in immiscible PLA/P[CL-co-LA] blends compatibilized by silica nanoparticles

From the literature, rubber-toughening of polymers is extensively described based on toughening and fracture mechanisms, as called crazing, shear yielding, cavitation or debonding.[52–56] In this respect, Stoclet et al.[57] characterized the complex plastic deformation behavior exhibited by PLA, considering both the elementary mechanisms and the drawing kinetics. As expected, the crazing mechanism is responsible for the brittle character of PLA. In contrast, rubber-modified PLA can undergo one or a combination of the most dissipative toughening mechanisms. This owes to the classical role of the dispersed rubbery phase acting as local stress concentrators at many sites throughout the material and initiating greater energy dissipative micromechanisms. Indeed, starting from its optimum microdomain average size around 0.9 µm, it might result that internal cavitation of rubber droplets (needing droplets of size higher than 200 nm) is easily achieved within the rubber-toughened PLA/P[CL-co-LA] binary blend.[58,59] This releases the hydrostatic strain energy and involves, to some extent, the dissipation of energy and emergence of other energy absorbing mechanisms for the main matrix, e.g. craze initiation (at such relatively high microdomain size, higher than 0.5 µm). The emergence of such energy dissipative micromechanisms initiated by the rubbery microdomains is further highlighted in Fig. 7. As aforementioned, the occurrence of crazing (and/or cavitation and fibrillation) might be considered as the prevailing toughening mechanism within these PLA-based materials. The contribution of the latter increases upon the morphological change (from droplets to ribbons and finally to interconnected and even co-continuous structures) as induced by the presence of nanosilica at the PLA/P[CL-co-LA] interface. This suggests us that craze initiation occurs at a low stress level from these peculiar morphologies. The high volume of cavitated matter and the capacity of the continuous rubber phase to stabilize the damaged microstructure allow achieving high energy dissipation. In this respect, the authors may postulate that the high capacity for energy dissipation is related to the interconnection of such peculiar microstructures, which promotes extensive plasticity in thinner matrix ligaments,

as compared to the previous binary blends. As a result, the contribution of dilatational toughening mechanisms increases with nanosilica content, enhancing the improved energy dissipation on impact stress conditions as another consequence (Fig. 7, middle). Besides craze growth, shear band nucleation further appears at an early stage in the presence of nanosilica, strengthening the ability of these materials to dissipate more energy before fracture (Fig. 7, bottom). It may be considered that the localization of silica nanoparticles at the PLA/P[CL-co-LA] interface strengthens the adhesion between polymer components, promoting interfacial cavitation under stress and releasing the triaxial state of stress. This therefore promotes shear band nucleation. Moreover, cavitational behavior became effective upon increasing deformation. Accordingly, this phenomenon favors the capacity of shear bands to stabilize craze growth, leading to improved energy dissipation before rupture. Even if the occurrence of shear banding is less pronounced than their cavitational counterpart, the combination of such energy dissipative micromechanisms leads to highly localized plastic deformation, endowing ultra-toughness of resulting materials. Scanning electron microscopy analysis along the draw axis on deformed samples confirms the presence of these specific deformations (Scheme 2). In the case of PLA/P[CL-co-LA] blends, macroscopic localized zones of micro-voids and micro-fibrils are observed after addition of a rubbery P[CL-co-LA] impact modifier within the PLA matrix and are related to crazing process on deformation. Further loading nanosilica within this PLA/P[CL-co-LA]-based blends imparted great ductility and impact strength to the materials, highlighting the uneven morphology change (from large droplets to small and long ribbons) and the formation of 'infinite' self-networked (and even co-continuous) structures mediated by silica nanoparticles as a more effective way to toughen glassy materials. These features therefore support the statement that higher stress levels are required from these peculiar structures to generate cavitation and crazes in these materials, involving a combination of shear yielding and multiple crazing mechanisms.

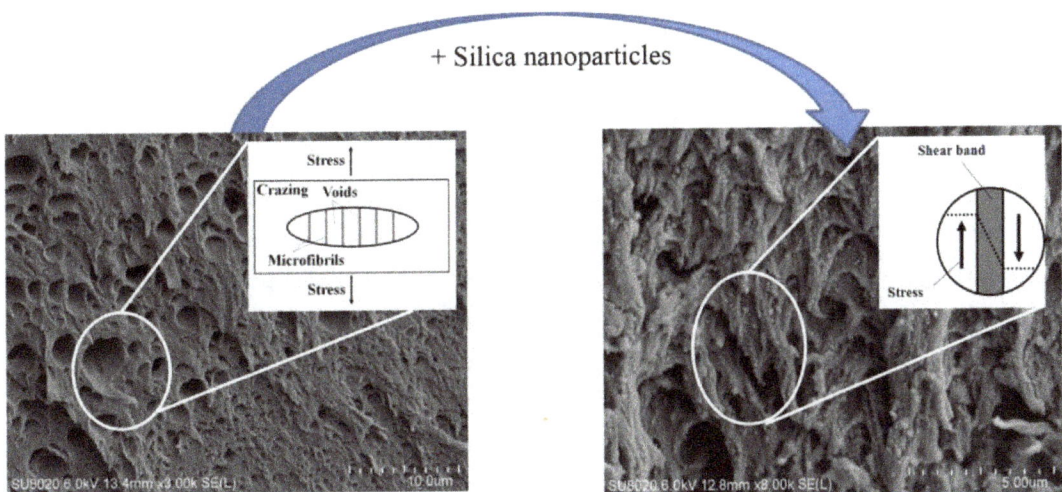

Scheme 2 Representation of brittle-to-ductile transition of immiscible PLA/P[CL-co-LA] blends after addition of silica nanoparticles

Conclusions

The compatibility of PLA/P[CL-*co*-LA] blends can be dramatically improved by the addition of silica nanoparticles, leading to the formation of self-networked (and even co-continuous) structures originated from rubbery impact modifier within the matrix. It was found that silica nanoparticles were mostly located at the PLA/P[CL-*co*-LA] interface, strengthening the interface of these peculiar morphologies. Using appropriate processing method and adjusting the affinity of silica nanoparticles through surface treatment, surface area and silica content, the conversion of rubbery spherical cavities into e.g. oblong structures in the presence of silica nanoparticles can be successfully reached, promoting, to some extent, filler-networks at 10 wt-% of TS-530, 5 wt-% of M-5 and 3 wt-% of H-5. The compatibilization mechanism of silica nanoparticles in PLA/P[CL-*co*-LA] blends and the progression to the uneven morphology change were proposed based on the coalescence and coarsening process. As reported, loading silica nanoparticles within PLA/P[CL-*co*-LA] blends leads to the increase in viscosity of resulting blends, impeding the coalescence of P[CL-*co*-LA] droplets while the migration and the nanosilica self-networking among the P[CL-*co*-LA] domains take part in the interconnected or even co-continuous organizations. Surprisingly, the co-addition of a P[CL-*co*-LA] rubbery impact modifier and silica nanoparticles in PLA through a two-step process increased significantly the material toughness as attested by the 18-fold increase recorded in impact strength with optimum stiffness/toughness balance. Accordingly, the resulting `infinite' network made of elongated rubbery domains mediated by the presence of silica nanoparticles affords a more efficient toughening strategy as compared to the individually dispersed spherical inclusions, which are often encountered in immiscible polymer blends. From these peculiar structures, it might result that a combination of shear yielding and multiple cavitational mechanisms provided large-scale deformation and high energy dissipation, determining the final mechanical performances of the as-produced PLA-based materials.

Conflicts of interest

The authors have no conflicts of interest to declare.

Acknowledgements

This research has been funded by the European Commission and Région Wallonne FEDER program in the frame of `Pôle d'Excellence Materia Nova' and OPTI²MAT program of excellence, by the Interuniversity Attraction Poles program initiated by the Belgian Federal Science Policy Office (PAI 6/27 and P7/05) and by FNRS-FRFC. J. Odent thanks F.R.I.A. for its financial support thesis grant. J.-M. Raquez is `chercheur qualifié' by the F.R.S.-FNRS. Jean-Michel Thomassin is a `Logistics collaborator' by the F.R.S.-FNRS. Franck Lauro and Jean-Michel Gloaguen are `Professeur des Universités' at UVHC and Université Lille1, respectively.

References

1. Z. H. Liu, X. D. Zhang, X. G. Zhu, Z. N. Qi, F. S. Wang, R. K. Y. Li and C. L. Choy: *Polymer*, 1998, **39**, (21), 5047–5052.

2. S. Wu: *Polym. Eng. Sci.*, 1990, **30**, (13), 753–761.
3. K. Cho, J. Yang, S. Yoon, M. Hwang and S. V. Nair: *J. Appl. Polym. Sci.*, 2005, **95**, (3), 748–755.
4. Y. Okamoto, H. Miyagi, M. Kakugo and K. Takahashi: *Macromolecules*, 1991, **24**, (20), 5639–5644.
5. J. Odent, J. -M. Raquez, E. Duquesne and P. Dubois: *Eur. Polym. J.*, 2012, **48**, (2), 331–340.
6. A. K. Vegt and J. J. Elmendorp: 'Blending of incompatible polymers', in 'Integration of fundamental polymer science and technology' (ed. L. A. Kleintjens, et al.), 381–389; 1986, Netherlands, Springer.
7. C. Koning, M. Van Duin, C. Pagnoulle and R. Jerome: *Prog. Polym. Sci.*, 1998, **23**, (4), 707–757.
8. G. Wildes, H. Keskkula and D. R. Paul: *Polymer*, 1999, **40**, (20), 5609–5621.
9. I. Fortelný: *Eur. Polym. J.*, 2004, **40**, (9), 2161–2166.
10. S. Lyu, T. D. Jones, F. S. Bates and C. W. Macosko: *Macromolecules*, 2002, **35**, (20), 7845–7855.
11. A. E. Nesterov and Y. S. Lipatov: *Polymer*, 1999, **40**, (5), 1347–1349.
12. M. L. Di Lorenzo and M. Frigione: 'Compatibilization criteria and procedures for binary blends: a review'; 1997, Tel Aviv, Israel, Freund.
13. J. K. Kim, S. Kim and C. E. Park: *Polymer*, 1997, **38**, (9), 2155–2164.
14. M. F. Díaz, S. E. Barbosa and N. J. Capiati: *Polymer*, 2005, **46**, (16), 6096–6101.
15. B. B. Khatua, D. J. Lee, H. Y. Kim and J. K. Kim: *Macromolecules*, 2004, **37**, (7), 2454–2459.
16. L. Ashabi, S. H. Jafari, B. Baghaei, H. A. Khonakdar, P. Pötschke and F. Böhme: *Polymer*, 2008, **49**, (8), 2119–2126.
17. X. -Q. Liu, W. Yang, B. -H. Xie and M. -B. Yang: *Mater. Des.*, 2012, **34**, (0), 355–362.
18. M. Y. Gelfer, H. H. Song, L. Liu, B. S. Hsiao, B. Chu, M. Rafailovich, M. Si, V. Zaitsev and J. Polym: *J. Polym. Sci. B Polym. Phys.*, 2003, **41**, (1), 44–54.
19. H. Yang, X. Zhang, C. Qu, B. Li, L. Zhang, Q. Zhang and Q. Fu: *Polymer*, 2007, **48**, (3), 860–869.
20. I. Kelnar, J. Rotrekl, J. Kotek, L. Kaprálková and J. Hromádková: *Eur. Polym. J.*, 2009, **45**, (10), 2760–2766.
21. H. -S. Lee, P. D. Fasulo, W. R. Rodgers and D. R. Paul: *Polymer*, 2005, **46**, (25), 11673–11689.
22. E. Moghbelli, H. -J. Sue and S. Jain: *Polymer*, 2010, **51**, (18), 4231–4237.
23. W. Wu and Z. D. Xu: *Acta Pharmacol. Sin.*, 2000, **1**, 99–104.
24. S. Xinqing, Q. Jinliang, H. Youqing, L. Yiqun, Z. Xiaohong, G. Jianming, T. Banghui and S. Zhihai: *Acta Polymerica Sinica.*, 2005, **1**, 142–148.
25. C. G. Ma, Y. L. Mai, M. Z. Rong, W. H. Ruan and M. Q. Zhang: *Compos. Sci. Technol.*, 2007, **67**, (14), 2997–3005.
26. P. R. Hornsby and K. Premphet: *J. Appl. Polym. Sci.*, 1998, **70**, (3), 587–597.
27. F. Sahnoune, J. M. Lopez Cuesta and A. Crespy: *Polym. Eng. Sci.*, 2003, **43**, (3), 647–660.
28. S. Chang, T. Xie and G. Yang: *J. Appl. Polym. Sci.*, 2006, **102**, (6), 5184–5190.
29. H. Yang, Q. Zhang, M. Guo, C. Wang, R. Du and Q. Fu: *Polymer*, 2006, **47**, (6), 2106–2115.
30. J. Jancar and A. T. Dibenedetto: *J. Mater. Sci.*, 1994, **29**, (17), 4651–4658.
31. Y. Long and R. A. Shanks: *J. Appl. Polym. Sci.*, 1996, **61**, (11), 1877–1885.
32. Y. S. Lipatov: *Prog. Polym. Sci.*, 2002, **27**, (9), 1721–1801.
33. A. E. Nesterov, Y. S. Lipatov and T. D. Ignatova: *Eur. Polym. J.*, 2001, **37**, (2), 281–285.
34. Y. S. Lipatov, A. E. Nesterov, T. D. Ignatova and D. A. Nesterov: *Polymer*, 2002, **43**, (3), 875–880.
35. K. Madhavan Nampoothiri, N. R. Nair and R. P. John: *Bioresour. Technol.*, 2010, **101**, (22), 8493–8501.
36. R. Auras, B. Harte and S. Selke: *Macromol. Biosci.*, 2004, **4**, (9), 835–864.
37. K. S. Anderson, K. M. Schreck and M. A. Hillmyer: *Polym. Rev.*, 2008, **48**, (1), 85–108.
38. S. Ishida, R. Nagasaki, K. Chino, T. Dong and Y. Inoue: *J. Appl. Polym. Sci.*, 2009, **113**, (1), 558–566.
39. J. Odent, P. Leclère, J. -M. Raquez and P. Dubois: *Eur. Polym. J.*, 2013, **49**, (4), 914–922.
40. J. Odent, Y. Habibi, J. -M. Raquez and P. Dubois: *Compos. Sci. Technol.*, 2013, **84**, (0), 86–91.
41. F. Gubbels, S. Blacher, E. Vanlathem, R. Jerome, R. Deltour, F. Brouers and P. Teyssie: *Macromolecules*, 1995, **28**, (5), 1559–1566.
42. C. B. Bucknall and D. Clayton: *J. Mater. Sci.*, 1972, **7**, (2), 202–210.
43. S. I. Naqui and I. M. Robinson: *J. Mater. Sci.*, 1993, **28**, (6), 1421–1429.
44. P. François, J. M. Gloaguen, B. Hue and J. M. Lefebvre: *J. Phys. III France*, 1994, **4**, (2), 321–329.

45. C. G'Sell, J. M. Hiver and A. Dahoun: *Int. J. Solids Struct.*, 2002, **39**, (13–14), 3857–3872.
46. A. C. Renouf-Glauser, J. Rose, D. F. Farrar and R. E. Cameron: *Biomaterials*, 2005, **26**, (29), 5771–5782.
47. Q. Zhang, H. Yang and Q. Fu: *Polymer*, 2004, **45**, (6), 1913–1922.
48. Y. -C. Ahn and D. R. Paul: *Polymer*, 2006, **47**, (8), 2830–2838.
49. Y. Li and H. Shimizu: *Polymer*, 2004, **45**, (22), 7381–7388.
50. F. Gubbels, R. Jerome, P. Teyssie, E. Vanlathem, R. Deltour, A. Calderone, V. Parente and J. L. Bredas: *Macromolecules*, 1994, **27**, (7), 1972–1974.
51. Y. -H. Wang, Y. -Y. Shi, J. Dai, J. -H. Yang, T. Huang, N. Zhang, Y. Peng and Y. Wang: *Polym. Int.*, 2013, **62**, (6), 957–965.
52. W. G. Perkins: *Polym. Eng. Sci.*, 1999, **39**, (12), 2445–2460.
53. R. M. Ikeda: *J. Appl. Polym. Sci.*, 1993, **47**, (4), 619–629.
54. P. A. O'Connell and G. B. McKenna: 'Yield and crazing in polymers', in 'Encyclopedia of polymer science and technology'; (ed Herman F. Mark); 627–681; 2002, New York, John Wiley & Sons, Inc.
55. I. Narisawa and A. Yee: 'Crazing and fracture of polymers', in 'Materials science and technology'; (ed. Edwin L. Thomas), 698–765; 2006, Weinheim, Wiley-VCH Verlag GmbH & Co. KGaA.
56. D. K. Mahajan and A. Hartmaier: *Phys. Rev. E*, 2012, **86**, (2), 021802.
57. G. Stoclet, J. M. Lefebvre, R. Séguéla and C. Vanmansart: *Polymer*, 2014, **55**, (7), 1817–1828.
58. A. M. Donald and E. J. Kramer: *J. Appl. Polym. Sci.*, 1982, **27**, (10), 3729–3741.
59. D. Dompas, G. Groeninckx, M. Isogawa, T. Hasegawa and M. Kadokura: *Polymer*, 1994, **35**, (22), 4750–4759.

17

Morphology of PA6 nanocomposites prepared by pressurized insertion of aqueous nanoparticle dispersions

Irene Hassinger*[1] and Martin Gurka[2]

[1]Rensselaer Polytechnic Institute, Troy, New York, USA
[2]Institute for Composite Materials, Kaiserslautern, Germany

Abstract Aqueous nanoparticle dispersions were introduced in extrusion processing using high enough pressure to build up a blend consisting of aqueous dispersion and polymer melt. Two mechanisms occurring during this process were investigated experimentally utilizing three different nanoparticles (TiO_2, $BaSO_4$, SiO_2) dispersed in water. The results were compared to a simplified theoretical model, describing the particle diffusion from the dispersing agent into the polymer melt and the competition of reduction of droplet size by shear forces of the mixing element and the build-up of bigger droplets by surface tension of polymer melt and dispersing agent. The smallest droplet size, occurring in the latter process, could be determined. The size was determined by the ratio of the viscosity of polymer melt and dispersing agent, their relative surface tensions and the shear stresses in the extruder. Such calculated smallest agglomerate size was not smaller than 100 nm in the pursued range. Furthermore, very dense agglomerates were formed. The SiO_2 nanoparticles did not undergo a diffusion mechanism ending up in big remaining agglomerates. As remarkably small nanoparticle agglomerates (<100 nm for TiO_2) and primary particles

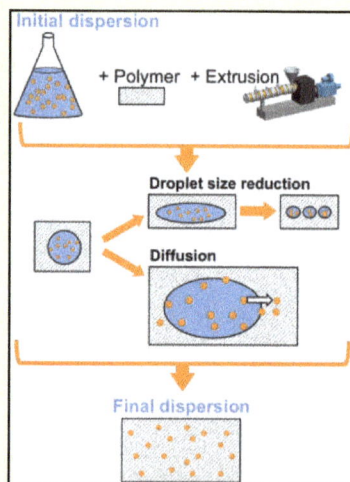

($BaSO_4$) could be found, it could be assumed that the diffusion mechanism also occurred. Thus, the pressurized injection of aqueous nanoparticle dispersions represents an adequate processing method for nanocomposites with $BaSO_4$ and TiO_2 fillers.

Keywords Review, Extrusion, Polyamide, Morphology, Dispersion mechanisms

Introduction

Fundamental changes of mechanical and thermal properties of polymers can be achieved by using nanofillers.[1–6] Nanoparticles have great impact on the abovementioned properties because of their large specific surface area. A strong dependence of the properties of the polymer nanocomposite (PNC) on:

- the chemical composition and structure of the polymer;
- the filler type, shape, size and size distribution;
- the filler content, polymer–filler interaction, filler orientation and dispersion state

can be found.[7–14]

To exploit the full potential of nanoparticles, they have to be well dispersed in the polymer matrix, maximizing the interface area between filler particles and surrounding matrix.

The advantage of thermoplastic nanocomposites is their easy processing, which allows cost-efficient and fast manufacturing in an industrial scale via e.g. extrusion.[15] Thermoplastic nanocomposites are traditionally made by melt-compounding of dry powders.[13] Agglomerates that exist already in the initial nanoparticle powders, therefore, have to be de-agglomerated during the processing.[13] Thus, the production of well-dispersed nanocomposites poses a challenge,[13] as the dispersion is limited, by achievable shear rate, residence time in the extruder, agglomerate strength and polymer–particle interaction (relative surface energies) and the viscosity of the polymer melt.

This paper is aimed at research if the application of nanoparticles in aqueous dispersions improves the nanoparticle dispersion in the PNC. Therefore, aqueous nanoparticle dispersions with an excellent *a priori* dispersity were utilized. A material processing route was developed that allowed the injection of these aqueous dispersions in the polymer melt during extrusion processing. Hence, the conservation of the good dispersion state of the initial aqueous dispersion during processing gives insight in the polymer–particle and the polymer–dispersion agent interaction.[16]

Nanoparticles that are not stabilized tend to agglomerate immediately. Thus, nanoparticle dispersions need to be stabilized by proper modification of their surface and choice of the right dispersion agent. Injecting the aqueous nanoparticle dispersions under pressure ensures a state where polymer and aqueous nanoparticle dispersion can be blended together in the extruder until the nanoparticles can transfer from the water in the polymer without agglomeration. In the following, the water can be evaporated and the nanoparticles remain in the polymer matrix.

Results dealing with non-polar polypropylene (PP) can be found in Refs. 16,17.

When nanoparticles remain in the agent they are synthesized in and they are used in this dispersion agent, the authors call the dispersion in situ nanoparticle dispersions. García et al.[18] used such dispersions in extrusion processing to produce thermoplastic PNC. Colloidal SiO_2 dispersions were used with a PA6 matrix. The dispersion agent water did not evaporate because the pressure in the mixing zone (30 bar) was higher than the evaporation pressure of water at the used temperature.[18] García et al.[5] also analyzed the difference of the dispersion quality of PNC that was produced via injection of an aqueous SiO_2 dispersion and via feeding of a SiO_2 powder. The dispersion quality of the colloidal SiO_2 from the aqueous dispersion was better than the dispersion that could be gained by dry powder compounding.[5]

From scanning electron microscopy (SEM), the authors analyzed the dispersion quality of three different fillers (TiO_2, $BaSO_4$, SiO_2) in a polyamide 6 matrix and compared the experimental results with the theoretical dispersion quality that can be achieved by different mechanisms, particle diffusion and droplet fragmentation during extrusion, that are described in the next paragraph.

Dispersion mechanisms

The blend of an aqueous nanoparticle dispersion and a polymer melt exists until the dispersion agent is evaporated during processing. Looking at a blend of an aqueous nanoparticle dispersion and a polymer melt, two mechanisms can occur to disperse the nanoparticles in the polymer: first, the fragmentation of droplets of dispersion or their growth, induced by shear and, second, the diffusion of dispersed nanoparticles from the dispersion agent into the polymer melt. These mechanisms are illustrated in the scheme in Fig. 1. The transfer of the particles in the polymer melt by diffusion requires time, while droplet fragmentation is controlled by the total shear energy during processing.

The shear stress in the extruder distributes the aqueous dispersion into smaller droplets. Thus, this blend can be treated like an emulsion. The tension, maintaining the droplet form is in opposition to the deforming shear stress τ_e. The relation of these two values determines to which extent a droplet is deformed in a shear field. This value is called Weber number.[19–21] The maintaining tension can be described by

the capillary pressure (p_c), which is proportional to the interfacial tension γ divided by the droplet size d_d[19–21]

$$p_c = \frac{4\gamma}{d_d} \qquad (1)$$

Equation (1) applies for spherical drops with a diameter d_d. The capillary pressure increases with decreasing drop size. Consequently, the Weber number (We) is[19–21]

$$We = \frac{\tau_e}{p_c} = \frac{\tau_e d_d}{4\gamma} \qquad (2)$$

The critical Weber number We_{crit} describes which minimal droplet size can be achieved at given shear stress.

The critical Weber number is only dependent on the relation of the viscosities of the dispersed and the continuous phase (η_d and η_c, respectively) and the kind of shear flow [19–21]. In laminar flow, the critical Weber number exhibits a minimum in the range between 0.1 and 1 ($0.1 < \frac{\eta_d}{\eta_c} < 1$). Beyond this, the critical Weber number is increasing again; therefore, the minimal droplet size is also increasing.[19]

If diffusion of particles into the polymer is not taking place during PNC production, the remaining agglomerate size in the PNC will be determined by the drop size achieved through droplet fragmentation during the mixing process because of shear stress. At the moment, when the dispersion agent is evaporated, the nanoparticles remaining from each droplet will form agglomerates immediately.

In order to get small agglomerates, the minimal droplet size should be as small as possible. Therefore,

- γ should be small,
- We_{crit} should be small (($\eta_c/10) < \eta_d < \eta_c$), and
- τ_e should be large.

Experimental

Materials

As matrix the authors used polyamide 6 (Ultramid B24 N 03 from BASF SE, Germany), well suitable for high-speed spinning of fibers.

The aqueous TiO_2 nanoparticle dispersion (trade name: Hombitec RM300 WP) was a top–down dispersion provided from Sachtleben Chemie GmbH, Germany, prepared by dispersion of dry particles in a dissolver. The primary particle size was 15 nm. The filler content was given at 17.3 vol.-% (45.5 wt.-%). The suspension contents additives for stabilization.

The $BaSO_4$ dispersions were kindly provided by MJR Pharmjet GmbH, Germany. They were produced via precipitation reaction in a microjet reactor. The particle size was found at 45–50 nm, and the particle concentration was 15 wt.-% (3.86 vol.-%). The aqueous dispersions were contaminated by sodium chloride, acetic acid and sodium hydroxide (NaOH).

In addition, commercially available SiO_2 dispersions from Evonik Nanoresins GmbH, Germany (trade name: Nanopol XP 20/0170), were used. The particle size was 20 nm, and the particle concentration in the aqueous dispersion was 24.43 vol.-% (41.56 wt.-%). The suspension contents additives for stabilization.

Compounding

The optimized extrusion parameters were found at a rotation speed of $n = 100 \text{min}^{-1}$ and a throughput of 9 kg h^{-1}.

(a)

(b)

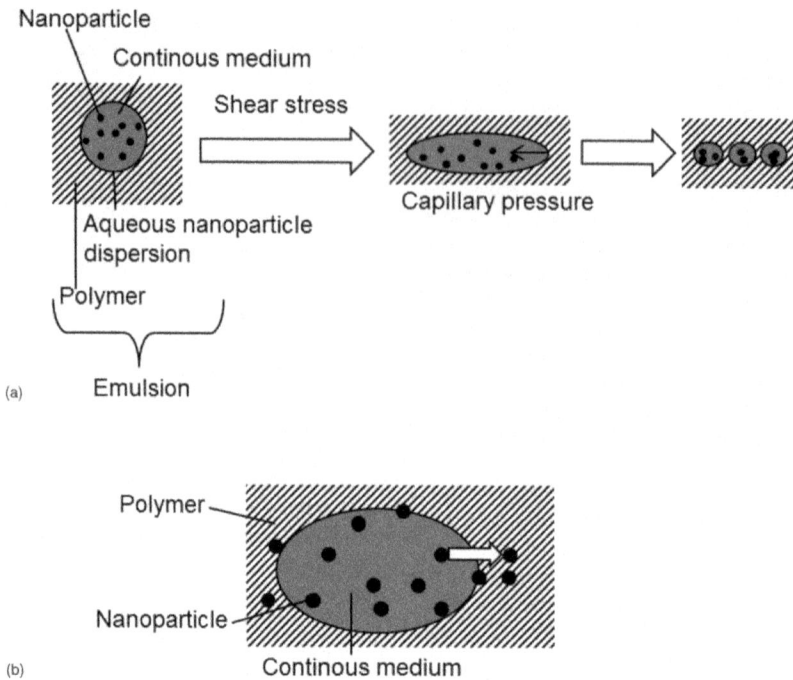

Figure 1 Scheme of mechanisms that occur when inserting aqueous nanoparticle dispersions under pressure: *a* droplet size reduction because of shear while extruding and *b* diffusion of nanoparticles from the dispersion agent in the polymer melt

$4\,kg\,h^{-1}$ was applied for the BaSO$_4$ composites with the highest filler concentration. The barrel temperatures were 65, 250, 220, 200, 200, 210, 210, 230, 230, and 230°C along the 10 heating zones (Fig. 2). Position E was cooled, in order to minimize the vapor pressure of the injected aqueous dispersion. The pressure reached was from 7.8 to 22 bar. Touchaleaume *et al.*[22] also reduced the temperature for about 20°C at the point where water was injected, compared to the other barrel zones (240°C).

The mixing screw had a diameter of $d_s=24.66\,mm$ and a flight land clearance of $s=0.34\,mm$. The screw configuration can be seen in Fig. 3. The first zone was for plasticizing, followed by four pressurized zones. Temperature and pressure were measured in the first zone, where the aqueous dispersion was added at a position with conveying screw elements. This zone was sealed against the plasticizing zone by two back-conveying elements. The extruder barrel was sealed from the first to the second pressurized zone by a back-conveying screw element and a blister. In the second pressurized zone, the material was dispersed by a kneading block. The third pressurized zone resembles this second zone. The sealing between pressurized zones number three and four resulted from a back-conveying element. Mixing was performed in pressurized zone four because of a kneading block and a teeth block. This was followed by the degassing via the side feeder and the vacuum degassing.

Analysis

The particle size and zeta-potential of the aqueous nanoparticle dispersions were analyzed via dynamic light

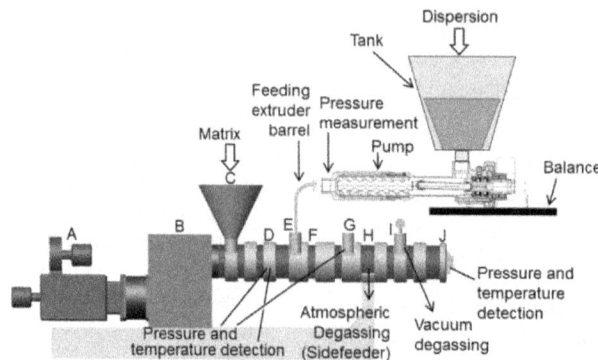

Figure 2 Schematic build-up of the extrusion processing for the pressurized insertion of aqueous nanoparticle dispersions in the polymer melt

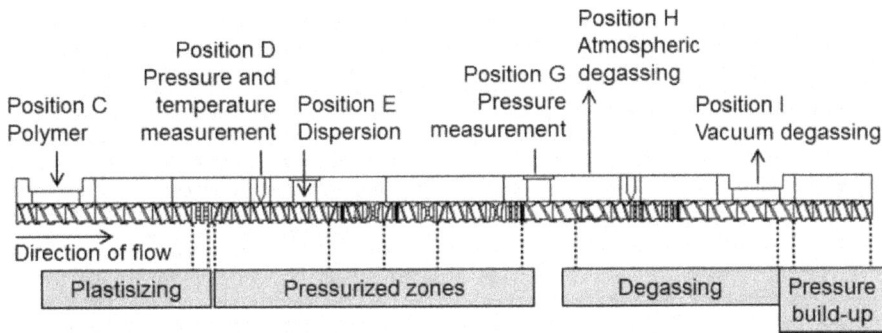

Figure 3 Schematic extruder screw configuration for the pressurized insertion of aqueous nanoparticle dispersions in the polymer melt

scattering in a Zetasizer Nano ZS from Malvern Instruments GmbH, Germany.

The surface tension of the dispersions and PA6 were determined via contact angle measurement with three–five single measurements on a PA6-suface for each material. For determination of the surface energy, the authors proceeded according to Neumann.[23-26]

Measurements in the high-pressure capillary rheometer (HKR, Göttfert Werkstoffprüfmaschinen GmbH, Germany) were conducted using a 2000 bar pressure sensor at 240°C to get the viscosity of the polymer. The applied shear rate was $500-8000\,s^{-1}$.

For SEM, a model Supra 40VP from Carl Zeiss, Germany, and a model JSM 6300 from Jeol Ltd., Japan, equipped with field emission cathode were used. All samples were sputtered with a coating of Pt/Pd for 60–80 s.

Results and discussion

Expected drop size, agglomerate size and effective impact time/theoretical consideration concerning agglomerate size

Equation (2) gives a value for the minimum drop size that can be created in the extruder, when polymer melt and aqueous nanoparticle dispersion are mixed under pressure, so that the water cannot evaporate.

Compared to the aqueous dispersion, the polymer melt exhibits more than 10^3 times higher viscosity. This is a way beyond the relation of $1 < \frac{\eta_p}{\eta_w} < 10$, where the Weber number exhibits a minimum, consequently, the critical Weber number exceeds the value of five.[19] To reach smaller Weber numbers, the viscosity of the polymer should not be more than 10 times larger than the viscosity of the aqueous dispersion. The effective shear stress can be calculated according to

$$\tau_e = \dot{\gamma} \qquad (3)$$

with the shear rate $\dot{\gamma}$ and the viscosity of the polymer. The shear rate can be determined according to Refs. 27,28 by

$$\dot{\gamma} = \frac{\pi(d_s - 2s)n}{s} \qquad (4)$$

Here, d_s is the screw diameter, s its flight land clearance and n the rotation speed of the extruder. With the given values,

the shear rate is $\dot{\gamma}=369\,s^{-1}$. The viscosity of the neat polymer at 230°C is 211 Pas, which results in a shear stress of $\tau_e=78\,kPa$. The interfacial tension of the aqueous dispersions from contact angle measurements was:

- $10\,mNm^{-1}$ for the TiO$_2$ dispersion,
- $16\,mNm^{-1}$ for the BaSO$_4$ dispersion, and
- $17\,mNm^{-1}$ for the SiO$_2$ dispersion.

Table 1 gives the smallest drop size that can be realized with these extrusion parameters.

The minimal theoretical drop size is in the range of 1.3–2.2 μm. The agglomerate size that can be calculated from this result, taking in account the nanoparticle concentration in the aqueous dispersion, is given in Table 1. It is obvious that the smallest agglomerate size that can be expected is not in the desired range, smaller than 100 nm. Thus, agglomerates smaller than 500 nm or perfectly dispersed primary particles cannot be achieved with this process. To produce smaller agglomerates or even primary particles, the diffusion of particles from the dispersion agent into the polymer must take place.

The following steps can improve the dispersion quality by reducing the drop size of the dispersion agent in the polymer melt:

- reducing the critical Weber number,
- increasing shear stress, and
- reducing the interfacial tension.

Adjusting the viscosities of polymer melt and dispersion agent in a range $1 < \frac{\eta_p}{\eta_w} < 10$ can reduce the critical Weber number. Therefore, the viscosity of the polymer should be reduced via i.e. temperature increase. This is limited by polymer degradation, occurring at elevated temperatures. Even a temperature increase to 270°C does not lower the viscosity of the polymer melt sufficient to significantly alter the critical Weber number. Furthermore, a viscosity decrease will result in a decreased shear stress, which increases the

Table 1 Smallest drop size and smallest agglomerate size that can be realized with these extrusion parameters for TiO$_2$, SiO$_2$ and BaSO$_4$ aqueous dispersions

	TiO$_2$	BaSO$_4$	SiO$_2$
Smallest drop size/μm	1.3	2.0	2.2
Smallest agglomerate size/nm	709	651	1395

achievable drop size. An easier method to reduce We_{crit}, therefore, is the increase of the viscosity of the aqueous dispersion. This can be achieved either by changing the dispersion agent, a higher particle concentration or addition of a viscosity modifier. The shear stress can be increased by higher rotation speed or screw elements with higher shearing effect. Adjusting the interfacial tension leads to a reduced capillary pressure and a promoted drop fragmentation.

Initial aqueous dispersions

The zeta-potential describes the characteristic, repulsive interactions of particles with the same charge in a dispersion. When the zeta-potential falls below a certain absolute value, it can lead to agglomeration of colloidal particles. The authors can assume that particles are stable beyond a zeta-potential of ±30mV (larger than 30mV or lower than 30mV) and do not agglomerate. How fast the agglomeration takes place is also dependent on the kinetic energy of the particles.

The particle-size distribution was gathered by dynamic light scattering. For the top–down dispersions with TiO_2 nanoparticles, most particles can be found as primary particles, with a size of 15nm. The measured zeta-potential is +29.7mV, being very close to the theoretical lower limit of +30mV for a well-stabilized dispersion.[29] The particle-size distribution of the *in situ* $BaSO_4$ dispersions reveals a primary particle size of 46.5nm. The majority of the particles can be found as primary particles. The zeta-potential with -36.4mV indicates very stable dispersions.[29] The maximum intensity of the particle-size distribution for *in situ* SiO_2 aqueous dispersions indicates a typical diameter of 20nm. The measured average zeta-potential of the dispersion was -34.6mV. For this reason, it can be assumed as very stable.[29] However, a second maximum lies at the zeta-potential of -25.9mV. This part of the particles is not permanent stable.

Fig. 4*a* illustrates the appearance of the dried dispersions in REM, and Fig. 4*b* indicates the energy dispersive X-ray spectrum (EDX) analysis. The remaining residues from the dispersion do not only consist of nanoparticles but also high amounts of sodium (Na) and chloride (Cl) can be found in the dispersions. The gold (Au) comes from the gold surface coating to reduce the charging in the electron microscope.

Structure of final polymer nanocomposites

The morphology of the nanocomposites was analyzed by SEM. The dispersion of the nanoparticles and the resulting

interface to the polymer was examined. Figure 5 shows the TiO_2 nanoparticle dispersion. The nanoparticles cannot be found as primary particles, but are present in small agglomerates. The larger agglomerates are isolated and impregnated with polymer. The nanoparticle–matrix interaction is reduced compared to nanocomposites, produced by nanoparticle powder incorporation (compare Ref. 30).

Primary nanoparticles can be found when $BaSO_4$ nanoparticle dispersions are used (Fig. 6). Nevertheless, nanoparticle agglomerates can be found too. The nanoparticles as well as the agglomerates are not well embedded in the matrix. The nanoparticle–matrix interaction is not very pronounced. Via SEM and EDX analysis, detected NaCl crystals can also be found in the fracture surfaces (Fig. 7). The EDX analysis of the fracture surface shows that NaCl crystals bigger than 1μm can be found in the material. Furthermore, there are regions in the surface consisting of sodium and oxygen that originate from NaOH. The mechanical properties of these composites will be determined by the NaCl crystals and not the nanoparticles.

Fig. 8 reveals very poor dispersion quality of the SiO_2 composites. The agglomerates show areas with high particle density and packing order (boxes). The very large agglomerates are not impregnated with matrix. Almost no small agglomerates or primary particles can be found in the composite. The particles are not connected with the matrix. Even though the *a priori* dispersity of the aqueous dispersion is very good, this status cannot be maintained during extrusion processing under pressure. Even though the pressures for the compounds (>17bar) are higher than the vapor pressure of water, large agglomerates are created.

Fig. 9 shows agglomerates >1μm in the polymer matrix detected with a backscattering electron detector. The TiO_2 and $BaSO_4$ nanoparticles are well dispersed compared to the SiO_2 nanoparticles. The achieved pressures for producing the TiO_2 and SiO_2 compounds are higher than the vapor pressure of water. The $BaSO_4$ compound was produced with a slightly lower pressure (11.2bar). Touchaleaume *et al.*[22] extruded PA6 with water under a pressure of 70–100bar to be certain that the water remains liquid in the extruder. This pressure is not reached in our extrusion process. Even the 30bars reached by García *et al.*[18] did not suffice for producing an agglomerate free PA6/SiO_2 composite.

The pressure during the extrusion processing can generally create agglomerates.[31,32] Shahabadi and Garmabi[33]

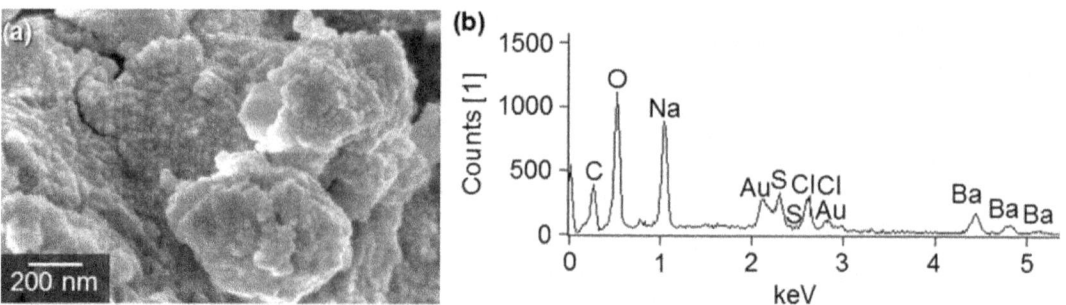

Figure 4 *a* SEM images of dried $BaSO_4$ nanoparticle dispersions and *b* their energy dispersive X-ray spectrum (EDX) analysis

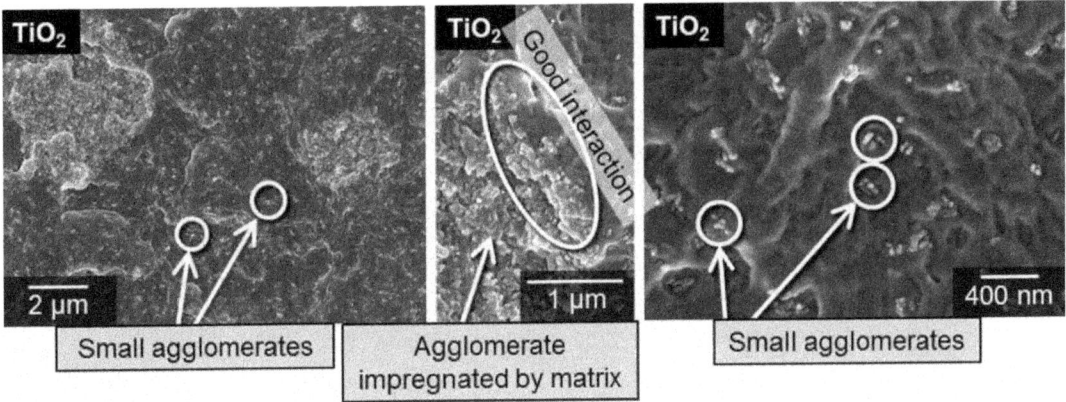

Figure 5 SEM images of the fracture surfaces of TiO$_2$ composites. Agglomerates and single nanoparticles and small agglomerates dispersed in the matrix can be found

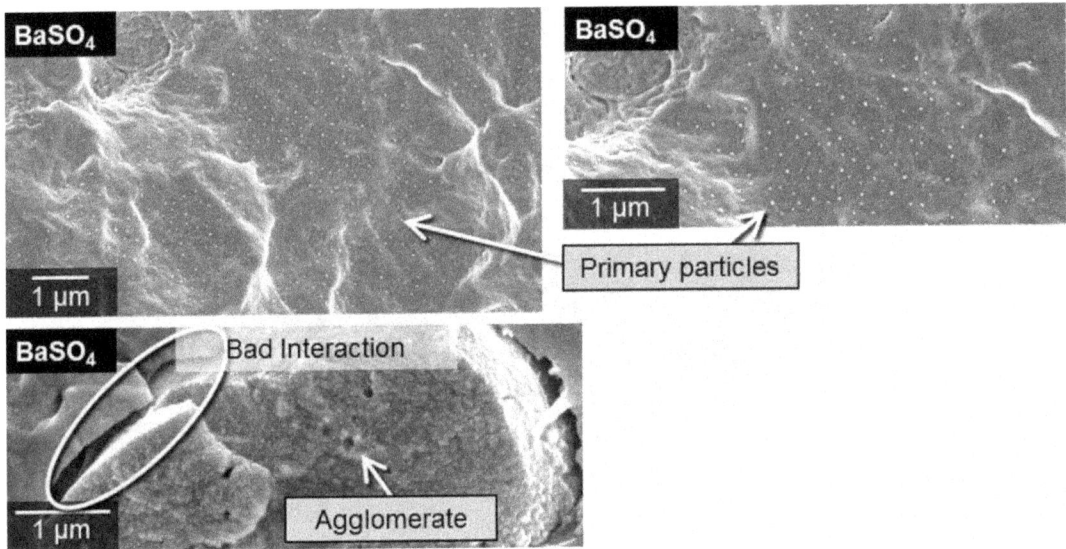

Figure 6 SEM images of the fracture surfaces of BaSO$_4$ composites. One agglomerate is shown and nanoparticles dispersed in the matrix

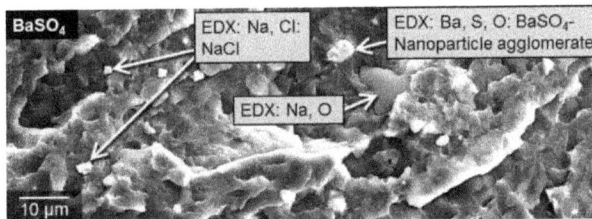

Figure 7 SEM image of the fracture surface of BaSO$_4$ composites with lower magnification. Additionally, their composition analyzed with EDX is given

proposed a mechanism where the drops of the dispersion are reduced in size via shear stress. When the contact time is sufficient, the nanoparticles diffuse from the water in the matrix.[34] When the dispersion agent water is evaporated and when the progression of the evaporation is faster than the diffusion and mixing of the particles with the polymer matrix, the remaining nanoparticles agglomerate. Agglomerates can be generated if the time for diffusion is not sufficient. In this process, the authors assume that agglomeration takes place when the particles are not stabilized and that re-agglomeration of particles that are already dispersed in the polymer matrix can be neglected.

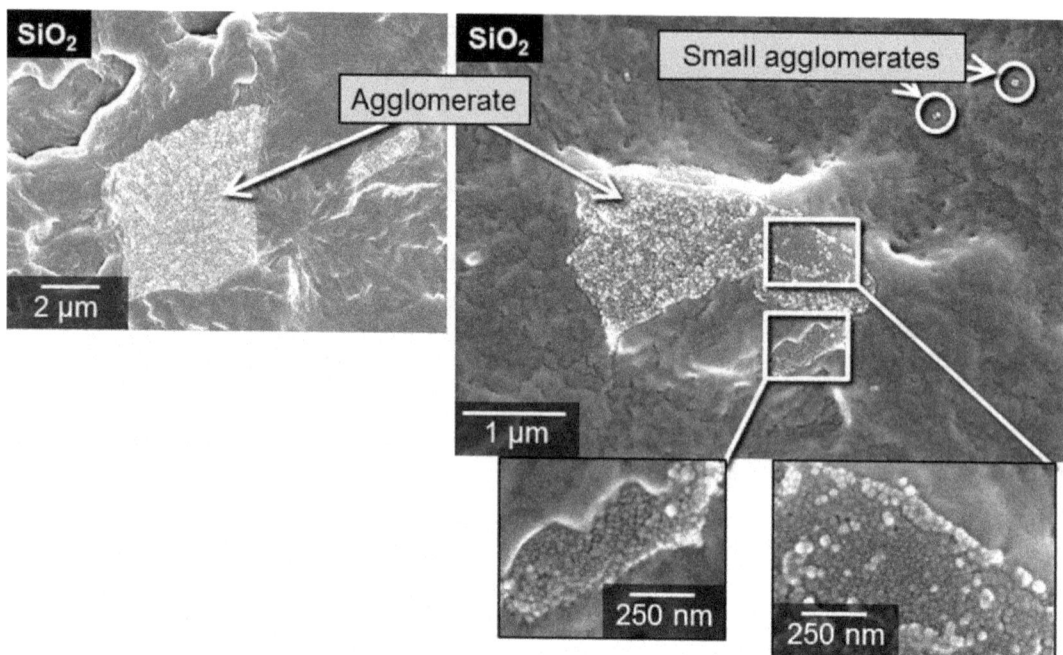

Figure 8 SEM images of the fracture surfaces of SiO₂ composites. Large agglomerates and few small agglomerates can be found

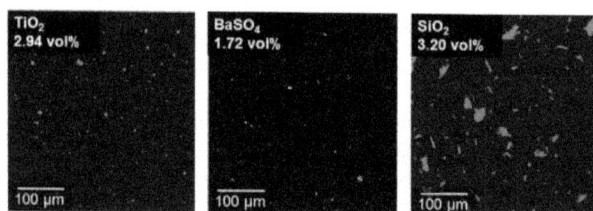

Figure 9 SEM images of the polished surface of TiO₂, BaSO₄ and SiO₂ composites pictured via backscattering electron detector

Fig. 10 gives a comparison of the theoretical calculated agglomerate size, when only the mechanism of drop break-up is present and the actual experiment results. As small agglomerates (TiO₂) and primary particles (BaSO₄) can be found in the matrix, diffusion also occurs. The TiO₂ agglomerates consist of few primary particles and have a size of ca. 100–500 nm. The residence time in the extruder is at minimum 86 s, which is obviously sufficient for the particles to diffuse. The SiO₂ particles occur in much larger agglomerates (>2 μm), which indicates that diffusion does not take place and coalescence might occur. The SiO₂ particles show the largest interfacial tension with a large particle concentration in the nanoparticle dispersion. This might lead to these large agglomerates with no diffusion occurring.

Touchaleaume *et al.*[22] supposed that 70–100 bar are needed to achieve a liquid aggregate state during extrusion. The achieved pressure was much smaller (<25 bar). It can be

Figure 10 Comparison of the theoretical determined (grey cycles with black frame) and the experimentally found agglomerate and particle size, respectively

assumed that this might be a reason for the occurrence of agglomerates. Yin et al.[17] and Liu et al.[16] showed agglomeration in spite of pressurization. García et al.[5,18] found agglomerates when injecting SiO_2 nanoparticle dispersions in PA6 and small agglomerates when inserting the particles in PP.

Conclusion

The pressurized injection of aqueous nanoparticle dispersions allows the generation of a blend of aqueous dispersion and polymer melt, but the production process does not allow a sufficient pressure build-up for all nanocomposites to be dispersed properly. There are two mechanisms that describe the behavior of the nanoparticle dispersions in the extruder: (a) drop size reduction via shear stress and (b) diffusion of particles from dispersion medium into the polymer. The Weber number indicates the smallest agglomerate size that can be gained through shear-induced drop reduction. The shear force induces a reduction of drop size to a theoretical minimum of 1.3–2.2 μm, which is not in the nanometer range. SiO_2 particles do not appear as single nanoparticles; thus, diffusion plays a minor role. Instead, the particles in the dispersion droplets agglomerate when water is evaporated. The agglomeration process advances faster than the diffusion. The particles in the agglomerates are very dense, which indicates a high agglomerate strength. TiO_2 particles show small agglomerates (<100 nm), which is smaller than the calculated smallest drop size found through shear-induced droplet break-up. $BaSO_4$ particles can be found as primary particles. It can be assumed that there are two mechanisms appearing: diffusion and droplet size reduction via shear stress. The $BaSO_4$ particles are present as primary particles in the matrix, and therefore, they can act as an interesting model system.

Conflicts of interest

The authors have no conflicts of interest to declare.

Acknowledgement

The authors gratefully acknowledge the financial support of the Federal German Ministry of Economics and Technology (BMWi) and the project management organization AIF Projekt GmbH for supporting this project included in the program Zentrales Innovationsprogramm Mittelstand (ZIM). Furthermore, the authors thank the project partner ViscoTec Pumpen-und Dosiertechnik GmbH.

References

1. K. Yano, A. Usuki, A. Okada, T. Kurauchi and O. Kamigaito: *J. Polym. Sci. A Polym. Chem*, 1993, **31**, 2493–2498.
2. A. Zhu, A. Cai, J. Zhang, H. Jia and J. Wang: *J. Appl. Polym. Sci.*, 2008, **108**, 2189–2196.
3. A. Ophir, A. Dotan, I. Belinsky and S. Kenig: *J. Appl. Polym. Sci.*, 2010, **116**, 72–83.
4. T. Hanemann and D. V. Szabó: *Materials*, 2010, **3**, 3468–3517.
5. M. García, G. van Vliet, S. Jain, B. A. G. Schrauwen, A. Sarkissov, W. E. van Zyl and B. Boukamp: *Rev. Adv. Mat. Sci*, 2004, **6**, 169–175.
6. J. X. Ren and R. Krishnamoorti: *Macromolecules*, 2003, **36**, 4443–4451.
7. R. Qiao, H. Deng, K. W. Putz and L. C. Brinson: *Journal of Polymer Science, Part B: Polymer Physics*, 2011, **49**, 740–748.
8. B. Pukánszky: 'Particulate filled polypropylene, structure and properties, polypropylene composites, in Polypropylene: an A-Z reference (polymer science and technology series)Karger-Kocsis', (ed. J. Karger-Kocsis), **2**, 575–580; 1999, Luxemburg, Springer.
9. X. Xu, B. Li, H. Lu, Z. Zhang and H. Wang: *Appl. Surf. Sci*, 2007, **254**, 1456–1462.
10. X. Xu, B. Li, H. Lu, Z. Zhang and H. Wang: *J. Appl. Polym. Sci.*, 2008, **107**, 2007–2014.
11. G. H. Michler and F. J. Balta-Calleja: 'Mechanical properties of polymers based on nanostructure and morphology'; 2005, Boca Raton, CRC Press.
12. R. P. Singh, M. Zhang and D. Chan: *J. Mater. Sci.*, 2002, **37**, 781–788.
13. N. F. Knör: 'Einfluss der Verarbeitungstechnologie und Werkstoffzusammensetzung auf die Struktur-Eigenschafts-Beziehungen von thermoplastischen Nanoverbundwerkstoffen', Dissertation, Technische Universität Kaiserslautern, Institut für Verbundwerkstoffe; 2009. [In: IVW Schriftenreihe Band93].
14. T. Villmow, P. Pötschke, S. Pegel, L. Häussler and B. Kretzschmar: *Polymer*, 2008, **49**, 3500–3509.
15. J. Metzger: *Nanotechnoloie in Kunststoffen: Innovationsmotor für Kunststoffe, ihre Verarbeitung und Anwendung*, 2009, **15**, 43–46.
16. Z. Liu, R. Yu, M. Yang, J. Feng, W. Yang and B. Yin: *Front. Chem. Eng. China*, 2008, **2**, 115–122.
17. C. -L. Yin, Z. -Y. Liu, W. Yang, M. -B. Yang and J. -M. Feng: *Colloid Polym. Sci.*, 2009, **287**, 615–620.
18. M. García, G. van Vliet, M. G. J. Cate, F. Chavez, B. Norder, B. Kooi, W. E. van Zyl, H. Verweij and D. H. A. Blank: *Polym. Adv. Technol.*, 2004, **15**, 164–172.
19. H. Schubert and H. Armbruster: *Chem. Ing. Tech.*, 1989, **61**, 701–711.
20. P. Walstra: *Chem. Eng. Sci.*, 1993, **48**, 333–349.
21. K. Köhler: 'Simultanes Emulgieren und Mischen, Dissertation Dissertation', Fakultät für Chemieingenieurwesen und Verfahrenstechnik des Karlsruher Instituts für Technologie (KIT); 2010.
22. F. Touchaleaume, J. Soulestin, M. Sclavon, J. Devaux, M. F. Lacrampe and P. Krawaczak: *Polym. Degrad. Stab.*, 2011, **96**, 1890–1900.
23. A. W. Neumann, R. J. Good, C. J. Hope and M. Sejpal: *J. Colloid Interface Sci.*, 1974, **49**, 291–304.
24. D. Li and A. W. Neumann: *Colloid Polym. Sci.*, 1992, **270**, 498–504.
25. K. Rieß: 'Plasmamodifizierung von Polyethylen', Dissertation, Mathematisch-Naturwissenschaftlich-Technische Fakultät (Ingenieurwissenschaftlicher Bereich) der Martin-Luther-Universität Halle, Wittenberg; 2001.
26. M. Feigl: 'Sol-Gel-Schutzschichten für Hochtemperaturlegierungen und Stähle', Dissertation, Fakultät für Natur- und Materialwissenschaften der Technischen Universität, Clausthal; 2011.
27. R. Prístavok: 'Analyse und Modellierung der Haftungsmechanismen bei der Beschichtung und Verklebung von Papierwerkstoffen', Dissertation, Fakultät für Maschinenbau, Verfahrens- und Energietechnik der Technischen Universität Bergakademie, Freiberg; 2006.
28. G. R. Kasaliwal: 'Analysis of multiwalled carbon nanotube agglomerate dispersion in polymer melts', Dissertation, Fachbereich Maschinenwesen an der Technischen Universität, Dresden; 2011.
29. C. Rauwendaal: 'Polymer extrusion'; 1986, München, Hanser.
30. T. Günther: 'Zum Fällungsprozess und Wachstum kugelförmiger SiO2-Partikel', Dissertation, Fakultät für Verfahrens- und Systemtechnik der, Otto-von-Guericke-Universität Magdeburg; 2008.
31. I. Hassinger and T. Burkhart: *J. Thermoplast. Compos. Mater.*, 2012, **1**, 573–590.
32. T. Lozano, P. G. Lafleur, M. Grmela and C. Thibodeau: *Polym. Eng. Sci.*, 2004, **44**, 880–890.
33. S. Shahabadi and H. Garmabi: *Polym. Lett.*, 2012, **6**, 657–671.
34. J. W. Ess and P. R. Hornsby: *Plast. Rubber Process. Appl.*, 1987, **8**, 147–156.

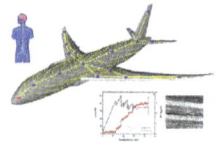

18

The use of carbon nanotubes for damage sensing and structural health monitoring in laminated composites

Han Zhang[1], Emiliano Bilotti[1,2], and Ton Peijs[*1,2]

[1]School of Engineering and Materials Science, and Materials Research Institute, Queen Mary University of London, Mile End Road, E1 4NS London, UK
[2]Nanoforce Technology Ltd., Queen Mary University of London, Joseph Priestley Building, Mile End Road, E1 4NS London, UK

Abstract The increasing use of fiber-reinforced plastics (FRPs) in industries such as aerospace, marine, and automotive, has resulted in a necessity to monitor the structural integrity of composite structures and materials. Apart from development of traditional non-destructive testing methods which are performed off-line, there is a growing need to integrate structural health monitoring (SHM) systems within composite structures. An interesting route toward multifunctional composite materials with integrated SHM capabilities is through the introduction of carbon nanotubes (CNTs) in fiber-reinforced composites as this provides not only integrated damage sensing capability, but may, at the same time, also lead to some additional mechanical reinforcement. Since the first use of CNTs for damage sensing in composite laminates, a significant number of studies have dealt with this topic, but a systematic understanding on the use of CNTs in FRPs for SHM is still lacking. Furthermore, a significant gap remains between results obtained in the laboratory and industrial applications. This review reports on the progress of this topic so far. The reviewed work had been categorized from model studies on single fiber composites to laminated composites under different loading conditions, as well as the development of reliable damage-sensing systems which could be transferred to real applications.

Keywords Carbon nanotubes, Damage sensing, Structural health monitoring, Fiber-reinforced plastics, Hierarchical composites, Nanoengineered composites

Introduction

The market demand for fiber-reinforced plastics (FRPs) has been continuously increasing over the last few decades,[1,2] especially for lightweight structural components where FRPs are replacing traditional metal materials due to their high specific strength and stiffness, good corrosion resistance, fatigue resistance, etc., in combination with advantages related to the manufacturing of complex integrated parts. However, the laminated nature of FRPs remains a challenge as this leads to relatively weak out-of-plane properties when compared to their excellent in-plane properties and may lead to failures such as delamination or matrix cracking which greatly limits their use. Furthermore, unlike in traditional materials, damage in composites is often non-visible which increases the difficulties to evaluate its effect on properties. Therefore, detection of various composite failure modes during usage, as well as monitoring

of health throughout service life has become increasingly important and necessary.

Due to the strict requirements on safety and reliability in aerospace applications (especially for civil aircraft), current composite parts and structures within aircraft are typically overdesigned to offset the possibility of unexpected internal failures, which in fact works against the weight reduction purpose of using them in the first place. A promising way to better utilize the potential of FRPs is by being able to detect and monitor the internal damage state, hence to set up proper safety thresholds and reduce the risk of catastrophic failures. The concept of structural health monitoring (SHM) is not new, but has become more integrated into the composites' industry in recent years.[3] Several health monitoring concepts have been proposed, including novel non-destructive testing (NDT) technologies, embedded sensors, and self-sensing material systems, etc. The ultimate goal of an SHM system is to mimic the function of the human nervous system, creating a fully integrated *in situ* health monitoring network for the next generation of

*Corresponding author, email t.peijs@qmul.ac.uk

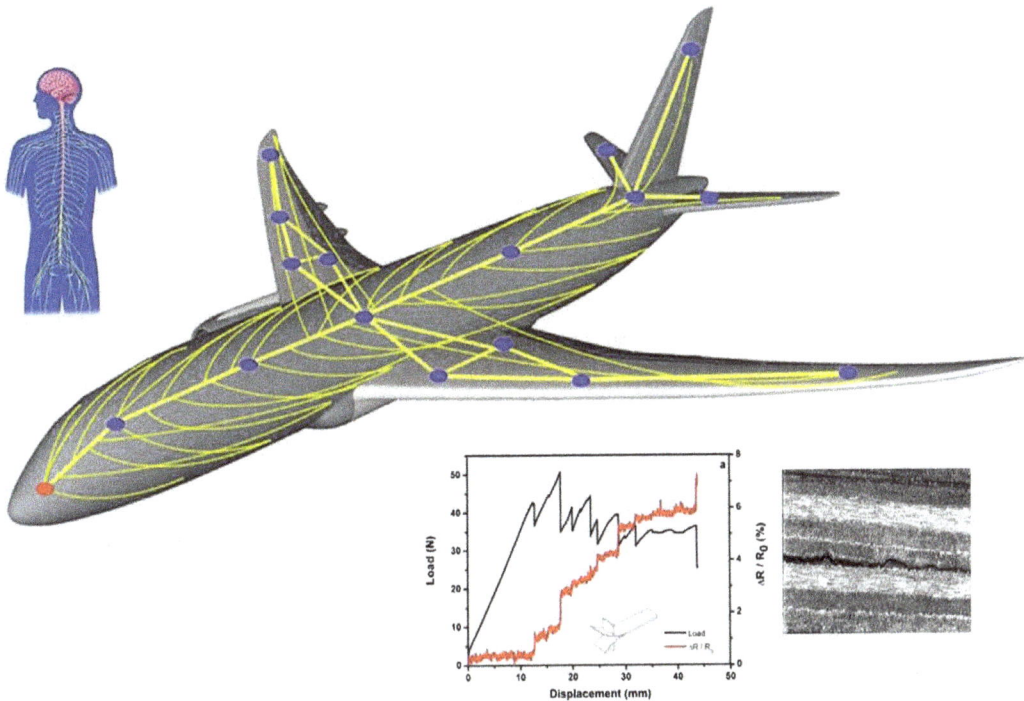

Figure 1 Illustration of a "smart aircraft" with fully embedded sensory network for SHM, similar to the human nervous system and integrated functionalities of damage detection. (Sensory network aircraft after Holger Speckmann Airbus®)

aircraft (Fig. 1). In this review, different concepts are introduced and briefly analyzed, with a special focus on the use of carbon nanotubes (CNTs) as integrated sensors for health monitoring in FRPs.

Various NDT methods have been applied in this field to examine and evaluate internal damage in composites. Acoustic emission (AE) is one of the most commonly used NDT methods for damage inspection and quality control. When an internal damage occurs, the released stress generates an elastic wave which can be recorded by a piezoelectric sensor on the specimen surface. By analyzing the collected information, the location as well as the possible failure modes can be identified.[4–7] Another commonly used NDT technology is ultrasonics which uses low-frequency acoustic pulses to detect the internal continuity of composite structures. Computed tomography has also been used more widely nowadays for inspection due to its accuracy in evaluating data, especially for complex engine components. Unfortunately, most of those NDT tests not only require special equipment to perform the damage detection, but also strongly depend on the experience of engineers to interpret data as well as involve additional assembly time, costs, and difficulties of monitoring the health of FRPs continuously. The downtime nature of these NDT methods, which often require that examinations are performed during out-of-service periods, increases the lifetime costs even further.

Hence, embedded sensors such as fiber optics or smart piezoelectric films or coatings have been introduced into FRPs to detect internal deformation and damage in-service. For instance, coated piezoelectric sensors either with or without CNTs,[8–14] could be used to detect the deformation of the

coated region through a change of electrical signals. This method provides the possibility to monitor the structural health during service life, avoiding complex downtime inspection and components assembly. However, it is worth noting that there is no structural contribution from most of those embedded sensors, and in certain cases, they may even lower the mechanical performance of components.[15] For example, the introduction of optical fibers in composite laminates may lead to early matrix cracking as the additional interfaces generated by these embedded sensors may weaken the composite in use, limiting their application in real structural components.

Self-sensing concepts which utilize electrical methods to detect the internal health status of composites have been introduced into this field for carbon fiber-reinforced plastics (CFRPs) using the intrinsically conductive carbon fibers (CFs) as damage sensors. The advantage of self-sensing methods is that all original mechanical properties are preserved while internal damage can be sensed and monitored.[16–22] By using the structural component itself as a sensor to detect damage, this concept overcomes issues related to the reduction of mechanical performance and durability as a result of the introduction of sensors in laminated composites. In contrast to electrically conductive CFRPs, insulating glass fiber-reinforced plastics (GFRPs) cannot be used for electrical damage sensing concepts, while various GFRPs applications like wind turbine blades still require advanced in-service damage detection systems. Apart from the highly anisotropic conductivities of CFs, the failure modes that can be self-sensed in CFRPs are mostly fiber breakage which normally occurs near the end of a component service life rather than early stage damage like

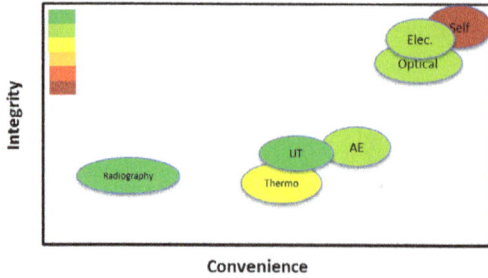

Figure 2   Comparisons between various SHM/NDT methods and their possibility of *in situ* damage detection in composites

matrix cracking or delaminations. Failure modes like matrix cracking or delaminations are harder to sense using electrical methods as the surrounding polymer matrix is typically insulating. In order to be able to apply the electrical sensing concept to insulating GFRPs, and to detect early matrix dominated failure modes in CFRPs, the polymer matrix needs to become electrically conductive. This can be achieved either through the introduction of conductive nanofillers in the polymer matrix or surface functionalization of reinforcing fibers with conductive nanofillers. Early studies on the use of CNTs as molecular sensors around the year 2000 guided the potential routes toward multifunctional composites,[23,24] while detailed reviews on CNT-based sensors and actuators were already published by Chou *et al.* around 2008.[25,26] An attempt to compare the strengths and limitations of several SHM/NDT methods is presented in Fig. 2 on the basis of convenience, integrity, and potential to detect different failure modes in FRPs, where integrity refers to the necessity to use less additional equipment and less interference with original material properties; and convenience refers to a realistic potential for *in situ* applications and dimensional limitations. Color index from top to bottom stands for the possibility to detect multiple failure modes ranging from high to low.

CNTs were first introduced into FRPs as sensors for SHM purposes by Fiedler *et al.* in 2004.[27] Since then, several studies exploring the use of CNTs for damage sensing in FRPs have been conducted, particularly in the last 5 years (Fig. 3). Because of their extraordinary mechanical and electrical properties, CNTs have become the nanofiller of choice for many multifunctional applications over the last decade.[28-34] With respect

to electrical properties their high aspect ratio guarantees a very low percolation threshold which does not significantly compromise the original resin properties. By introducing CNTs into insulating polymer matrices, various matrix-dominated failure modes can be sensed and detected, providing a promising route for next-generation SHM systems. The use of CNTs in FRPs also provides the possibility of in-service health monitoring, without complicated equipment and intensive labor requirements. In-service damage detection can greatly save aircraft or wind turbine downtime, in other words, providing more in-service time, improving efficiency and reducing costs.

Although a good amount of research has been conducted in this field, a systematic understanding of the use of CNTs in FRPs for damage sensing and health monitoring purposes is lacking, not to mention the gap that still exists between laboratory successes and real industrial applications in structural components. In the present paper, the use of CNTs for damage sensing in composites is reviewed, with the aim to contribute to a better understanding and systematic comparison of the field. Apart from this, special efforts have been made to analyze the existing knowledge from an end-user point of view, to classify various works and data, and list requirements from industry (i.e. SHM in aircraft), trying to reduce the gap between laboratory data and real applications.

Model studies on single fiber composites

Single fiber composites are often chosen as model systems to study interfacial properties of composites and for this reason, they have also been used to study the effect of CNT sensory networks within interfacial regions. Various research works have been conducted to understand how conductive interfacial networks behave during loading which is an important aspect of damage sensing in FRPs, especially for loading situations where interfacial debonding is likely to occur.

Park *et al.*[35,36] have combined an electrical sensing method together with traditional AE, in order to obtain damage sensing results. In this study, CNTs were dispersed in epoxy using a solvent and after solvent evaporation, composites with CF positioned at the mid-plane were manufactured, as illustrated in Fig. 4.[36] During tensile loading along the fiber direction, fiber breakage occurred and electrical sensing signals as well as AE signals were collected. For the specimen with extremely

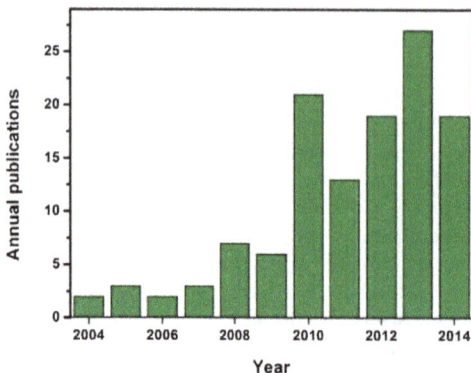

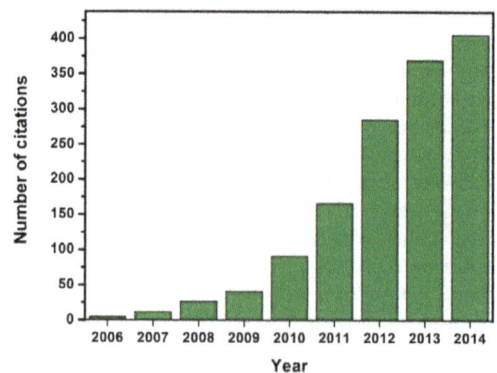

Figure 3   Number of annual journal publications and citations on the use of CNTs for damage sensing in composite materials

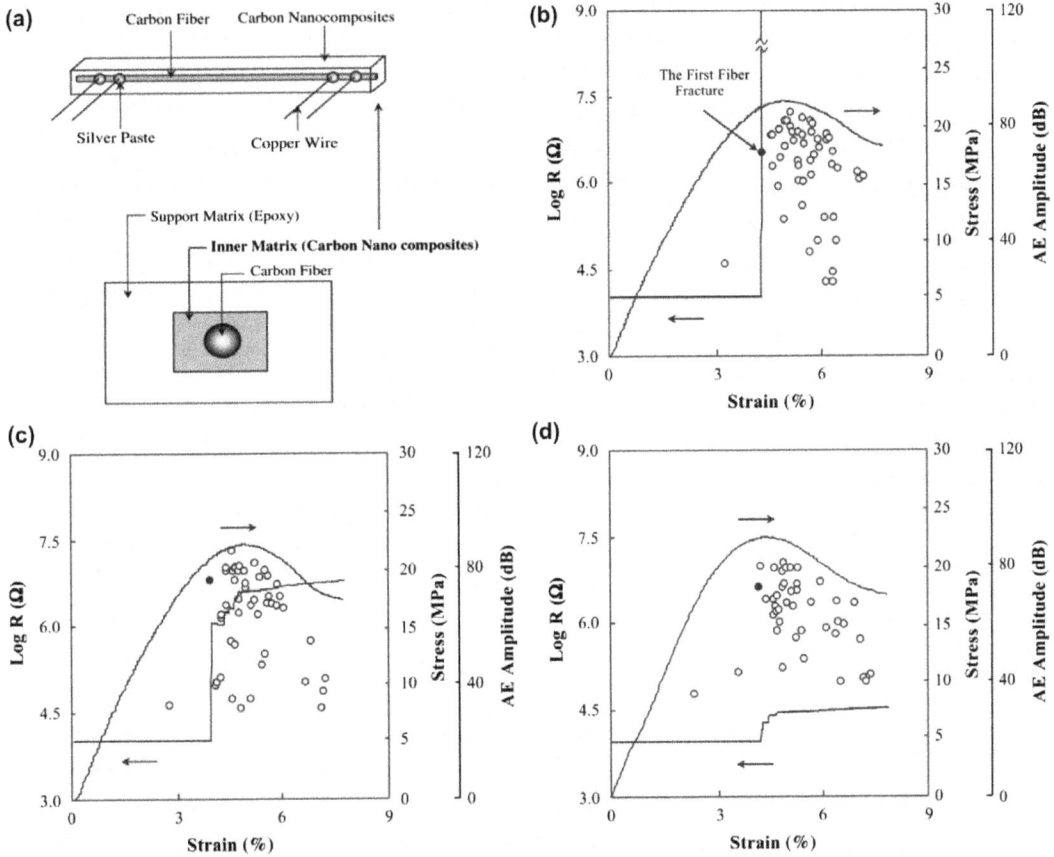

Figure 4　Park *et al.*'s single carbon fiber model composite results: *a* experimental setup of double-matrix single carbon fiber composites; *b* electrical sensing signals suddenly jump with fracture of carbon fiber in the insulating matrix; *c* electrical sensing signals reveal more information after fiber breakage in CNT-filled matrix; *d* reduced sensing signals with higher CNT concentrations

Source: Reproduced with permission from Ref. 36

low CNT content (0.1 wt.%), electrical signals were obtained until the first fiber fracture when electrical resistance suddenly jumped to insulating level (Fig. 4*b*). This jump was due to the low amount of CNTs which could not build up a stable electrically conductive network. The insulating nature of the surrounding matrix made that only the CF could transfer electrons and once broken, this also shuts off the pathway for electrons, leading to an immediate increase in resistivity. Specimen with relatively high CNT loadings (0.5 wt.%) possessed a percolating CNT network within the matrix. Therefore, after initial fiber fracture, the electrical signals still progressively increased with increasing strain, revealing more internal damage information which could be confirmed by AE data (Fig. 4*c*). This was attributed to the existence of a percolating network of CNTs in the matrix that maintains conductivity after the first fiber breakage. At higher CNT loadings (2.0 wt.%), the measured resistance change during the test was obviously less than for the 0.5 wt.% specimen, although the sensing signals still increased progressively after the first fiber break (Fig. 4*d*). This was due to more possible contact points from this percolated CNT network while with increasing strain and deformation electrons can still pass through the specimen via remaining CNT pathways.

These model studies showed some promise of the use of CNTs into FRPs. Even in combination with conductive CF, CNTs showed their potential for damage sensing, not to mention their potential benefits through additional multi-scale mechanical reinforcement. After the establishment of a percolating network, larger amounts of CNTs reduce the sensitivity of the electrical damage sensing signal, which was also confirmed by other sensing studies based on CNT networks.[37–40] Apart from the concentration, the solvent used to facilitate CNT dispersion at the manufacturing stage also affected the sensing results. Poor solvents such as water gave a lower electrical resistivity change than acetone and 2-propanol as they lead to poor dispersion and poor sensory networks.[41] Therefore, both CNT loading and dispersion state are key factors that need to be addressed for optimized damage sensing behavior.

With the aim of highlighting the potential of percolated CNT networks as sensory systems, insulating glass fiber (GF) was also used in similar model composite studies. Zhang *et al.*[42] used both electrophoretic deposition (EPD) and dip-coating to produce multifunctional GF interfaces to be evaluated in single fiber fragmentation tests. EPD showed improved dispersion of CNTs on GF surfaces compared to dip-coating,

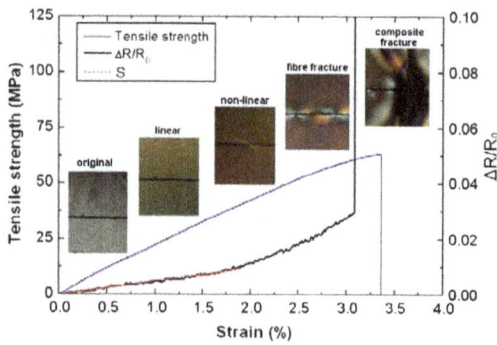

Figure 5 *In situ* **damage sensing on single glass fiber with CNT coating via EPD. Different stages of fiber under tensile loading are shown in the graph together with the corresponding electrical sensing signals**
Source: Reprint with permission from Ref. 42

while a more than 30% increment in interfacial shear strength compared to the reference system was reported. The EPD specimens were chosen to perform *in situ* electrical resistance measurements, as shown in Fig. 5. The electrical resistivity curve showed three stages with a linear, non-linear, and abrupt change, indicating the possibility of using this multifunctional layer as a mechanical sensor. It is worth mentioning that a large scatter in electrical resistivity was measured in this work for specimens with similar CNT loadings, while dip-coating led

to a reduction in single fiber tensile strength. These studies showed that in order to be successful in SHM applications, stable and repeatable sensing signals, as well as no degradation of original material properties are key factors.

Based on these results, Gao *et al.*[13] improved the dip-coating technique for CNTs using a surfactant and controlled pH, generating a multifunctional interface to sense various stimuli. Single GFs with deposited CNTs at the surface were characterized for electrical properties, piezoresistivity, humidity, and temperature sensitivity (Fig. 6). Although the strength and other mechanical performances of these CNT-modified GF interfaces were not measured in this research work, and the use of surfactant, pH control, as well as the complicated manufacturing might limit scale-up, these model studies did unveil the possibility of using CNTs directly onto insulating fibers such as GFs to produce multifunctional composites.

Apart from depositing CNTs directly onto carbon or glass fiber surfaces for sensing functionalities, polymer/CNT sizings and coatings have also been studied as interface sensors.

Rausch and Mäder[10,43] introduced an alternative route to incorporate CNTs into FRPs at the fiber/matrix interface by sizing or coating a CNT/polymer film former onto GF. For their CNT/GF/PP model composites, 0.5 wt.% CNT/PP film was coated onto GF surfaces, followed by an annealing process to achieve a more homogeneous coating (Fig. 7). The CNT/polymer coating enabled the localization of the sensory network within the interfacial region rather than being distributed throughout the matrix. However, for such a coating

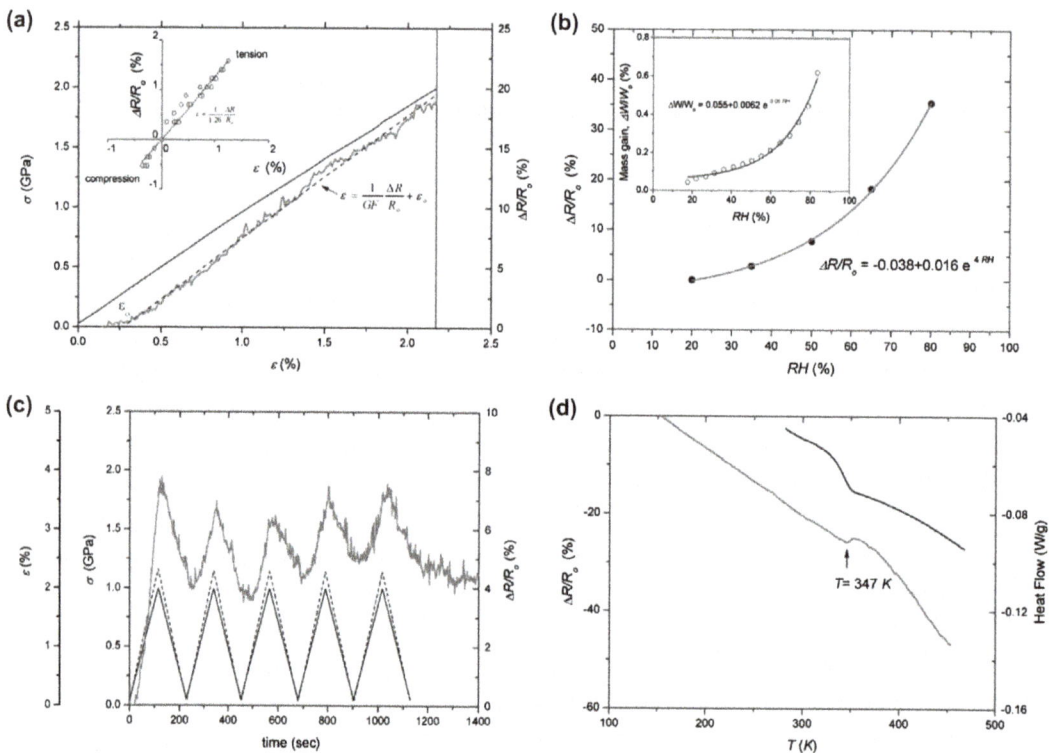

Figure 6   Single glass fiber with CNT coating as sensor for various stimuli: *a* sensing signals under static tensile stress until failure; *b* under cyclic tensile loading; *c* sensing signals with relative humidity changes; *d* signal changes with temperature
Source: Reprint with permission from Ref. 13

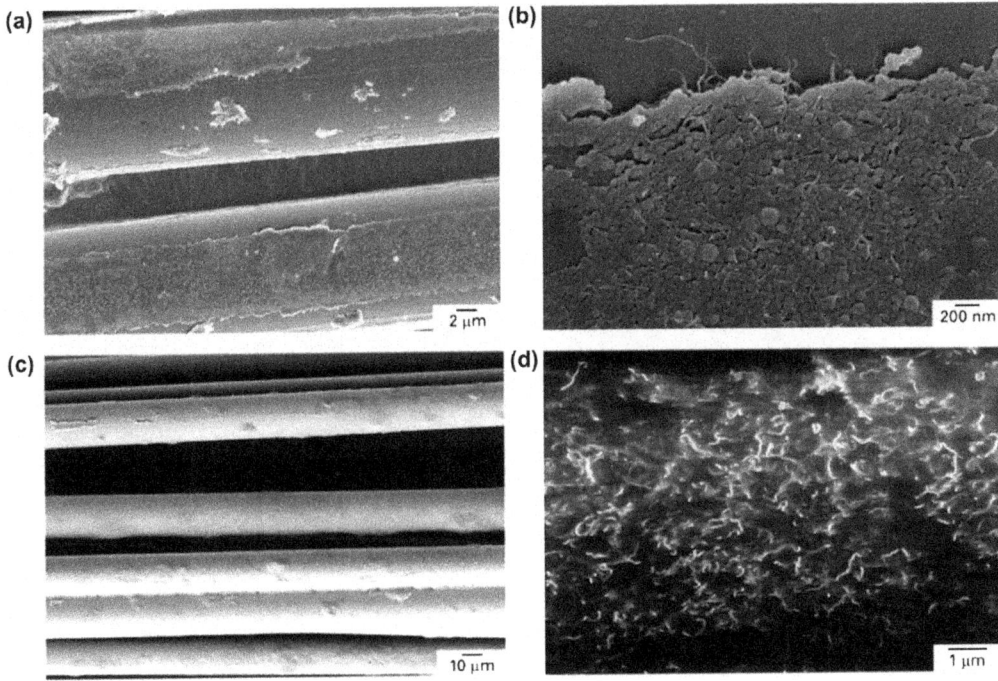

Figure 7   SEM images of glass fibers with CNT coatings, solid coating content on glass fiber is 5.5 wt.%.: *a* coated glass fiber; *b* higher magnification of coated glass fiber with CNTs dispersed in PP particles of the film former; *c* glass fiber after annealing at 200 °C for 15 min; *d* charge contrast SEM image of CNT coating after annealing
Source: Reprint with permission from Ref. 43

to be effective, it needs to be continuous and homogeneous (>10 wt.%). During tensile loading, interface failure as well as GF breakage could be identified (Fig. 8). The sensitivity of such a multifunctional coating could be adjusted by changing the coating thickness and CNT content in the coating. Both thicker coatings and higher CNT loadings in the coating led to a reduction in sensitivity of the coating. Similar effects were also found in other studies.[35–39,41]

Although problems like strength reduction of the original microfiber as well as issues related to complex additional processing steps exist, these single fiber model studies have shown great potential of the use of CNT networks as sensors for the detection of interfacial damage in both conductive and insulating reinforcing fibers. Localized CNT sensory networks on insulating GFs can introduce multifunctionality in their composites, regardless of polymer or interface.

Model studies on laminates

After the initial idea of using CNTs as multifunctional sensors for FRPs,[23,24,27] numerous studies based on GF/CNT (or CF/CNT) hybrid multi-scale composites have been conducted[44–60] to further explore the potential of CNT damage sensing methodologies in laminated composites. It is well established that damage in composite laminates starts from matrix-dominated failure modes such as matrix cracking and delamination, followed by fiber breakage which normally occurs near the end of the composites' lifetime. The introduction of CNT networks

within an insulating matrix acts like a neuron sensory networks for FRPs, enabling the detection of microcrack initiation and propagation, with the possibility of identifying the damage stage under certain conditions. Both static and cyclic loading conditions have been studied and are reported below, while also different manufacturing processes such as calendering or solution-based processes as well as specific specimen preparation methods are briefly described for each work.

Static loadings

Thostenson and Chou[49] used high shear mixing methods (calendering and three-roll mill) to disperse MWCNTs in epoxy resin, creating a percolating network for the onset of damage detection and evolution in GFRPs. During their specifically designed mechanical tests, interlaminar delamination and transverse microcracking was evaluated in unidirectional and cross-ply laminates, respectively. In the delamination tests (Fig. 9a), the electrical resistance increased significantly when delamination started and then progressively increased until the specimen failed. On the other hand, in the case of microcracking detection (Fig. 9b), the electrical resistance increased stepwise after crack initiation. These differences in sensing signal response opened up the possibility to identify different failure modes by using conductive CNT networks. The authors also compared the initial undamaged and damaged electrical resistance, and showed that it provided a more sensitive route to evaluate self-healing efficiency compared to traditional stiffness measurement since the stiffness of

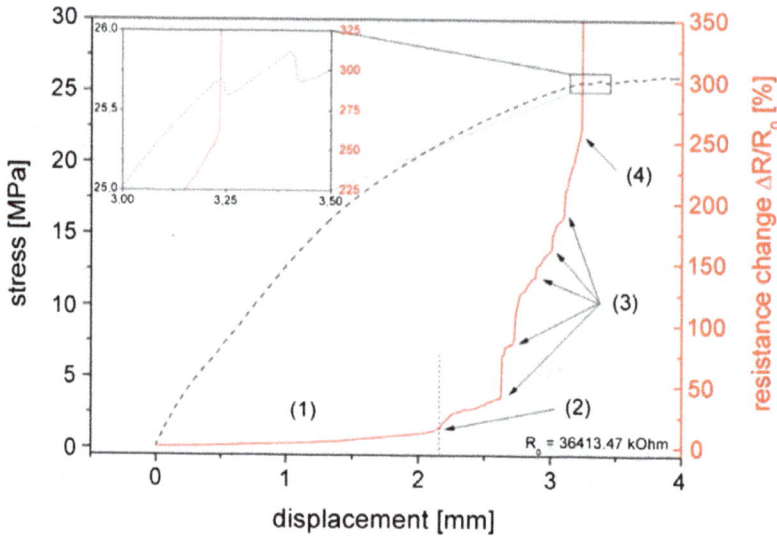

Figure 8 Damage sensing results of CNT/PP-coated glass fiber under tensile loading, where a step-change in resistance represents the occurrence of interface failures
Source: Reprint with permission from Ref. 43

cross-ply laminates is dominated by the 0° plies rather than the 90° plies, meaning that the residual stiffness after matrix cracking remains quite high. It is worth noting that in these studies the electrical resistance returned to nearly its initial value after the first damage (Fig. 9c), which was believed to be due to the specific design of the cross-ply laminate with the 0° plies in the outer layers.

Böger et al.[51] also found that the first matrix crack during tensile testing induced an obvious change in resistivity, which could be used to identify damage initiation. The introduction of CNTs did improve the fatigue behavior of laminates, but unfortunately no sensing properties were measured during these fatigue tests.

The electrical damage sensing method has another advantage over traditional techniques like AE as was shown by Gao et al.[61] Here, calendering was used to disperse CNTs into vinyl ether matrix, creating an electrically conductive polymer matrix which was combined with AE to detect damage in GFRP. Excellent correlation was found between electrical signals and AE cumulative counts during initial loading. However, during reloading, AE did not register new events (Kaiser effect), whereas electrical resistivity still increased (Fig. 10). This result highlighted another advantage of the electrical sensing method. Since the resistance change during elastic deformation in loading and reloading could not be attributed to damage but mainly to strain and crack opening, this part of the resistivity change was deducted in order to obtain the resistivity change due to damage, as defined by sensing parameter $\Delta R/L - \Delta R_E/L = \Delta R_D/L$.[61]

The dispersion of CNTs in the polymer matrix is another important aspect for both damage sensing and mechanical properties of FRPs. Gao et al.[62] compared the effect of CNT loading and dispersion in an epoxy matrix on damage sensing properties under static and incremental tensile loading of [0/90$_2$/0] GF laminates by using both three-roll milling of CNTs

with and without sizing agent. By using a sizing agent, the dispersion of CNTs in epoxy resin became non-uniform which lead to a reduction in mechanical performance compared to uniformly dispersed specimen via three-roll milling. During tensile testing, the resistance change per unit length was also found to be much less for non-uniformly dispersed specimen. This was not only due to agglomeration of sized CNTs, but also the results of higher CNT concentrations in these specimens.[62]

Gao et al's research showed the effect of CNT dispersion and concentration on sensing properties. In the case of CNTs, a uniform dispersion is preferred to create conductive pathways for electrons during loading. However, it is worth mentioning that for structural applications where the critical failure mode or damage prone region is often known, the localization of CNTs within these regions can be much more effective and efficient. In the case of CFRPs based on conductive CFs, research results showed that the presence of CNTs within the polymer matrix not only improved the electrical conductivity in transverse and through-thickness direction of composite laminates, but also enhanced the sensitivity to damage through more obvious resistance changes.[45] This can be attributed to the more sensitive neuron-like CNT percolated network in between CFs, as explained elsewhere.

Cyclic loadings

Apart from static testing conditions, cyclic loading was performed with the aim of evaluating damage progression as well as identifying the failure stages during the composites' usage.

Thostenson and Chou[63] examined microstructural damage evolution under cyclic loading with CNTs dispersed in epoxy via a calendering method. Cracks initiated within the 90° mid-plies (Fig. 11) with the applied load increasing until crack saturation in these plies, and with CNT pullout being observed at the crack surface. For cyclic sensing (Fig. 12), the strains of

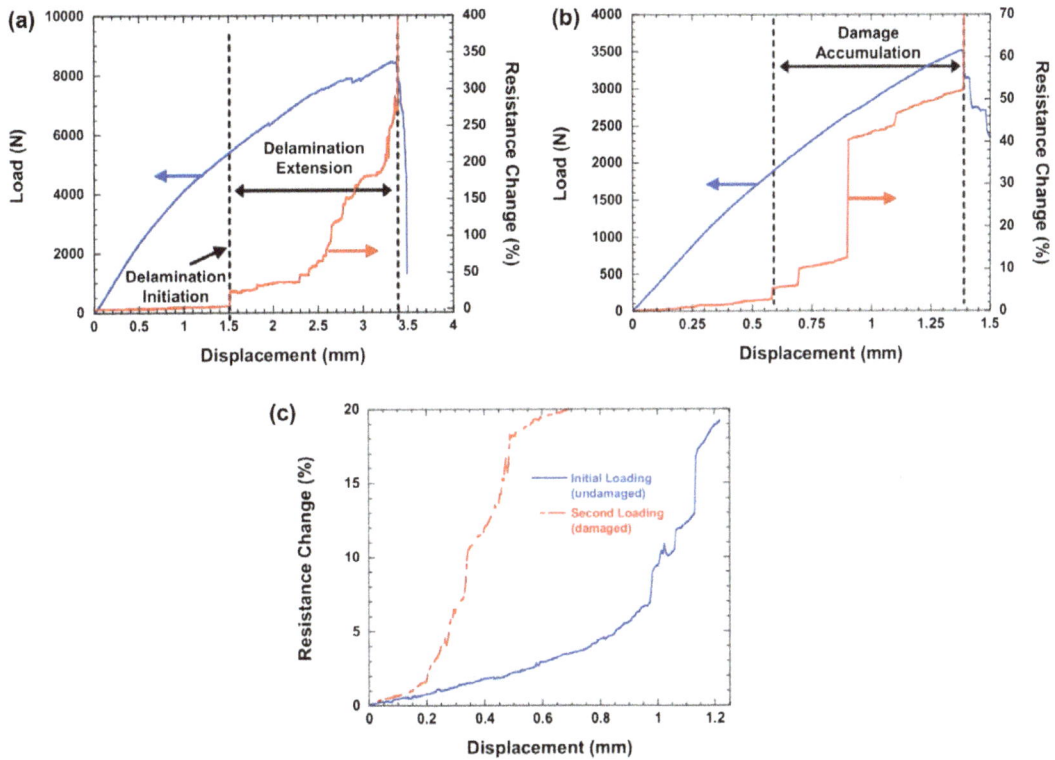

Figure 9 *In situ* damage sensing results for: *a* mid-ply delamination; *b* microcracking; *c* initial and second loading conditions
Source: Reprint with permission from Ref. 49

the first two cycles were lower than the critical first ply failure strain of transverse cracking and therefore the resistance followed the strain curve like in the case of strain sensing.[63] As strain increased, cracks were initiated and the resistance clearly started to deviate from the strain response, leaving a permanent resistance change after unloading. After crack initiation, for each incremental strain applied, three types of sensing signals were captured (Fig. 12*b*): (i) crack reopening, (ii) elastic deformation (which did not lead to new crack formation), and (iii) new crack accumulation, although this slope change was only clearly shown in the fifth cycle and not that obvious in later cycles which might be due to a too large applied strain. Twenty five cycles of smaller applied strain were also tested, and according to the authors, all reloading curves exhibited the same shape (Fig. 12).

In another incremental cyclic tensile loading study performed by Gao *et al.*,[48] three damage stages including crack initiation, transverse microcracking, and delamination were reported. Transverse crack densities as well as residual elastic modulus were studied in order to confirm those three damage stages. Delamination at the final stage was promoted by the specific specimen lay-up [0/90/90/0].

Similar to quasi-static conditions, differences between electrical sensing method and AE have been confirmed for incremental cyclic loadings. AE data together with electrical sensing data are shown in Fig. 13[61] and indicate that AE can only detect new crack formation during each incremental

loading cycle, while electrical sensing shows a more continuous increase throughout the test.

The possibility of using CNT networks to identify different damage stages was successfully shown in composite laminates. However, transferring these results to real composite applications remains a great challenge. For example, issues of how to adapt the differences in specimen sizes and lay-ups from model studies to real components and the ability to correlate sensing signals to failure modes remain a challenge.

However, initial results have indicated that CNT networks that establish themselves as a nerve-like sensory network within composite laminates are able to sense deformation and damage initiation and propagation, potentially providing a useful tool for health monitoring of composite structures. Moreover, initial work suggests that by analyzing the data, different failure modes and damage stages can potentially be identified.

Damage sensing in standardized tests

In the previous sections, it was shown that the use of CNTs in FRPs, especially in the case of insulating fibers like GF, can lead to hierarchical composites with damage sensing capability. Various stages of failure initiation and propagation, under both static and dynamic loadings, have been identified and analyzed. However, in order to replace (or combine with) existing embedded sensor SHM technologies in FRPs, improved sensitivity and reliability of signals remains a challenge. Most

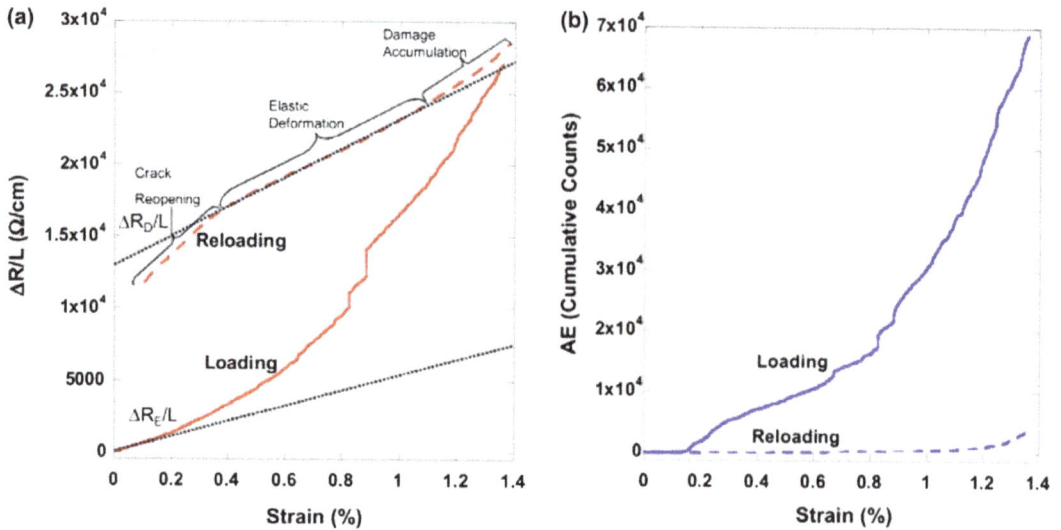

Figure 10   Comparison between electrical sensing method and AE counts for initial loading: *a* and reloading *b*
Source: Reprint with permission from Ref. 61

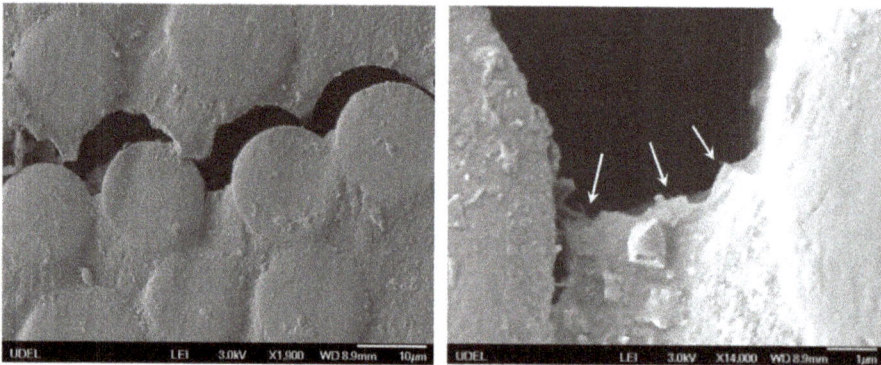

Figure 11   SEM images of crack formation in 90° ply (left) and CNTs within epoxy matrix (right)
Source: Reprint with permission from Ref. 63

of the research in this area is performed on model systems or specially designed tests on laminates aiming to promote specific failure modes to prove the concept. Few works have tried to transfer existing know-how to realistic composite applications. The first step to bridge this gap would be to apply the technique to standard composite tests which are closely related to realistic loading conditions. For instance, delaminations are one of the most common failure modes in laminated composites and can be evaluated under ASTM D5528, while damage sensing under impact conditions can be evaluated under ASTM D7136. It is necessary to demonstrate stable and repeatable sensing signals under these standardized test conditions before trying to implement this technology into real composite applications. In the following section, up-to-date damage sensing studies for various composite tests are introduced and analyzed, with the aim to build up a systematic understanding on how to transfer existing knowledge on CNT damage sensing to real components. After a systematic characterization of sensing in standardized test conditions, the electrical sensing method could be adapted to different applications with specific failure tendencies.

Impact

Impact damage is one of the major concerns in health monitoring of FRPs.[47,64] Normally, the electrical sensing signals are correlated to the impact energy during data analysis. Yesil *et al.*[65] compared different treated CNTs for dispersion as well as impact damage sensing properties. Better dispersion gave better electrical sensing signals which correlated to the impact energy applied. Similar findings were reported in other research works.[62,66] In another research by Arronche *et al.*,[67] two-probe and four-probe measurements were compared for sensing impact damage, with the four-probe method showing better reliability as it eliminates contact resistance effects.

It is worth mentioning that the impact studies listed below were not performed under ASTM D7136 for drop-weight

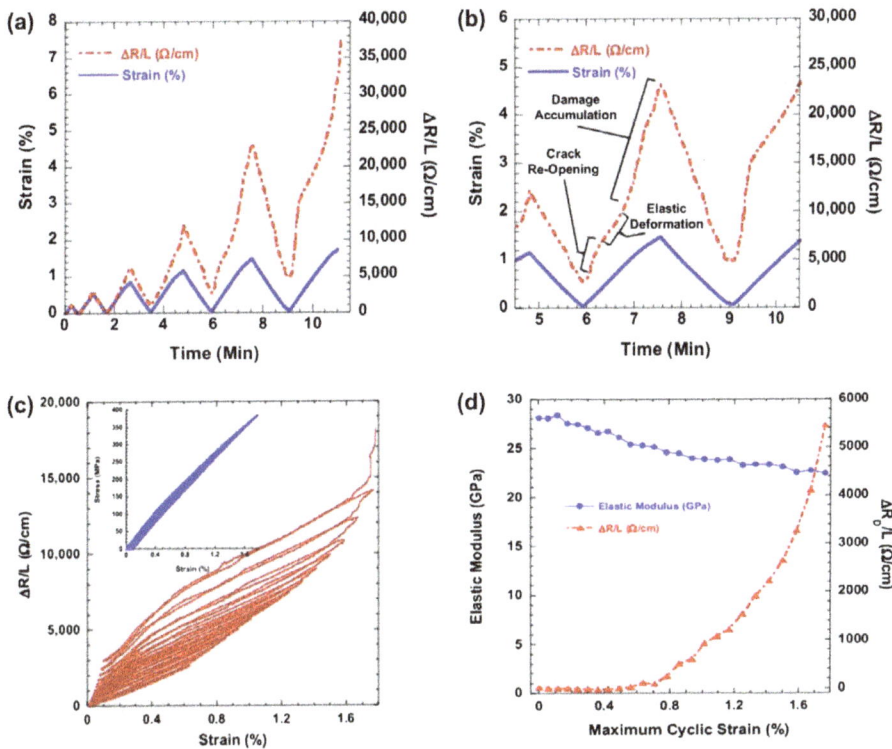

Figure 12 Damage initiation and evolution study under incremental cyclic loading in 90° mid-ply of cross-ply laminated composites: *a* resistance response under cyclic loading; *b* focused view of resistance response of fifth cycle; *c* substantial hysteresis response from resistance–strain curve; and *d* elastic modulus and resistance changes with maximum cyclic strain
Source: Reprint with permission from Ref. 63

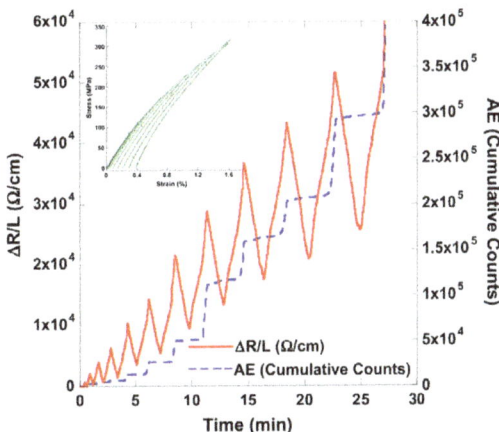

Figure 13 Comparison between electrical sensing and AE under incremental cyclic loading conditions, showing good correlation between electrical signals and AE counts
Source: Reprint with permission from Ref. 61

impact, although the potential of using electrical signals for impact damage sensing was shown. Gao et al.[47] examined electrical sensing signals during repeated impact loadings, in combination with AE to confirm the damage (Fig. 14). Monti et al.[68] also observed an increase in electrical resistivity for

specimen after impact, which again was due to breakdown of internal CNT networks.

In order to apply the electrical sensing method to larger plates, Naghashpour et al.[69] performed electrical grid mapping to detect internal damage in large composite laminates. Both high-velocity and low-velocity impact was applied to initiate damage, and real-time detection was achieved.

Some other CNT-related impact damage sensing studies including electrical impedance tomography,[70] or sensing skins,[71] etc. are not discussed here, as these works are not the main topic for this review paper.

Fatigue

It is well known that the fatigue life of composites is strongly affected by matrix-dominated properties.[50,72,73] In fatigue, residual strain is often used as an indicator for the internal health of a composite laminate. Hence, by comparing the resistance change and number of cycles together with residual strain, the internal health condition of laminates can be revealed by electrical signals.

Yesil et al.[65] followed ASTM D3039 to produce tensile specimen for fatigue testing, with various treated CNTs being introduced in the GFRP laminates. Among their results, the largest sensing signals were obtained for treated CNTs which gave better dispersion and interaction within the matrix (Fig. 15). Generally, it was found that electrical resistance increased

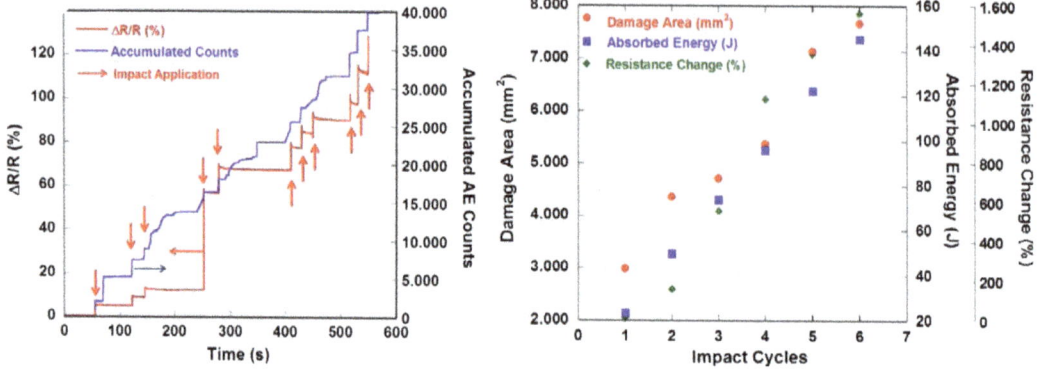

Figure 14 *In situ* impact damage sensing using percolated CNT networks in composite laminates: in comparison with AE results (left) and damage areas measured by ultrasonic C-scan (right)
Source: Reprint with permission from Ref. 47

with decreasing stiffness during fatigue testing as a result of internal damage accumulation process. The existence of CNTs has also shown increased high cycle fatigue life compared to the reference.[51]

Flexural loading and short beam shear loading

Thostenson and Chou[49] used two different spans to promote different failure modes in three-point bending tests (Fig. 16). For short beam shear conditions, i.e. a short span-to-depth ratio of 4, electrical signals increased by several orders close to the point of shear failure, indicating rather abrupt delamination. Subsequent reloading did establish some new electrical contacts but still resulted in a much greater resistance than its initial value. For a longer span-to-depth ratio of 8, i.e. more toward pure bending conditions, a more incremental resistance change was observed, which corresponds to matrix cracking and damage accumulation.

Boger et al.[50] followed ASTM D2344 for short beam shear testing of GF/CNT hybrid composites. During this interlaminar shear strength test, in-plane electrical resistance remained more or less constant until a sudden increase due to composite failure. In this study, through-thickness resistivity showed no obvious change during loading (Fig. 17). In contrast, Zhang et al.[74] found that monitoring through-thickness resistance was the most effective way to detect interlaminar shear failure, as the sensing signals simultaneously changed with increasing load, providing a tool to detect early stage matrix damage. Moreover, the sensitivity of the measured electrical signals was also higher for their through-thickness measurements compared to in-plane measurement (Fig. 18).[74]

Delamination

Delamination is one of the most common failure modes in composites as out-of-plane properties remain one of their weaknesses due to their laminated nature. Various studies have been conducted with the aim of improving delamination resistance through the addition of CNTs, while at the same time introducing multifunctionalities for *in situ* health monitoring.[66,75–77]

Zhang et al. applied various routes to deliver CNTs into FRPs through interleaves for damage sensing purposes, including direct spraying deposition[66] for both insulating GF and conductive CF preforms. With this simple but versatile spray coating technique, an extremely low CNT concentration (~0.01 wt.%) could be introduced to create a percolating network for internal health monitoring under Mode-I test conditions. Fine CNT networks positioned on the surface of GFs as well as in between fibers were observed. Internal crack propagation was correlated to measured electrical signal changes, and a good correlation was reported (Fig. 19).

Zhang et al. extended this method to CF prepregs, and obtained good sensing signals.[66] Compared to the CF reference laminate, the introduction of CNTs within the interfacial regions successfully improved the stability of the sensing signals. This reduction in scatter in sensing signals, which is essential for industrial applications, was attributed to changes in sensing mechanism going from physical contacts between conductive CFs to a tunneling mechanism between CNTs in percolated networks in resin-rich regions with improved stability and consistency, as illustrated in Fig. 20.

Overview of CNT damage sensing and other on-line SHM methods

The current review has demonstrated the potential of CNT networks for damage sensing from model studies to standardized composite coupon tests. Table 1 lists the majority of research studies related to damage sensing in fiber composites using CNTs. The materials used, manufacturing processes applied, as well as main results are reported. The table also indicates whether the damage sensing studies were performed using standardized test methods. Both top-down (mixing CNTs into polymer matrices) as well as bottom-up (CNT coatings, growing CNTs onto fibers etc.) approaches have been used to produce these multifunctional composites. Moreover, it is interesting to note that only less than half of the reported sensing work was performed using standardized ASTM/ISO tests.

For on-line SHM, in general, two main approaches are followed to integrate sensors into composites: (i)

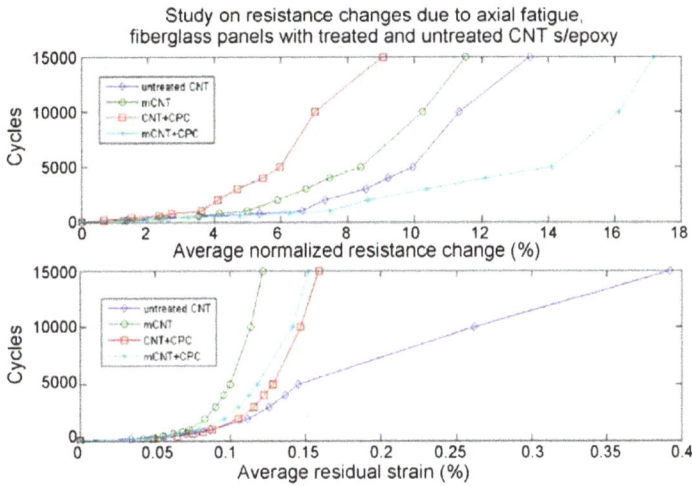

Figure 15   Resistance change under axial fatigue vs. number of cycles (top), and average residual strain vs. number of cycles under axial fatigue (bottom)
Source: Reprint with permission from Ref. 65

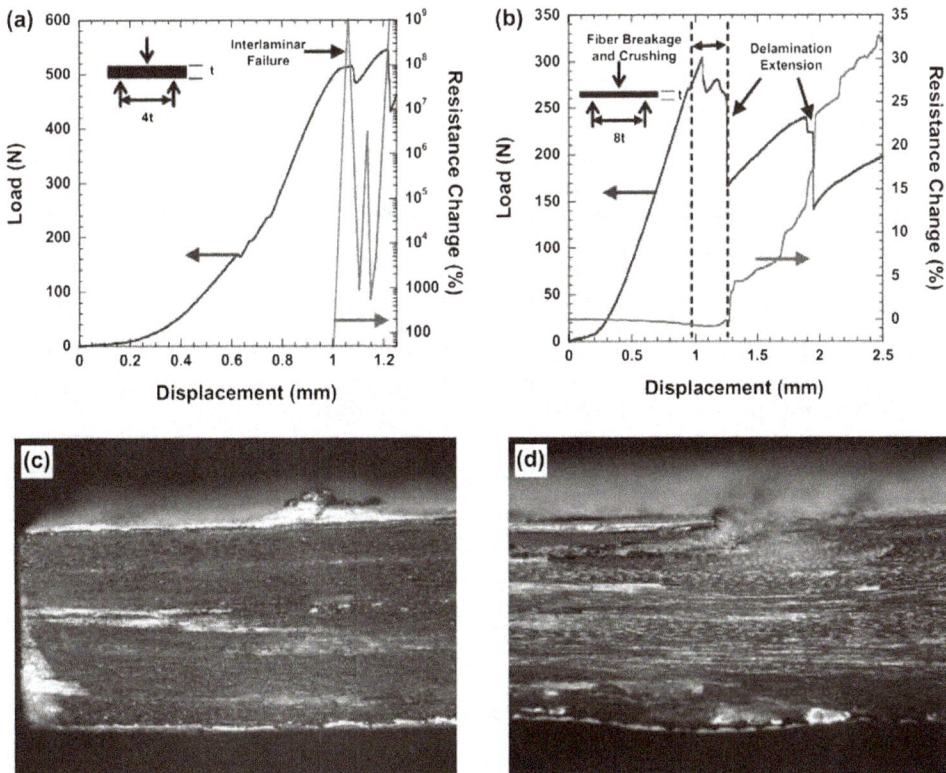

Figure 16   *In situ* damage sensing results at different span-to-depth ratios of 4 *a* and 8 *b*, together with their optical microscope images *c* and *d*, respectively
Source: Reprint with permission from Ref. 49

surface-mounted sensors as in the case of piezoelectric, eddy current sensors, and comparative vacuum monitoring (CVM), and (ii) embedded sensors like fiber Bragg grating sensors or

CNTs. For industrial applications, especially for civil aircrafts where safety is always paramount, surface-mounted sensors are often favored in the short term due to their simplicity and

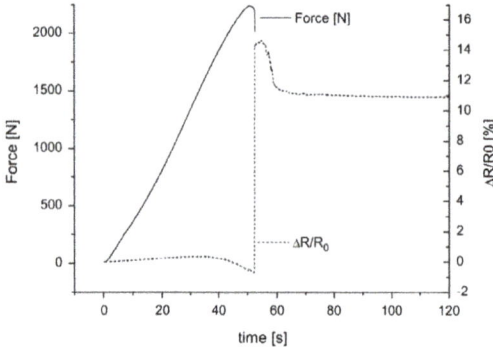

Figure 17   Resistance change during ILSS test of GF/CNT hybrid composites
Source: Reprint with permission from Ref. 50

more importantly, the preservation of the original material properties. From an application point of view, novel hybrid CFRP-CNT composites are regarded as a new material system which needs to go through a long validation program before it can be certified for use in aircraft. Sensors that are attached to the outer surface without affecting the components' integrity are easier to be approved for use in current aircraft. However, surface-mounted sensors have their limitations as they have to face limited space for sensor attachments and strict operating environments, not to mention the additional equipment involved which adds more weight to the aircraft. Moreover, such sensors have difficulties to detect internal damage in composite laminates. Despite these disadvantages, nanocarbon based surface coatings are expected to be among the first CNT based sensing systems for composite structures, because of advantages over piezoelectrical sensors such as lightweight and added multifunctionality through increased electrical conductivity for improved lighting strike protection (LSP).

On the other hand, regardless of the certification process, embedded sensing systems like CNT-based damage sensing technologies can be integrated into the composite components without requiring additional space or equipment, greatly saving space and weight of the complete system. In the long term, a self-sensing composite system with reliable SHM

functionalities is definitely the ultimate goal for the aviation industry. Hence, for the medium- to long-term on-line SHM, the embedded or integrated sensory networks in composite materials rather than surface-mounted sensors are favored. However in order to be successful, a couple of important aspects need to be addressed from a materials point of view:

- The integrated sensor system cannot affect components' integrity and structural properties.
- The system needs to be integrated or as compact as possible, with minimum number of components involved.
- The system needs to have clear and reliable sensing signals to identify different failure modes, and good sensitivity to determine the level of damage.

Clearly, compared to other on-line SHM methods, the here presented CNT sensing technologies have advantages of integration and lightweight, preserved (or even improved) mechanical and electrical performance, potential to detect different matrix-dominated failures, as well as long-term durability for on-line damage monitoring. However, at its current stage, the technology has also some limitations and challenges, such as (i) how to position electrical circuits to build detection networks, (ii) how to decouple sensing signals from the CNT network from signals originating from the naturally conductive CFs, and (iii) how to deal with the increase in viscosity and inhomogeneous dispersion of CNTs in polymer resins and in laminate manufacturing. For the first two issues, previous SHM know-how based on existing methods could provide guidance of how to establish successful sensory networks, while combinations with other NDT or SHM methods might be required to improve its sensitivity to different failure modes. Regarding manufacturing issues, innovative methods of CNT deposition like spray coating, electrophoresis deposition, or catalytic growth of CNTs onto fiber preforms provide interesting solutions to localize the CNT networks at desired positions without affecting resin viscosity or creating filtration problems arising from flow of a CNT-filled resin through the fiber preform.

The study of damage sensing using CNTs (or other nanocarbons) as integrated sensing materials into FRPs is still at an early stage but has already shown great promise of both the fundamental understanding of sensing mechanisms, as well as the potential to apply this technology in composite

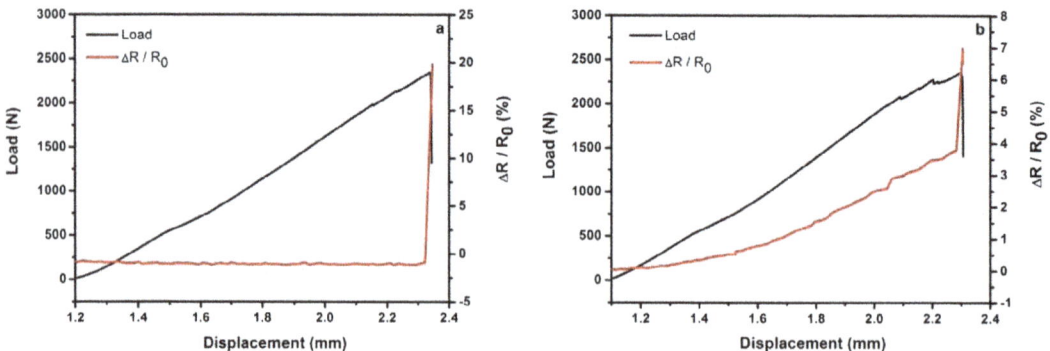

Figure 18   Resistance change during ILSS test of CNT-modified CFRP composites forin-plane (left) and through-thickness (right) measurements[74]

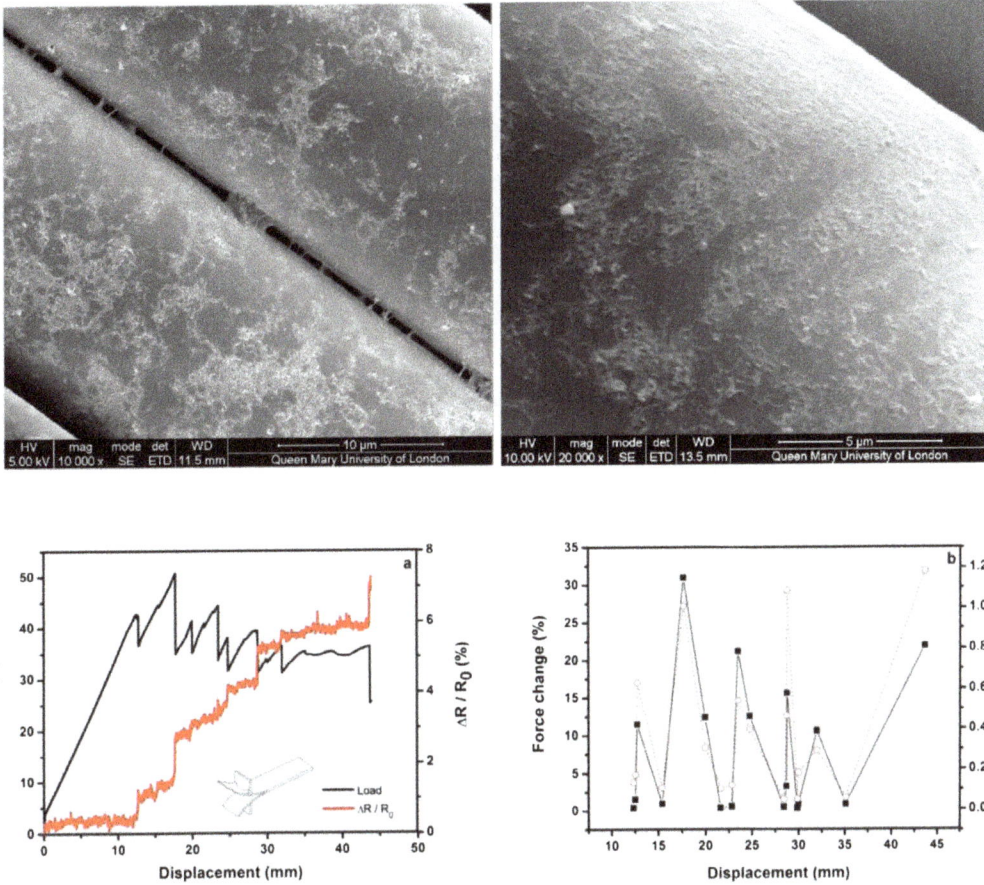

Figure 19　Spray deposited CNT network on glass fiber fabrics, and *in situ* damage sensing in hybrid CNT/GF laminates: *a* electrical sensing data with load curve; *b* correlation between electrical sensing signals and external loads[78]

components for on-line health monitoring. However, before this technology reaches maturity, it probably needs to be used in combination with other traditional SHM or NDT methods to deliver reliable results. For instance, in order to utilize the CNT sensing method to detect various failure modes under both static and cyclic loading conditions, a combination with AE may be needed to determine those failure modes, while it can also help to decouple the effects from the conductive CFs. However, in the long term, the use of CNTs in FRPs possesses huge potential for creating integrated self-sensing materials and structures.

Conclusions and perspectives

Fiber-reinforced composites have been used extensively to replace traditional materials like metals in many fields. To improve the reliability of these materials, integrated on-line health monitoring systems have been the topic of numerous studies for many years. In the last decade, the use of CNTs to create sensory networks that are able to detect various stages of internal damage and failure modes in composite materials has been extensively studied. Percolated networks of electrically conductive CNTs allow electrons to pass through the

material or structure and provide the possibility to monitor internal deformation and damage in composites using an electrical resistivity method. Recent advances in both model composites as well as composite laminates have been discussed. The effect of CNT concentration and dispersion on electrical sensing behavior has also been compared and analyzed.

It is worth noting that different manufacturing and dispersion processes can have a significant effect on the feasibility of industrial scale-up. Each processing route has its pros and cons, which might be suitable for a certain application. For instance, direct mixing of CNTs into low-viscosity epoxy resin may seem easy from a logistics point of view, but may result in poor dispersion, CNT alignment along the resin flow direction, and filtering effects by fiber preforms during liquid molding processes or too high viscosities for composite processing. To achieve good CNT dispersion (and localization), *in situ* growth of CNTs directly onto the reinforcing fiber surface or electrophoresis deposition can be performed. Unfortunately, such nanofunctionalization processes are relatively complex and costly procedures that could limit industrial applications, while there are still remaining issues regarding the preservation of fiber strength. Spray coating is an easy industrial-scalable process which has shown

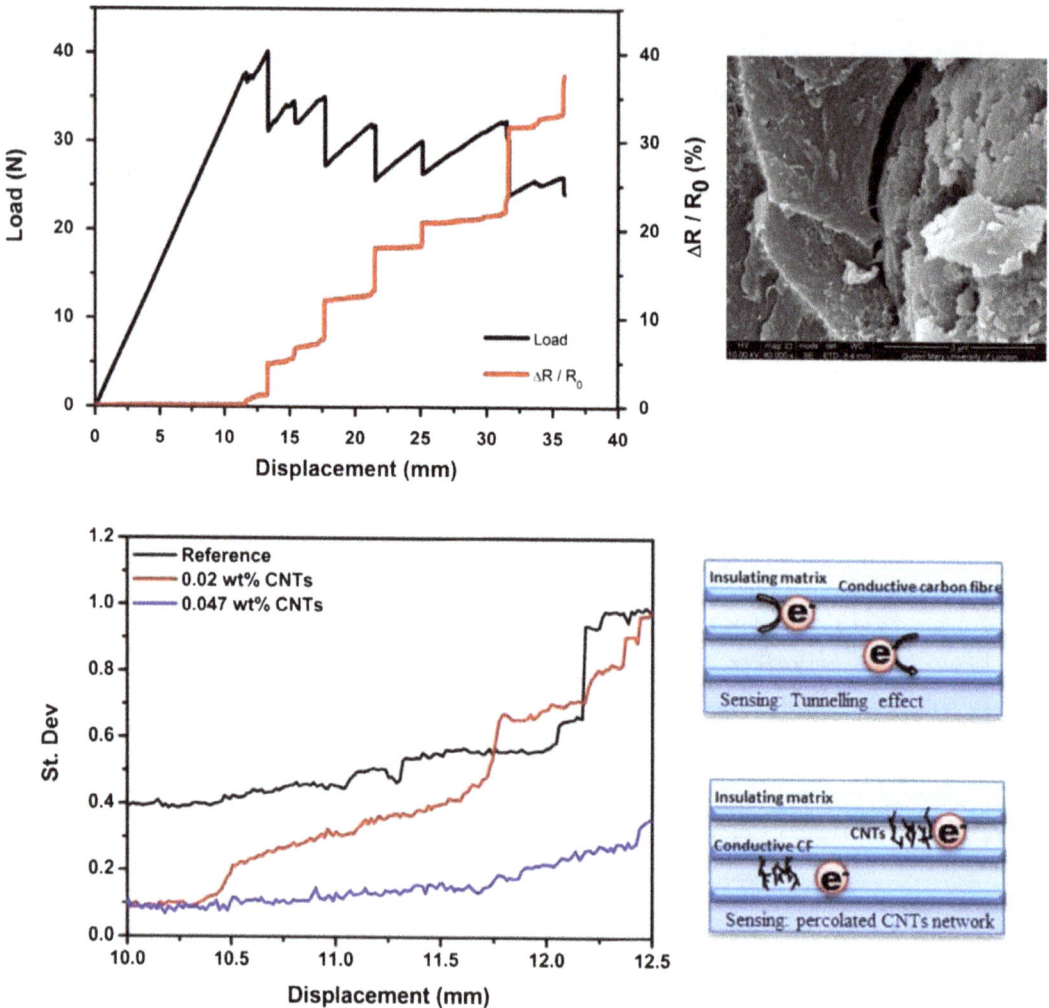

Figure 20 *In situ* damage sensing results of CNT-modified carbon fiber prepregs under Mode-I test conditions
Source: Reprint with permission from Ref. 66

to create good CNT dispersion and localization. However, when spraying nanomaterials, care needs to be taken that safety precautions are taken and that solvents are recovered in order to reduce the risk on personnel and environment. Apart from technological issues related to dispersion and localization of CNTs in composites, easy scale-up, relatively simple procedures, safety and environmental issues are among the most important aspects when considering technology transfer from laboratory to an industrial environment.

SHM requires a multidisciplinary approach, involving topics like structural mechanics, damage sensing and monitoring, electronic engineering (to process and manage signals), software engineering and statistics (to interpret data), and multi-scale composite manufacturing (to localize sensors). To obtain the best SHM results, a combination of different methods may be required. Lab-scale damage sensing concepts for composites based on CNTs to detect damage has now been

around for a couple of years. Since then a number of studies have examined the potential of CNT sensing in composites, as well as interpreted obtained data to identify different failure modes. After some promising results which confirmed the potential of CNTs as an effective nanomaterial to detect damage in laminated composites, the question remains, however, how to transfer this technology into real industrial components. To progress this research area and reduce the gap to industrial applications, research needs first to demonstrate reliable and repeatable sensing data under standardized test conditions together with the development of reliable and cost-effective manufacturing methods for such multifunctional composites in an industrial environment. Subsequently, these CNT damage sensing concepts need to be tested in real components, while the information obtained from these self-sensing materials should be implemented into damage models to improve lifetime predictions of composite structures.

Table 1 Summarized results of composites damage sensing works based on CNT networks

Fiber/matrix	Nanofillers (loadings) and location	Manufacturing processes	Detected failure and damages	Main results	Standard test* (ASTM or ISO)	References
GF/epoxy	MWCNTs (0.5 wt.%) in epoxy matrix	Three-roll mill	Delamination; microcracking	Showed possibility of using CNTs to identify different failure modes	No	Thostenson and Chou[49]
GF/epoxy	MWCNTs on GF surface	Electrophoretic deposition (EPD)	Microcracking	Coated single GF as *in situ* sensor for tracking microcracks	No	Zhang[14]
GF/epoxy	CNTs (0.5 wt.%) in epoxy matrix	Calendering	Crack initiation during static and incremental tensile loading	Used CNT networks to identify different failure stages	No	Thostenson[48, 63]
GF/vinyl ester	CNTs (0.5 wt.%) in matrix	Calendering	Crack initiation during static and incremental tensile loading	Combined with AE, it showed some advantages of the electrical sensing method	No	Gao LM[61]
GF/epoxy	MWCNTs (0.3 wt.%) in epoxy matrix	Calendering	Crack initiation during tensile loading	Used CNTs to identify damage initiation	No	Böger[51]
GF/epoxy	SWCNT on GF surface	Spray coating	Microcrack initiation and propagation under tensile loading	Embedded SWCNT-coated fiber sensor provided *in situ* information on resin curing and deformation under loading	No	Luo[58]
GF/epoxy	MWCNTs (1.0 wt.%) in matrix	Three-roll mill	Localization of failure under tensile loading	Localized failure by placing electrodes at different positions	No	Nofar[73]
GF/epoxy	MWCNTs (0.5 wt.%) in epoxy	Three-roll mill and sizing agent	Sensing under tensile loading	Compared effect of dispersion and CNT localization on sensing data	No	Gao LM[62]
CF/epoxy	CNTs (0.1/0.5/1.0 wt.%) in epoxy	Calendering	Sensing under tensile loading	Improved sensing signals for CNT-modified specimen	No	Kostopoulos[45]
CF/epoxy	MWCNTs (0.5 wt.%) in epoxy matrix	Sonication	Sensing under tensile loading	Improved sensitivity for CNT-modified CFRPs compared to neat CFRPs	Yes (Tensile) ASTM D3039	Grammatikos[79]
GF/PP	MWCNTs (0.5 wt.%) in polymer film coated on GF	Sizing or coating of CNT/polymer film onto GF	Interface strain sensing under tensile loading	Alternative route utilizing aqueous CNT/polymer coating on GF for sensing	No	Rausch and Mäder[10, 43, 80]
CF/epoxy	CNTs (0.1 to 2.0 wt.%) in epoxy	Solution-based processing	Fiber breakages in dual-matrix composites	Use of CNT networks to detect fiber breakage in model sample	No	Park JM[35, 36]
CF/epoxy	CNTs (0.5 wt.%) in epoxy	Solution-based processing	Fiber breakages in dual-matrix composites	Showed effect of CNT dispersion on sensing results	No	Park JM[41]
GF/epoxy	MWCNTs (0.3 wt.%) in epoxy	Three-roll mill	Sensing during interlaminar shear strength testing (ILSS)	In-plane sensing signals gave sudden increase at failure, while through-thickness signals showed no obvious changes during testing	Yes (ILSS) ASTM D2344	Böger[50]
GF/epoxy	CNTs in PVA fiber	CNT/PVA fiber strain sensor	Strain sensing under tensile and three-point bending test conditions	Use of a CNT/PVA fiber as embedded strain sensor in GFRP	Yes (Tensile) ASTM D3039	Alexopoulos[8, 11, 81]
GF/epoxy	MWCNTs in epoxy matrix	Dispersed in epoxy using sizing agent	Impact damage	Showed potential of electrical sensing method for impact damage	No	Gao LM[47]
CF/epoxy	CNTs (0.1–0.5 wt.%) in epoxy matrix	Mechanical stirring and sonication	Impact damage	Showed increased resistance after impact damage	No	Monti[68]
GF/epoxy	MWCNTs (0.5 wt.%) in epoxy matrix	As received	Impact damage	Four probes measurement provided better sensing data compared to two probes method	Yes (Impact) ASTM D7136	Arronche[67]

Material	CNT details	Processing	Damage/Application	Description	Standard testing	Reference
GF/epoxy	CNTs (0.1–1.0 wt.%) in epoxy matrix	Three-roll mill	Open hole and impact damage	Applied electrical sensing method to large composite plates	No	Naghashpour[69]
GF/epoxy	MWCNTs in epoxy	Solution-based processing	Sensing under tensile, cyclic tensile fatigue, and impact	Compared effect of sizing and treatment of CNTs with improved dispersion leading to better sensing signals	Yes (Impact) ASTM D7136	Yesil S[65]
GF/epoxy	MWCNTs (0.5 wt.%) on GF surface	Dip coating	Single glass fiber with coated CNTs as multifunctional sensor	First time coating of CNTs onto GF for sensing rather than mixing into matrix	No	Gao SL[13]
GF/epoxy	MWCNTs on GF	Spray coating	Delamination and interlaminar shear	First time use of spray coating to deposit CNTs in GFRPs, introduced percolated network for damage sensing with extremely low CNT content	Yes (DCB, SBS) ASTM D5528, ASTM D2344	Zhang[78]
CF/epoxy	MWCNTs (0.047 wt.%) on CF prepregs	Spray coating	Delamination	Showed correlation between electrical sensing signals and force drop, and improved sensing stability	Yes (DCB) ASTM D5528	Zhang[66]
GF/epoxy	CNT thread	Embedded fiber	Matrix cracking	Use of CNT thread as embedded sensor in GFRPs for damage detection	No	Hehr[67]
CF/epoxy	MWCNTs (0.08 wt.%) on CF	Spray coating	Interlaminar shear	Demonstrated progressive damage sensing with through-thickness measurements	Yes (SBS) ASTM D2344	Zhang[74]
CF/epoxy	CNT (0.5 wt.%)	Sonication	Fatigue	Correlation between electrical sensing results and AE signals under fatigue mechanical testing	No	Grammatikos[82]
GF/epoxy	CNT/Al_2O_3 (0.5 wt.%)	Three-roll mill	Sensing under tensile loading	Use of CNT/Al_2O_3 as conductive filler in epoxy to detect damage	Yes (Tensile) ASTM D3039	Li[59]
CF/epoxy	CNT growth on GF	Embedded fiber	Sensing under tensile loading	Use of fuzzy glass fiber as strain sensor in CF prepreg	No	Sebastian[60]

*Standard testing in this table indicated that both specimen dimension and test conditions were according to test standard

Conflicts of interest

No potential conflict of interest was reported by the authors.

References

1. M. Holmes: 'US demand for fibre reinforced plastic composites to rise'; 2013. http://www.materialstoday.com/composite-industry/news/us-demand-for-fibre-reinforced-plastic-composites/
2. S. Black: Compos. World, 2012, vol. March 2012. http://www.compositesworld.com/articles/carbon-fiber-market-gathering-momentum
3. C. Boller: Int. J. Syst. Sci., 2000, 31, 1333–1349.
4. P. J. deGroot, P. A. M. Wijnen and R. B. F. Janssen: Compos. Sci. Technol., 1995, 55, 405–412.
5. H. Jeong and Y. S. Jang: Compos. Struct., 2000, 49, 443–450.
6. O. Ceysson, M. Salvia and L. Vincent: Scr. Mater., 1996, 34, 1273–1280.
7. M. Surgeon and M. Wevers: Ndt & E Int., 1999, 32, 311–322.
8. N. D. Alexopoulos, C. Bartholome, P. Poulin and Z. Marioli-Riga: Compos. Sci. Technol., 2010, 70, 260–271.
9. B. R. Loyola, V. La Saponara and k. J. Loh: J. Mater. Sci., 2010, 45, 6786–6798.
10. J. Rausch and E. Mäder: Compos. Sci. Technol., 2010, 70, 1589–1596.
11. N. D. Alexopoulos, C. Bartholome, P. Poulin and Z. Marioli-Riga: Compos. Sci. Technol., 2010, 70, 1733–1741.
12. Y. Shimamura, K. Kageyama, K. Tohgo and T. Fujii. Cyclic behavior of electrical resistance type low stiffness, large strain sensor by using carbon nanofiber/flexible epoxy composite, In: 'Fracture and strength of solids, Vii', (ed. A. K. Ariffin, S. Abdullah, A. Ali, A. Muchtar, M. J. Ghazali and Z. Sajuri, Pts 1 and 2, vol. 462–463; 2011, pp. 1200–1205.
13. S.-I. Gao, R.-C. Zhuang, J. Zhang, J.-W. Liu and E. Maeder: Adv. Funct. Mater., 2010, 20, 1885–1893.
14. J. Zhang, J. Liu, R. Zhuang, E. Maeder, G. Heinrich and S. Gao: Adv. Mater., 2011, 23, 3392–3397.
15. T. Takeda, Y. Shindo, T. Fukuzaki and F. Narita: J. Compos. Mater., 2014, 48, 119–128.
16. D. D. L. Chung: Carbon, 2012, 50, 3342–3353.
17. S. Wang and D. D. L. Chung: Carbon, 2006, 44, 2739–2751.
18. K. Schulte and C. Baron: Compos. Sci. Technol., 1989, 36, 63–76.
19. I. Weber and P. Schwartz: Compos. Sci. Technol., 2001, 61, 849–853.
20. M. Kupke, K. Schulte and R. Schüler: Compos. Sci. Technol., 2001, 61, 837–847.
21. R. Schueler, S. P. Joshi and K. Schulte: Compos. Sci. Technol., 2001, 61, 921–930.
22. N. Muto, Y. Arai, S. G. Shin, H. Matsubara, H. Yanagida, M. Sugita and T. Nakatsuji: Compos. Sci. Technol., 2001, 61, 875–883.
23. J. R. Wood, M. D. Frogley, E. R. Meurs, A. D. Prins, T. Peijs, D. J. Dunstan and H. D. Wagner: J. Phys. Chem. B, 1999, 103, 10388–10392.
24. J. R. Wood, Q. Zhao, M. D. Frogley, E. R. Meurs, A. D. Prins, T. Peijs, D. J. Dunstan and H. D. Wagner: Phys. Rev. B, 2000, 62, 7571–7575.
25. C. Li, E. T. Thostenson and T. W. Chou: Compos. Sci. Technol., 2008, 68, 1227–1249.
26. T.-W. Chou, L. Gao, E. T. Thostenson, Z. Zhang and J.-H. Byun: Compos. Sci. Technol., 2010, 70, 1–19.
27. B. Fiedler, F. H. Gojny, M. H. G. Wichmann, W. Bauhofer and K. Schulte: Ann. Chim. Sci. Matér., 2004, 29, 81–94.
28. T. W. Ebbesen, H. J. Lezec, H. Hiura, J. W. Bennett, H. F. Ghaemi and T. Thio: Nature, 1996, 382, 54–56.
29. M. M. J. Treacy, T. W. Ebbesen and J. M. Gibson: Nature, 1996, 381, 678–680.
30. R. H. Baughman, A. A. Zakhidov and W. A. de Heer: Science, 2002, 297, 787–792.
31. S. Iijima: Nature, 1991, 354, 56–58.
32. S. Iijima and T. Ichihashi: Nature, 1993, 363, 603–605.
33. S. Iijima, C. Brabec, A. Maiti and J. Bernholc: J. Chem. Phys., 1996, 104, 2089–2092.
34. E. T. Thostenson, Z. F. Ren and T. W. Chou: Compos. Sci. Technol., 2001, 61, 1899–1912.
35. J. M. Park, D. S. Kim, J. R. Lee and T. W. Kim: Mater. Sci. Eng. C-Biomimetic Supramol. Syst., 2003, 23, 971–975.
36. J.-M. Park, D.-S. Kim, S.-J. Kim, P.-G. Kim, D.-J. Yoon and K. L. DeVries: Compos. Part B Eng., 2007, 38, 847–861.
37. E. Bilotti, H. Zhang, H. Deng, R. Zhang, Q. Fu and T. Peijs: Compos. Sci. Technol., 2013, 74, 85–90.
38. E. Bilotti, R. Zhang, H. Deng, M. Baxendale and T. Peijs: J. Mater. Chem., 2010, 20, 9449–9455.
39. R. Zhang, H. Deng, R. Valenca, J. H. Jin, Q. Fu, E. Bilotti and T. Peijs: Compos. Sci. Technol., 2013, 74, 1–5.
40. H. Zhang, E. Bilotti, W. Tu, C. Y. Lew and T. Peijs: Eur. Polym. J., 2015, 68, 128–138.
41. J.-M. Park, P.-G. Kim, J.-H. Jang, Z. Wang, J.-W. Kim, W.-I. Lee, J. G. Park and K. L. DeVries: Compos. Part B Eng., 2008, 39, 1170–1182.
42. J. Zhang, R. Zhuang, J. Liu, E. Mäder, G. Heinrich and S. Gao: Carbon, 2010, 48, 2273–2281.
43. J. Rausch and E. Mäder: Compos. Sci. Technol., 2010, 70, 2023–2030.
44. E. T. Thostenson and T.-W. Chou: Compos. Sci. Technol., 2008, 68, 2557–2561.
45. V. Kostopoulos, A. Vavouliotis, P. Karapappas, P. Tsotra and A. Paipetis: J. Intell. Mater. Syst. Struct., 2009, 20, 1025–1034.
46. Q. An, A. N. Rider and E. T. Thostenson: ACS Appl. Mater. Interfaces. 2013, 5, 2022–2032.
47. L. M. Gao, T. W. Chou, E. T. Thostenson, Z. G. Zhang and M. Coulaud: Carbon, 2011, 49, 3382–3385.
48. L. M. Gao, E. T. Thostenson, Z. Zhang and T. W. Chou: Adv. Funct. Mater., 2009, 19, 123–130.
49. E. T. Thostenson and T.-W. Chou: Adv. Mater., 2006, 18, 2837.
50. L. Böger, M. H. G. Wichmann, L. O. Meyer and K. Schulte: Compos. Sci. Technol., 2008, 68, 1886–1894.
51. L. Böger, J. Sumfleth, H. Hedemann and K. Schulte: Compos. Part A Appl. Sci. Manuf.: 2010, 41, 1419–1424.
52. G. Pandey, M. Wolters, E. T. Thostenson and D. Heider: Carbon, 2012, 50, 3816–3825.
53. k. J. Kim, W.-R. Yu, J. S. Lee, L. Gao, E. T. Thostenson, T.-W. Chou and J.-H. Byun: Compos. Part A Appl. Sci. Manuf., 2010, 41, 1531–1537.
54. A. S. Wu, A. M. Coppola, M. J. Sinnott, T. W. Chou, E. T. Thostenson, J. H. Byun and B. S. Kim: Compos. Sci. Technol., 2012, 72, 1618–1626.
55. Y. Song, A. Hehr, V. Shanov, N. Alvarez, N. Kienzle, J. Cummins, D. Koester and M. Schulz: Smart Mater. Struct., 2014, 23, 075008-1–075008-12.
56. Z. Zhang, H. Wei, Y. Liu and J. Leng: Struct. Health Monit., 2015, 14, 127–136.
57. A. Hehr, M. Schulz, V. Shanov and Y. Song: Struct. Health Monit., 2014, 13, 512–524.
58. S. D. Luo, W. Obitayo and T. Liu: Carbon, 2014, 76, 321–329.
59. W. K. Li, D. L. He, Z. M. Dang and J. B. Bai: Compos. Sci. Technol., 2014, 99, 8–14.
60. J. Sebastian, N. Schehl, M. Bouchard, M. Boehle, L. Li, A. Lagounov and K. Lafdi: Carbon, 2014, 66, 191–200.
61. L. Gao, E. T. Thostenson, Z. Zhang and T.-W. Chou: Carbon, 2009, 47, 1381–1388.
62. L. Gao, T.-W. Chou, E. T. Thostenson and Z. Zhang: Carbon, 2010, 48, 3788–3794.
63. E. T. Thostenson and T.-W. Chou: Nanotechnology, 2008, 19, 215713-1–215713-6.
64. J. K. Kim and M. L. Sham: Compos. Sci. Technol., 2000, 60, 745–761.
65. S. Yesil, C. Winkelmann, G. Bayram and V. La Saponara: Mater. Sci. Eng. A, 2010, 527, 7340–7352.
66. H. Zhang, Y. Liu, M. Kuwata, E. Bilotti and T. Peijs: Compos. Part A Appl. Sci. Manuf., 2015, 70, 102–110.
67. L. Arronche, V. La Saponara, S. Yesil and G. Bayram: J. Appl. Polym. Sci., 2013, 128, 2797–2806.
68. M. Monti, M. Natali, R. Petrucci, J. M. Kenny and L. Torre: J. Appl. Polym. Sci.: 2011, 122, 2829–2836.
69. A. Naghashpour and S. Van Hoa: Nanotechnology, 2013, 24, 455502-1–455502-9.
70. B. R. Loyola, T. M. Briggs, L. Arronche, k. J. Loh, V. La Saponara, G. O'Bryan and J. L. Skinner: Struct. Health Monit., 2013, 12, 225–239.
71. k. J. Loh, T.-C. Hou, J. P. Lynch and N. A. Kotov: J. Nondestr. Eval., 2009, 28, 9–25.
72. C. S. Grimmer and C. K. H. Dharan: J. Mater. Sci., 2008, 43, 4487–4492.
73. M. Nofar, S. V. Hoa and M. D. Pugh: Compos. Sci. Technol., 2009, 69, 1599–1606.
74. H. Zhang, Y. Liu, E. Bilotti and T. Peijs: Adv. Compos. Lett., 2015, 24, 83–88.
75. F. H. Gojny, M. H. G. Wichmann, U. Köpke, B. Fiedler and K. Schulte: Compos. Sci. Technol., 2004, 64, 2363–2371.
76. C. B. Kim, S. W. Park and D. G. Lee: Compos. Struct., 2008, 86, 69–77.
77. M. Arai, Y. Noro, K. I. Sugimoto and M. Endo: Compos. Sci. Technol., 2008, 68, 516–525.
78. H. Zhang, M. Kuwata, E. Bilotti and T. Peijs: J. Nanomater, 2015, 2015, 785834-1–785834-7. doi:10.1155/2015/785834
79. S. A. Grammatikos and A. S. Paipetis: Compos. Part B Eng., 2012, 43, 2687–2696.
80. J. Rausch and E. Mäder: Mater. Technol., 2011, 26, 153–158.

19

Probing polymer chain constraint and synergistic effects in nylon 6-clay nanocomposites and nylon 6-silica flake sub-micro composites with nanomechanics

Jian Chen[1], Ben D. Beake[*2], Gerard A. Bell[2], Yalan Tait[3], and Fengge Gao[3]

[1]Jiangsu Key Laboratory of Advanced Metallic Materials, School of Materials Science and Engineering, Southeast University, Nanjing 211189, China
[2]Micro Materials Ltd., Willow House, Ellice Way, Yale Business Village, Wrexham, LL13 7YL, UK
[3]School of Science and Technology, Nottingham Trent University, Clifton campus, Nottingham, NG11 8NS, UK

Abstract In this study, we report that a synergistic effect exists in the surface mechanical properties of nylon 6–clay nanocomposites (NC) that can be shown by nanomechanical testing. The hardness, elastic modulus, and nanoindentation creep behavior of nylon 6 and its nanocomposites with different filler loading produced by melt compounding were contrasted to those of model nylon 6 sub-microcomposites (SMC) reinforced by sub-micro-thick silica flakes in which constraint cannot occur due to the difference in filler geometry. Polymer chain constraint was assessed by the analysis of nanoindentation creep data. Time-dependent creep decreased with increasing the filler loading in the NC consistent with the clay platelets exerting a constraint effect on the polymer chains which increases with filler loading. In contrast, there was no evidence of any reduced time-dependent creep for the SMC samples, consistent with a lack of constraint expected due to much lower aspect ratio of the silica flakes.

Keywords Nanomechanics, Creep compliance, Chain constraint, Nanoindentation, Modulus

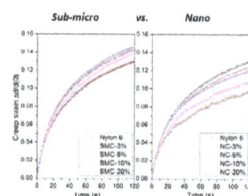

Introduction

After the initial reports into clay/polymer nanocomposites, [1–6] the following years of extensive research have gradually recognized the strength and weakness of these materials. In terms of the mechanical reinforcement, this technology has advantages over traditional micro-sized fiber and filler technology at low filler contents.[7] However, as the filler content exceeds 5 wt.%, the efficiency of reinforcement of the clay/polymer nanocomposite technology can rapidly decline due to the difficulties in the exfoliation of the multi-layered structure of clay and the onset of filler agglomeration to produce small stacks of clay platelets. Although attempts have been made to produce nanocomposites with a well-ordered structure of reinforcement using pre-exfoliated clay through layer-by-layer assembly techniques,[8] the brittleness of the composites produced by this method is a barrier to many potential applications.

The mechanical properties of composites depend on the interaction of many factors, such as the filler content, its spatial distribution, and the crystallinity of the matrix, resulting in complex behavior.[9,10] In clay/nylon 6 nanocomposites, the glass transition temperature is usually unchanged or very close to that of the unfilled nylon.[6,11] It is not yet clear to what extent the improvements in stiffness in clay/polymer nanocomposites are predominantly due to the incorporation of a much stiffer clay filler or whether the filler improves the stiffness and creep resistance by interaction with the polymer matrix exerting a constraining effect on the polymer chains. To maximize their potential, ideally NC would be designed with synergistic benefits over the base polymer, but as yet, there have been no reports that have clearly shown that such an enhancement exists and contributes to the improved surface mechanical properties. Vlasveld *et al.*[12] applied the Halpin–Tsai model originally designed for semi-crystalline polymers finding that no additional stiffening from constraint factors was necessary to explain the modulus enhancements in the clay/nylon 6 nanocomposites they studied. Understanding the relative significance of these different enhancements in mechanical properties with different processing conditions is a key step in the development of improved fabrication and processing routes capable of delivering polymer nanocomposites with optimized mechanical properties.

A recent review has highlighted that the current literature provides apparently conflicting information on the synergistic effects of polymer nanocomposites on their mechanical properties.[13] It is notable that many of the more remarkable increases in modulus in nanocomposite systems occur when the elastic modulus of the bulk polymer is very low. Nanomechanical testing provides a convenient route to testing the mechanical behavior of novel polymer nanocomposites[14-18] since only very small volumes are required. As a surface-sensitive technique, it can provide information about surface mechanical properties more relevant to applications involving surface contact loading (e.g. sliding wear) than bulk tensile testing. In addition to the determination of mechanical properties, such as hardness (H) and elastic modulus (E), it is possible to assess time-dependent behavior[14,19-21] and the response of the composites to high-strain rate impact and repetitive contact so that fatigue can be directly studied at the nanoscale.[22] In the majority of studies[13,] there has been a focus on reporting H and E with the time-dependent behavior being an unwanted complication. However, with suitable experimental design[14,19-21] the creep behavior of the polymeric materials can be investigated at the same time as the determination of hardness and elastic modulus.

An investigation of the extent of homogeneity of the load–displacement curves and analytical treatment of the time-dependent behavior has the potential to provide information on the mechanical homogeneity and the extent of filler dispersion on finer length scales than accessible by more macro-scale techniques. Nanoindentation creep in NC has been studied previously with a range of behavior being reported.[14,23,24] The creep of layered silicate/PA6 nanocomposites has been reported to show reduced creep compared to unfilled nylon.[9,25] However, for example, Shen and co-workers found that higher nanoindentation creep in nylon 66 nanocomposites than for the base polymer.[14] Using X-ray diffraction and optical microscopy, they found a reduction in the crystal size and degree of crystallinity, concluding that morphological changes were probably the main reasons for the greater creep. Seltzer and co-workers have recently investigated the nanoindentation creep behavior of skin and core regions of injection molded samples of nylon 6/organoclay nanocomposites with 1.1–4.5 vol. % of organoclay. They found that for their samples while organoclay does improve the indentation creep resistance of nylon 6, the enhancement was solely due to a decrease in the initial creep compliance at zero time as the time-dependent creep was actually increased, i.e. in the surface/near-surface region of their NC samples probed by nanoindentation, there was no evidence of the organoclay imparting a constraining effect on the polymer chains.

We have investigated whether the finding of Seltzer et al. is more generally applicable by studying the nanomechanical behavior and creep response of a set of NC samples with a wider range of filler loadings from 3 to 20 wt.%. As the

properties of the composites strongly depend on the filler size, composites with fillers in the sub-micro range have been investigated. Bonderer et al.[26] demonstrated that comprehensive reinforcement of strength, stiffness, and toughness can be achieved at relative high filler loading by applying sub-micro-thick fillers. Data from the NC samples were contrasted to a control set of nylon 6–single-layered sub-micro silica flake composites produced with the same wt.% filler loadings. During the processing, these filler in the SMC exhibited a strong reduction in aspect ratio and therefore these samples were expected to behave quite differently from the NC samples, behaving as a model system where appreciable polymer chain constraint does not occur.

In addition to charactering the properties of the NC and SMC in terms of differences in their hardness, modulus, and crystallinity, in this study, we have (i) investigated the relative importance of the initial creep compliance and time-dependent creep contributions to the total time-dependent deformation, (ii) used two different analytical methods to confirm our findings, and (iii) questioned whether nanomechanical testing can reliably produce indirect but useful information about the effectiveness of the dispersion. The results provide an improved understanding on the effects of filler size, aspect ratio, crystallinity, and content of the NC and SMC on their hardness, modulus, and creep behavior and their suitability for different applications.

Experimental

Sample preparation

The nylon 6 used for the NC and SMC composites was a high molecular weight, high viscosity nylon 6 with commercial name Grilon F50. The NC were produced by compounding with 3, 5, 10, and 20 wt.% montmorillonite clay modified with quaternary ammonium cations (commercially named as 93A) and the SMC samples were produced by compounding with 3, 5, 10, and 20 wt.% sub-micron-sized silica flake. Before composite processing, fillers and the polymer were dried under vacuum at 120 °C for 4–6 h in a vacuum oven. Then, the dried fillers were pre-blended with the polymer at different filler contents (0, 3, 5, 10, and 20 wt.%). The pre-blended materials were compounded in melt extrusion using a Prism Eurolab 16-mm twin-screw with 40/1 L/D ratio. Extruded samples were cooled in a water bath and chopped into pellets using a pelletizer. The feeding rate, screw speed, and processing temperature applied were 15–20%, 400 rpm, and 235 °C, respectively. The silica flakes were initially 350 nm thick with average aspect ratio (flake equivalent diameter to thickness ratio) of 1750 although a significant reduction of flake size occurred in the composites during processing as shown by SEM analysis of the original silica flakes before composite processing and the flakes obtained from the composites with 3wt.% and 20wt.% filler loading, respectively, by burning off the polymer matrix. During processing, the average aspect ratio of the flakes decreased from the original 1750 to 28 in the 3 wt.% SMC and 17 in the 20 wt.% SMC. The size breakdown may be caused by mechanical collision either between flakes or between flakes and the extrusion equipment during the melt processing but for our purposes is convenient in providing a range of SMC samples whose properties can be contrasted to the NC.

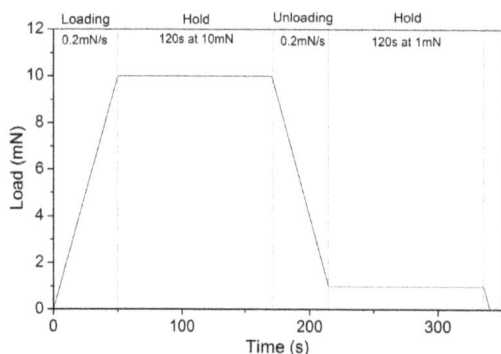

Figure 1   Load vs. time protocol in the nanoindentation tests. The hold at peak load is for creep assessment and the further hold after 90% unloading is for creep recovery assessment

Crystallinity measurements

The crystallinity measurements were carried out using DSC analysis. The procedure involved: (1) heating up from −20 to 270 °C at 50 °C/min, (2) holding at 270 °C for 5 min, (3) cooling from 270 to −20 °C at 10 °C/min, (4) heating up from −20 to 270 °C at 10 °C/min, and (5) cooling from 270 to −20 °C at 50 °C/min. The first cycle of heat/cooling was aimed to remove the thermal history of the materials. The curves obtained during the temperature ramp at 10 °C/min were used to calculate the crystallinity of the materials using Equation 1:

Eqn. 1
$$crystallinity = \Delta H_m \big/ (1 - f) \Delta H_c^0$$

where ΔH_m and ΔH_c^0 represent the melt enthalpy of samples and crystallization enthalpy of 100% crystalline polymer, respectively. f is the filler mass fraction of the composite. ΔH_c^0 of 100% crystalline nylon 6 was taken as 190 J/g.

X-ray diffraction analysis of the clay and NC samples

The structure of the clay and its composites formed was characterized using X-ray diffractometry. The equipment used was an X'Pert X-ray diffractometer with Cu Kα with wavelength 1.54 nm as radiation source. Prior the experiment, the powdered clay sample was compressed into dense square-shaped tablets that could fit into the equipment sample holder. For composite samples, the materials were injection molded into square plates with dimensions 15 mm × 15 mm × 1 mm. The molded plate sample was inserted into the sample holding position directly without using a sample holder. During the experiment, the generator voltage and X-ray tube current applied were 45 kV and 40 mA, respectively. The scanning range was from 0.991° to 120° with step size 0.00835.

Nanoindentation testing

A NanoTest system produced by Micro Materials Ltd., Wrexham, UK was used for the nanoindentation testing. The system has a very high thermal stability enabling nanoindentation creep measurements to be performed without thermal drift. The experiments were conducted in an environmental enclosure controlled at 27.0 ± 0.2 °C. The humidity was ~65% RH. All samples were molded flat surfaces without polishing. The nanoindentation tests were load controlled with a Berkovich diamond indenter with the loading protocol shown in Fig. 1. The initial (contact) load was 10 μN. All the samples were loaded from this initial load to a peak load of 10 mN at a fixed loading rate of 0.2 mN s⁻¹. The load was held constant at the peak load for 120 s to record the creep response before unloading at the same rate. There was an additional 120 s hold after 90% unloading which can be used to assess the instrument drift and the sample creep recovery. For all the experiments, the rate of creep recovery was found to be almost two orders of magnitude greater than the underlying instrumental signal drift so that the second hold period was not used for thermal drift correction. A matrix of 20 indentations spaced 100 μm apart was performed on each of the samples. The area function for the diamond indenter was determined from indentations into fused silica, although it was found that the indenter essentially indistinguishable from an ideal Berkovich geometry at the penetration depths (maximum depth ~2–3 μm) reached at 10 mN on all the samples.

The elastic modulus was determined by power law fitting to the unloading slope of the indentation curve.[27] The reduced elastic modulus, E_r, was converted to the elastic modulus, E, using the Poisson ratio for nylon 6, $v = 0.39$. The indentation creep behavior was analyzed in the instrument software using Equation 2.

Eqn. 2
$$\Delta d \big/ d(0) = \left[A \big/ d(0) \right] \cdot \ln(Bt + 1)$$

where A and B are the constants, $d(0)$ is the initial depth in the hold period, and Δd is the increase in depth during the hold period. $\Delta d/d(0)$ is the dimensionless indentation creep strain. Fitting the experimental data with Equation 2 has the benefit of (i) not requiring a particular constitutive model such as linear viscoelasticity to be assumed, therefore, allowing indentation with sharp indenters in the viscoelastic–viscoplastic regime[19,20,] (ii) being an excellent quantitative fit to the experimental creep curves enabling subtle differences in behavior to be uncovered. Equation 2 has proved successful in interpreting loading rate effects and in determination of the strain rate sensitivity, a dimensionless creep parameter that exhibits some correlation with tan delta,[20,28,29] and in determination of glass transition temperatures occurring below 0 °C.[19]

Results

Crystallinity and XRD

The crystallinity varied with filler loading in the NC and SMC samples as shown in Fig. 2a. The crystallinity of the SMC decreased significantly with only 3 wt.% loading of silica flakes and the further increase in the filler content had little effect. The crystallinity of the NC, however, increased slightly when the filler content reaching 3wt.% and then reduced significantly at higher loading. At 10 wt.%, the crystallinity of the NC was ~16%, lower than that of the SMC with the same loading.

Fig. 2b shows the XRD patterns of the organoclay 93A, nylon GF50, and its composites with 3, 5, 10, and 20 wt % organoclay 93A produced by melt processing. The organoclay clay 93A before added into the polymer has clear distinctive (001) peak at 2θ 3.81° which is equivalent to 2.316 nm interlayer spacing of silicate layers in the clay structure. However, the peak is not visible in all composite samples even in the enlarged chart in Fig. 2c.

Nanoindentation

Hardness and elastic modulus

Typical nanoindentation curves for the composites are shown in Fig. 3. The variation in elastic modulus with filler loading from analysis of such curves is shown in Fig. 4. It can clear be seen that while the NC and SMC samples were stiffer than the base polymer the enhancement in stiffness was greater for the NC samples.

The hardness and elastic modulus data are summarized in Table 1. The hardness of the nylon 6 is also increased by the incorporation of both fillers. The relative increase in hardness for the NC over the base polymer is less than the corresponding relative increase in elastic modulus. For example, incorporation of 20 wt.% clay increases the elastic modulus of nylon 6 by 150%, while the corresponding increase in hardness is only around 50%. There is greater variation in the H and E data for the composites with the sub-micron filler, especially at the higher loading. The highest average hardness was for the SMC with 20 wt.% clay.

Nanoindentation creep

The strain rate sensitivity are summarized in Table 2. Illustrative examples of the creep strain during the hold at peak load are shown in Fig. 5. While the NC show reduced creep strain compared to the matrix polymer, the SMC samples showed increased values which did not vary with filler loading.

Equation 2 was found to be an excellent fit to the creep data of NC, SMC, and the base nylon 6. The creep parameters, creep strain rate sensitivity $A/d(0)$, creep extent A, and time constant $1/B$, were calculated as shown in Table 2 and Fig. 6a–c. The $A/d(0)$, A and $1/B$ parameters were not significantly different from those of the base polymer for the SMC, but in contrast, there was a clear decrease in all these parameters for the NC.

Discussion

As shown in Table 2, all the NC and SMC showed higher hardness and elastic modulus than the base polymer. However, the extent of this improvement was different for the NC and SMC samples. In this discussion, we aim to deconvolute the separate influences of accompanying changes in the crystallinity of the composites and the type and dispersion of the fillers on the homogeneity and enhancements in mechanical properties of the composites. A key question is whether the enhancements in stiffness are explainable without constraint. Previously at the bulk scale Vlasveld et al. reported that when a Halpin–Tsai model originally designed for semi-crystalline polymers is used that no additional stiffening from constraint factors was necessary to explain the modulus enhancements in the clay/nylon 6 nanocomposites, and similarly at the nanoscale, Seltzer and co-workers reported there was no evidence for constraint in the surface mechanical behavior.

Decreasing crystallinity has an adverse effect on hardness and elastic modulus.[30] Consistent with this, the mechanical properties of the lower crystallinity nylon 6 studied here are lower than those previously reported for nylon 6 with higher crystallinity.[31] For the NC, it has been reported that low loading levels (1–3 wt.%) of nanofillers can act as nucleation sites for spherulitic crystallization so that the activation energy of crystallization is lower than the matrix polymer [14]. The small increase in crystallinity observed for the 3 wt.% NC in Fig. 2 is consistent with this explanation. At higher loading levels (10–20 wt.%), the activation energy of crystallization of the nanocomposites is higher than the neat polymer as more clay particles tend to obstruct the mobilization of the polymer molecular chains, retarding crystallization and crystal growth.[32] The observed decrease in crystallinity (~30% at 10 wt.%) is that determined by Shen and co-workers by DSC.[23]

For the SMC, the crystallinity decreased to ~18% at 3 wt.% filler and then increased slightly at 5 wt.%. Further increase in filler content has a little effect on the crystallinity as shown in Fig. 2. The sharp reduction in crystallinity in the 3 wt.% SMC appears to be due to similar reasons as NC with higher filler loading because of reduced mobility of the polymer molecules to crystallize. From this point of view, a further increase in filler content should further decrease the crystallinity due to the increased barriers to prevent crystallization. However, the progressive decrease in the aspect ratio with the increase in filler content of the SMC makes the shape of the filler more and more "roundish" as evidenced in SEM and optical microscopic studies[33] and progressively less able to influence the chain mobility of polymer molecules. As a result from these two opposing effects, further increasing filler loading did not result in a significant change in the crystallinity of the composites.

Displaying the data as relative modulus enhancement (modulus of the composites/modulus of the matrix) enables comparison between studies with different MW (and hence different stiffness) nylon 6 base materials.[6] The enhancement in stiffness, approximately doubling at 10 wt. %, is consistent with the results reviewed by Cho et al.,[6] and later by Vlasveld and co-workers,[12,25] Liu et al.,[10] Shen et al.[23,] and Diez-Pascual et al.[13] The efficiency of hardness and modulus enhancement begins to saturate at high loading levels as has been reported in previous work due to the decrease in crystallinity and a reduction in the efficiency of the exfoliation. Application of the Halpin–Tsai model can provide evidence of the effectiveness of the reinforcement with different filler loading. Following the approach taken in reference 12, the relative moduli of the fully amorphous and crystalline phases were used to estimate the elastic modulus contribution of the matrix in the nanocomposites. The Halpin–Tsai model was then used with the amorphous matrix component adjusted for the presence of surfactant to determine the effective aspect ratios of the fillers taking the density of clay = 2.8 g cm^{-3} and for silica flakes ρ = 2.2 g cm^{-3}, surfactant density = 0.93 g cm^{-3}, E (clay) = 172 GPa, and E (silica flakes) = 70 GPa. The effective aspect ratios ranged from ~210 at 3 wt.% to ~110 at 20 wt. %, consistent with the typical values of ~100–300 for layered silicate nanocomposites. The decrease in reinforcement effectiveness at higher clay, due to reduced exfoliation, is in good agreement with the literature.[12] In contrast, the values for the SMC are much lower (aspect ratios 10–60) and show little clear trend with wt.% silica flake, providing clear evidence for a marked difference in reinforcement efficiency between them and the NC.

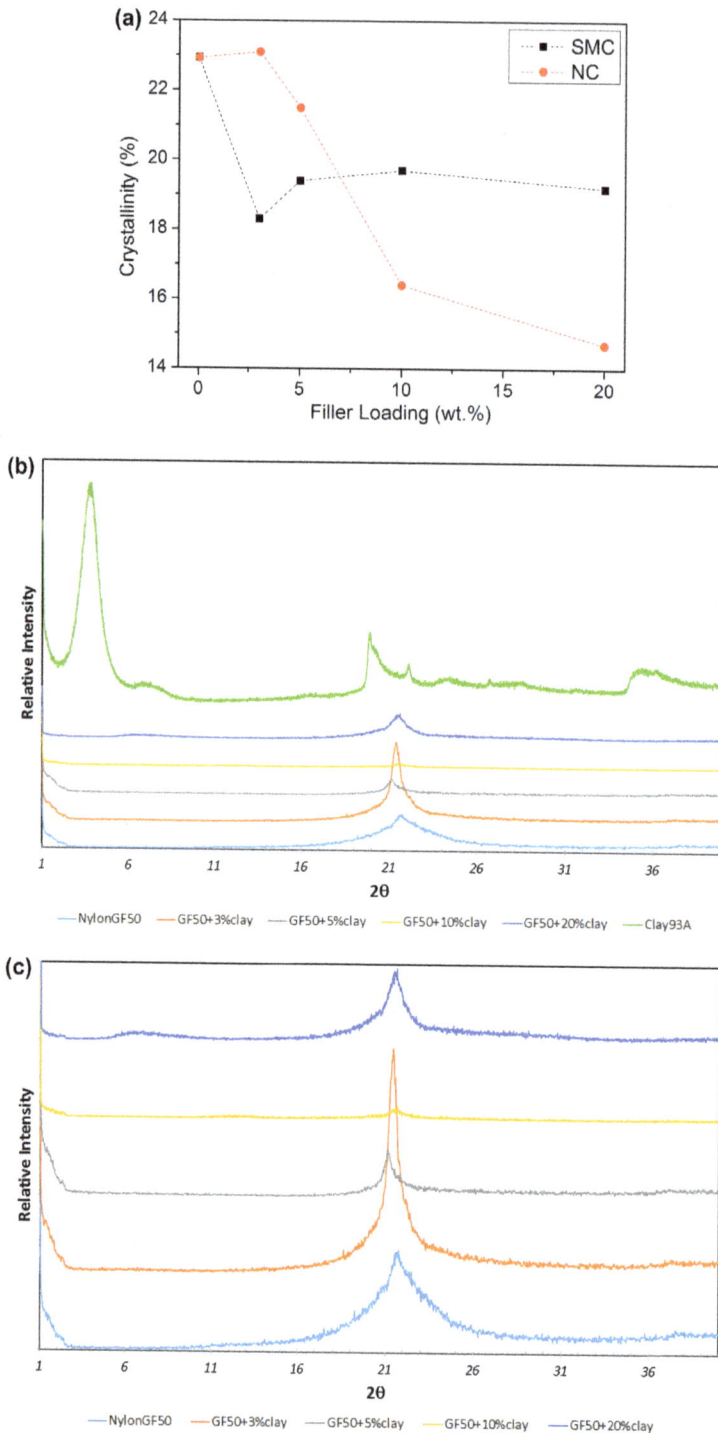

Figure 2   *a* Variation in crystallinity with wt.% filler loading for NC and SMC *b* XRD patterns of the original nylon GF50, and its composites with 3, 5, 10, and 20wt% organoclay 93A, and clay 93A *c* enlarged view of *b*

The *H* and *E* of the composite with 5 wt.% clay were found to be lower than those of the composites with 3 wt.% clay. In a previous study involving nylon 6/organoclay NC[22,] it was found that the hardness and modulus of NC with 3 and 5% organoclay were almost identical. Shen and co-workers observed anomalous creep effects at similar loading levels on nylon 6,6 NC[14] and PEO NC samples have also shown similar behavior at 5 wt.%.[16] In all these cases, the reasons are

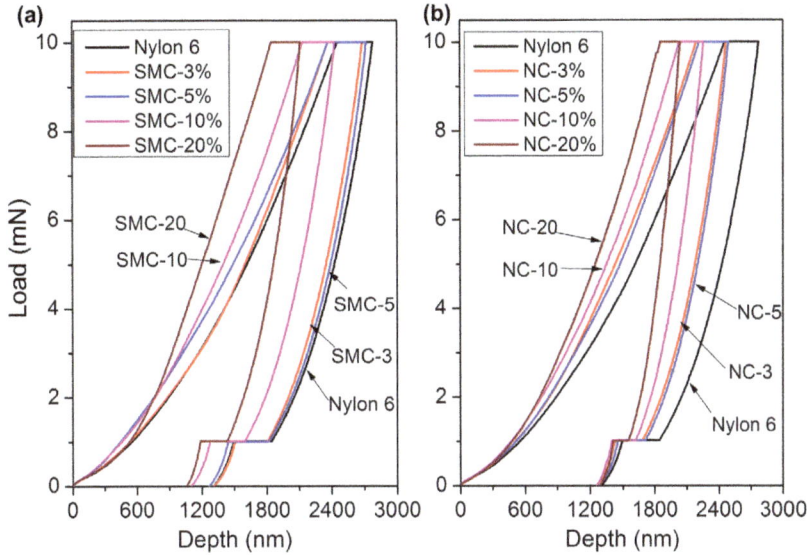

Figure 3 Typical nanoindentation curves for nylon 6 and *a* SMC and *b* NC

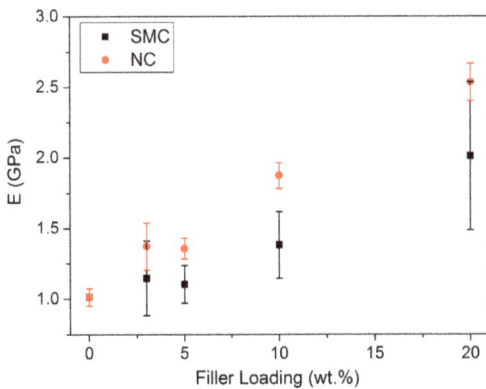

Figure 4 Variation in elastic modulus with filler loading for *a* SMC and *b* NC

not completely clear but it is conceivable that the structural change from the high extent of clay exfoliation state at a low content of 3 wt.% to an intercalated structure dominated state at 5 wt.% could be associated with this phenomenon since it is well known that highly exfoliated structure with less extent of aggregation of clay in the composites is more effective in enhancing mechanical properties due to the increase in the number of effective reinforcing particles and the elimination

of the weak van der Waals force between the silicate layers within a clay filler. The lower crystallinity in the 5 wt. % NC sample may also be a factor. The XRD data in Fig. 2*b,c* show that the clay (001) peak is not clearly present in all the NC samples. This may be a limitation of the XRD technique in the effective characterization of layered structured materials when number of layers in the structure is reduced. It is reasonable to believe that a high extent of exfoliation and significant chopping down of the layered structure has occurred in all the composite materials. Of course, this is not to say perfectly exfoliated nanocomposites have been achieved and it is likely that some intercalated structures and stacked layers with lower number of silicate layers are present that could not be detected by the XRD technique. This result is quite different from those obtained from nylon materials with medium and low molecular weight. A significant (001) peak is observed in for composites when clay loading level exceeds 5 wt % when using the same clay with medium and low molecular weight nylon 6 in the melt processing under the same experimental conditions in our laboratory. However, the nylon 6 used in the current study, Grilon F50, is a high molecular weight polymer. It is well known that the efficiency of clay exfoliation and interaction is higher when increasing molecular weight of nylon.[34]

The elastic modulus of polymeric materials can be overestimated by nanoindentation due to the continuing pronounced time-dependent deformation during unloading. This more

Table 1 Hardness and elastic modulus

Sample	d_{max} /nm	H/MPa	E/GPa
Nylon 6	2779 ± 106	83 ± 7	1.01 ± 0.06
SMC-3%	2688 ± 401	94 ± 27	1.15 ± 0.26
SMC-5%	2723 ± 197	86 ± 12	1.10 ± 0.13
SMC-10%	2419 ± 233	124 ± 23	1.38 ± 0.24
SMC-20%	2071 ± 333	157 ± 52	2.01 ± 0.54
NC-3%	2460 ± 189	105 ± 19	1.37 ± 0.17
NC-5%	2505 ± 74	98 ± 6	1.36 ± 0.07
NC-10%	2195 ± 76	124 ± 10	1.88 ± 0.09
NC-20%	2052 ± 101	133 ± 13	2.53 ± 0.13

Table 2 Creep and strain rate sensitivity

Sample	Strain rate Sensitivity $A/d(0)$ *	Creep/nm	$(1/B)$/s *	A/nm *
Nylon 6	0.043 + 0.001	319 ± 17	6.0 + 0.3	105 ± 5
SMC-3%	0.042 + 0.005	308 + 54	5.8 + 1.1	100 + 19
SMC-5%	0.046 + 0.004	330 + 34	6.2 + 0 .5	110 + 10
SMC-10%	0.045 + 0.005	287 + 51	6.2 + 0.7	95 + 14
SMC-20%	0.050 + 0.009	268 + 54	6.2 + 1.0	89 + 17
NC-3%	0.039 + 0.003	272 + 28	5.0 + 0.5	85 + 7
NC-5%	0.037 + 0.001	269 + 15	4.7 + 0.3	82 + 5
NC-10%	0.033 + 0.001	219 + 8	4.2 + 0.3	65 + 2
NC-20%	0.030 + 0.002	187 + 7	4.5 + 0.5	56 + 3

*Obtained by fitting the experimental creep data to Equation 2.

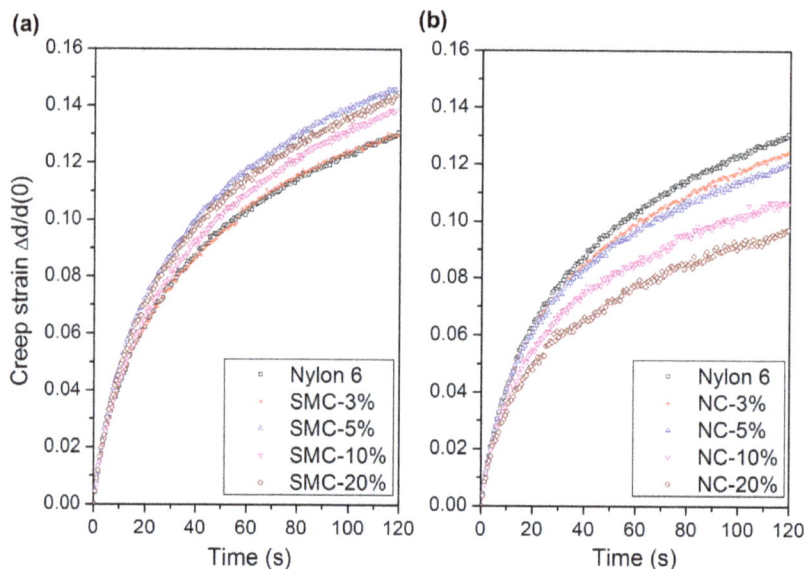

Figure 5 Typical creep strain curves for nylon 6 and *a* SMC and *b* NC

commonly occurs when (i) the holding period is too short, (ii) and/or the loading rate is too high, (iii) and/or the indentation depth is lower than ~ 200 nm.[17,35,36] To avoid any of these artifacts influencing the measurements in this study the load vs. time protocol has been designed to (i) load sufficiently slowly (50 s), (ii) hold for a long time at peak load (120 s), and (iii) indent sufficiently deep (2–3 μm).

The significant increases in E in the composites over the nylon 6 were accompanied by a smaller rise in H. In a previous study involving nylon 6/organoclay NC[22] it was found that although the stiffness of 3 and 5 wt. % organoclay was over 20% greater than nylon 6 the hardness of the NC did not change. Consequently, the H/E in the composites is lower than in the matrix polymer. The change in H/E may have important implications for tribological applications where performance often is correlated with parameters such as H/E and H^3/E^2, with higher H^3/E^2 commonly resulting in a higher load for non-elastic deformation, and higher wear resistance to wear as has been achieved by cross-linking ultra-high molecular weight polyethylene (UHMWPE) for replacement of hip prostheses.[37] The tribology and wear of polymers and polymer-based composites is complicated by viscoelastic recovery and brittleness,[38–40] with Brostow and co-workers reporting a direct correlation between a brittleness index and scratch

recovery.[40] Although wear resistance may improve by adding silicate layers to nylon 6, this has been associated with an increase in brittle modes of deformation such as the formation of brittle cracks during scratching.[41] A reduction in H/E when compared to matrix polymers has been reported previously for nylon 6 with 3–5% organoclay and for PEO with 7–15% organoclay.[22] In the current study, the decrease in H/E is more marked for the NC than the SMC as shown in Fig. 7. As the hardness is determined from the unloading curve after creep it is important to consider the influence of the reduced creep on its determination. Increased hardness in NC in comparison to nylon 6 is partially due to the reduction in creep. At 3 and 5% filler, the increased hardness of the NC is consistent with their greater creep resistance. For the higher loadings, the more rapid increase in hardness on the SMC samples is more likely to be connected to the increase in crystallinity with creep acting to minimize the difference.

Indentation creep analysis

With high-thermal stability instrumentation and suitable experimental design, the analysis of nanoindentation creep data can provide detailed information on the ability of the fillers to constrain the polymer chains in the composites.

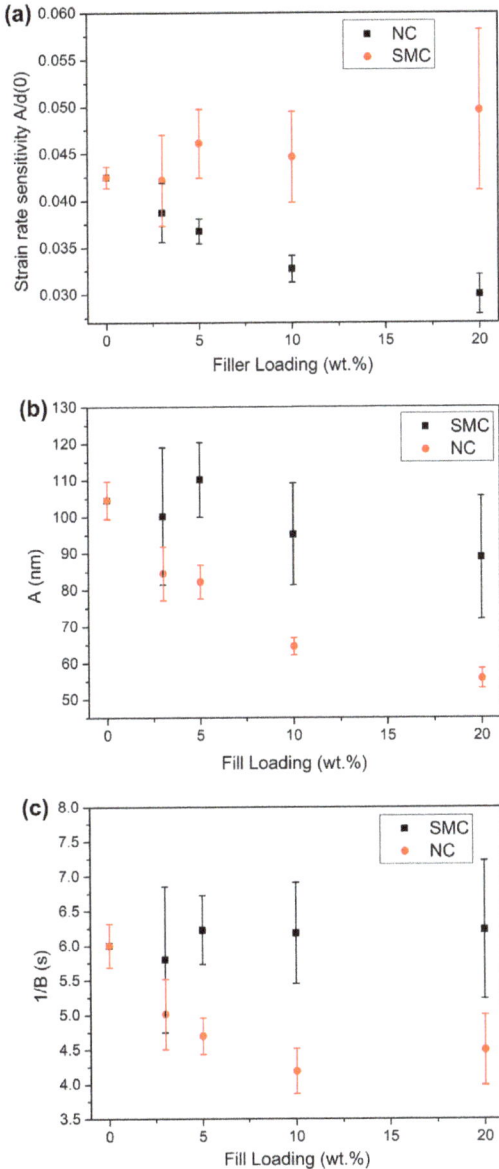

(a)

(b)

(c)

Figure 6 Variation in the *a* strain rate sensitivity *A/d(0)*, *b* creep extent parameter *A*, and *c* time constant *1/B*

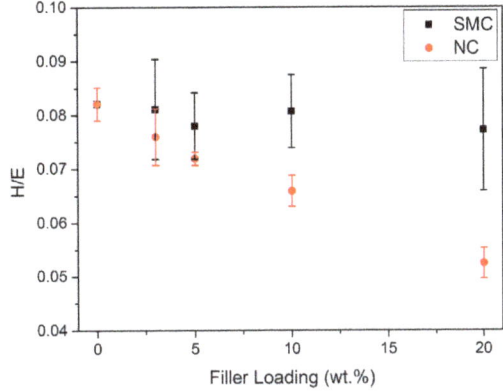

Figure 7 Variation in *H/E* with filler loading for *a* SMC and *b* NC

The creep strain rate sensitivity, *A/d(0)*, and time constant *1/B*, are sensitive to viscoelastic behavior.[19,29] The experimental creep depth data are closely fitted by the logarithmic Equation 2 enabling subtle changes in behavior to be determined. Tests at room and elevated temperatures on a range of amorphous and semi-crystalline polymers have shown that minimum values of the time constant *1/B* are observed close to the glass transition region.[29] More recently[19,] the treatment has been applied to analyze sub-ambient temperature nanoindentation creep data of atactic polypropylene and observe a minimum in time constant *1/B* in the vicinity of the glass transition region more clearly than the maximum in *A/d(0)* at the same temperature. The SMC showed much improved *H* and *E* than the matrix but without any clear variation in creep parameters in the composite as compared with the nylon 6 as shown in Table 2 and Fig. 6, consistent with no synergistic constraint in the SMCs.

In contrast, the creep parameters for the clay nanocomposites decrease compared nylon 6 and continue to fall as filler loading increases. Previous studies on the creep of layered silicate/PA6 nanocomposites have shown reduced creep compared to unfilled nylon.[9,25] Interestingly, Seltzer and co-workers[9] investigated the creep properties of their organoclay-enhanced PA6 nanocomposites using both nanoindentation and cantilever bending. In nanoindentation, the nanocomposites showed increased creep resistance which is in agreement with our findings on NC and SMC. To separate out the relative effects of initial creep compliance and time-dependent creep, the data were normalized by the initial creep compliance. To enable comparison with their work, the time-dependent creep compliance, $(J_{app}-J_0)/J_0$, was also determined using their proposed method according to Equation 3.[21] Eqn. 3

$$(J_{app} - J_0)/J_0 = \left(d/d(0) \right)^2 - 1$$

where *d* is the penetration depth. The time-dependent creep is shown in Fig. 8 which shows the same trends as the creep strain curves as shown in Fig. 5. This finding contrasts with the same analysis of the time-dependent creep during nanoindentation by Seltzer *et al.*[9] who found that it was larger for all their NC sample than for the matrix. This result is similar to that

Nanoindentation of NC is commonly either load or strain rate-controlled, with a load ramp terminated at a set indenter penetration depth. However, it has been shown[20] that the analysis of indentation creep is aided by the choice of a *constant loading time* rather than the use of (i) a depth-terminated load history, (ii) constant loading rate experiments to different peak loads, or (iii) constant strain (proportional loading), as in all of these cases the changing time taken to reach the peak load influences the creep behavior, making a robust analytical comparison between samples more problematic. By employing the same 50 s loading ramp for all the samples, the current study has made it possible to observe small but clear changes in strain rate sensitivity and time constant.

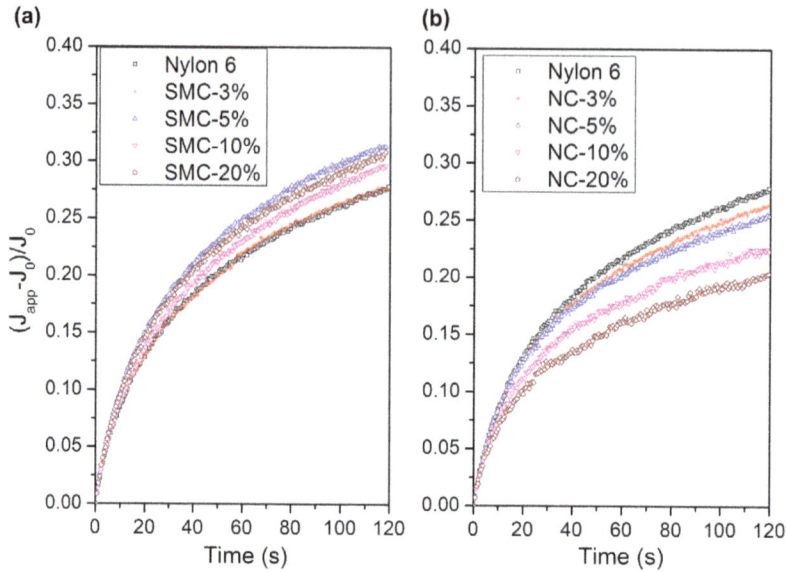

Figure 8 Variation in time-dependent creep with filler loading for *a* SMC and *b* NC

reported herein for the SMC composites (where no constraint was expected or found) and implies differences in the surface structure of the NC produced by Seltzer *et al*. In contrast to their behavior in nanoindentation, in bulk measurements by cantilever bending[9,] they observed reduced time-dependent behavior in the composites and therefore suggested that the organoclay imparts a constraint effect on the PA 6 molecular chains, restricting their mobility in the bulk compared to the surface.

In our analysis of the nanoindentation creep on NC, reduced initial creep compliance *and* time-dependent creep were found to contribute to the improvement in creep resistance. The strain rate sensitivity approach (Equation 2) also clearly shows this difference in the time-dependent component. As a further confirmation, we reanalyzed indentation creep data in our laboratory on stiffer nylon 6 NC samples with 3–5% organoclay (hardness and modulus data previously reported in ref[22]). The analysis also revealed reduced strain rate sensitivity (Equation 2) and time-dependent creep (Equation 3) in the nanocomposites in comparison to nylon 6. The analysis of the data shows that some constraint does indeed occur. Smaller but still significant rises in stiffness can be observed (e.g. ~40% increase for 10 wt. % SMC and ~100% increase at 20 wt. % SMC) for systems without constraint.

Conclusions

The results show that NC are more effective in enhancing stiffness and creep behavior in nylon 6 than in the SMC composites despite the larger decrease in crystallinity that occurs for the former. The analysis of nanoindentation creep data appears to be a very effective approach for assessing the presence of surface/near-surface constraint of the polymer by clay platelets or silica flakes. The creep analysis showed that both reduced initial creep compliance and time-dependent creep contribute to the improved creep resistance of NC implying that the clay has a beneficial constraining effect on the polymer chains

in the nanocomposites. This new finding is in contrast to the creep behavior of nylon 6–silica flake SMC studied here and to that of nylon 6/organoclay nanocomposites reported in[9] where improved indentation creep was found to only be due to the improved initial creep compliance, and constraint did not occur.

The methodology applied in this study has provided strong evidence that constraint does occur in nylon 6–clay nanocomposites and could prove a valuable tool in their future optimization. It is also easily applicable to other systems such as CNT/polymer nanocomposites.

Acknowledgements

Jian Chen would like to thank the National Natural Science Foundation of China (11204031 and 11472080), the Natural Science Foundation of Jiangsu Province of China (BK20141336), and the State Key Laboratory for Mechanical Behavior of Materials (2014603).

Conflicts of interest

The authors have no conflicts of interest to declare.

References

1. Y. Kojima, A. Usuki, M. Kawasumi, A. Okada, Y. Fukushima, T. Kurauchi and O. Kamigaito: *J. Mater. Res.*, 1993, **8**, 1185–1189.
2. A. Usuki, Y. Kojima, M. Kawasumi, A. Okada, Y. Fukushima, T. Kurauchi and O. Kamigaito: *J. Mater. Res.*, 1993, **8**, 1179–1184.
3. A. Usuki, M. Kawasumi, Y. Kojima, A. Okada, T. Kurauchi and O. Kamigaito: *J. Mater. Res.*, 1993, **8**, 1174–1178.
4. R. A. Vaia and E. P. Giannelis: *Macromolecules*, 1997, **30**, 7990–7999.
5. M. Kawasumi, N. Hasegawa, M. Kato, A. Usuki and A. Okada: *Macromolecules*, 1997, **30**, 6333–6338.
6. J. W. Cho and D. R. Paul: *Polymer*, 2001, **42**, 1083–1094.
7. F. Gao: *Mater. Today*, 2004, **7**, 50–55.
8. P. Podsiadlo, A. K. Kaushik, E. M. Arruda, A. M. Waas, B. S. Shim, J. Xu, H. Nandivada, B. G. Pumplin, J. Lahann, A. Ramamoorthy and N. A. Kotov: *Science*, 2007, **318**, 80–83.

9. R. Seltzer, Y. Mai and P. M. Frontini: *Composites Part B: Engineering Nanomechanics and Nanocomposites: Mechanical Behavior*, 2012, **43**, 83–89.

10. T. X. Liu, Z. H. Liu, K. X. Ma, L. Shen, K. Y. Zeng and C. B. He: *Compos. Sci. Technol.*, 2003, **63**, 331–337.

11. D. Vlasveld, J. Groenewold, H. Bersee and S. J. Picken: *Polymer*, 2005, **46**, 12567–12576.

12. D. Vlasveld, J. Groenewold, H. Bersee, E. Mendes and S. J. Picken: *Polymer*, 2005, **46**, 6102–6113.

13. A. M. Díez-Pascual, M. A. Gómez-Fatou, F. Ania and A. Flores: *Prog. Mater. Sci.*, 2015, **67**, 1–94.

14. L. Shen, I. Y. Phang, L. Chen, T. X. Liu and K. Y. Zeng: *Polymer*, 2004, **45**, 3341–3349.

15. J. Chen, X. L. Guo, Q. Tang, C. Y. Zhuang, J. S. Liu, S. Q. Wu and B. D. Beake: *Carbon*, 2013, **55**, 144–150.

16. B. D. Beake, S. Chen, J. B. Hull and F. Gao: *J. Nanosci. Nanotech.*, 2002' **2**, 73–79.

17. D. Tranchida, S. Piccarolo, J. Loos and A. Alexeev: *Macromol.*, 2007, **40**, 1259–1267.

18. A. Dasari, Z. Z. Yu and Y. W Mai: *Mater. Sci. Eng. R.*, 2009, **63**, 31–80.

19. J. Chen, G. A. Bell, H. S. Dong, J. F. Smith and B. D. Beake: *J. Phys. D: Appl. Phys.*, 2010, **43**, 425404.

20. B. Beake: *J. Phys. D: Appl. Phys.*, 2006, **39**, 4478–4485.

21. R. Seltzer and Y. Mai, *Eng. Fract. Mech.*, 2008, **75**, 4852–4862.

22. B. D. Beake, S. R. Goodes, J. F. Smith and F. Gao: *J. Mater. Res.*, 2004, **19**, 237–247.

23. L. Shen, I. Y. Phang, T. X. Liu and K. Y. Zeng: *Polymer*, 2004, **45**, 8221–8229.

24. L. Shen, W. C. Tjiu and T. Liu: *Polymer*, 2005, **46**, 11969–11977.

25. D. P. N. Vlasveld, H. E. N. Bersee and S. J. Picken: *Polymer*, 2005, **46**, 12539–12545.

26. L. J. Bonderer, A. R. Studart and L. J. Gauckler: *Science*, 2008, **319**, 1069–1073.

27. W. C. Oliver and G. M. Pharr: *J. Mater. Res.*, 1992, **7**, 1564–1583.

28. A. Gray and B. D. Beake: *J. Nanosci. Nanotech.*, 2007, **7**, 2530–2533.

29. A. Gray, D. Orecchia and B. D. Beake: *J Nanosci. Nanotech.*, 2009, **9**, 4514–4519.

30. B. D. Beake and G. J. Leggett: *Polymer*, 2002, **43**, 319–327.

31. G. A. Bell, D. M. Bieliński and B. D. Beake: *J Appl. Poly. Sci.*, 2008, **107**, 577–582.

32. W. R. Broughton: 'Characterization of nanosized filler particles in polymeric systems: a review', NPL Report MAT12; 2008, Teddington, UK, NPL.

33. Y Li, PhD thesis: 'The development of sub-micro filler enhanced composites', Nottingham Trent University, UK, Nov 2007.

34. T. D. Fornes, P. J. Yoon, H. Keskkula and D. R. Paul: *Polymer*, 2001, **42**, 9929–9940.

35. C. A. Tweedie, G. Constantinides, K. E. Lehman, D. J. Brill, G. S. Blackman and k. J. Van Vliet, Adv. Mater., 2007, **19**, 2540.

36. D. Tranchida, S. Piccarolo, J. Loos and A. Alexeev: Appl. Phys. Lett., 2006, **89**, 171905.

37. P. A. Williams, K. Yamamoto, T. Masaoka, H. Oonishi and I. C. Clarke: Tribol. Trans., 2007, **50**, 277–290.

38. W. Brostow, J. –L. Deborde, M. Jaklewicz and P. Olszynski: J. Mater. Ed., 2003, **25**, 119–132.

39. N. K. Myshkin, M. I. Petrokovets and A. V. Kovalev: Tribol. Int., 2005, **38**, 910–921.

40. W. Brostow, V. Kovacevic, D. Vrsaljko and J Whitworth: J. Mater. Ed., 2010, **32**, 273–290.

41. D. Aravind and Z. Yu: Nanotechnol., 2008, **19**, 55708.

20

Nanocomposite toughness, strength and stiffness: role of filler geometry

Israel Greenfeld and H. Daniel Wagner*

Department of Materials and Interfaces, The Weizmann Institute of Science, Rehovot 76100, Israel

Abstract The toughness, strength and stiffness of nanocomposites are considered for the general case of hollow fillers with arbitrarily shaped cross-sections. The particular cases of nanotubes, thin wall general cylindrical fillers, and thin ribbons are examined. The toughness is expressed by the energy dissipated when the filler pulls out from the matrix during the composite fracture, taking into consideration the filler critical length. The study reveals how the properties of nanocomposites can be optimized by modulating the filler shape and dimensions, as well as the mechanical properties of the material and interface. The tradeoffs between toughness, strength and stiffness are analyzed in view of their different and sometimes opposite dependence on the material and geometric parameters. It is shown that when the filler is shorter than its critical length, typical of most current nanotubes, the toughness, strength and stiffness can be improved simultaneously by reducing the filler cross-sectional aspect ratio (wall thickness divided by diameter). The mechanical performance of composites reinforced by carbon nanotubes and microfibers is compared for several possible filler packing conformations, demonstrating the high potential of nanoreinforcement.

Keywords Nanocomposite, Toughness, Strength, Stiffness, Nanotube, CNT, CNTF, WSNT

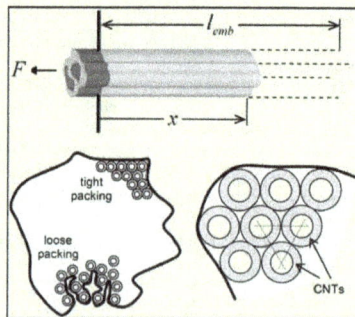

Introduction

Reinforcement at the nanoscale, using fillers like carbon nanotubes (CNTs), tungsten disulfide nanotubes (WSNTs) and molybdenum disulfide nanoplatelets (MSNPs), has the potential for improving the composite strength, stiffness and toughness compared to microscale reinforcement. Nanofillers could potentially be used in standard composite applications, as well as in confined spaces, thin polymer nanofilms[1,2] and electrospun polymer nanofibers.[3-5] Nanofillers such as carbon nanotubes possess strength and stiffness an order of magnitude higher than traditional microfillers like carbon microfibers.[6,7] Furthermore, the ratio of the filler's interfacial area to volume ($R/R^2 \sim R^{-1}$), which determines its load bearing capacity, becomes extremely large for nanofillers, 10^2-10^3 higher than microfillers. At the same time, nanofillers are often short and hollow, have arbitrarily shaped cross-sections, and form complex arrays and networks.[8]

It is the aim of the present study to generalize the modeling of the mechanical properties of hollow fillers, to those having an irregular cross-sectional shape, of which nanofillers are a typical example. Specifically, we demonstrate that the cross-sectional shape of a reinforcing filler has a significant influence on the overall mechanical performance of the composite, particularly so for nanocomposites in view of the effects of the filler's small size and diverse shapes and structures. For the purpose of this study, we assume uniform dispersion of the filler in the matrix and unidirectional alignment, both optimal conditions for enhanced performance.

We focus on the structural toughening properties of nanoreinforcement, for which there is growing evidence of improved mechanical performance.[3,9-11] The major mechanism contributing to toughening of reinforced composites is the energy dissipated when the filler is pulled out from the matrix through the fracture surface.[12] The pullout energy of nanotube reinforced composites was modeled by Wagner,[13,14] based on the classic models by Cottrell[15] and Kelly–Tyson[16] developed for solid fibers. We expand these models to encompass fillers with arbitrarily shaped cross-sections, including the distinct class of thin wall fillers. A starting point for the model is the filler critical length, the length above which the filler breaks rather than pulls out.

Also, the classic models by Piggott[17] and others for the strength and stiffness of composites are expanded for such

*Corresponding authors, email daniel.wagner@weizmann.ac.il and green_is@netvision.net.il

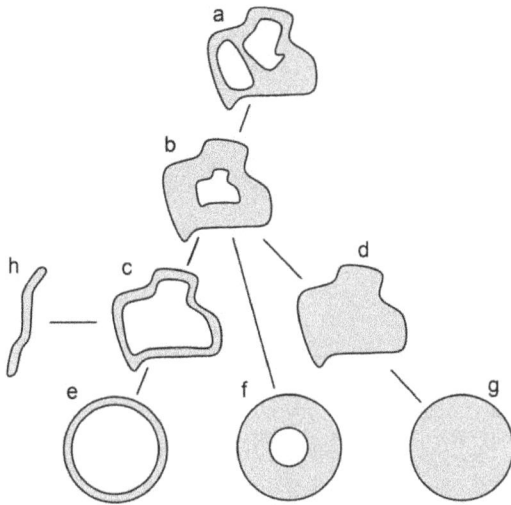

Figure 1 Categories of cross-sectional shapes of cylindrical fillers: *a* arbitrarily shaped outline and cavities (e.g. aligned NT bundle); *b* arbitrary shape, single cavity; *c* arbitrary shape, thin wall; *d* arbitrary shape, solid; *e* circular, thin wall (e.g. single wall nanotube); *f* circular, thick wall (e.g. multiwall nanotube); *g* circular, solid (e.g. fiber); *h* open shape, thin or thick wall (e.g. flat ribbon, graphene)

arbitrarily shaped fillers. The overall mechanical performance of the composite structure is presented as a tradeoff between the toughness, strength and stiffness, since these properties depend differently (and sometimes oppositely) on parameters such as the filler's length, cross-sectional shape, strength, stiffness and volume fraction, as well as the matrix and interface strength and stiffness. Finally, the performance of nano- and microreinforcements is compared for several composite structures with different filler packing schemes.

Effect of filler shape on critical length

The cross-section of an elongated reinforcing nanofiller may have diverse shapes, as illustrated in Fig. 1. In the most general case, a single filler has the form of a hollow *general cylinder* (a solid bounded by a right ruled closed surface),[18] whose cross-section has arbitrarily shaped outline and multiple cavities (Fig. 1*a*); a spun bundle of aligned hollow nanotubes is an example of such a filler (Appendix). An arbitrarily shaped filler may be solid (Fig. 1*d*), or have a thick wall (Fig. 1*b*) or a thin wall (Fig. 1*c*). A filler may have a circular cross-section, and its core may be solid as in a fiber (Fig. 1*g*), or hollow with a thin or thick wall, as in a single wall (Fig. 1*e*) or multiwall (Fig. 1*f*) nanotube. Finally, a thin or thick ribbon (Fig. 1*h*) is another category of a cylindrical filler.

We therefore seek to generalize our recently proposed toughness model[13,14] of a unidirectional filler, based on the pullout mechanism, to a wider class of cross-sectional shapes. We first focus on the critical length of a filler having an arbitrarily shaped cross-section.

Consider a hollow filler of the form of a general cylinder embedded in a matrix, with a material cross-sectional area *a* and an outer perimeter *p*. The longitudinal tensile stress in the filler $\sigma(x)$, and the interfacial stress between the filler and

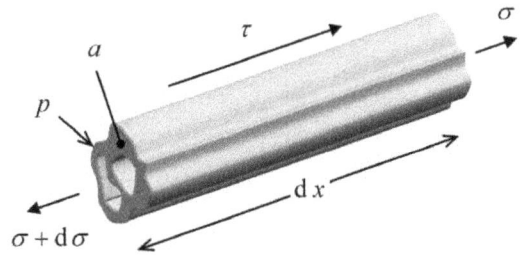

Figure 2 Tensile stress σ and interfacial stress τ acting on differential element dx of hollow filler with arbitrarily shaped cross-section, having perimeter p and material area a

the matrix $\tau(x)$, acting on a differential element of the filler (Fig. 2), are expressed by

$$p\tau(x)\,dx = a[\sigma(x)+d\sigma]-a\sigma(x) = a\,d\sigma \qquad (1)$$

where x is the distance from the filler edge. When the filler is long and its cavities are narrow, the matrix column embedded internally inside the filler is constrained by the relatively small displacement of the stiff filler, resulting in high matrix stresses close to the filler ends, leading to local breaking of the matrix. Therefore, the contribution of the internal matrix column to the stress in the filler is negligible. Furthermore, for nanofillers, the matrix material (e.g. polymer) is not likely to penetrate the cavities at all.

The CKT model[15,16] assumes that when the tensile load is applied, the interfacial stress reaches a constant value $\tau(x)=\tau_i$, corresponding to the yield strength of the matrix or the bonding strength between the filler and the matrix, whichever is lower. More specifically, according to Mallick,[19] this assumption is valid for a ductile matrix, which yields under a high interfacial stress and flows plastically with little or no strain hardening. Such an elastic perfectly plastic matrix maintains a constant yield strength, equal to the interfacial shear stress τ_i. By comparison, shear lag based models predict that in the filler central length the stress transfer is elastic rather than plastic, whereas in regions close to the filler edges the matrix may yield or debond.[17,20] For a ductile matrix, with a yield strength lower than the filler–matrix bonding strength, when the external stress is gradually increased, the matrix yielded regions expand from the filler edges toward the center, and the resulting shear lag stress profile approaches that of CKT. In case it is the filler–matrix bonding strength which is lower than the matrix yield strength, the interfacial shear stress τ_i is the friction stress caused by the compression applied by the matrix on the filler.

Integrating equation (1) with a constant interfacial stress, we obtain

$$\sigma(x)=\tau_i x\frac{p}{a}, \quad \sigma(x)\leq\sigma_f \qquad (2)$$

where σ_f is the filler ultimate strength. Thus, the tensile stress in the filler $\sigma(x)$ grows linearly with the distance from its edges, until σ_f is reached and the filler breaks. This occurs when the filler is longer than a critical length l_c, obtained by substituting $x=l_c/2$ into equation (2)

$$l_c = \frac{2\sigma_f}{\tau_i}\frac{a}{p} \qquad (3)$$

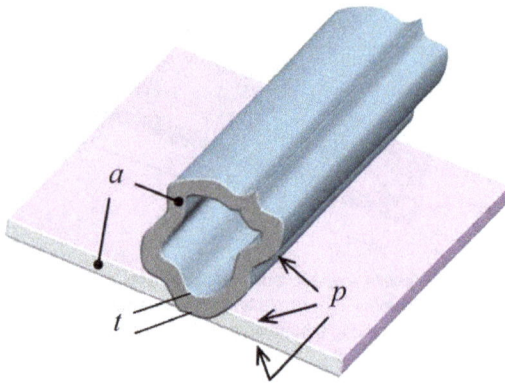

Figure 3 Analogy between thin wall arbitrarily shaped cylinder and thin ribbon: cylinder is spread out flat and forms ribbon with thickness t, width p, total perimeter $2p$ (both sides) and cross-sectional area a; ribbon has half critical length of cylinder, assuming double sided bonding to matrix, and its critical length is independent of its width

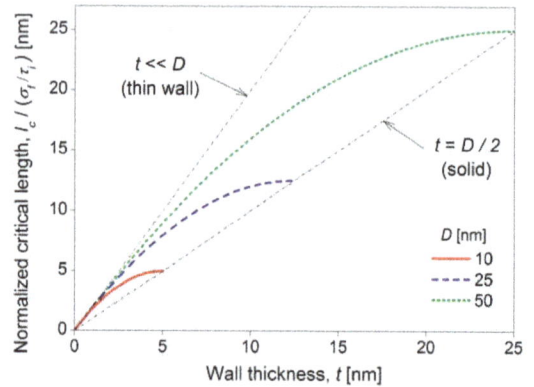

Figure 4 Normalized critical length $l_c/(\sigma_f/\tau_i)$ vs. wall thickness t and outer diameter D of circular hollow tube (equation (7)): multiply by $\sim 10^3$ to obtain the order of l_c for a CNT; bounds for thin wall and solid cylinders are marked; lowest possible wall thickness of CNT is that of single wall that is 0.34 nm

In the case of a thin wall hollow filler having a wall thickness t (Figs. 1c and 3), the ratio $a/p \cong t$, and therefore

$$l_c^{thin} \cong \frac{2\sigma_f t}{\tau_i} \qquad (4)$$

regardless of the cross-section shape, whether circular, elliptic or warped.

Note that the thin wall critical length does not depend on the filler lateral size (i.e. width) but only on its thickness. This outcome is clarified when spreading out the general cylinder into an equivalent flat ribbon (Fig. 3), for which $a/p \cong t/2$ and the critical length is $l_c = \sigma_f t/\tau_i$, independent of the ribbon width p. The critical length of the equivalent ribbon is half that of the closed shape because its bonding to the matrix is double-sided. Hence, possibly, the use of ribbons (e.g. graphene) may be beneficial when a shorter critical length is desired.

For a circular hollow tube (e.g. a nanotube), with an outer diameter D, inner diameter d and wall thickness $t = (D-d)/2$, we define the following function of the cross-sectional aspect ratio t/D

$$A_{tD} \equiv \frac{a}{a_{tot}} = \frac{(D^2-d^2)}{D^2} = 4\frac{t}{D}\left(1-\frac{t}{D}\right) \qquad (5)$$

This function, which represents the ratio between the filler's material area a and its total area a_{tot} (the area bounded by the filler perimeter p), ranges from $4t/D \cong 0$ for a thin wall tube ($t \ll D$) to 1 for a solid fiber ($t = D/2$). The ratio a/p for circular cylinders is therefore

$$\frac{a}{p} = \frac{\pi(D^2-d^2)}{4\pi D} = t\left(1-\frac{t}{D}\right) = \frac{D}{4}A_{tD} \qquad (6)$$

The corresponding critical length from equation (3) is

$$l_c = \frac{2\sigma_f t}{\tau_i}\left(1-\frac{t}{D}\right) = \frac{\sigma_f D}{2\tau_i}A_{tD} = l_c^{solid}A_{tD} \qquad (7)$$

which reduces to the known expression $l_c^{solid} = \sigma_f D/(2\tau_i)$ for a solid fiber, and to $2\sigma_f t/\tau_i$ of equation (4) for a thin wall tube.

A tube with thicker wall, larger diameter and higher tensile strength has a longer critical length, and vice versa (Fig. 4). For example, the scale of the critical length of a carbon nanotube is two to three orders of magnitude lower than that of a carbon fiber (using equation (7) with 10^1 higher σ_f, 10^{-3} smaller D and $A_{tD}<1$). When the filler length is longer than its critical length (a typical case in microscale reinforcement), it is desirable to increase the critical length in order to optimize toughness. On the other hand, as will be shown later (see also Ref. 13), when the filler length is shorter than its critical length (a typical case in nanoscale reinforcement), it is desirable to reduce the critical length, in other words to decrease the wall thickness, diameter and tensile strength and increase the interfacial strength.

The aspect function A_{tD}, depicted in Fig. 5, is useful for comparing the critical length (and as will be seen, the pullout energy) of a hollow tube to a solid fiber having the same external diameter. The higher the aspect ratio (thicker tube wall and/or smaller diameter), the longer the critical length of the tube with respect to a solid fiber, and vice versa. Note that $l_c \leq l_c^{solid}$ in all cases.

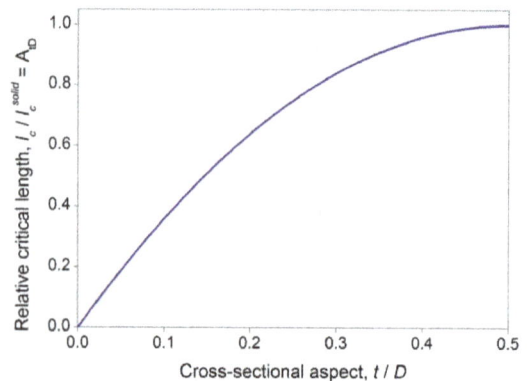

Figure 5 Relative critical length l_c/l_c^{solid} (or aspect ratio function A_{tD}) versus cross-sectional aspect ratio t/D of circular hollow tube (equations (7) and (5))

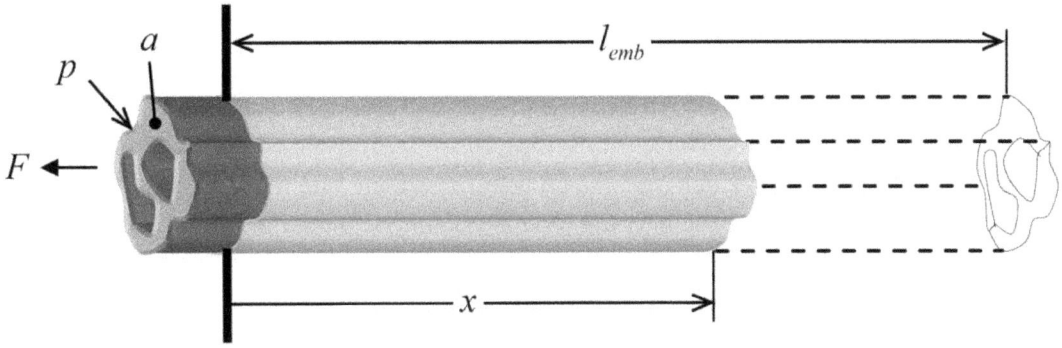

Figure 6 Force F pulling out hollow filler that has initial embedded length l_{emb} in matrix: filler has arbitrarily shaped cross-section with perimeter p and material area a

Effect of filler shape on pullout energy

The energy absorbed during pulling out of a unidirectional elongated filler from an embedding soft matrix is generally the major contributor to toughening of a composite structure. The following analysis generalizes the toughness model[13] to an arbitrarily shaped hollow filler, representative of a wide class of cross-sectional shapes (Fig. 1), and expands on thin wall hollow filler (Fig. 1c) and open filler (Fig. 1h), as well as circular hollow filler (Fig. 1e and f), with a solid fiber (Fig. 1g) as a particular case.

The force needed to pull out a hollow filler embedded length x in a soft matrix, when the filler has the form of a general cylinder with perimeter p, is $F(x) = px\tau_i$ (Fig. 6). The interfacial stress τ_i is the yield strength of the matrix or the interfacial bonding, and is assumed constant as previously described. Thus, the work for a complete pullout of a filler that is initially embedded l_{emb} in the matrix is

$$W(l_{emb}) = \int_0^{l_{emb}} F(x)\,dx = \frac{p\tau_i l_{emb}^2}{2} \tag{8}$$

As discussed in the previous section, when the filler is long and its cavities are narrow, the matrix column embedded internally (if any) in the filler's cavities does not pull out and does not transfer any load from the applied force, since it breaks at the filler ends during the onset of pullout (this break dissipates significantly less energy compared to the pullout energy).

The pullout work for N unidirectional such fillers, whose embedded length is evenly distributed from 0 to l_{emb}, is

$$\langle W \rangle = N \frac{\int_0^{l_{emb}} W(l_{emb})\,dl_{emb}}{\int_0^{l_{emb}} dl_{emb}} = N \frac{p\tau_i l_{emb}^2}{6} \tag{9}$$

Consider a composite's representative volume element (RVE), whose total cross-sectional area is A, encompassing the matrix area, the filler cumulative material area Na, and any cavities inside or outside the filler. Since the volume fraction of the filler in that RVE is $V_f = Na/A$, the number of fillers intersecting the fracture surface is on average

$$N = \frac{V_f A}{a} \tag{10}$$

independent of the filler length.

Using equations (9) and (10), the pullout energy area density (in short, pullout energy) is given by

$$G = \frac{\langle W \rangle}{A} = \frac{V_f \tau_i l_{emb}^2}{6} \frac{p}{a} \tag{11}$$

which can be rewritten in terms of the critical length by substituting the ratio p/a from equation (3)

$$G = \frac{V_f \sigma_f l_{emb}^2}{3 l_c} \tag{12}$$

independent of the cross-section shape and whether it is hollow or not.

In the case of a filler whose length l is shorter than its critical length, for example as typical of most current nanotubes, the maximum embedded length is $l_{emb} = l/2$, and therefore

$$G = \frac{V_f \sigma_f l^2}{12 l_c}, \quad l < l_c \tag{13}$$

Note that this expression is universal for any cross-sectional shape. For a thin wall filler with a wall thickness t, we substitute l_c from equation (4)

$$G_{thin} \cong \frac{V_f \tau_i l^2}{24 t}, \quad l < l_c \tag{14}$$

independent of the cross-section shape and lateral size. In the case of a thin wall ribbon-shaped filler (e.g. graphene), assuming the same interfacial strength (which may not necessarily be the case), the pullout energy will increase by a factor of 2. Substituting l_c for a circular cross-section from equation (7) into equation (13)

$$G = \frac{V_f \tau_i l^2}{24 t(1 - t/D)} = \frac{V_f \tau_i l^2}{6D} A_{tD}^{-1}, \quad l < l_c \tag{15}$$

which reduces to $G_{solid} = V_f \tau_i l^2/(6D)$ for a solid fiber (using equation (5) with $t = D/2$, leading to $A_{tD} = 1$).

The pullout energy in this filler length domain is invariant with respect to the filler ultimate strength σ_f. However, when σ_f is very high, the pullout energy will not grow indefinitely with the filler length l because of fracture mechanics considerations arising from inherent flaws in the filler. Consequently, the filler will tend to break at the flaw nearest to the matrix fracture plane, where the filler stress is highest, and, according to Piggott,[17] the pullout energy will

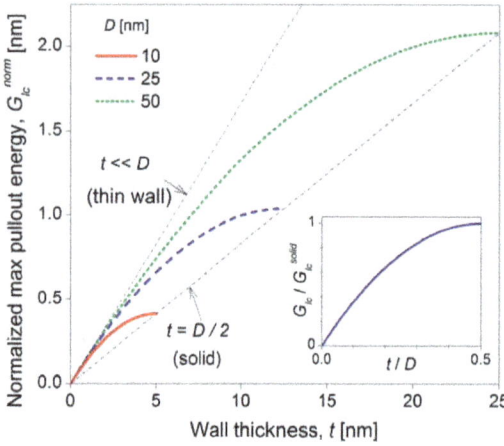

Figure 7 Maximum pullout energy G_{lc} versus wall thickness t and outer diameter D of circular hollow tube (equation (21)): normalized energy $G_{lc}^{norm} = G_{lc}/(V_f \sigma_f^2/\tau_i)$, and relative energy G_{lc}/G_{lc}^{solid} (inset)

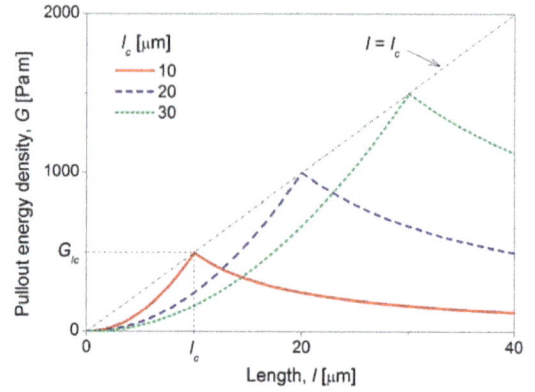

Figure 8 Pullout energy G as function of filler length l, for three values of critical length l_c (equations (13), (16) and (19)): $V_f \sigma_f/12 = 50$ MPa; holds for any cross-sectional shape

saturate at $G_{sat} = 2V_f \tau_i L^2/D$, where L is the filler fragment pullout length. The length L is inversely proportional to the density of flaws along the filler, derived from the flaw statistical distribution. Note the similarity between the expressions of G_{sat} and G_{solid}, obtained by substituting $l = 2 \times 3^{1/2} L$.

In the case of a filler that is longer than its critical length, we observe two possible types of events: if the filler is embedded more than $l_c/2$ beyond the fracture surface in both directions, its stress exceeds the ultimate strength and it will break at the fracture surface,[15] dissipating a relatively low energy compared to pullout; if the filler is embedded less than $l_c/2$ in one of the two directions, it will pull out without breaking. Thus, only the latter event contributes to the pullout energy, and its probability of occurrence is l_c/l. Since the embedded length in that type of event ranges from 0 to $l_c/2$, we substitute $l_{emb} = l_c/2$ into equation (12), and adjust by the probability of occurrence

$$G = \frac{V_f \sigma_f (l_c/2)^2}{3l_c} \frac{l_c}{l} = \frac{V_f \sigma_f l_c^2}{12l}, \quad l > l_c \tag{16}$$

Note that this expression is universal for any cross-sectional shape. For a thin wall filler having a wall thickness t, we substitute l_c from equation (4)

$$G_{thin} \cong \frac{V_f \sigma_f^3 t^2}{3\tau_i^2 l}, \quad l > l_c \tag{17}$$

independent of the cross-section shape and lateral size. In the case of a thin wall ribbon shaped filler (e.g. graphene), the pullout energy will degrade by a factor of 4 (because of the term l_c^2 in equation (16)). Substituting l_c for a circular cross-section (equation (7))

$$G = \frac{V_f \sigma_f^3 t^2}{3\tau_i^2 l} \left(1 - \frac{t}{D}\right)^2 = \frac{V_f \sigma_f^3 D^2}{48\tau_i^2 l} A_{tD}^2, \quad l > l_c \tag{18}$$

which reduces to $G_{solid} = V_f \sigma_f^3 D^2/(48\tau_i^2 l)$ for a solid fiber.

The maximum achievable pullout energy G_{lc} is obtained when the filler's length is equal to its critical length, or $l = l_c$

(in fact, l should be slightly shorter than l_c in order to ensure no filler breaking). Substituting $l_{emb} = l_c/2$ into equation (12)

$$G_{lc} = \frac{V_f \sigma_f l_c}{12} \tag{19}$$

which is universal for any cross-sectional shape. For a thin wall hollow filler with arbitrarily shaped cross-section, this expression reduces to (using equation (4))

$$G_{lc}^{thin} \cong \frac{V_f \sigma_f^2 t}{6\tau_i} \tag{20}$$

For a circular hollow tube (e.g. a nanotube), the maximum pullout energy is (using equation (7))

$$G_{lc} = \frac{V_f \sigma_f^2 D}{24\tau_i} A_{tD} = G_{lc}^{solid} A_{tD} \tag{21}$$

which reduces to $G_{lc}^{solid} = V_f \sigma_f^2 D/(24\tau_i)$ for a solid fiber. The dependence of G_{lc} on the cross-sectional geometry is depicted in Fig. 7. The maximum pullout energy cannot grow indefinitely with σ_f, and, as previously described, is bounded by the saturation energy G_{sat}, which depends on the density of filler flaws.

The pullout energies for the two filler length domains are depicted in Fig. 8. Similar plots are known in the literature for solid fibers, for example see Ref. 21; however, the current analysis demonstrates the universality of such plots with respect to any arbitrarily shaped hollow cross-sections.

Figure 8 demonstrates how the pullout energy can be increased when $l < l_c$: the preferred method, which yields the highest possible pullout energy for a given l_c, would be to increase l until it approaches l_c and the energy reaches its peak. If that is not possible, l_c can be reduced until it approaches l, achievable by (equation (7) and Fig. 4) using a filler with smaller wall thickness and/or lateral width, and/or by increasing the interfacial strength (e.g. by chemical functionalization of the filler). In the latter method, the energy will also reach a peak, but a lower peak than in the first method. For example (Fig. 8), suppose we have $l = 10$ μm and $l_c = 20$ μm: in the first method the energy will rise from 250 Pa m to 1000 Pa m by increasing l to 20 μm, while in the second method it will rise only to 500 Pa m by decreasing l_c to 10 μm.

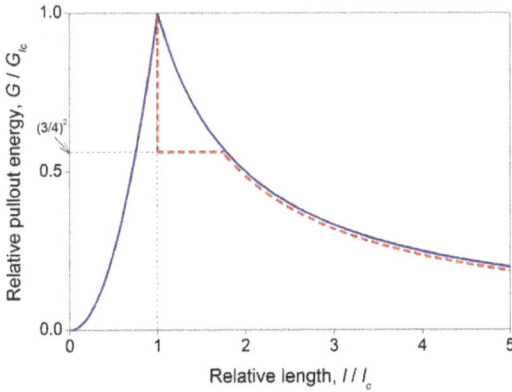

Figure 9 Universal curve depicting relative pullout energy G/G_{lc} versus relative filler length l/l_c (equation (22)): holds for any cross-sectional shape; dashed curve illustrates case of fillers somewhat longer than l_c described in text

Combining the two length domains (equations (13) and (16)), and the maximum achievable pullout energy G_{lc} defined by equation (19), we obtain

$$G = \frac{V_f \sigma_f}{12} \begin{cases} l^2/l_c \\ l_c^2/l \end{cases} = G_{lc} \begin{cases} (l/l_c)^2, & l<l_c \\ (l/l_c)^{-1}, & l>l_c \end{cases} \quad (22)$$

which holds for any cross-sectional shape. So, the pullout energy when $l>l_c$ is higher for longer l_c and shorter l. By contrast, when $l<l_c$, the pullout energy is higher for shorter l_c and longer l. A universal pullout energy curve, obtained by normalizing G by G_{lc}, is depicted in Fig. 9.

Cottrell[15] points out an additional domain, where the filler length is longer than its critical length but not much longer. In the domain $l \gg l_c$, when a filler breaks, the nearby fillers will become overloaded and will also break, and therefore the crack will propagate at a cross-sectional plane. However, when $l>l_c$ (only slightly longer), the filler will not tend to break at the fracture plane but rather at its highly stressed length center, creating a 'new' composite with filler fragments of length $1/2\, l_c < l < l_c$ (fragments longer than l_c will rebreak). The pullout energy that corresponds to this case is given by equation (13), and will remain constant upon increasing the filler length (since the length remains constant on average after fragmentation), until, when $l \gg l_c$ it will gradually drop in accordance with equation (16). See illustration in Fig. 9.

Since the average fragment length in this domain is $l=3/4\, l_c$, the pullout energy will be, according to equation (13), lower by a factor of $(3/4)^2 \cong 0.56$ compared to a filler that is slightly shorter than its critical length, or $l \cong l_c$. From a toughness perspective, since the pullout energy decreases from its peak for any filler length in the range $l>l_c$, whether somewhat longer or much longer than l_c, it is always preferable to use fillers that are slightly shorter than their l_c.

The effect of the aspect ratio t/D of a circular hollow tube can be summarized from equations (15) and (18)

$$G = G_{solid} \begin{cases} A_{tD}^{-1}, & l<l_c \\ A_{tD}^2, & l>l_c \end{cases} \quad (23)$$

depicted in Fig. 10. So, when $l>l_c$, the higher the aspect ratio (thicker tube wall), the higher the pullout energy. In other

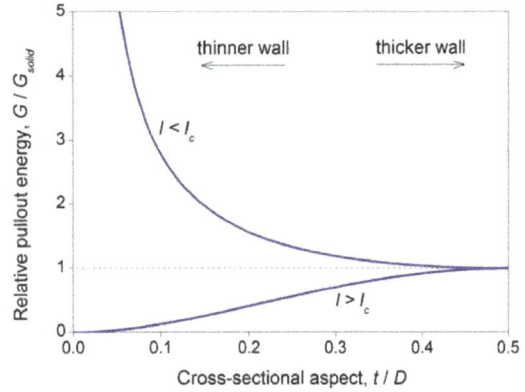

Figure 10 Dependence of relative pullout energy G/G_{solid} on cross-sectional aspect ratio t/D of circular hollow tube, when tube length l is longer (lower curve) and shorter (upper curve) than critical length l_c (equation (23) combined with equation (5)): maximum achievable G in domain $l<l_c$ is bounded by minimal possible wall thickness, and so long as l_c is not reduced below l

words, hollow tubes in this domain dissipate less pullout energy compared to solid fibers. By contrast, when $l<l_c$, the lower the aspect ratio (thinner tube wall), the higher the pullout energy. In other words, hollow tubes in this domain dissipate more pullout energy than solid fibers, and more so at lower values of the aspect t/D (i.e. shorter l_c according to equation (7)). We can keep on reducing t/D in order to increase the pullout energy, so long as l_c does not become smaller than l, switching to the domain $l>l_c$. Clearly, the wall thickness cannot be reduced below the width of a single layer of atoms as in a single wall nanotube.

The combined effect of the diameter and aspect ratio for a circular hollow tube can be expressed by rewriting equation (22) in the following form

$$G = G_{max} \begin{cases} (l_c/l)^{-1}, & l<l_c \\ (l_c/l)^2, & l>l_c \end{cases} \quad (24)$$

where l_c is given in equation (7), and the term $G_{max} = V_f \sigma_f l/12$ indicates the maximum possible pullout energy that can be achieved for a given tube length l by matching l_c to it. When $l \ll l_c$ (Fig. 11) as is typical for most current nanotubes, the effect of reducing the diameter at a given tube length has a dramatic positive impact on the pullout energy. This size effect occurs earlier (i.e. at a larger diameter) for lower aspect ratios (thinner wall thickness). In other words, the maximum pullout energy for a given tube length can be achieved by gradually reducing the external diameter, while preferably keeping a thin wall, effectively diminishing the critical length toward the tube length. If the diameter is reduced beyond the point where $l=l_c$, the trend will be reversed and the pullout energy will gradually decrease.

Effect of filler shape on strength and stiffness

A change in a design parameter, such as the filler's length or cross-sectional shape, in order to improve one material

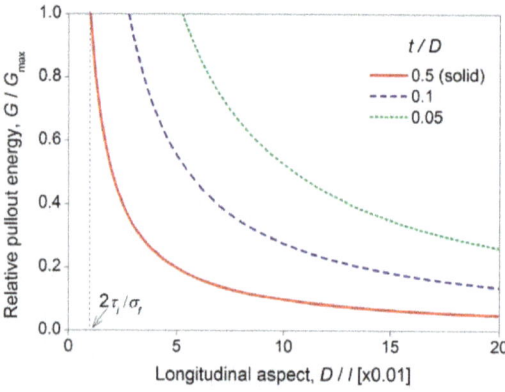

Figure 11 Relative pullout energy G/G_{max} versus longitudinal aspect ratio D/l for three aspect ratios t/D of circular hollow tube (equation (24) combined with equation (7)): plotted for domain $l<l_c$, with $2\tau_i/\sigma_f=0.01$

property, does not necessarily result in improving the other. Specifically, we are interested in studying the tradeoffs between the composite strength, stiffness and toughness, with respect to the filler geometry. We now use existing composites strength and stiffness models (for an example see Ref. 17) and generalize them to an arbitrarily shaped hollow filler. We then compare their strength and stiffness parametric trends to those of the toughness model.

Composite strength

Consider a composite's RVE, whose total cross-sectional area is A. The matrix has an ultimate strength σ_m, containing evenly dispersed unidirectional filler with ultimate strength σ_f. The filler is a general cylinder with length l, material cross-section area a and perimeter p (Fig. 2). When a sufficiently high load is applied on the composite, the interfacial stress is assumed to reach a constant value τ_i as previously described, corresponding to the matrix yield strength or the interfacial bonding strength. The stress in the filler, as a function of the distance x from the filler end, can be derived by substituting the ratio p/a from equation (3) into equation (2)

$$\sigma(x) = \tau_i x \frac{p}{a} = \frac{2\sigma_f x}{l_c}, \quad \sigma(x) \leq \sigma_f \quad (25)$$

Note that this formulation is universal with respect to the filler's cross-sectional shape and lateral size. At distances longer than $l_c/2$ from the filler end, the filler stress will reach its ultimate strength σ_f.

When an ultimate load F is applied, a crack develops in the matrix as a result of small debonding cracks at the filler ends. Consequently, fillers that are embedded a short length beyond the fracture plane will pull out, while those that are embedded longer will break at the fracture plane. So, except for shortly embedded filler sections, both matrix and filler are simultaneously exploited to their ultimate strength. The force borne by the matrix surface bonded to the filler end is neglected, since it is much lower than the pullout force.

When the filler length is shorter than its critical length $l<l_c$, the maximum stress in the filler occurs at $x=l/2$, where $\sigma(l/2) < \sigma_f$, and the filler's contribution to the overall stress is by means of pullout. Since the matrix fracture plane can fall

anywhere between the filler ends, the average stress in the filler at the fracture surface is $1/2\,\sigma(l/2)$, and the longitudinal strength of the composite is

$$\sigma_c = \frac{F}{A} = \frac{F_{filler}^{pullout} + F_{matrix}^{ultimate}}{A}$$
$$= 1/2\frac{Na}{A}\sigma(l/2) + \frac{A_{matrix}}{A}\sigma_m, \quad l<l_c \quad (26)$$

where N is the average number of fillers intersecting the fracture surface, and A_{matrix} is the material cross-sectional area of the matrix. Using N from equation (10) and $\sigma(l/2) = \sigma_f l/l_c$ from equation (25)

$$\sigma_c = 1/2\frac{l}{l_c}V_f\sigma_f + V_m\sigma_m, \quad l<l_c \quad (27)$$

where V_f and V_m are the filler and matrix volume fractions, respectively. Note that V_m is slightly less than $(1-V_f)$ for hollow fillers (see more on this further on). This equation is invariant with respect to the cross-sectional shape and lateral size, in other words it is a generalization of standard fiber reinforcement models such as in Ref. 17 to arbitrarily shaped fillers. As for the pullout energy in this length domain, a higher l/l_c improves the composite strength. Furthermore, the same value of the maximum composite strength can be achieved either by increasing l toward l_c or decreasing l_c toward l, contrary to the maximum pullout energy (G_{lc} from equation (19)), which decreases when l_c is decreased. The maximum contribution of the filler to the composite strength is $1/2V_f\sigma_f$.

For a thin wall hollow filler with a wall thickness t, we substitute l_c from equation (4)

$$\sigma_c^{thin} \cong 1/4\frac{\tau_i}{\sigma_f}\frac{l}{t}V_f\sigma_f + V_m\sigma_m, \quad l<l_c \quad (28)$$

independent of the cross-section shape and lateral size. In the case of a thin wall ribbon shaped filler (e.g. graphene), the contribution of the filler to the composite strength will improve by a factor of 2 as a result of the duplication of the interfacial area, as for the pullout energy in this length domain.

For a circular hollow tube, we substitute l_c from equation (7)

$$\sigma_c = \left(\frac{\tau_i}{\sigma_f}\frac{l}{D}A_{tD}^{-1}\right)V_f\sigma_f + V_m\sigma_m, \quad l<l_c \quad (29)$$

depicted in Figs. 13 and 14. For a solid fiber, substitute $A_{tD}=1$. Note that σ_f can be eliminated from equations (28) and (29) (it appears in both the numerator and denominator), since the filler ultimate strength is not reached in the domain $l<l_c$.

When the filler length is longer than its critical length $l>l_c$, at a distance longer than $l_c/2$ from the filler ends the fracture stress is σ_f for a fraction of $(l-l_c)/l$ of the fillers, while at shorter distances from the ends the filler's contribution to the overall stress is through pullout and its average stress is $1/2\,\sigma_f$ for a fraction of l_c/l of the fillers. Thus, the composite longitudinal strength is

$$\sigma_c = \frac{F}{A} = \frac{F_{filler}^{pullout} + F_{filler}^{ultimate} + F_{matrix}^{ultimate}}{A}$$
$$= 1/2\frac{Na}{A}\frac{l_c}{l}\sigma_f + \frac{Na}{A}\frac{l-l_c}{l}\sigma_f + \frac{A_{matrix}}{A}\sigma_m, \quad l>l_c \quad (30)$$

Using N from equation (10)

$$\sigma_c = \left(1 - 1/2\frac{l_c}{l}\right)V_f\sigma_f + V_m\sigma_m, \quad l > l_c \tag{31}$$

invariant with respect to the cross-sectional shape and lateral size. Note that, contrary to the pullout energy in this length domain, a higher l/l_c improves the composite strength, and the maximum strength is obtained for long fillers ($l \gg l_c$), for which equation (31) reduces to the known mixture rule, $\sigma_c = V_f\sigma_f + V_m\sigma_m$. Thus, the maximum contribution of the filler to the composite strength is $V_f\sigma_f$, while the minimum is $1/2V_f\sigma_f$.

For a thin wall hollow filler with a wall thickness t, we substitute l_c from equation (4)

$$\sigma_c^{thin} \cong \left(1 - \frac{\sigma_f}{\tau_i}\frac{t}{l}\right)V_f\sigma_f + V_m\sigma_m, \quad l > l_c \tag{32}$$

independent of the cross-section shape and lateral size. For a very thin filler wall (small t), this equation reduces to the same simple mixture rule as for very long fillers, implying that for a given volume fraction, a short hollow filler can deliver the same composite strength as a long solid filler, while maintaining toughness by keeping its length slightly below the critical length. We elaborate on such length-thickness tradeoffs further on. In the case of a thin wall ribbon-shaped filler (e.g. graphene), the contribution of the filler to the composite strength will improve (by a factor smaller than 2) as a result of the duplication of the interfacial area, contrary to the pullout energy in this length domain.

For a circular hollow tube, we substitute l_c from equation (7)

$$\sigma_c = \left(1 - 1/4\frac{\sigma_f}{\tau_i}\frac{D}{l}A_{tD}\right)V_f\sigma_f + V_m\sigma_m, \quad l > l_c \tag{33}$$

depicted in Figs. 13 and 14. For a solid fiber, substitute $A_{tD} = 1$. Note that reducing the cross-section aspect t/D (lower A_{tD}) has the same effect on improving the composite strength as reducing the longitudinal aspect D/l.

As mentioned before, V_m is slightly lower than $(1 - V_f)$ when the filler contains cavities. Thus, a correction factor is needed for hollow fillers if the matrix material (e.g. polymer) does not penetrate the filler's cavities, particularly when the longitudinal aspect ratio l/D is high or when the internal cavity is narrow or closed at its ends (e.g. in CNT). In such cases, the filler cavities are void and do not share the load. Thus, using equations (5) and (10)

$$V_m = 1 - \frac{Na_{tot}}{A} = 1 - \frac{V_f}{A_{tD}} \tag{34}$$

where a_{tot} is the total area bounded by the filler perimeter p, and the term on the right is for a circular hollow tube. Consequently, for a given volume fraction, while a thinner tube wall (smaller A_{tD}) improves the contribution of the filler to the composite's strength, it weakens the contribution of the matrix. Similarly, V_m should also be corrected for gaps between closely-packed fillers, which are not penetrated by the matrix or are too small to share the load, as in the case of NT bundles (Appendix).

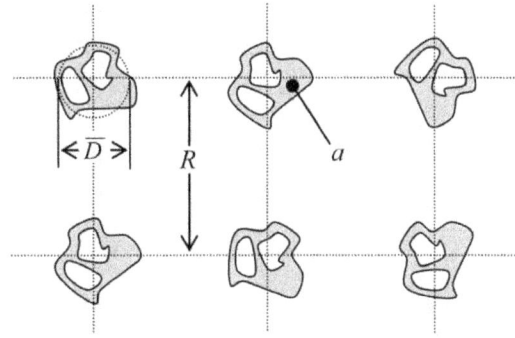

Figure 12 Square packing of hollow fillers with arbitrarily shaped cross-section: R is mean distance between fillers, a is filler material cross-section area and $\bar{D}$ is filler mean lateral size

Composite stiffness

The theoretical model for the elastic modulus of a composite with unidirectional reinforcing fibers was developed by Cox, using the shear lag theory.[17,20] We adjust the model for a filler with an arbitrarily shaped cross-section

$$E_c = \left[1 - \frac{\tanh(\beta l/2)}{\beta l/2}\right]V_fE_f + V_mE_m, \quad \beta^2 = \frac{2\pi G_m}{E_f a \ln(2R/\bar{D})} \tag{35}$$

where E_f and E_m are the tensile moduli of the filler and the matrix respectively, G_m is the matrix shear modulus, a is the filler material cross-section area, $\bar{D}$ is the filler mean lateral dimension and R is the mean lateral distance between adjacent fillers (Fig. 12). As in the strength model, V_m is slightly less than $(1 - V_f)$ for slender hollow fillers, as calculated by equation (34).

Note that R is a mean value for the distance between fillers that protrude through an arbitrary cross-section of the composite. Assuming square lateral packing, a single cell $R \times R$ has an area R^2, and contains four portions of a filler with an average cumulative area a. Hence, the filler volume fraction is $V_f = a/R^2$, and therefore

$$\ln\left(\frac{2R}{\bar{D}}\right) = 1/2\ln\left(\frac{4a}{V_f\bar{D}^2}\right) \tag{36}$$

Contrary to the pullout energy, when extending the length of the filler (very large l), the bracketed term in equation (35) has an upper limit value of 1, and the composite's elastic modulus reaches its maximum (for a given volume fraction), expressed by the known mixture rule, $E_c = V_fE_f + V_mE_m$. Thus, the maximum contribution of the filler to the composite stiffness is V_fE_f.

For a thin wall hollow filler with a wall thickness t, we approximate the filler perimeter by $\pi\bar{D}$, and therefore the cross-section material area can be estimated by $a \cong \pi\bar{D}t$. Substituting into equation (35) and using equation (36), we arrive after rearrangement of terms at the following estimation

$$\left(\frac{\beta l}{2}\right)_{thin} \cong \frac{l}{\bar{D}}\left(\frac{G_m}{E_f}\right)^{1/2}\left[\frac{t}{\bar{D}}\ln\left(\frac{4\pi}{V_f}\frac{t}{\bar{D}}\right)\right]^{-1/2} \tag{37}$$

Note that, unlike the toughness and strength, the stiffness of a thin wall filler is dependent on the cross-section mean

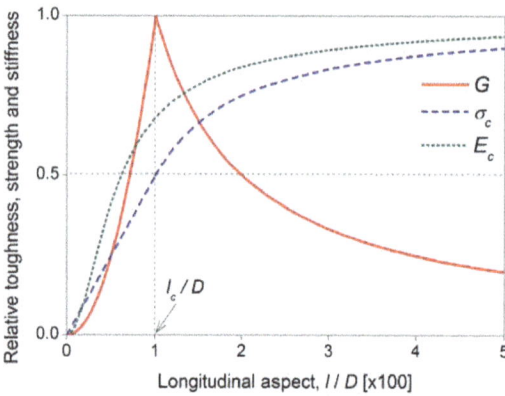

Figure 13 Relative toughness G/G_{Ic} (equation (22)) combined with equation (7)), strength $\sigma_c/(V_f\sigma_f)$ (equations (29) and (33)) and stiffness (elastic modulus) $E_c/(V_fE_f)$ (equations (35) and (39)) versus longitudinal aspect ratio l/D of circular hollow tube: toughness and strength curves are plotted for $l_c/D = \sigma_f A_{tD}/(2\tau_i) = 100$; thinner wall tube will shift l_c/D point to left (i.e. aspect function A_{tD} will be smaller, see equation (5)); stiffness curve is plotted for $G_m/E_f = 10^{-3}$ and $A_{tD} \ln(\pi A_{tD}/V_f) = 4$; thinner wall tube will stretch graph upward; strength and stiffness curves account only for filler's contribution to composite (first term in equations (29), (33) and (35))

Figure 14 Relative toughness G/G^{solid} (equation (23)), strength $\sigma_c/\sigma_c^{solid}$ (equations (29) and (33)) and stiffness E_c/E_c^{solid} (equations (35) and (39)) versus cross-sectional aspect ratio t/D of circular hollow tube: aspect function A_{tD} is given by equation (5), and is equal to 1 for solid fiber; strength curve for $l > l_c$ is plotted for $l/l_c^{solid} = 2(l/D)(\tau_i/\sigma_f) = 1$; higher values will squeeze graph downward, with minimum at 1; lower values will stretch it upward, with maximum at the $l < l_c$ curve; stiffness curve is plotted for $(l/D)(G_m/E_f)^{1/2} = 1$ and $V_f = 0.05$; higher values will squeeze graph downward (with minimum at 1), while lower values will stretch it upward; strength and stiffness curves account only for filler's contribution to composite (first term in equations (29), (33) and (35))

size $\overline{D}$, which appears in the longitudinal and cross-sectional mean aspect ratios, $l/\overline{D}$ and $t/\overline{D}$ respectively. For a very thin filler wall (small values of $t/\overline{D}$), the term $\beta l/2$ of equation (37) becomes very large, and equation (35) reduces to the same simple mixture rule as for very long fillers. This outcome implies that, for a given volume fraction, a short hollow filler can deliver the same composite stiffness as a long solid filler, without compromising the toughness if its length is kept slightly below the critical length. More on such tradeoffs is discussed further on.

In the case of a circular hollow tube, we replace (with the help of equation (5)) $\overline{D} = D$ and $a = \pi D^2 A_{tD}/4$, thus

$$\ln\left(\frac{2R}{\overline{D}}\right) = \frac{1}{2}\ln\left(\frac{\pi A_{tD}}{V_f}\right) \tag{38}$$

Substituting into equation (35)

$$\frac{\beta l}{2} = 2\frac{l}{D}\left(\frac{G_m}{E_f}\right)^{1/2}\left[A_{tD}\ln\left(\frac{\pi A_{tD}}{V_f}\right)\right]^{-1/2} \tag{39}$$

which reduces to the known expression for a solid fiber when $A_{tD} = 1$. Equation (35) with equation (39) is depicted in Figs. 13 and 14. Note that reducing the cross-section aspect t/D (lower A_{tD}) has the same effect on improving the composite stiffness as increasing the longitudinal aspect l/D. When the cross-section aspect is increased (higher A_{tD}), the composite's elastic modulus monotonically diminishes until it reaches a minimum at a solid fiber. Hence, when high stiffness is desired (at a given volume fraction), a thin wall filler is preferable to a solid fiber.

Comparison to toughness

Reinforcement design is the result of a tradeoff between the desired composite performance criteria, mainly its tough-

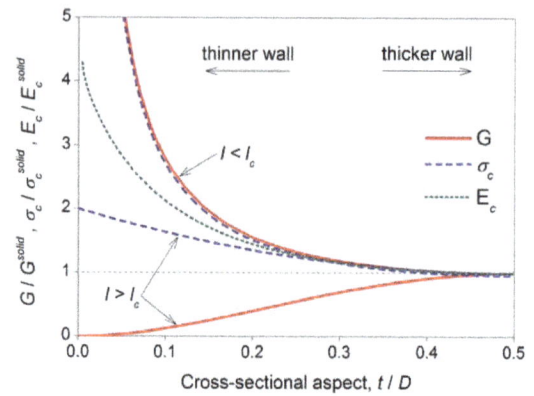

ness, strength, and stiffness (elastic modulus), which are determined by the filler's length, cross-sectional shape, mechanical properties and volume fraction, as well as by the matrix and interface mechanical properties. The following description highlights some of these dependencies, for a fixed filler volume fraction.

The plot in Fig. 13 summarizes the composite's toughness, strength and stiffness performance as a function of the longitudinal aspect ratio l/D of a circular hollow tube. In the domain $l < l_c$, a common case for nanoreinforcement, the performance parameters improve simultaneously when the ratio l/l_c is increased, either by increasing l or decreasing l_c. By contrast, in the domain $l > l_c$, a typical case for microreinforcement, when the strength and stiffness improve, the toughness degrades. However, the strength and stiffness in the domain $l < l_c$ are much lower compared to the domain $l > l_c$, and do not approach their optimal values. For example, when the toughness is at its optimum ($l = l_c$), the composite strength is only half its maximum possible value.

Furthermore, when $l \ll l_c$ as is typical for nanotubes, reducing the diameter has a dramatic positive impact on all three performance parameters. By gradually reducing the external diameter, the toughness increases toward its maximum value, while the strength and stiffness increase as well. This size effect occurs earlier for lower aspect ratios (thinner wall thickness). However, when $l > l_c$, the toughness trend with respect to the diameter is reversed, while strength and stiffness keep improving.

The effect of the cross-sectional aspect ratio t/D of a circular hollow tube is depicted in Fig. 14. In the domain $l < l_c$, the performance parameters improve simultaneously when t/D is decreased (thinner tube wall). By contrast, in the domain $l > l_c$, when the strength and stiffness improve, the toughness degrades. When the tube's longitudinal aspect ratio l/D is increased to high values, the strength and stiffness level off (Fig. 13) at a maximum value, and cannot be further improved by modifying the cross-sectional aspect ratio.

Geometric tradeoffs

We observe that the equations describing the composite toughness (equations (15) and (18)), strength (equations (29) and (33)) and stiffness (equations (35) and (39)) for circular hollow fillers contain functions of both the longitudinal aspect l/D and the cross-sectional aspect t/D, whereas the common models for circular solid fillers (e.g. fibers) contain l/D only. Hence, where hollow fillers are concerned, both geometrical aspect ratios play a significant role in determining the overall mechanical performance of the composite. In fact, the overall performance is the result of a tradeoff between the two aspect ratios, which can be visualized by writing equal performance expressions for G, σ_c and E_c, based on the aforementioned equations

$$\left.\begin{array}{c} (l^2/D)_G^{l<l_c} \\ (l^{1/2}/D)_G^{l>l_c} \\ (l/D)_{\sigma c} \end{array}\right\} \sim A_{tD} \tag{40}$$

$$(l/D)_{Ec} \sim [A_{tD}\,\ln(\pi\,A_{tD}/V_f)]^{1/2}$$

where A_{tD} is the aspect function of t/D defined in equation (5). The dependencies in equation (40), depicted in Fig. 15, were derived by expressing l/D as functions of t/D, where all other parameters are assumed constant. Although, in the case of G, the longitudinal aspect is expressed as l^2/D ($l<l_c$) and $l^{1/2}/D$ ($l>l_c$), the trend of l/D is maintained and it acts in the same direction for both l domains.

Figure 15 demonstrates the performance invariance with respect to size scaling (e.g. from microscale to nanoscale, and *vice versa*) obtained by modulating the geometric parameters D, t and l. It can be seen, for all three performance parameters, that the longitudinal aspects l^n/D are monotonically increasing functions of t/D. Consequently, a desired performance change obtained by increasing (or decreasing) l/D can be matched by increasing (or decreasing) t/D instead, of course within the minimum and maximum wall thickness boundaries. This means, for example, that the overall performance of short hollow fillers, typical of nanofillers, can match and even exceed that of long solid microfillers. In the case of G, t/D can be decreased so long as the resulting G_{lc} (equation (21)) does not drop below the desired G.

The following example, comparing the composite strength of two alternative fillers with the same tensile and interfacial strength and same volume fraction, but with a completely different size scale, may help clarify the issue. First, consider a solid fiber with $D=1\ \mu m$ and $l=1\ mm$, which delivers a desired composite strength. Its aspect ratios are $t/D=0.5$ and $l/D=1000$, denoted by point F on Fig. 15 (l/D is normalized to 1). Now, exchange this filler by a hollow single wall nanotube, with $D=20\ nm$ and $t=0.34\ nm$, so that $t/D=0.017$. Its corresponding length is $l=1.3\ \mu m$ ($l/D=67$),

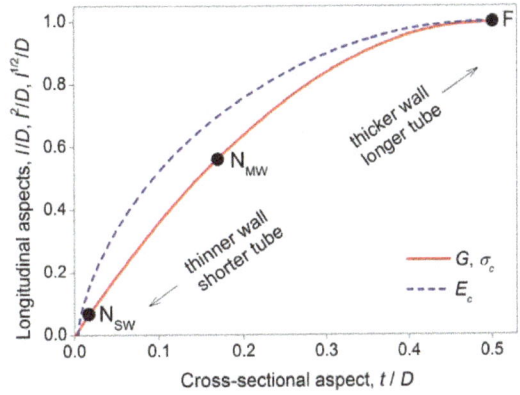

Figure 15 Equal composite performance curves (or lines of isoperformance) of the toughness G, strength σ_c and stiffness E_c, depicted in aspect ratios space – longitudinal aspect l^n/D versus cross-sectional aspect t/D, for a circular hollow tube (equation (40)): longitudinal aspects are l/D for σ_c and E_c, l^2/D for G when $l<l_c$, and $l^{1/2}/D$ for G when $l>l_c$, normalized to a maximum value 1; E_c curve is plotted for $V_f=0.05$; points F, N_{MW} and N_{SW} denote solid fiber, multiwall nanotube and single wall nanotube examples respectively, discussed in the text

denoted by point N_{SW} in Fig. 15. Optionally, we exchange the filler by a multiwall nanotube, with $D=20\ nm$ and $t=3.4\ nm$ (10 single atom layers), so that $t/D=0.17$ and $l=11.3\ \mu m$ ($l/D=560$), denoted by point N_{MW} in Fig. 15. All three filler types deliver the same composite strength (however, since nanotubes are typically stronger than fibers, they will in fact deliver higher composite strength, as discussed in the next section). Similar calculations, carried out for the toughness and stiffness using equation (40) and Fig. 15, show similar geometrical tradeoffs.

Nanotoughness compared to microtoughness

The potential toughness, strength and stiffness performance of nanoreinforcement using multiwall CNT fillers, is compared to microreinforcement using carbon microfibers (CFs), expanding our previous work.[10,13] We focus on the following four types of gedanken experiments illustrated in Figure 16b–e, compared to the reference carbon microfiber configuration in Fig. 16a, assuming the same filler volume fraction for both nanotubes and microfibers.

First, we consider uniformly dispersed unidirectional long carbon nanotubes, of length $l=l_c$ (Fig. 16c), a somewhat optimistic assumption in view of current technological feasibility. The length of the reference carbon microfibers is assumed to be $l=l_c$ as well. Using the optimal pullout energy $G_{lc}=V_f\sigma_f^2 D/(24\tau_i)$ from equation (21), the toughness performance ratio between CNT and CF reinforced composites is estimated by

$$\frac{G_{NT}}{G_{CF}} = \frac{G_{lcNT}}{G_{lcCF}} = \frac{\sigma_{NT}^2 D_{NT}\tau_{iCF}}{\sigma_{CF}^2 D_{CF}\tau_{iNT}} \approx \left(\frac{\sigma_{NT}}{\sigma_{CF}}\right)^2 \frac{D_{NT}}{D_{CF}} \sim 1 \tag{41}$$

assuming the same interfacial strength for both filler types, $\tau_{iCF} \cong \tau_{iNT}$. In this equation, the following order of magnitude

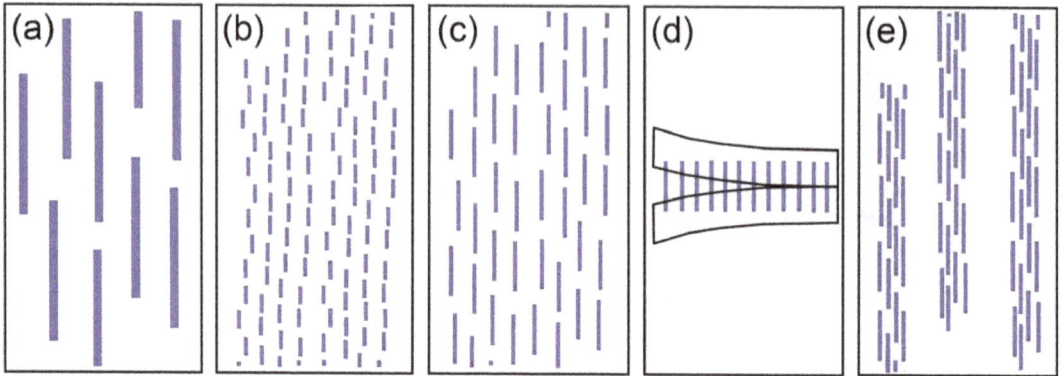

Figure 16 Gedanken experiments of unidirectionally aligned nanoreinforcing fillers: *a* uniformly dispersed long CF (reference configuration); *b* uniformly dispersed short NT; *c* uniformly dispersed long NT; *d* interlaminar reinforcement by long NT; *e* uniformly dispersed long NT fibers (NTF)

ratios were used: $\sigma_{NT}/\sigma_{CF} \sim 10$ and $D_{NT}/D_{CF} \sim 0.01$. This estimate finds the microfibers and nanotubes equivalent in their composite toughness. Note that the use of graphene has the potential of improving the nanoreinforcement by a factor of 2. The strength and stiffness performance ratios for this structure are estimated by $\sigma_{NT}/\sigma_{CF} \sim 10$ (equation (27)) and $E_{NT}/E_{CF} \sim 10$ (equation (35)) respectively, demonstrating a potential advantage for the nanotube reinforced composite, even though the toughness is not improved.

As already noted, most current CNTs are shorter than their critical length, and therefore for such cases the assumption $l=l_c$ used in the previous structure (equation (41)) should be replaced by $l<l_c$ (Fig. 16b). Using a nanotube length of order $l_{NT}/l_{cNT} \approx 0.1$ for this structure, the toughness ratio estimate is reduced by the factor $(l_{NT}/l_{cNT})^2 \sim 0.01$ (equation (22)) with respect to equation (41), favoring microfibers over nanotubes. If, however, we use single wall instead of multiwall nanotubes, the critical length can possibly be reduced so that $l_{NT}/l_{cNT} \approx 1$, and consequently, the toughness will decrease (in accordance with equation (21)) only by the factor $A_{tD} \sim 0.1$ (assuming $t/D \approx 0.025$)

$$\frac{G_{NT}}{G_{CF}} = \frac{G_{lcNT}^{thin}}{G_{lcCF}} = \frac{G_{lcNT}^{solid}}{G_{lcCF}} A_{tD} \sim 0.1 \qquad (42)$$

where we used $G_{lcNT}^{solid}/G_{lcCF} \sim 1$ from equation (41). The strength and stiffness performance ratio for this structure are estimated by $\sigma_{NT}/\sigma_{CF} \sim 10$ (equation (27)) and $E_{NT}/E_{CF} \sim 10$ (equation (35)) respectively, demonstrating a potential advantage for the nanotube reinforced composite, even though the toughness is degraded.

A compact nanotubes packing was demonstrated by Garcia et al.,[22] using reinforcement by a forest of aligned carbon nanotubes transplanted in the interlaminar region in laminates (Fig. 16d). Analysis of a similar structure was carried out by Wagner et al.,[13,14] considering a large number of parallel nanotubes with their total volume equivalent to that of a microfiber, yielding

$$\frac{nG_{NT}}{G_{CF}} = \frac{nD_{NT}\tau_{iNT}l_{cNT}^2}{D_{CF}\tau_{iCF}l_{cCF}^2} \approx \frac{\sigma_{NT}}{\sigma_{CF}} \sim 10 \qquad (43)$$

where $n = (D_{CF}^2/D_{NT}^2)(l_{cCF}/l_{cNT})$ is the number of nanotubes, and assuming $l=l_c$ for both the nanotubes and the

microfiber. This structure is advantageous to nanotubes. The strength performance ratio for this structure is estimated by $(\sigma_{NT}/\sigma_{CF})(l_{cCF}/l_{cNT}) \cong D_{CF}/D_{NT} \sim 100$, demonstrating a significant potential advantage for the nanotube reinforced composite, as well as toughness improvement.

The fourth structure compared here is a composite reinforced by long bundles of aligned compactly packed carbon nanotubes, or nanotube fibers (NTFs) (Fig. 16e), discussed in Appendix. The toughness performance ratio is (equations (49) and (21))

$$\frac{G_{NTF}}{G_{CF}} = \frac{G_{lcNTF}}{G_{lcCF}} \approx \left(\frac{\sigma_{NT}}{\sigma_{CF}}\right)^2 \frac{D_{NTF}}{D_{CF}} \approx \left(\frac{\sigma_{NT}}{\sigma_{CF}}\right)^2 \sim 100 \qquad (44)$$

using $D_{NTF}/D_{CF} \sim 1$, and assuming $l=l_c$ for both the nanotube fibers and the carbon fibers. This structure offers a substantial potential advantage over carbon microfibers. The strength and stiffness performance ratios for this structure are estimated by $\sigma_{NT}/\sigma_{CF} \sim 10$ (equation (27)) and $E_{NT}/E_{CF} \sim 10$ (equation (35)) respectively, advantageous as well to the nanotube fibers. These estimates suppose perfect compact packing of the nanotubes in each bundle, and efficient stress transfer from the bundle's boundary to its core, both somewhat compromised in practice due to gaps between the nanotubes and matrix penetration into the bundle (Appendix).

Based on the above estimates, the relative toughness, strength and stiffness of the four nanostructural types of Fig. 16b–e, with respect to the reference carbon microfiber configuration of Fig. 16a, are illustrated in Fig. 17. From a mechanical viewpoint, the nanocomposite structures of Fig. 16c–e are most advantageous, showing a potential simultaneous improvement in toughness, strength and stiffness.

As a final comment, we address the potential performance of composites reinforced by inorganic nanotubes such as WSNTs. Recent studies[23,24] have shown that, although WSNTs are intrinsically weaker than CNTs, WSNT reinforced composites have a comparable and sometimes higher toughness compared to CNT reinforced composites, without compromising other mechanical properties. Among the reasons cited in these studies is the presence of sulfide and oxysulfide functional groups, which increases the

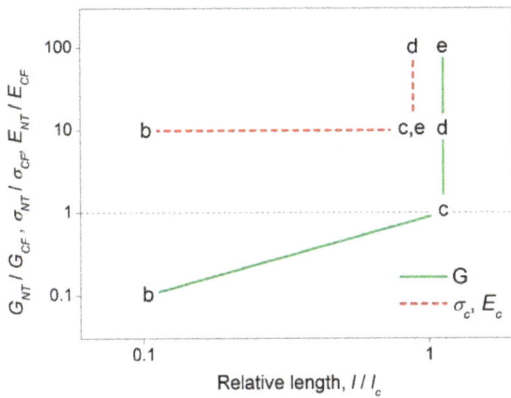

Figure 17 Nanoreinforcement potential – carbon nanotubes (NTs) compared to carbon microfibers (CFs): relative toughness G_{NT}/G_{CF}, strength σ_{NT}/σ_{CF} and stiffness E_{NT}/E_{CF}, versus relative length l/l_c; letters refer to structural types defined in Fig. 16

cross-linking within the matrix and the interfacial strength between the WSNTs and the matrix, and generates a more uniform and controlled dispersion of the WSNTs in the matrix, in contrast to the tendency of CNTs to aggregate. These preliminary results should be substantiated and expanded, for example by means of experimental data on the interfacial strength between WSNTs and various matrix materials.

Conclusion

The present analysis generalizes the modeling of the mechanical properties of composites comprising a soft matrix and unidirectional uniformly dispersed hollow fillers with arbitrarily shaped cross-sections, typical of nanoreinforced composites. The generalized model applies to a wide variety of filler types, including CNT, WSNT, MSNP and CNT fibers (CNTF). Of particular interest in this work is the evaluation of the composite toughness, strength and stiffness dependence on the filler size and shape characteristics, such as diameter, wall thickness and length. The toughness was expressed by the energy dissipated when the filler pulls out from the matrix during the composite fracture, taking into consideration the filler critical length.

It is shown that the critical length and the pullout energy of a thin wall filler, such as a single wall carbon nanotube (SWCNT), depend on the wall thickness rather than on the external shape and lateral size of the filler. This outcome leads to an analogy between a thin wall filler and a thin flat ribbon such as graphene, which has a double sided bonding area and hence half the critical length. The use of graphene instead of nanotubes may therefore offer improvement of the composite toughness in those cases where a shorter critical length is desired.

A convenient function of the aspect ratio between the wall thickness and the outer diameter is used to express the effect of the wall thickness on the critical length and pullout energy of hollow tubes. This function, termed the 'aspect function', rises monotonically with the aspect ratio (from a thin wall tube to a solid fiber). A tradeoff exists between the

filler cross-sectional aspect ratio (wall thickness/diameter) and longitudinal aspect ratio (length/diameter), such that a desired performance can be achieved by modifying either of them. In this way, for example, thin wall short fillers can match the performance of long solid fillers at the same volume fraction.

Two filler length categories are characterized with respect to the filler critical length, regardless of the cross-sectional shape: fillers shorter than their critical length, whose pullout energy is inversely dependent on the critical length, and fillers longer than their critical length, whose pullout energy is proportional to the square of the critical length. The maximum pullout energy is achieved when the filler length is slightly below the critical length. Nanotubes typically belong to the short filler category, and therefore, to achieve a high pullout energy, their critical length should be shortened. This can be achieved by selecting nanotubes with smaller diameter and/or wall thickness (e.g. switching from MWCNT to SWCNT), keeping the filler volume fraction constant, as well as by increasing the interfacial strength (e.g. by using a stronger matrix or by functionalizing the nanotubes[11]). These dependencies reverse when the filler is longer than the critical length.

The parametric trends regarding the toughness of short fillers coincide with those for the composite strength and stiffness, such that decreasing the critical length benefits all. By contrast, in the domain of long fillers, the trends are reversed, and an increase in the critical length benefits the toughness while degrading the strength and stiffness, and vice versa.

Comparison of the reinforcement performance of CNT to carbon microfibers, when the filler is uniformly and unidirectionally dispersed in the matrix at the same volume fraction, shows that nanotubes have the potential of simultaneously achieving up to two orders of magnitude better toughness, strength and stiffness, depending on the structural configuration. Specifically, nanotubes are most advantageous when their length approaches their critical length, and when they are compactly packed in structures such as aligned forests embedded in the interlaminar region of laminates, or long woven bundles in the form of fibers (CNTF). An interesting outcome of the analysis is that aligned agglomerations of nanotubes that form bundles may be advantageous in toughness compared to unbundled nanotubes.

The full potential of nanocomposites is yet to be achieved. Challenges are still high in forming desired superstructures of nanofillers in the matrix, with the intention of exploiting their utmost properties. Current technology runs into difficulties in obtaining homogenous blends of nanotubes in a polymer solution, achieving unidirectional alignment of nanotubes in a matrix, or reaching a very high filler volume fraction as in biological tissues. However, the present analysis demonstrates how the properties of nanocomposites can be optimized by modulating the filler shape and dimensions, as well as material and interface properties. In future work, the analysis will be expanded to randomly oriented nanofillers.

Conflicts of Interest

The authors have no conflicts of interest to declare.

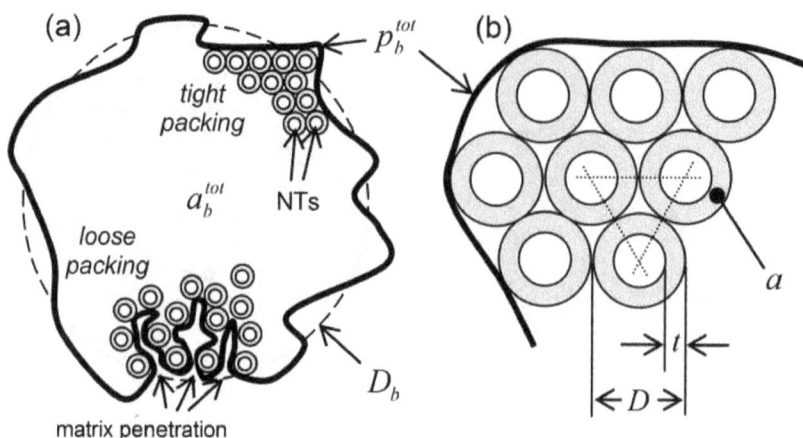

Figure A1 Cross-sectional view of nanotube bundle, illustrating *a* tight and loose packing and *b* magnified section of tight packing: *D*, *t* and *a* are diameter, wall thickness and material cross-sectional area of single NT respectively; p_b^{tot} is total outer perimeter of bundle, and a_b^{tot} is its total area (area bounded by perimeter p_b^{tot}); D_b is outer diameter of bundle with circular cross-section

Acknowledgements

The authors would like to acknowledge support from the INNI Focal Technology Area program 'Inorganic nanotubes (INT): from nanomechanics to improved nanocomposites', as well as from the G.M.J. Schmidt Minerva Centre of Supramolecular Architectures at the Weizmann Institute. This research was also made possible in part by the generosity of the Harold Perlman family. HDW is the recipient of the Livio Norzi Professorial Chair in Materials Science.

Appendix

Aligned nanotube bundles: example of complex geometry

Ultra long aligned CNT bundles, commonly termed CNT fibers or CNTF, synthesized by spray pyrolysis of ferrocene/xylene solutions, can reach several microns in diameter and contain 10^5–10^6 multiwall CNTs in a cross-section.[25,26] Within the same category of structures are bundles that contain a smaller amount of CNTs, which sometimes form spontaneously as a result of aligned agglomeration of CNTs in a polymer solution, for example under extensional flow.[3,4] The latter phenomenon is usually perceived as undesirable from a mechanical viewpoint, but this is not always the case as will be shown further on. The cross-sectional geometry of an aligned bundle is an example of a filler with arbitrarily shaped outline and multiple cavities, as presented in Fig. 1a. The following analysis applies the methods developed in this work to such bundles.

The calculation of the mechanical properties of aligned bundles embedded in a soft matrix should take into consideration the actual complex cross-sectional geometry of the bundle. Figure A1 illustrates the cross-section of a bundle for two types of packing: tightly packed nanotubes wherein the gaps between adjacent nanotubes are void of matrix material, and loosely packed nanotubes wherein the matrix material penetrates inside the bundle, thus increasing

the effective interfacial area and affecting the stress transfer between outer and inner nanotubes.

Assuming the nanotubes in a bundle are tightly packed in a hexagonal array (Fig. A1b), the nanotube-matrix interface exists only at the external surface of the outermost nanotubes exposed to the matrix, and the stress in the bundle propagates inwards via interactions between adjacent nanotubes. Thus, the geometry is defined by the total outer perimeter of the bundle p_b^{tot} and the total area a_b^{tot} (the area bounded by the perimeter p_b^{tot}, including material and all forms of internal voids).

A correction factor should be applied to the perimeter, to account for the actual contact area between the matrix and the circular profile of each nanotube. For a nanotube with an external diameter D, the factor is equal to half its perimeter, $1/2\,\pi D$, divided by its width D, and therefore the actual perimeter is

$$p_b \cong \frac{1/2\pi D}{D} p_b^{tot} \cong \frac{\pi}{2} p_b^{tot} \tag{45}$$

Observing a representative area element indicated by a triangle in Fig. A1b, its material area fraction is equal to the material area of three 60° nanotube sections, $3/6a$ (*a* is the material cross-sectional area of a single nanotube), divided by the triangle area $3^{1/2}/4\,D^2$. Thus, the bundle material cross-sectional area is given by

$$a_b \cong \frac{1/2\,a}{3^{1/2}/4\,D^2} a_b^{tot} \cong \frac{\pi\,A_{tD}}{2 \times 3^{1/2}} a_b^{tot} \tag{46}$$

where A_{tD} is the cross-sectional aspect function of a single nanotube, defined in equation (5), which ranges from $4t/D \cong 0$ for a thin wall tube ($t \ll D$) to 1 for a solid fiber ($t = D/2$). If the nanotube wall thickness is unknown, the bundle's actual material cross-sectional area can be derived by $a_b = \rho_l/\rho_{NT}$, where ρ_l is the linear density of the bundle, and ρ_{NT} is the nanotube material density.

Substituting the bundle's actual perimeter p_b from equation (45) and its actual material cross-sectional area a_b from equation (46) into equation (3), we obtain the bundle's critical length

$$l_{cb} = \frac{2\sigma_f}{\tau_i} \frac{a_b}{p_b} \cong \frac{2\sigma_f}{3^{1/2}\tau_i} A_{tD} \frac{a_b^{tot}}{p_b^{tot}} \qquad (47)$$

regardless of the bundle cross-sectional shape. For a bundle with a circular cross-section having a diameter D_b (Fig. A1a), $a_b^{tot}/p_b^{tot} = \frac{1}{4}D_b$, and therefore, the critical length is

$$l_{cb} \cong \frac{\sigma_f D_b}{2 \times 3^{1/2}\tau_i} A_{tD} \qquad (48)$$

which reduces to $l_{cb}^{solid} = \sigma_f D_b/(2 \times 3^{1/2}\tau_i)$ for solid fillers, and to $l_{cb}^{thin} \cong 2\sigma_f t D_b/(3^{1/2}\tau_i D)$ for thin wall tubes. Note the similarity of this equation to equation (7).

Since a compact nanotube bundle can be made very long, owing to the efficient transfer of stress between adjacent nanotubes, the maximum achievable pullout energy can be reached (assuming $l_b = l_{cb}$) for a given volume fraction V_f (equation (19))

$$G_{lcb} = \frac{V_f \sigma_f l_{cb}}{12} = \frac{V_f \sigma_f^2 D_b}{24 \times 3^{1/2}\tau_i} A_{tD} \qquad (49)$$

This energy is higher by a factor of $1/3^{1/2}D_b/D$ compared to the maximum pullout energy of unbundled uniformly dispersed nanotubes (equation (21)) with the same volume fraction and cross-sectional shape, each with a length $l = l_c$. This factor can be of order 10^2–10^3 for a large bundle, and is reflected in equation (44) and Fig. 17. The reason for this dramatic advantage in toughness is that the critical length of a bundle l_{cb} is much longer than that of an individual nanotube l_c, because of the high ratio $a_b/p_b \sim D_b$ in equation (47).

If the fillers composing the bundle are solid

$$G_{lcb}^{solid} = \frac{V_f \sigma_f^2 D_b}{24 \times 3^{1/2}\tau_i} \qquad (50)$$

similar to equation (21), and if they are thin-wall with a wall thickness t

$$G_{lcb}^{thin} \cong \frac{V_f \sigma_f^2 t}{6 \times 3^{1/2}\tau_i} \frac{D_b}{D} \qquad (51)$$

Under the same comparison assumptions as before ($l_b = l_{cb}$, $l = l_c$ and same V_f), the bundle's ultimate strength will be equivalent to uniformly dispersed nanotubes (equations (27) and (31)). The same is true for the stiffness, assuming the longitudinal aspect ratio is about the same for the bundle and a single nanotube ($l_b/D_b \cong l/D$) (equations (35) and (39)). Moreover, since bundles can be made very long, it is possible to reach a length l_b much higher than l_{cb}, thus approaching the maximum achievable strength and stiffness (refer to Fig. 13), a difficult goal to achieve in unbundled nanotubes.

When the nanotubes in a bundle are loosely packed (Fig. A1a), the matrix material partially penetrates the bundle. This phenomenon may occur in practice as a result of, for example, electrostatic charge that builds up on the nanotubes during processing or chemical treatment of the bundle. Consequently, the nanotube–matrix effective interfacial area is expanded, and the stress propagation in the bundle is modified, leading to a reduction in the effective cross-sectional area that carries the load.

Defining the effective interfacial perimeter $\langle p_b \rangle$ and the effective cross-sectional area $\langle a_b \rangle$, the effective critical length can be written as in equation (3)

$$\langle l_{cb} \rangle = \frac{2\sigma_f}{\tau_i} \frac{\langle a_b \rangle}{\langle p_b \rangle} = \frac{\sigma_f \langle D_b \rangle}{2\tau_i} \qquad (52)$$

where $\langle D_b \rangle$ is the effective diameter of a bundle with a circular cross-section. Correspondingly, the effective pullout energy of the bundle, $\langle G_{lcb} \rangle = V_f \sigma_f \langle l_{cb} \rangle/12$. For a given bundle, $\langle D_b \rangle$ can be assessed by obtaining the bundle's $\langle l_{cb}^{test} \rangle$ from a fragmentation test, so that $\langle D_b \rangle \cong 2\tau_i \langle l_{cb}^{test} \rangle/\sigma_f$. The effective diameter $\langle D_b \rangle$ is evidently smaller than D_b of a tightly packed bundle (a rough estimate is $\langle D_b \rangle/D_b \sim 0.1$), and therefore the maximum achievable effective pullout energy is proportionally lower, but is still much higher than unbundled uniformly dispersed nanotubes with the same volume fraction.

While the composite strength and stiffness of tightly packed bundles were shown to be the same as for uniformly dispersed nanotubes with the same volume fraction, this is not the case for loosely packed bundles. The less efficient stress transfer between the nanotubes in a bundle means that the effective nanotube volume fraction is lower than V_f, and therefore the composite strength and stiffness will be degraded to some extent in comparison to uniformly dispersed nanotubes.

In summary, aligned tightly packed bundles have the potential of offering a significant improvement in composite toughness with respect to unbundled nanotubes, while the strength and stiffness remain unimpaired. Penetration of matrix material into the bundle may somewhat degrade the toughness, but it is still expected to remain much higher than unbundled nanotubes. Since the discussion in this appendix is generalized for any type of aligned bundle, these conclusions hold true also for small bundles, formed by agglomeration during the composite production process. This means that, from a toughness perspective, aligned agglomeration is potentially advantageous compared to uniformly dispersed nanotubes.

References

1. N. Satyanarayana, K. S. S. Rajan, S. K. Sinha and L. Shen: *Tribol. Lett.*, 2007, **27**, (2), 181–188.
2. K. L. White and H. J. Sue: *Polymer*, 2012, **53**, (1), 37–42.
3. X. M. Sui and H. D. Wagner: *Nano Lett.*, 2009, **9**, (4), 1423–1426.
4. Y. Dror, W. Salalha, R. L. Khalfin, Y. Cohen, A. L. Yarin and E. Zussman: *Langmuir*, 2003, **19**, (17), 7012–7020.
5. Y. Ji, C. Li, G. Wang, J. Koo, S. Ge, B. Li, J. Jiang, B. Herzberg, T. Klein, S. Chen, J. C. Sokolov and M. H. Rafailovich: *EPL*, 2008, **84**, (5), 56002.
6. B. G. Demczyk, Y. M. Wang, J. Cumings, M. Hetman, W. Han, A. Zettl and R. O. Ritchie: *Mater. Sci. Eng. A*, 2002, **A334**, (1–2), 173–178.
7. M. F. Yu, O. Lourie, M. J. Dyer, K. Moloni, T. F. Kelly and R. S. Ruoff: *Science*, 2000, **287**, (5453), 637–640.
8. I. Levchenko, Z.-J. Han, S. Kumar, S. Yick, J. Fang and K. Ostrikov: 'Large arrays and networks of carbon nanotubes: morphology control by process parameters', in 'Syntheses and applications of carbon nanotubes and their composites', (ed. S. Suzuki), 2013, Philadelphia, PA, InTech.
9. H. D. Wagner: *Nat. Nanotechnol.*, 2007, **2**, (12), 742–744.
10. M. H. G. Wichmann, K. Schulte and H. D. Wagner: *Compos. Sci. Technol.*, 2008, **68**, (1), 329–331.
11. N. Lachman and H. D. Wagner: *Composites Part A*, 2010, **41A**, (9), 1093–1098.
12. A. G. Evans: *J. Am. Ceram. Soc.*, 1990, **73**, (2), 187–206.
13. H. D. Wagner, P. M. Ajayan and K. Schulte: *Compos. Sci. Technol.*, 2013, **83**, 27–31.
14. H. D. Wagner: *Chem. Phys. Lett.*, 2002, **361**, (1–2), 57–61.
15. A. H. Cottrell: 'Strong solids', *Proc. R. Soc. Lond. A*, 1964, **282A**, 2–9.

16. A. Kelly and W. R. Tyson: *J. Mech. Phys. Solids*, 1965, **13**, (6), 329.

17. M. Piggott: 'Load bearing fibre composites', 475; 2002, New York, Kluwer Academic Publishers.

18. E. W. Weisstein: 'Generalized Cylinder', From MathWorld – A Wolfram Web Resource [viewed Nov 2014]; Available from: http://mathworld.wolfram.com/GeneralizedCylinder.html.

19. P. K. Mallick: 'Fiber-reinforced composites: materials, manufacturing, and design', xvii; 2008, Boca Raton, FL, CRC Press.

20. H. L. Cox: *Br. J. Appl. Phys.*, 1952, **3**, 72.

21. H. D. Wagner and A. Lustiger: *Compos. Sci. Technol.*, 2009, **69**, (7–8), 1323–1325.

22. E. J. Garcia, B. L. Wardle and A. J. Hart: *Composites Part A*, 2008, **39A**, (6), 1065-1070.

23. G. Lalwani, A. M. Henslee, B. Farshid, P. Parmar, L. J. Lin, Y. X. Qin, F. K. Kasper, A. G. Mikos and B. Sitharaman: *Acta Biomater.*, 2013, **9**, (9), 8365–8373.

24. M. Shtein, R. Nadiv, N. Lachman, H. D. Wagner and O. Regev: *Compos. Sci. Technol.*, 2013, **87**, 157–163.

25. Z. Yang, X. Chen, H. Nie, K. Zhang, W. Li, B. Yi and L. Xu: *Nanotechnology*, 2008, **19**, 085606.

26. C. Li, G. J. Fang, L. Y. Yuan, N. S. Liu, L. Ai, Q. Xiang, D. S. Zhao, C. X. Pan and X. Z. Zhao: *Nanotechnology*, 2007, **18**, (15), 155702.

Evaluation of fracture behavior of polyethylene/CaCO₃ nanocomposite using essential work of fracture (EWF) approach

Meymanat S. Mohsenzadeh[1], Mohammad Mazinani[1] and Seyed Mojtaba Zebarjad*[2]

[1]Department of Materials Engineering, Faculty of Engineering, Ferdowsi University of Mashhad, Mashhad, Iran
[2]Department of Materials Engineering, Faculty of Engineering, Shiraz University, Shiraz, Iran

Abstract The fracture behavior of two series of polyethylene (PE) films, both the unreinforced and reinforced with different amounts of nanoscale calcium carbonate (CaCO₃) particles was investigated in this research work by means of the essential work of fracture (EWF) approach

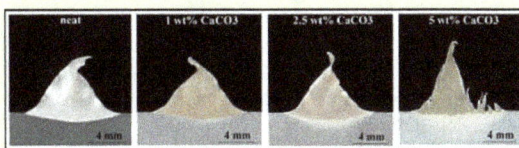

under the loading condition of mode I. For this purpose, EWF test was conducted on deeply double edge notched tension (DDENT) specimens; and the essential and non-essential components of fracture work, were estimated using the plots representing the variation of the specific total work of fracture (w_f) versus the ligament length of the specimen. The results showed that the crack resistance for the PE/CaCO₃ nanocomposites was lower than that of the unreinforced PE, and for the case of nanocomposite samples, the total plastic energy absorbed during fracture increased with increasing CaCO₃ content up to 2.5 wt-% and then decreased for nanocomposite samples containing 5 wt-%CaCO₃. The plastic zone size decreased with increasing CaCO₃ content, whereas the specific plastic work per unit volume increased with increasing weight percent CaCO₃. The observed trends were rationalized using the fracture and deformation mechanisms of nanocomposites.

Keywords Polyethylene, Calcium carbonate, Nanocomposite, Essential fracture work, Stress whitening zone

Introduction

Incorporation of mineral fillers in plastics is a common practice with the final goal to typically reduce the overall costs and to improve their physical and mechanical properties such as stiffness and dimensional stability.[1] The enhancement of these properties depends upon the geometrical parameters of dispersed phase, e.g. the particles size, shape, distribution and orientation, as well as the filler concentration. To this, one may add the compatibility between the polymer matrix and the filler which plays an important role in controlling the properties of the composite. This is because it promotes interaction between the polymer matrix and the filler.[4] It is believed that the addition of inorganic fillers causes the composites to become brittle through decreasing their impact energy. Most of the studies on modification of semi-crystalline polymers filled with rigid particles have reported a significant decrease in the composite toughness compared to the neat polymer.[1] However, several studies have demonstrated that an

increase in impact resistance of polypropylene and polyethylene has been observed by adding an appropriate amount of these rigid particles.[1-4] The reason of this toughness increase is that for a specific volume fraction, the thickness of interparticle matrix ligaments becomes less than a critical threshold value and a structure with reduced plastic resistance percolates throughout the composite. Consequently, a significant improvement in the toughness could be achieved through void formation and concomitant matrix deformation. Calcium carbonate (CaCO₃) is one of the most commonly inorganic fillers used in thermoplastics mainly due to its availability in readily usable form as well as its low cost.[5] The mechanical properties of composites reinforced with CaCO₃ particles are determined by the control of their volume fraction and size,[2,3] and mixing condition[6] as well as adding suitable surface modifiers.[7]

Nanocomposite materials have received a great deal of attention in recent years because of the remarkable enhancements that have been successfully made on their physical and mechanical properties by the addition of only a limited amount of filler materials when compared with conventional composites.[7,8,23] The extent to which the

*Corresponding author, email mojtabazebarjad@shirazu.ac.ir

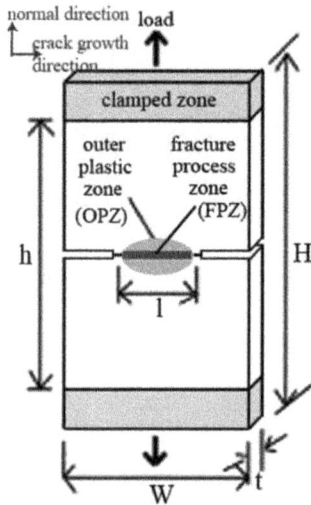

Figure 1 DDENT specimen used for EWF experiments representing dimensions and different zones within specimen developed during loading

properties of nanocomposites are improved depends largely on how the nanoparticles are dispersed in the matrix. Homogeneous dispersion of the nanoparticles in the matrix is usually very difficult to achieve, due to the strong tendency of these ultrafine particles to agglomerate. In a few cases, a significant decrease in toughness of nanocomposite materials has been reported due to inappropriate dispersion of nanoparticles in the matrix.[2,3,7]

Toughness is one of the most important mechanical properties of materials that should be taken into account in engineering applications. Linear elastic fracture mechanics is adopted for characterization of the toughness of materials displaying only elastic deformation or for materials with only small scale yielding near the crack tip.[9] Indeed, for ductile materials in which a large plastic zone is developed at the crack tip, linear elastic fracture mechanics is no longer valid. Conventional Charpy and Izod impact tests are commonly used to determine the toughness of these materials.[1–4,7] However, impact toughness is not an appropriate parameter for materials design. Alternatively, the essential work of fracture (EWF) technique has been recently accepted globally as a reliable method for the characterization of ductile fracture behavior of polymeric materials, especially those in the form of films and sheets.[10–18,28] EWF has been employed for evaluating the fracture toughness of polymeric materials that are toughened by rubber particles,[11,12] or for those containing different fillers and reinforcing components.[13–15] Recently, the fracture toughness of nanocomposites with a polymer matrix has been evaluated by the EWF method.[16–18]

The EWF concept was initially proposed by Broberg[19] and then developed by Cotterell, Reddell and Mai.[20,21] In this approach, the total fracture work (W_f) dissipated in a precracked specimen is divided into two parts, i.e. the work required to fracture the material in its inner process zone (W_e), and that consumed during various deformation mechanisms in the plastic zone (W_p) (Fig. 1). Under

prevailing plane-stress conditions, W_f can be written as

$$W_f = W_e + W_p = w_e l t + \beta w_p l^2 t \tag{1}$$

$$w_f = w_e + \beta w_p l \tag{2}$$

where w_f is the specific total fracture work, w_e and w_p are respectively the specific essential work of fracture and the specific plastic work, l is the ligament length, t is the sample thickness and β is the shape factor corresponding to the size of plastic zone. The term w_e is considered to be a material property and thus, characterizes the material resistance against cracking under plane stress conditions.

The EWF approach has been applied to polypropylene–calcium carbonate (PP-CaCO$_3$) composites by Gong et al.[14] It has been found that addition of CaCO$_3$ into the PP matrix reduces both w_e and βw_p values. However, an optimum amount of maleic anhydride grafted polypropylene (PP-g-MA) could improve fracture toughness of the composite. Tjong and Bao[18] have reported a simultaneous reduction in w_e and βw_p of high density polyethylene with the addition of organoclay. Incorporation of elastomers has resulted in an increase in both parameters. For poly(propylene-block-ethylene)–1 wt-% organoclay (MMT) nanocomposites, a simultaneous reduction in w_e and βw_p with respect to neat poly(propylene-block-ethylene has also reported.[22] It has been found that the finely and uniformly dispersed boehmite alumina (BA) nanoparticles could increase the essential work of fracture of the nanocomosites.[23–25] Pedrazzoli et al.[23] have reported a significant enhancement (up to 64%) with incorporation of the BA into linear low density polyethylene. In the study of Yang et al.,[17] it has been found that the addition of TiO$_2$ nanoparticles into polyamide 66 improves the essential work for the crack initiation ($w_{e,ini}$) at the expense of lowering the non essential work term ($\beta_{ini}w_{p,ini}$). $w_{e,ini}$ is the specific essential work of resistance to crack initiation and $\beta_{ini}w_{p,ini}$ is the specific plastic work dissipated during crack initiation, while w_e and βw_p are the specific essential and plastic work of fracture defined for whole fracture process. Similar results were obtained with addition of the BA into polypropylene.[26] Although the fracture properties of polyethylene–CaCO$_3$ nanocomposites have been greatly investigated, the evaluation of their fracture behavior using the EWF method has not been reported. In the present study, the fracture behavior of a medium density polyethylene (MDPE) and its composites reinforced with the CaCO$_3$ nanoparticles have been studied by using the EWF approach. The influence of CaCO$_3$ content on the main parameters of the EWF analysis, the development of plastic zone as well as the absorption of the plastic energy has been also investigated.

Experimental procedure

Materials

MDPE, HP3840 UA, was used in this investigation as the matrix material for composite specimens. This material was used in powder form having a melt flow index of 4.2 g/10 min (measured at 190°C with the application of 2.16 kg) and a density of 0.937 g cm^{-3}. The precipitated calcium carbonate (CaCO$_3$) powder with an average particle size of

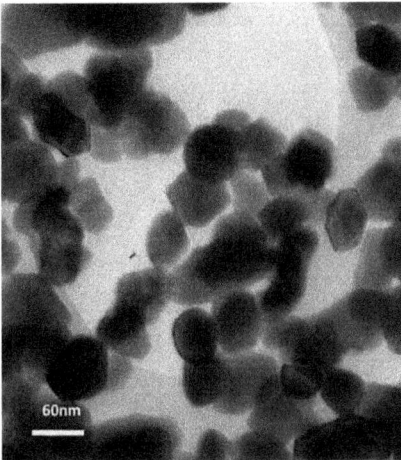

Figure 2 Transmission electron micrograph showing nanosized $CaCO_3$ particles

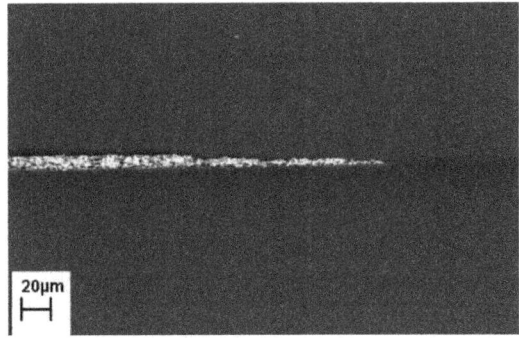

Figure 3 Scanning electron micrograph of region near crack tip

about 70 nm was obtained from Solvay Company (Suresnes CDX, France), under a trade name of Socal 312. Figure 2 shows the transmission electron micrograph of the $CaCO_3$ nanoparticles used in this study. As it can be seen in this TEM image, particles have an irregular shape.

Preparation of test specimens

The nanocomposite samples were prepared by mixing the MDPE and $CaCO_3$ powders in a mixer mill (Retsch MM400) with preselected powders' weight ratios. Mixing was carried out at a frequency of 20 Hz for 10 min. The sample names according to their compositions are: MDPE (0 wt-%$CaCO_3$), MDPE-1 (1 wt-%$CaCO_3$), MDPE-2.5 (2.5 wt-%$CaCO_3$) and MDPE-5 (5 wt-%$CaCO_3$). The mixtures were then compression molded into the films of 0.3 mm thick with the application of 30 kPa pressure at 170°C. Deeply double edge notched tensile (DDENT) specimens with a total length of $H=110$ mm (leaving an effective length of $h=60$ mm between grips) and a width of $W=60$ mm were cut from the films. The notches were made using a sharp scissors. The subsequent fine sharpening of the notches was done by slowly pushing a razor blade into their bottoms. The geometry of the notches produced in this way was observed by a scanning electron microscope (SEM LEO 1450VP). Figure 3 shows the scanning electron micrograph of the region near the crack tip. As it can be seen, a good quality of the crack was obtained. From our SEM observations, the crack tip radius seems to be 300 nm. The ligament lengths of the specimens were confined by ESIS TC4 test protocol.[28] At least 25 specimens with the ligament lengths ranging between 5 and 15 mm were prepared for each type of the composites. The ligament length (l) and thickness (t) of each sample were measured with the help of an optical microscope and a micrometer respectively.

Essential work of fracture test

Tensile tests were conducted on the DDENT specimens using a Zwick (Z 250) universal tensile testing machine at room temperature under a constant crosshead speed of 5 mm min^{-1}. The displacement of samples under loading was measured through an extensometer with a gauge length

of 25 mm. The load–displacement curves were recorded accordingly, and the energy absorbed in the specimens during the tensile failure was determined by the calculation of the area under the load–displacement curves. The specific work of fracture was plotted against the ligament length. The specific essential work (w_e) and the specific non-essential work (βw_p) were determined for each group of samples through the calculation of the intercept and the slope of the linear relationship, which has been fitted to their corresponding experimental data points respectively. The maximum load for each specimen was used to check the stress criterion.[28] According to the ESIS-TC4 test protocol available for the EWF analysis,[28] a predetermined stress criterion was applied to each test specimen. This criterion ensures that the ligament region is entirely in plane stress conditions and that full ligament yielding occurs before the onset of crack growth stage. This criterion has been developed on the basis of the evaluation of the net section stress

$$\sigma_{max} = \frac{F_{max}}{lt}$$

where F_{max} is the maximum peak load. For all the test specimens, an average value of σ_{max}, denoted by σ_m, was calculated, and a specimen for which $\sigma_{max} < 0.9\sigma_m$ or $\sigma_{max} > 1.1\sigma_m$, was rejected from the list of specimens possessing valid experimental results.

The variation in the βw_p is caused by that of both β, the dimensionless shape factor which depends on plastic zone size, and w_p, the specific energy absorption per unit volume in the plastic zone.[29] Thus, In order to explain the observed trend in βw_p parameter, the variations of β and w_p were investigated for these materials.

Results and discussion

Load–displacement curves

The photographs of the fractured MDPE specimen and different nanocomposites are shown in Fig. 4. As can be seen in these photographs, all the specimens have fractured in a fully ductile manner. Furthermore, the well defined yielding of the regions neighboring the initial ligaments of the specimens can be observed in this series of photographs. The effect of $CaCO_3$ content on the load–displacement curves of DDENT specimens with the same ligament length is shown in Fig. 5a. It is clearly seen that the maximum stress has slightly

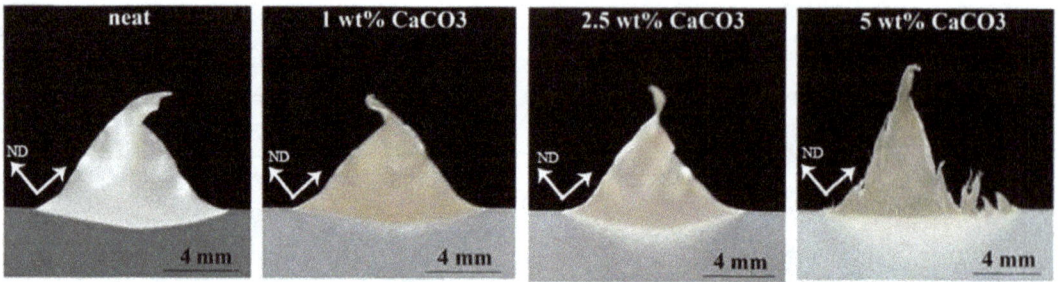

Figure 4 Photographs showing fractured specimens during EWF fracture experiments: MDPE and composites with 1, 2.5 and 5 wt-%CaCO₃; ND, normal to the fracture surface

decreased when 1 wt-%CaCO₃ is added to the MDPE. This shows that the addition of such nanoparticles with non-continuous type of spatial distribution and a very limited load transfer capability within the continuous PE matrix has no considerable reinforcing effect and, indeed, lowers the overall load bearing capacity of the composite through introducing a certain number of nonhomogeneity sites inside the matrix material. Obviously, the lowering effect of nanoparticels on load bearing capacity of the composite samples is more significant when more CaCO₃ particles (2.5 and 5 wt-%) are added to their matrices. However, the total displacement experienced by the composite has been increased by 10.3% when 1 wt-%CaCO₃ nanoparticles is added to the PE matrix. This experimental evidence clearly shows that the resistance of the composite material with 1 wt-%CaCO₃ against the overall crack growth as an important stage of the whole ductile fracture phenomenon associated with a significant energy adsorption has been improved. Although such behavior is also observed for the case of 2 wt-%CaCO₃ composite (with a 3.9% increase), the effect is less significant when compared with 1 wt-%CaCO₃ composite. However, the presence of 5 wt-%CaCO₃ in the composite has decreased the total displacement to an extent almost equal to that of the MDPE.

The load–displacement curves for the MDPE–1 wt-%CaCO₃ composite with various ligament lengths are shown in Fig. 5b. It can be seen in this figure that irrespective of the initial ligament length, the shapes of all curves are similar. This

satisfies one of the basic requirements for the application of EWF analysis. Similar plots were obtained for MDPE specimens and those with 2.5 and 5 wt-%CaCO₃ nanoparticles. The mentioned similarity of the load–displacement curves was checked for each series of the experiments, and the results showed the same trend for all the test conditions.

The visual observation of fracture mechanism during the experiments and the type of overall load–displacement response of the specimens show that the ligament region has been fully yielded before the crack propagation stage indicating that one of the necessary conditions for the EWF testing to be valid has been satisfied.

The graphs showing the variation of maximum stress versus the ligament length for the tested specimens are represented in Fig. 6a and b. The specimens for which $\sigma_{max} < 0.9\sigma_m$ or $\sigma_{max} > 1.1\sigma_m$ were rejected from the list of specimens possessing valid experimental results.

Determination of EWF parameters

Figure 7 shows the specific work of fracture (w_f) as a function of the ligament length of the test samples for the MDPE and MDPE–CaCO₃ nanocomposites. The results of quantitative measurements of EWF parameters for all the specimens are summarized in Table 1. From the values listed in this table, it can be clearly seen that the values of specific essential work of fracture (w_e) for composite samples with different amounts of CaCO₃ nanoparticles are lower than that of the pure MDPE specimen, indicating that the essential compo-

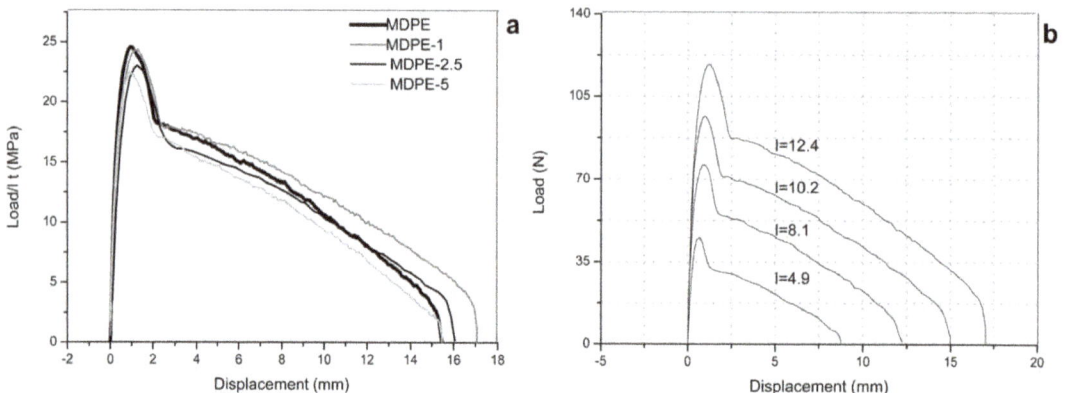

Figure 5 *a* specific load–displacement curves for all materials in this investigation with ligament length of 12 mm and *b* load–displacement curves of composite samples with 1 wt-%CaCO₃ having different ligament lengths (ligament is in mm)

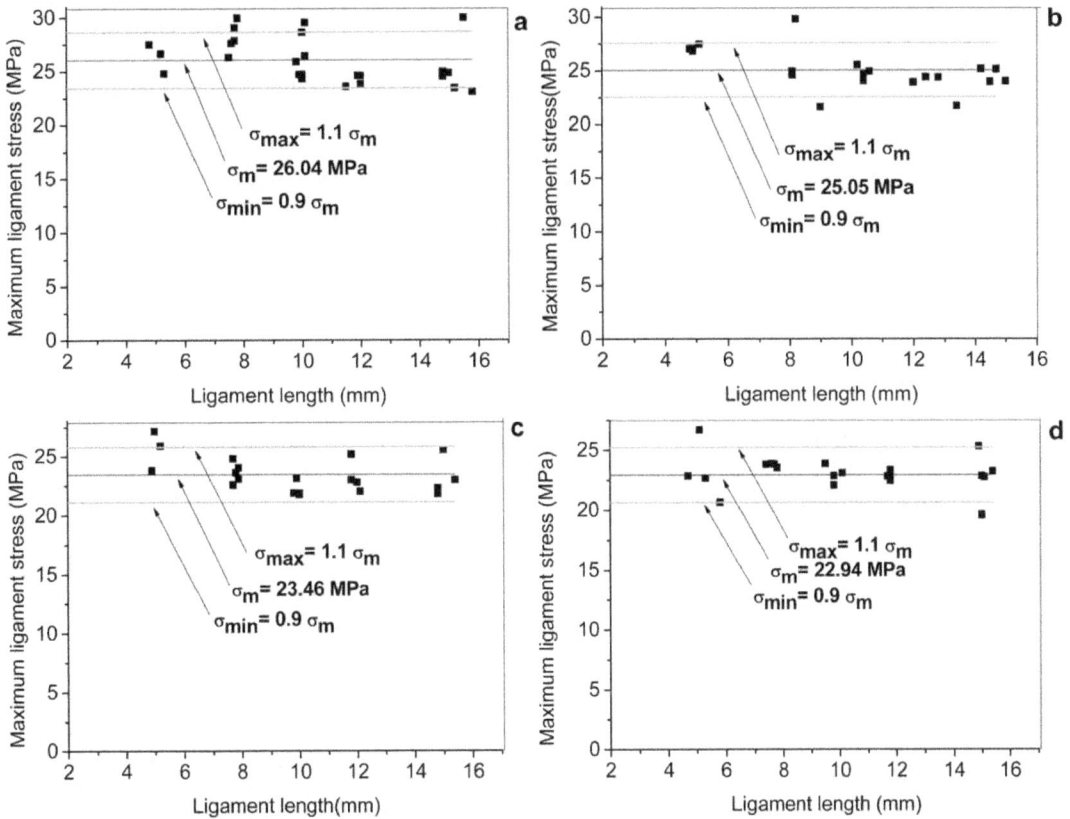

Figure 6 Stress criterion for quantitative results of EWF approach. *a* MDPE material; *b* MDPE–1 wt-%CaCO₃; *c* MDPE–2.5 wt-%CaCO₃; *d* MDPE–5 wt-%CaCO₃

nent of the total energy needed for the occurrence of fracture phenomenon decreases as the $CaCO_3$ nanoparticles are added to MDPE. Decrease in the values of essential fracture work when nanoparticles are present in the MDPE material, and as their relative quantity increases in composite specimens, clearly indicates that $CaCO_3$ nanoparticles have a detrimental effect on this component of the total energy of fracture. To accurately describe this influence of $CaCO_3$ nanoparticles, one should think of the predominant mechanism by which the essential component of the whole fracture is probably controlled. As pointed out previously, nanoparticles with a minimal load transfer efficiency during straining exhibit a limited strengthening effect and instead, introduce a considerable number of nanosize stress concentration sites within the MDPE matrix of composite samples, which finally leads to the formation of voids in the matrix. Consequently, the dissipated work for the creation of two new surfaces reduces. Accordingly, one may consider the void nucleation to be the controlling mechanism which determines the amount of essential work of fracture.

The essential work w_e considerably decreases for MDPE-5 compared to that of neat MDPE. This effect can be related to the poor dispersion of $CaCO_3$ nanoparticles in MDPE matrix. When agglomerates are present in the matrix, voids created by decohesion are not stable and grow rapidly to larger sizes.[4] This may lead to easier crack growth in the matrix

with the final result of lowering the essential component of the total fracture work. Figure 8 shows the images taken from the ligament region in MDPE-5 wt-%CaCO₃ composite during different stages of loading. As can be observed in this series of macrographs, the plastic deformation zones clearly distinguished as white regions have been developed at both crack tips along the ligament length (Fig. 8a and b). At the later stages of loading when the load reaches its maximum value, the plastic zones approach one another such that they overlap to form a single zone extended over the entire ligament length (Fig. 8c). The cracks propagate afterwards in a stable manner and accordingly, the extent of crack opening increases gradually (Fig. 8d–f). Some macrovoids are observable in Fig 8f mostly in the precrack tip regions experiencing highly localized stresses. These macrovoids may be caused by coalescence of microvoids formed by agglomerates.

As can be seen in Table 1, the βw_p parameter has increased from 10.8 to 13 MJ m^{-3} by the addition of 1 wt-%CaCO₃ nanoparticles into the MDPE matrix. Although βw_p decreases when 2.5 wt-%CaCO₃ is added (13–11.48 MJ m^{-3}), the βw_p parameter for this specimen is still greater than that of the MDPE (10.8 MJ m^{-3}). More importantly, the βw_p value has decreased considerably when relatively a high amount of CaCO₃ (5 wt-%) is present in the material, i.e. a big drop from 13 to 4.21 MJ m^{-3}. This experimental evidence is of crucial importance since it has

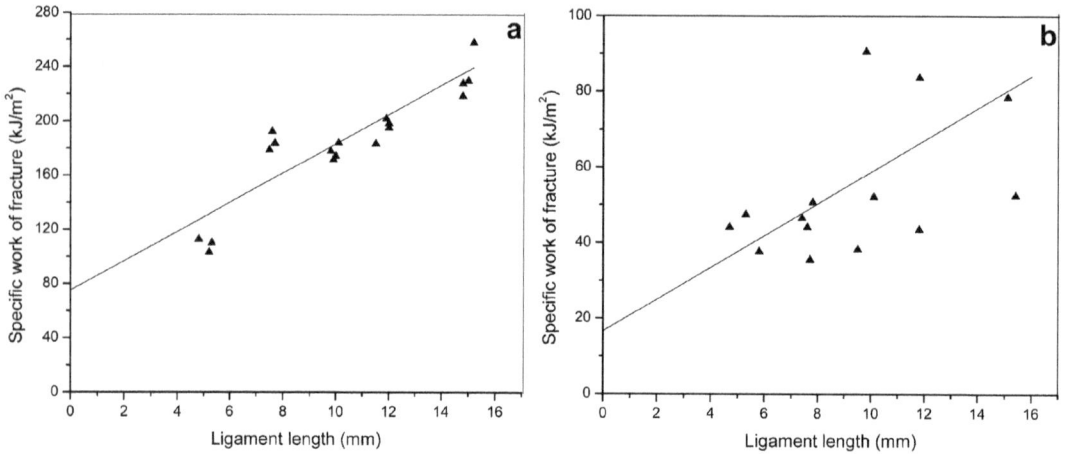

Figure 7   Variation of specific work of fracture w_f as function of ligament length for different samples. *a* MDPE; *b* composite samples with 5 wt-%CaCO$_3$ nanoparticles

been shown that the micron sized CaCO$_3$ particles to decrease the βw_p parameter in PP–CaCO$_3$ composite materials.[14]

Development of plastic zone

The load–displacement curve for the composite sample with 2.5 wt-%CaCO$_3$ is shown in Fig. 9, together with different images taken from the ligament region of the test specimen representing the development of the plastic deformation zones. As can be observed in this figure, in the early stages of loading corresponding to the first region of the load–displacement curve, the plastic deformation zones appear to be initiated from both the notch tips. These localized regions have been extended towards the middle of the ligament length at the later stages of loading (Fig. 9a). The plastic deformation in DDENT specimens can be visually examined through the stress whitening phenomenon, which occurs as a result of the light scattering. The light scattering phenomenon takes place in such composite samples as a result of the occurrence of various deformation processes such as matrix crazing, matrix shear yielding and filler/matrix debonding.[7]

At maximum load (point b), two plastic zones have met each other at the mid-way region along the ligament length. Beyond this point, the plastic zones on both sides have overlapped and accordingly, the ligament region has necked down giving rise to a significant load drop in the load–displacement curve (region c). After point d, the load decreases more slowly representing that the newly generated crack tips propagate from both sides of the specimen towards the centerline in a stable manner, perpendicular to the tensile loading direction (Fig. 9e and f). The final fracture

Table 1   EWF parameters obtained experimentally in this research

Sample	w_e/kJ m^{-2}	βw_p/MJ m^{-3}	β	w_p/MJ m^{-3}	R^2
MDPE	75.3	10.8	0.092	117.4	0.75
MDPE-1	50.9	13	0.075	173.3	0.85
MDPE-2.5	43	11.5	0.066	173.9	0.73
MDPE-5	16.6	4.2	0.02	210.5	0.48

of the specimen has happened when the crack tips from both sides have met at the centerline region of the ligament (point g). This experimental evidence was observed for all the remaining specimens investigated in this research.

The development of macroscopic plastic deformation zones in the nanocomposite samples was different from that of the MDPE. A double plastic zone was observed in the nanocomposite samples (Figs. 4 and 9). The double plastic zone is schematically illustrated in Fig. 10. The whole plastic zone in this figure consists of an intense outer plastic zone (IOPZ) near the fracture process zone and a diffuse outer plastic zone (DOPZ) slightly away from it, depending upon the intensity of stress whitening in the sample. The contrast in intensity observed in the plastic zone indicates that different stages of deformation have taken place during fracture experiment.[29] The concentration of localized stress near the fracture process zone is higher than that in the region away from this region. Beyond the fracture process zone, DOPZ was observed where the stress concentration is comparatively lower. This is probably the reason for the formation of double plastic zone in these nanocomposite samples. As can be clearly seen in Fig. 4, the size of DOPZ depends on the CaCO$_3$ content in the composite samples. The higher CaCO$_3$ content has resulted in a larger DOPZ since the number of stress concentration sites increases considerably within the matrix with increasing filler content. At approximately the end of the linear elastic region, the DOPZ has been formed without having consumed a noticeable amount of plastic work. On the other hand, the formation of the IOPZ has taken place during the whole fracture experiment during which it has expanded into the previously formed DOPZ.[30] The formation of double plastic zone has been previously reported for the case of ethylene–propylene block copolymer,[29,30] polypropylene–EPDM rubber blends[31] and PP/CaCO$_3$/PP-g-MAH composites.[15] The occurrence of this phenomenon has been attributed to the elongation and coalescence of microvoids, which increase the intensity of stress whitening in the IOPZ.[30]

In order to calculate the specific non-essential work of fracture w_p, the plastic zone shape factor β was first

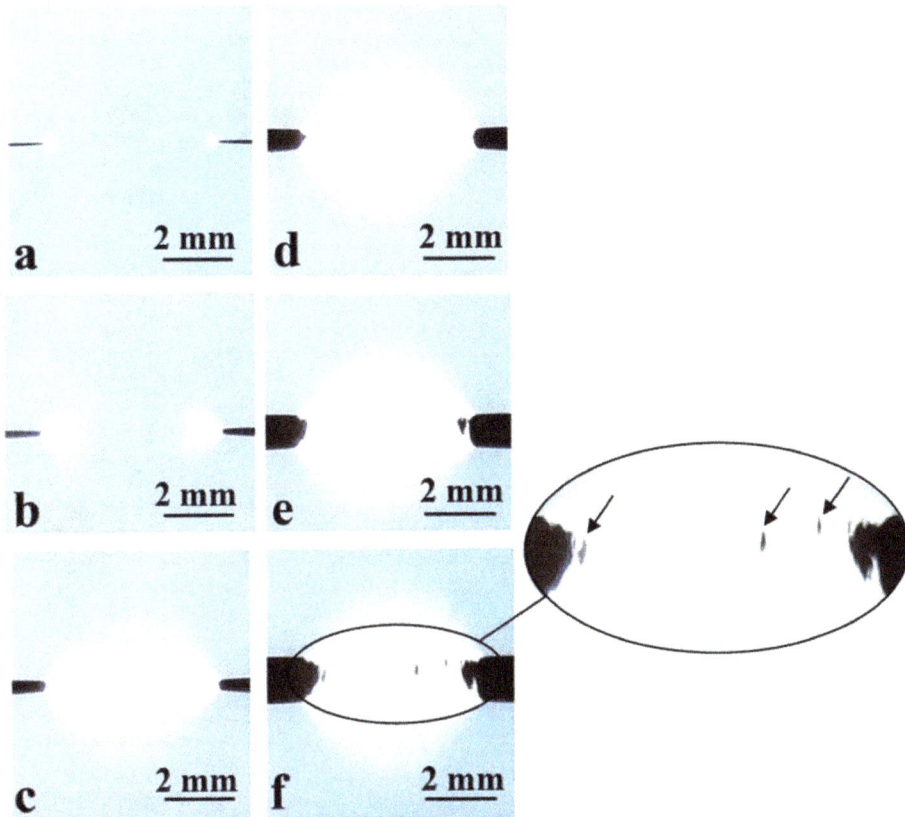

Figure 8 Progressive development of fracture process for MDPE-5 (ligament length of the specimen is 5 mm)

determined. Since the plastic deformation zone in all specimens was found to be parabolic in shape, the relation between β, the ligament length l and the height of the plastic zone h, i.e. $\beta = 2/3(h/l)$ was used.[29] The variations of h as a function of the ligament length for the MDPE sample as well as the composite specimens are shown in Fig. 11. The values of β parameter for these specimens measured from the curves in Fig. 11 are given in Table 1. It should be noted that in the presence of a double plastic zone in test specimens, the height of IOPZ (Fig. 10) was considered for the calculation of β. The main reason for this assumption is that the energy dissipated during the formation of DOPZ (Fig. 10) was lower than that consumed for the formation of IOPZ.[29] As it can be observed in Table 1, the β parameter decreases when $CaCO_3$ nanoparticles are added to the MDPE matrix, as well as when the $CaCO_3$ content in the composite samples increases from 1 to 2.5 wt-%. However, a significant decrease in the β value has been resulted for the case of MDPE–5 wt-%$CaCO_3$ nanocomposite. As mentioned earlier, the presence of agglomerated particles in this composite sample produces relatively non-stable voids that lead to a faster crack propagation stage. This may diminish the area of IOPZ with the final result of a sharp reduction of β value.

The experimental results in Table 1 show that the specific plastic work of fracture (w_p) increases with the addition of 1–5 wt-%$CaCO_3$ nanoparticles. The increase of w_p, irrespective of the plastic zone size (i.e. β parameter), with the

presence of $CaCO_3$ nanoparticles within the MDPE matrix, indicates that a considerable amount of plastic work has been done in the material by the addition of these particles. Additional energy absorption mechanism(s) has (have) to be invoked in order to explain the higher values of w_p for composite samples when compared to the MDPE. Kim *et al.* have investigated the deformation mechanisms that occurs in particle-filled semi-crystalline polymers.[31] The results of their investigation have showed that voids are nucleated around the particles in such composite systems as a result of the decohesion of their interface with the matrix, leading to a change in the stress state at the regions within the matrix adjacent to the particles with the final result of the matrix plastic deformation in the so called interparticle ligament regions. Accordingly, one may speculate that the enhancement of w_p in these composite samples is caused by the plastic deformation of localized ligaments between the particles within the matrix following the particles–matrix decohesion.

The addition of 1 wt-%$CaCO_3$ nanoparticles into the MDPE has produced a composite material with higher non-essential work of fracture (βw_p) than the matrix alone. To understand the reason for increasing the βw_p parameter for this composite, one should look at the change of individual parameters β and w_p with the addition of 1 wt-%$CaCO_3$ into the MDPE matrix. According to the results in Table 1, although the presence of this amount of nanoparticles has

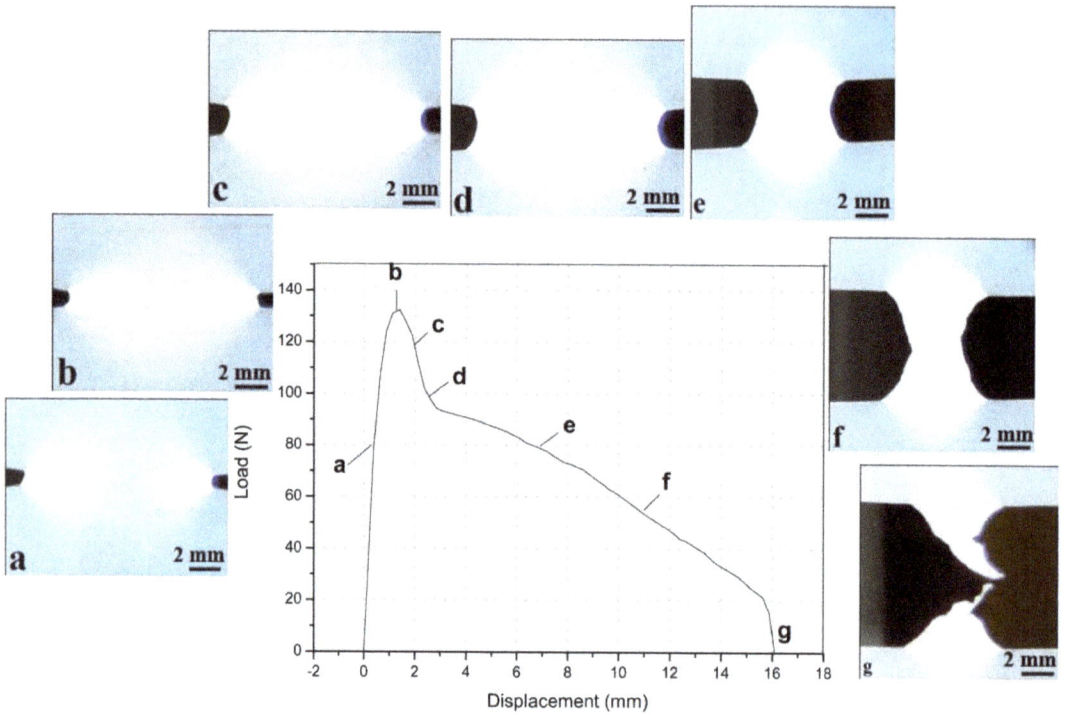

Figure 9 Load–displacement curve for MDPE-2.5 together with photographs of fracture process taken during loading of DDENT specimen with ligament length of 11.8 mm

an undesirable effect on the shape factor β (indicated as a decrease from 0.092 to 0.075) corresponding to the plastic zone size in the composite sample, the plastic fracture work (w_p) has increased considerably so that a greater value of βw_p has been obtained for this composite sample (13 MJ m^{-3} compared to 10.8 MJ m^{-3} for the MDPE). However, the addition of greater amounts of nanoparticles has continuously decreased the βw_p parameter in such a way that it has finally reached its minimum value of 4.21 MJ m^{-3} for the case of 5 wt-%CaCO$_3$ composite.

Conclusion

The EWF approach was successfully used to characterize the fracture behavior of the MDPE as well as that of the composite samples when 1, 2.5 and 5 wt-%CaCO$_3$ nanoparticles are added to MDPE. It was shown the CaCO$_3$ nanoparticles added into the MDPE matrix in all quantities to decrease the specific essential work of fracture w_e. This detrimental effect of CaCO$_3$ nanoparticles was attributed to the introduction of a certain number of stress concentration sites within the matrix and the fact that the situation

becomes much more severe when greater amounts of nanoparticles are added to the composite giving rise also to increase the chance of particles agglomeration during the manufacture of test specimens. Also the results showed that βw_p values increased with increasing CaCO$_3$ content up to 2.5 wt-%, and then decreased for the nanocomposite containing 5 wt-%CaCO$_3$.

The specific plastic work per unit volume w_p increased when the CaCO$_3$ nanoparticles were added to the MDPE matrix, while β decreased with increasing amount of CaCO$_3$, which revealed that the specific energy absorption per unit volume in the plastic zone could be markedly improved at the

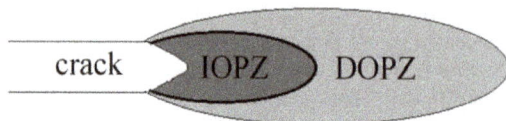

Figure 10 Schematic illustration of double plastic zone (DPZ) consisting of intense outer plastic zone (IOPZ) and diffuse outer plastic zone (DOPZ)

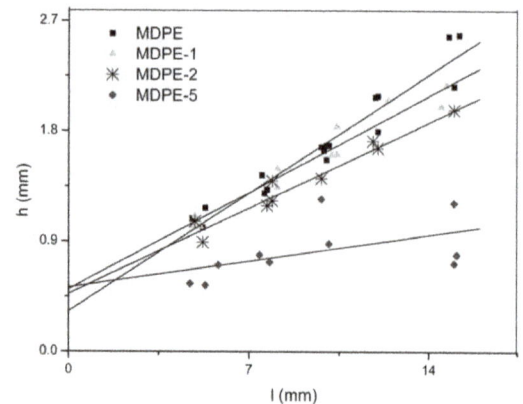

Figure 11 Linear relation of height of plastic zone h and ligament length l for determination of β

cost of restricted development of plastic deformation zones. The reduction of plastic zone size for MDPE-5 was more significant, which resulted in the sharp reduction of βw_p.

Conflicts of interest

The authors have no conflicts of interest to declare.

References

1. Z. Bartczak, A. S. Argon, R. E. Cohen and M. Weinberg: *Polymer*, 1999, **40**, 2347.
2. Y. S. Thio, A. S. Argon, R. E. Cohen and M. Weinberg: *Polymer*, 2002, **43**, 3661.
3. W. C. J. Zuiderduin, C. Westzaan, J. Huétink and R. J. Gaymans: *Polymer*, 2003, **44**, 261.
4. C. M. Chan, J. Wu, J. X. Li and Y. K. Cheung: *Polymer*, 2002, **43**, 2981.
5. R. N. Rothon: 'Particulate-filled polymer composites'; 2003, Shrewsbury, Rapra Technology Limited.
6. R. Gedron and D. Binet: *J. Vinyl Addit. Technol.*, 1998, **4**, 54.
7. A. Lazzeri, S. M. Zebarjad, M. Pracella, K. Cavalier and R. Rosa: *Polymer*, 2005, **46**, 827.
8. G. Levita, A. Marchetti and A. Lazzeri: *Polym. Compos.*, 1989, **10**, 39.
9. 'Plastics. Determination of fracture toughness (Glc and Klc) – linear elastic fracture mechanics (LEFM) Approach', ISO 13586, ISO, Geneva, Switzerland, 2000.
10. T. Bárány, T. Czigány and J. Karger-Kocsis: *Prog. Polym. Sci.*, 2010, **35**, 1257.
11. A. Pegoretti and T. Ricco: *Eng. Fract. Mech.*, 2006, **73**, 2486.
12. O. Okada, H. Keskkula and D. R. Paul: *Polymer*, 2000, **41**, 8061.
13. S. C. Tjong, S. A. Xu, R. K. Y. Li and Y. W. Mai: *Compos. Sci. Technol.*, 2002, **62**, 2017.
14. G. Gong, B. H. Xie, W. Yang, Z. M. Li, W. Q. Zhang and M. B. Yang: *Polym. Test.*, 2005, **24**, 410.
15. G. Gong, B. H. Xie, W. Yang, Z. M. Li, S. M. Lai and M. B. Yang: *Polym. Test.*, 2006, **25**, 98.
16. B. K. Satapathy, R. Weidisch, P. Potschke and A. Janke: *Compos. Sci. Technol.*, 2007, **67**, 867.
17. J. L. Yang, Z. Zhang and H. Zhang: *Compos. Sci. Technol.*, 2005, **65**, 2374.
18. S. C. Tjong and S. P. Bao: *Compos. Sci. Technol.*, 2007, **67**, 314.
19. K. B. Broberg: *J. Mech. Phys. Solids*, 1975, **23**, 215.
20. B. Cotterell and J. K. Reddel: *Int. J. Fract.*, 1977, **13**, 267.
21. Y. W. Mai and B. Cotterell: *Int. J. Fract.*, 1986, **32**, 105.
22. J. Karger-Kocsis, V. M. Khumalo, T. Bárány, L. Mészáros and A. Pegoretti: *Compos. Interfaces*, 2013, **20**, 395.
23. D. Pedrazzoli, R. Ceccato, J. Karger-Kocsis and A. Pegoretti: *Express Polym. Lett.*, 2013, **7**, 652.
24. F. Tuba, V. M. Khumalo and J. Karger-Kocsis: *J. Appl. Polym. Sci.*, 2013, **129**, 2950.
25. T. Turcsan, L. Meszaros, V. M. Khumalo, R. Thomann and J. Karger-Kocsis: *J. Appl. Polym. Sci.*, 2014, **131**, 40447.
26. D. Pedrazzoli, F. Tuba, V. M. Khumalo, A. Pegoretti and J. Karger-Kocsis: *J. Reinf. Plast. Compos.*, 2014, **33**, 252.
27. E. Clutton: 'Essential work of fracture', in 'Fracture machanics testing methods for polymers'. (ed. D. R. Moor *et al.*), Vol. 28, 177; 2001, Amsterdam, ESIS Publication.
28. D. Ferrer-Balas, M. L. Maspoch, A. B. Martinez, E. Ching, R. K. Y. Li and Y. W. Mai: *Polymer*, 2001, **42**, 2665.
29. D. Ferrer-Balas, M. L. Maspoch and Y. W. Mai: *Polymer*, 2002, **43**, 3083.
30. A. van der Wall and R. J. Gaymans: *Polymer*, 1999, **40**, 6067.
31. G. M. Kim and G. H. Michler: *Polymer*, 1998, **39**, 5699.

Bioactive nanocomposites for dental application obtained by reactive suspension method

Oussama Boumezgane[1,2], Federica Bondioli[3] ⓘD, Sergio Bortolini[4] ⓘD, Alfredo Natali[4], Aldo R. Boccaccini[5], Elena Boccardi[5] and Massimo Messori[1,2]* ⓘD

[1]Dipartimento di Ingegneria "Enzo Ferrari", Università di Modena e Reggio Emilia, via P. Vivarelli 10/1, Modena, Italy
[2]Consorzio INSTM, Unità di ricerca di Modena e Reggio Emilia, via Giusti 9, Firenze, Italy
[3]Dipartimento di Ingegneria Industriale, Università di Parma, Parco Area delle Scienze 181/A, Parma, Italy
[4]Dipartimento Chirurgico, Medico, Odontoiatrico e di Scienze Morfologiche con Interesse Trapiantologico, Oncologico e di Medicina Rigenerativa, Università di Modena e Reggio Emilia, via del Pozzo 71, Modena, Italy
[5]Department of Materials Science and Engineering, University of Erlangen-Nuremberg, Cauerstraße 6, Erlangen, Germany

Abstract Hydroxyapatite (HA) filled poly(methyl methacrylate)/ poly(hydroxyethyl methacrylate) (PMMA/PHEMA) blends were prepared by reactive suspension method: HA was synthesized by co-precipitation process directly within a HEMA solution and the so-obtained suspension was polymerized in the presence of PMMA. HA particles were obtained in form of nanorods with a length

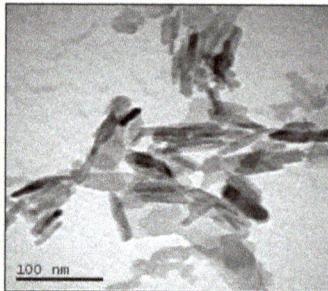

100 nm

Reactive suspension method: synthesis of hydroxyapatite within hydroxyethyl methacrylate as suspending medium and polymerizable monomer.

Hydroxyapatite nanoparticles of acicular shape (length 50-200 nm and diameter 10-30 nm, average aspect ratio of 6).

of 50–200 nm and a diameter of 10–30 nm. A significant increase in glass transition temperature was observed in the nanocomposites with respect to the unfilled polymer blends. Dynamic-mechanical thermal analysis showed a significant increase in the storage modulus in the nanocomposites measured in the rubbery region. This increase was unpredicted by Mooney's predictive equation and was attributed to the presence of cross-linking points due to the *in situ* generated HA particles. An increase in the elastic modulus was also observed at room temperature in compression and three-point bending tests. The presence of HA in the polymer blends resulted in an important decrease in the water sorption values. The bioactivity of the nanocomposites was verified by the precipitation of HA layer on the surface after soaking in simulated body fluid.

Keywords Nanocomposites, Nanoparticles, Polymer-matrix composites, Bioactivity, Hydroxyapatite

Introduction

Oral health has become increasingly important during the last decades. In this respect, dental restorative materials have received particular attention by numerous researchers, especially in the area of extensively used acrylic-based – resins. A large number of investigations have been performed to improve the properties of composite resins for dental applications, such as abrasion resistance, mechanical, and anti-bacterial properties.[1,2]

Acrylic materials, consist of a solid component (poly (methyl methacrylate) powder, PMMA) and a liquid component (methyl methacrylate monomer), are one of the most frequently and extensively used dental restorative materials, due to their satisfactory aesthetic properties, easy operation, low cost, and good stability in the oral environment.[2] The most common use of this material includes the fabrication of denture base, temporary crowns, and temporal seal of cavities.[3]

One method of improving the mechanical properties of an acrylic-based denture base is by adding preformed nanoparticles (*ex situ* approach) into the liquid phase before mixing with

*Corresponding author, email massimo.messori@unimore.it

the PMMA powders. The most common nanoparticle fillers used are silica, zirconia, titania, alumina, and hydroxyapatite.[4–6] The use of nanoparticles can improve the resin properties such as wear resistance, modulus, and flexural strength even at very low mass fraction values. However, these properties are greatly influenced by the quality of the nanofiller dispersion in the polymeric matrix. Because of the very high surface area and surface charge of inorganic nanoparticles, agglomerates are generally formed, and thus the homogeneous dispersion of nanofillers in the organic matrix becomes difficult at high filler concentration. In fact, particles, with a size less than 100 nm, tend to agglomerate into larger clusters to minimize their surface energy. This agglomeration affects materials performance by the inclusion of voids that act as preferential sites for crack initiation and failure.[7]

A common method to limit agglomeration phenomena is the chemical modification of particles using suitable coupling agents, which decrease the surface energy of the particles and improve the compatibility with the organic matrix. The main drawback of this approach is the time-consuming steps of chemical modification and purification of particles. An alternative approach is represented by the synthesis of the nanoparticles directly within the organic matrix (*in situ* approach).

In this respect, the aim of this study is to realize an innovative nanocomposite material using PMMA in form of powder and hydroxyethyl methacrylate (HEMA) as liquid phase, reinforced with *in situ* generated hydroxyapatite ($Ca_{10}(PO_4)_6(OH)_2$, HA) nanoparticles. HA is an osteogenic and osteoconductive inorganic phase,[8] similar to the bone mineral, and confers its bioactivity to polymer-based composites promoting bone regeneration.[9–15]

In the present work, the so-called 'reactive suspension method'[16] is used. The *in situ* synthesis in a HEMA solution of HA by a co-precipitation process, starting from calcium nitrate tetrahydrate and ammonium dihydrogen phosphate as precursors, is proposed as a promising strategy for the achievement of homogeneous hybrid materials having higher degree of phase interaction between the polymer matrix and the inorganic filler. This procedure should avoid extensive particles agglomeration typically seen in polymer/HA composites obtained by mechanical incorporation of preformed HA powders into the polymer melt or solution, causing non-homogeneous materials, and it was already proved to be an effective approach for the preparation of homogeneous PCL/silicate glasses composites,[17,18] poly(propylene fumarate)/HA composites,[19] PCL/HA composites,[16] and polyacrylic acid/calcium phosphate ceramics composites.[20]

In this study, the structure and morphology of the obtained PMMA/HEMA/HA composites are investigated. The mechanical properties are determined in flexural and compression mode, while the bioactivity of the composites is studied by immersion in simulated body fluid (SBF).

Materials and methods

Materials

Calcium nitrate tetrahydrate ($Ca(NO_3)_2\cdot4H_2O$), HEMA, 3-(trimethoxysilyl)propyl methacrylate (MSDS), and dibenzoyl peroxide (BPO) were purchased from Sigma–Aldrich (Milan, Italy). Ammonium phosphate ((NH_4)$_2HPO_4$), ammonium hydroxide (NH_4OH), and ethanol were purchased from Carlo Erba (Italy). PMMA powder was obtained from Lang Dental Manufacturing Co. under the trade name of Jet Kit. Reagent-grade NaCl, NaHCO$_3$, KCl, $K_2HPO_4\cdot3H_2O$, $MgCl_2\cdot6H_2O$, Na_2SO_4, 1 M HCl, and Tris buffer ($(CH_2OH)_3CNH_2$), used for SBF preparation, were purchased from Sigma–Aldrich.

Synthesis of hydroxyapatite, $Ca_{10}(PO_4)_6(OH)_2$

The reagents used to synthesize HA were $Ca(NO_3)_2H_2O$ and (NH_4)$_2HPO_4$[21] (Table 1). $Ca(NO_3)_2H_2O$ was dissolved in HEMA to obtain a 0.259 M solution and (NH_4)$_2HPO_4$ was dissolved in distilled water to obtain a 0.156 M aqueous solution. 50 mL of (NH_4)$_2HPO_4$ aqueous solution was added drop wise into 50 mL of $Ca(NO_3)_2$/HEMA solution under vigorous stirring at 65 °C over a period of approximately 1 h. Vigorous stirring was maintained for 1 h and the pH was periodically monitored and leveled at value 9 by addition of NH_4OH.

0.025 mmol of silane coupling agent MSDS[22] per m² of HA surface (see the evaluation of specific surface area of synthesized particles in Table 1) were then added to stabilize the suspension, avoiding the precipitation of HA nanoparticles, and increasing the compatibility between the organic phase (HEMA) and the inorganic nanoparticles.

To characterize the HA powder, the suspension was filtrated and the powder was washed with ethanol by centrifugation to solubilize and remove the organic solvent. After the drying step, the powder was accurately weighed in order to verify the reaction yield that is 100% (Table 1).

Nanocomposites preparation

PMMA powder was added to the HEMA/HA suspension in two different weight ratios, 1:2 and 1:3, respectively, under vigorous stirring, after the addition of 1 phr of BPO as radical initiator. The obtained nanocomposites formulations are listed in Table 2. Unfilled HEMA/PMMA samples were also prepared in the absence of HA, as reference materials.

The mixtures were then polymerized by thermal curing at 60 °C for 1 h with a post-curing at 100 °C for 1 h.

Powder and composites characterization

The structural characterization of the dried powders was performed by X-rays diffraction analysis (XRD) using an X'Pert PRO diffractometer (PANalytical), operating in the 10–90 2theta range with step size 0.01 ° and step time of 1 s.

Table 1 Composition of the synthesized powder and specific surface area (SSA) value (HA content expressed in parts per hundred of resin, phr)

Code	$Ca(NO_3)_2H_2O$ (g)	(NH_4)$_2HPO_4$ (g)	Theoretical HA (g)	Obtained HA (g)	HEMA (g)	HA content in HEMA (phr)	SSA (m²/g)
HA	12.22	4.12	5.20	5.47	50.00	10.94	117.02

Table 2 Composition of the prepared samples

Code	PMMA/HEMA weight ratio	HA in PMMA-HEMA matrix (phr)
1:2 HA	0.5	7
1:3 HA	0.33	8

The particle morphology was examined by transmission electron microscopy (TEM, JEM 2010, Jeol, Japan). Dried powders were dispersed in *n*-butanol and a drop of the so-obtained suspension was placed on a copper grid (200 mesh) covered with PELCO® support films of Formvar (thickness of 30–60 nm), followed by drying.

To determine the amount of the silane coupling agent added to the HA suspension, the specific surface area of the powder was measured by the Brunauer, Emmett, and Teller method (Gemini 2360 apparatus, Micromeritics, Norcross, GA, USA) after degassing under vacuum at 150 °C.

In order to investigate the presence of residual organic groups on the particle surface, FT-IR analysis was performed on the obtained powder. The analysis was carried out in the attenuated total reflectance mode (ATR) with an Avatar 330 spectrometer (Thermo Nicolet, Germany). A minimum of 64 scans with a resolution of 1 cm^{-1} was performed.

Finally, simultaneous thermogravimetry and differential thermal analysis (TG-DTA) was performed, on the modified HA powder, with a Netzsch STA 429 CD with a heating rate of 20 °C·min^{-1} up to 1000 °C in air atmosphere.

The nanocomposites were characterized by differential scanning calorimetry (DSC) to measure the glass transition temperature (T_g) of composites and, analyzing the effect of the HA nanofiller on the glass transition temperature of composites. The test was carried out using a scanning rate of 25 °C min^{-1} from 0 to 200 °C and the T_g value was assumed as the mean value of the energy jump of the thermograms (average value between the onset and the endpoint of the glass-transition range).

Dynamic-mechanical thermal analysis (DMTA) was carried out on DMA Q800 TA instrument in the temperature range between −30 and +150 °C with a heating rate of 3 °C min^{-1} using a single cantilever clamp. The storage modulus E′ and tanδ were measured and the T_g value was assumed as the maximum of the tanδ curve.

Three-point bending test and compression test were performed using an Instron series 5500 dynamometer for flexural and compression tests on the sample with the highest content of HA (1:3 HA) and on the corresponding unfilled 1:3 sample.

Three-point bending test was carried out following the ISO 4049 standard procedure[23] for the flexural strength test. Five prismatic samples with $25 \times 2 \times 2$ mm^3 were prepared. The samples were stored in water at 37 °C for 24 h just before the measurement. A crosshead speed of 0.75 mm·min^{-1} was applied until sample fracture.

Compression test was performed following the ISO 604 standard.[24] Five prismatic samples $10 \times 10 \times 4$ mm^3 were prepared for both 1:3 HA sample and unfilled 1:3 sample. The samples were stored in water at 37 °C on the day before the measurement. A crosshead speed of 2 mm·min^{-1} was applied until sample fracture.

For water sorption test, the samples were cut in small pieces with prismatic shape, and then were placed in a desiccator at 37 °C until constant mass (m_1) was reached. The dried samples were stored in water at 37 °C for 7 days and weighted to get the wet mass (m_2) and then placed in a desiccator until a dried constant mass was reached (m_3).

The water sorption value was calculated by the formula:

$$W_{sp} = \frac{(m_2 - m_3)}{V} \tag{1}$$

where W_{sp} is the water sorption value in µg·mm^{-3} and V the sample volume.[23]

In vitro tests were carried out to assess the bioactivity of the nanocomposites. The SBF is an aqueous solution with inorganic ion composition very similar to human blood plasma; it is a protein-free solution with a pH of 7.4 prepared in the laboratory according to the procedure developed by Kokubo and Takadama,[25] and utilized by many other authors.[26–28] Proper quantities of the reagents NaCl, NaHCO$_3$, KCl, K$_2$HPO$_4$·3H$_2$O, MgCl$_2$·6H$_2$O, CaCl$_2$, Na$_2$SO$_4$, tris(hydroxymethyl) aminomethane, were dissolved in deionized water and the solution was buffered at pH 7.4 at 36.5 °C by adding tris and 1 M HCl, with the help of a magnetic stirrer, according to the concentrations given in;[29] then, the pH of the solution was adjusted to 7.4 by addition of 1 M hydrochloric acid. The samples were cut in rectangular shape and immersed into SBF solution with a ratio SBF/material of 255 ml/g[26] maintained at body temperature (37 °C) in a rotating incubator. The samples were tested for four different periods of 1 h, 24 h, 7 days, and 28 days; in the case of 7 and 28 days, the SBF solution was changed every 3–4 days in order to better mimic *in vitro* the *in vivo* behavior of the material.[30] The samples were removed from SBF solution at the end of each treatment period, washed with deionized water and dried at room temperature.

To analyze the ability to form an apatite layer, the surface of the samples after SBF immersion was observed with an electron scanning microscopy (SEM) equipped with an energy-dispersive X-ray spectroscopy (EDS) using electron and ion beam microscope Zeiss Auriga 60. Infrared spectroscopy analysis was also carried out in ATR using an FT-IR machine (Nicolet 6700, Thermo Scientific Germany). The analysis was performed under the following conditions: spectral range between 4000 and 530 cm^{-1}; window material, CsI; 32 scans at a resolution of 4 cm^{-1}. Finally, X-ray diffraction (XRD) using an X'Pert PRO diffractometer (PANalytical), operating in the 10–90 2theta range with step size 0.01 ° and step time of 1 s, was carried out to determine the phase composition of the crystallization on the surface of the samples.

Results and discussion

HA nanoparticles characterization

Figure 1 reports the XRD pattern of the HA powder obtained with Ca/P in 1.67 stoichiometric ratio, such as that of HA present in natural bone tissues. The peaks in the XRD pattern were those characteristics of pure HA and closely matched with the JCPDS 09-432 of stoichiometric calcium hydroxyapatite, indicating the absence of other calcium phosphate phases.[31]

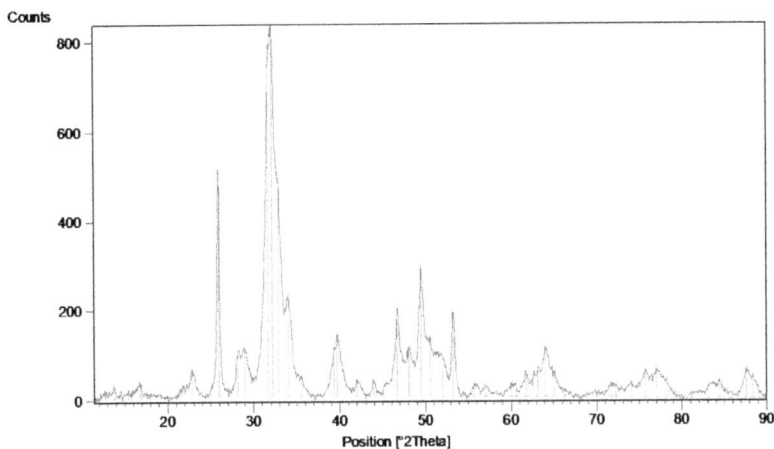

Figure 1   XRD pattern of HA powder (the reported bars indicate the JCPDS 09–432 phase)

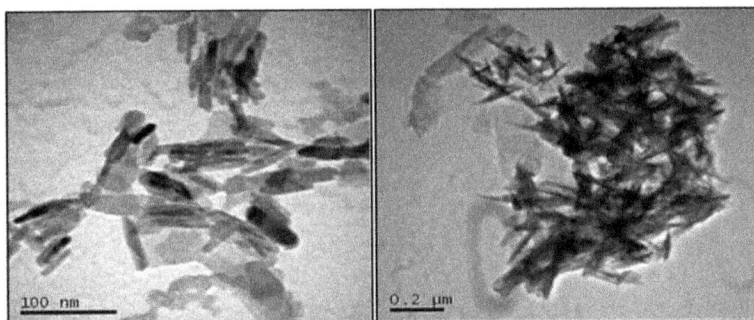

Figure 2   TEM images of HA powder

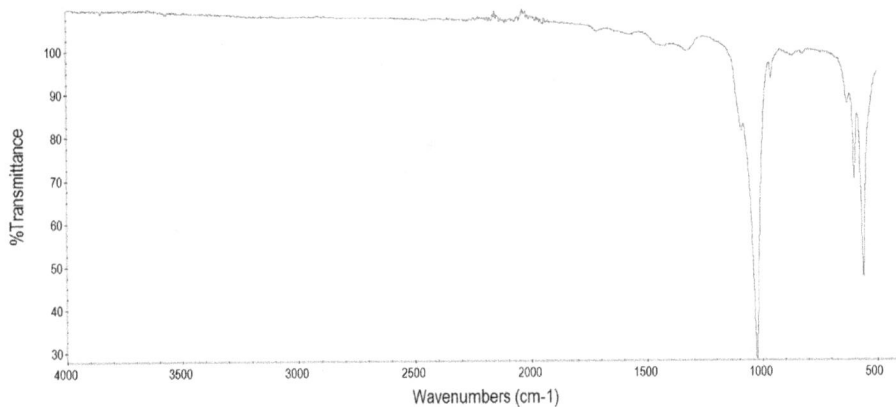

Figure 3   IR spectrum of HA powder

TEM micrographs of the synthesized HA powder are reported in Fig. 2, showing particles of acicular shape with an average length between 50 and 200 nm and an average diameter between 10 and 30 nm, with an average aspect ratio of about 6. The particles tend to form roughly spherical agglomerates of about 100 nm to decrease their high surface energy.

In Table 1, the powder SSA value is reported which is necessary to evaluate the content of silane coupling agent that was added to control the suspension stability. The powder showed a value of SSA of 117 $m^2 g^{-1}$, indicating the high surface/volume ratio of the nanoparticles that could be estimated around 100×10^{-6} m^{-1}. As mentioned in paragraph 2.2, 7.9 g of silane coupling agent were then added to the HEMA/HA suspension.

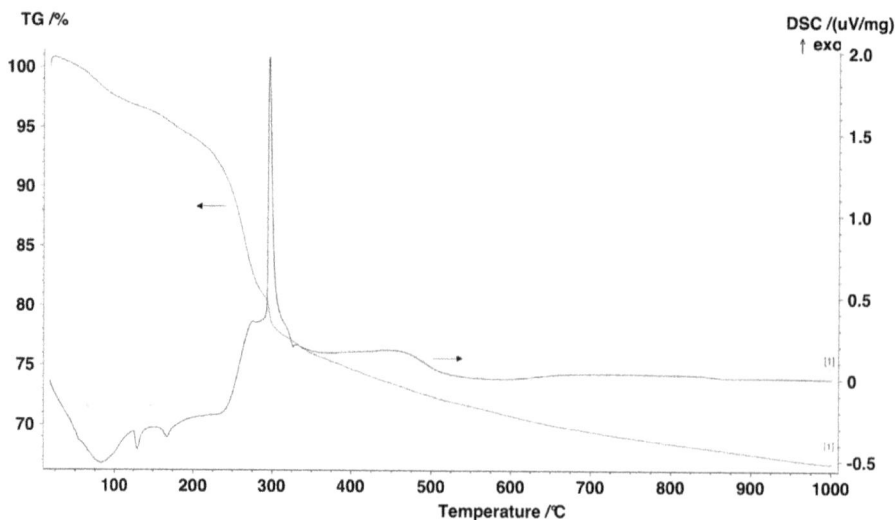

Figure 4   TG/DTA curves of HA powder

Table 3   Thermal and dynamic-mechanical properties (E': conservative modulus, tanδ$_{max}$: maximum value of loss factor)

Sample	T_g (°C) by DSC	E' elastic region 29 °C (GPa)	E' rubbery region 145 °C (MPa)	E' Mooney region 145 °C $s = 1.35$	T_g (°C) by DMTA (tanδ$_{max}$)
Unfilled 1:2	66	2.27	1.4	1.4	98
1:2 HA	76	4.21	9.4	2.1	109
Unfilled 1:3	68	3.91	1.2	1.2	102
1:3 HA	83	3.09	10.0	1.9	103

Figure. 3 shows the IR spectra of the obtained powders. The formation of HA is indicated by the presence of an intense peak centered at around 1000–1100 cm^{-1}, since the phosphate bands v3 fall in the region 1092–1048 cm^{-1} (corresponding to triply degenerated anti-symmetric P–O stretching). Moreover, the v1 at 963 cm^{-1} (corresponding to non-degenerate symmetric P–O stretching), v4 at 603 and 571 cm^{-1} (corresponding to triply degenerated O–P–O bending)[32] are present. Finally, a weak broad band can be observed in the range between 3100 and 3600 cm^{-1} that corresponds to the presence of OH groups.[33] The small peak observed between 1200 and 1350 cm^{-1} could be due to the silane coupling agent as the Si-O-CH$_3$ vibrational state falls at 1207 cm^{-1} (Si-O bending).[34]

TG-DTA analysis (Fig. 4) performed on the same sample confirmed the absence of secondary calcium phosphate phases; the curves revealed the endothermic loss of water molecules at 100 °C and the exothermic crystallization of the amorphous HA phase at 280 °C associated with the elimination of some volatile components present as traces in initial reagents.[22] The exothermic weight loss at 300 °C might be due to combustion of the organic residue probably associated with the silane coupling agent. Finally, the weight loss observed in the range of 400–1000 °C is assumed as the gradual dehydroxylation in the HA powder. These results assure that, during the curing step (1 h at 60 °C with post cure at 100 °C for another hour) of the PMMA/HEMA/HA composites, the HA powder is stable and does not undergo any structural changes.

Nanocomposites characterization

The glass transition temperatures, evaluated by DSC analysis (Table 3), were significantly affected by the presence of HA, as a strong increase was observed in the nanocomposites in comparison with the respective unfilled matrix for both HA contents. This effect is attributed to a stiffening effect due to interactions between the polymer matrix and the filler in the interfacial region.[35]

DMTA results (Figs. 5 and 6) showed that the nanocomposites dynamic-mechanical properties, such as storage modulus, loss modulus and tanδ, were affected by the HA content. This is an expected behavior due to the addition of rigid fillers to polymer matrix that contrasts the movement of the polymer chains leading to a damping decrement and a shift of T_g values to higher temperatures.[36]

The glass transition temperature (T_g) at maximum value of loss factor, tanδ, and storage modulus E' at the elastic region and rubbery region obtained by DMTA are also reported in Table 3. In the low temperature region (below T_g), the storage modulus does not follow a systematic trend with composition showing the highest value in the case of 1:2 HA sample and, surprisingly, a higher value for unfilled 1:3 sample with respect to the corresponding filled one (1:3 HA). On the contrary, a significant increase in storage modulus E' in the rubbery region was observed in both nanocomposites, as they presented a value of E' almost 10 times

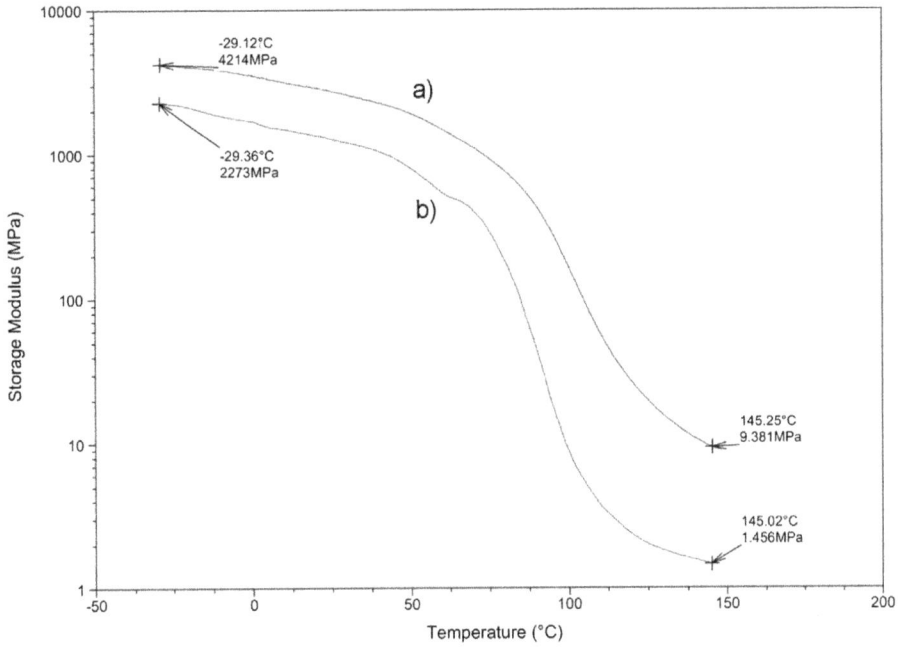

Figure 5   Conservative modulus (E') trace of a 1:2 HA and b unfilled 1:2 (chosen as representative)

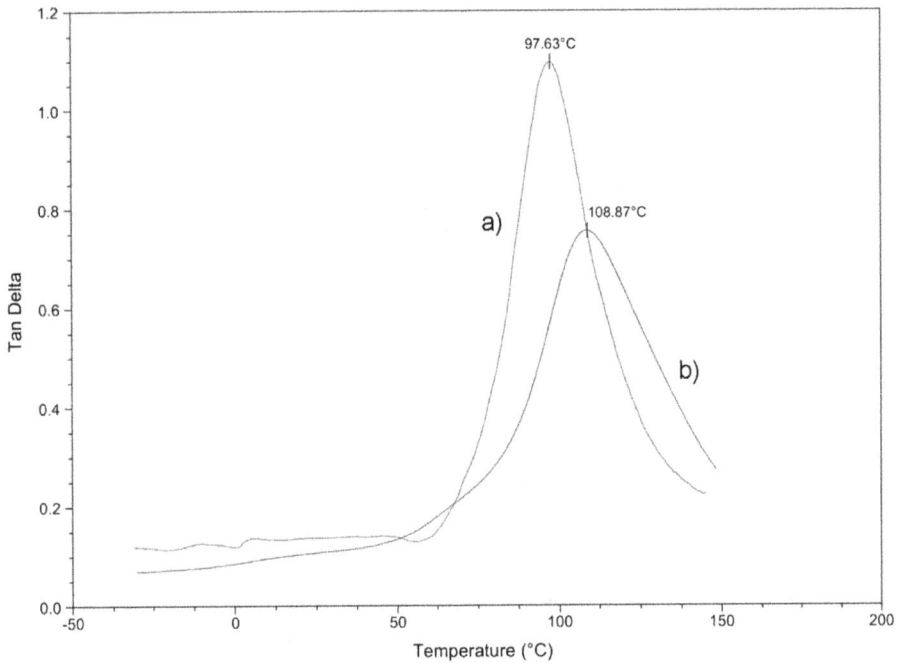

Figure 6   Loss factor (tanδ) trace of a 1:2 HA and b unfilled 1:2 (chosen as representative)

higher than the respective unfilled samples. As expected, lower values of tanδ (loss factor) in the nanocomposites were also observed even at high temperature indicating a lower damping in the nanocomposites due to the interaction between the filler and the polymer matrix.

In order to compare the experimental results with data predicted by the well-known models for composite materials, Mooney's empirical equation for non-spherical particles[37] was applied to study the dependence of the elastic modulus of filled polymers on the nanoparticles content.

Table 4 Flexural and compression modulus as experimentally obtained and calculated by Equation (2)

	E_c experimental (MPa)	E_c Mooney (MPa)	E_f experimental (MPa)	E_f Mooney (MPa)
Unfilled 1:3	226±87	226	347±122	347
1:3 HA	256±56	361	662±197	552

Table 5 Experimental and calculated composite strength (σ)

	σ experimental (MPa)	σ Nielsen (MPa)	σ Pukanszky* (MPa)
Unfilled 1:3	15±5.33	15	15
1:3 HA	8.2±1.1	13.5	15.1

*calculated assuming a good adhesion: $B = 3.5$.

Table 6 Water sorption values

Sample	W_{sp} (µg·mm^{-3})
Unfilled 1:2	297
1:2 HA	290
Unfilled 1:3	369
1:3 HA	291

$$\frac{E_c}{E_m} = \exp\left(\frac{2.5V_p + 0.407(p-1)^{1.508}V_p}{1 - sV_p}\right) \quad (2)$$

where Ec and Em are the Young's modulus of composite and matrix, Vp is the particle volume fraction, p is the aspect ratio of the filer, s is a crowding factor for the ratio of the apparent volume occupied by the particle to its own true volume. The value of s lies between 1.35 and 1.91, where the minimum value of s comes from the least dense packing factor, which is the sphere packaging factor ($s = 3\sqrt{2}/\pi \cong 1.35$), while the

maximum value of s comes from the densest cubic factor ($s = \frac{6}{\pi} \cong 1.91$)

As shown in Table 3, the values of the obtained storage modulus, in the rubbery region, are higher than the data predicted by Mooney's equation and the crowding factor had no significant effect due to the very low filler content.

The high increment of the storage modulus obtained in the nanocomposites has been also observed in previous studies such as;[36] considering that in the rubbery region, the modulus value is principally governed by the cross-linking density, an increase in the latter could be attributed to the presence of

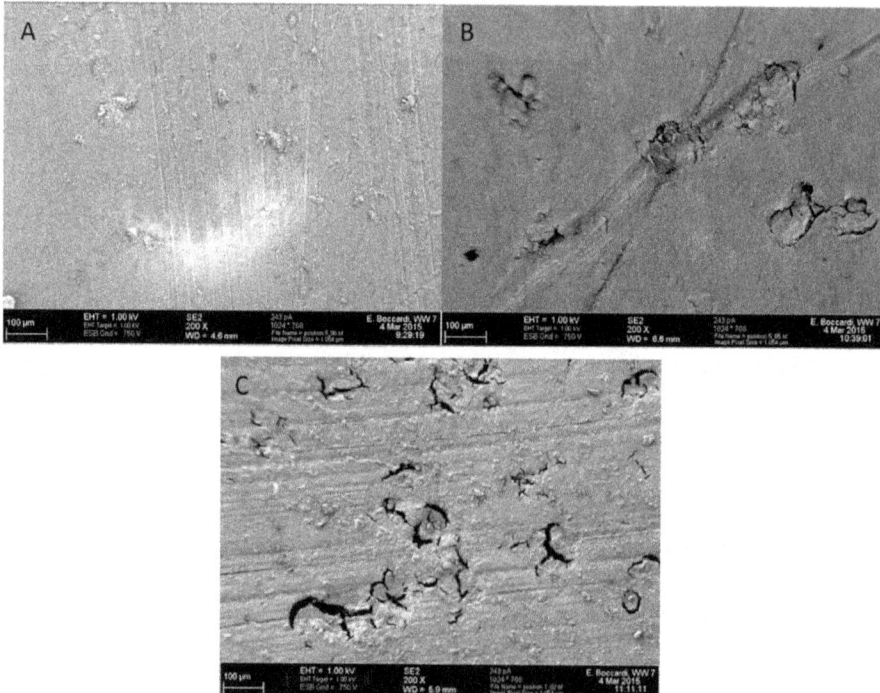

Figure 7 Surface morphology of the samples: (A) unfilled, (B) 1:2 HA, (C) 1:3 HA

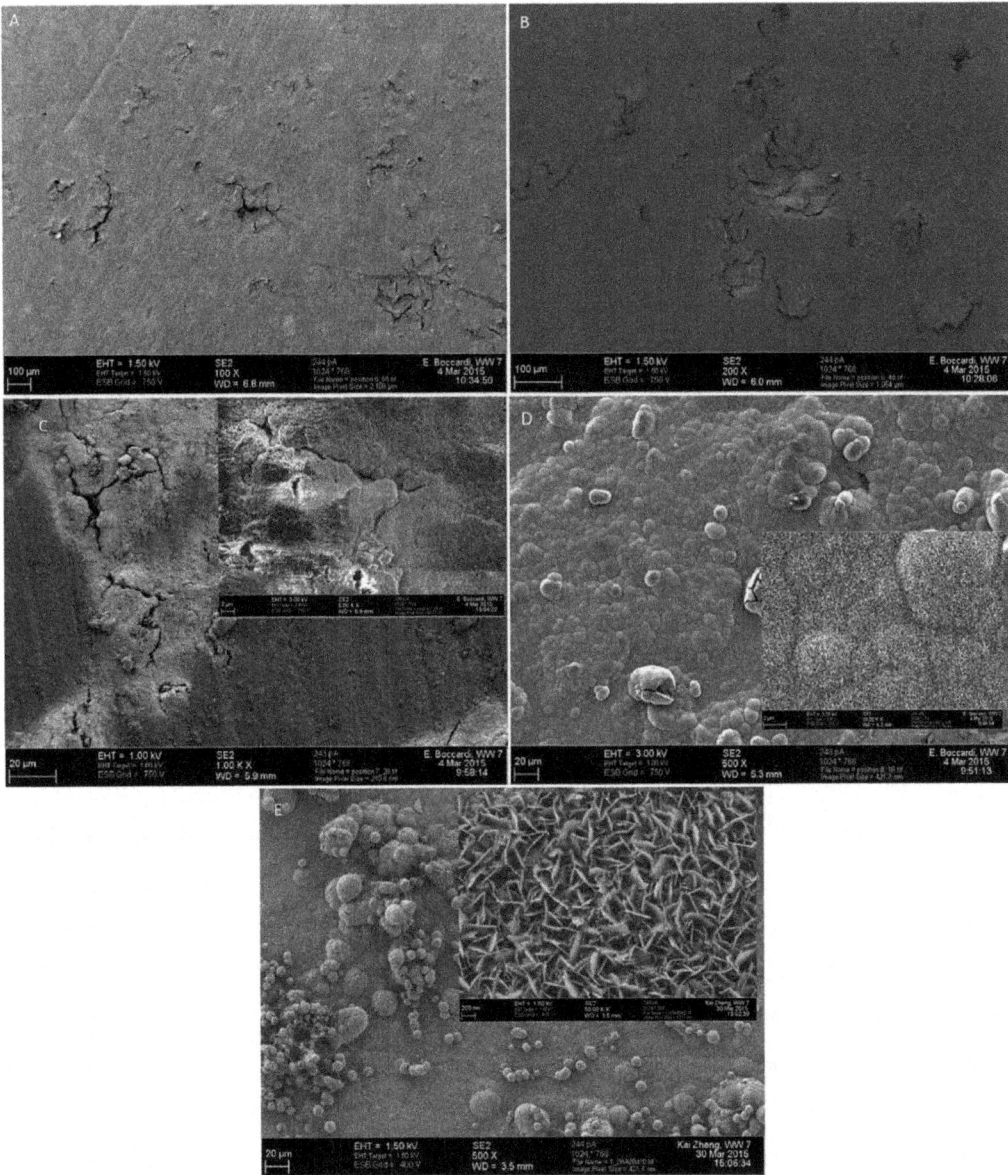

Figure 8   SEM images of 1:2 HA after SBF treatment: (A) untreated; (B) 1 h; (C) 24 h; (D) 7 d; (E) 28 d

the *in situ* generated HA, therefore, the HA nanoparticles could act not only as rigid reinforcing filler but also as cross-linking points.

In Table 4, the results of the flexural and compression tests on 1:3 filled and unfilled samples are reported. The flexural test showed an increment of the flexural modulus for nano-composite by 52% that is an expected result since in the composites the deformations are strongly obstructed by the nanofiller. However, the nanocomposite presented a brittle fracture before the yield point was reached and a lower value of the flexural strength, probably due to the presence of certain agglomeration that acts as stress concentrations. The compression tests gave a value of compression modulus of

the nanocomposite greater than that of the unfilled of almost 12%.

The values of the E_f and E_c were also compared with data predicted using Equation (2). As it can be seen in Table 4, for the flexural test, the increment obtained is greater than the increment predicted by the Mooney's equation while in the case of the compression test, the theoretic and the experimental results are very close considering the experimental error, probably because the effect of the filler is more evident in case of flexural deformation.

For the flexural strength, two different empirical equations were applied, considering the cases of good and poor interfacial adhesion.

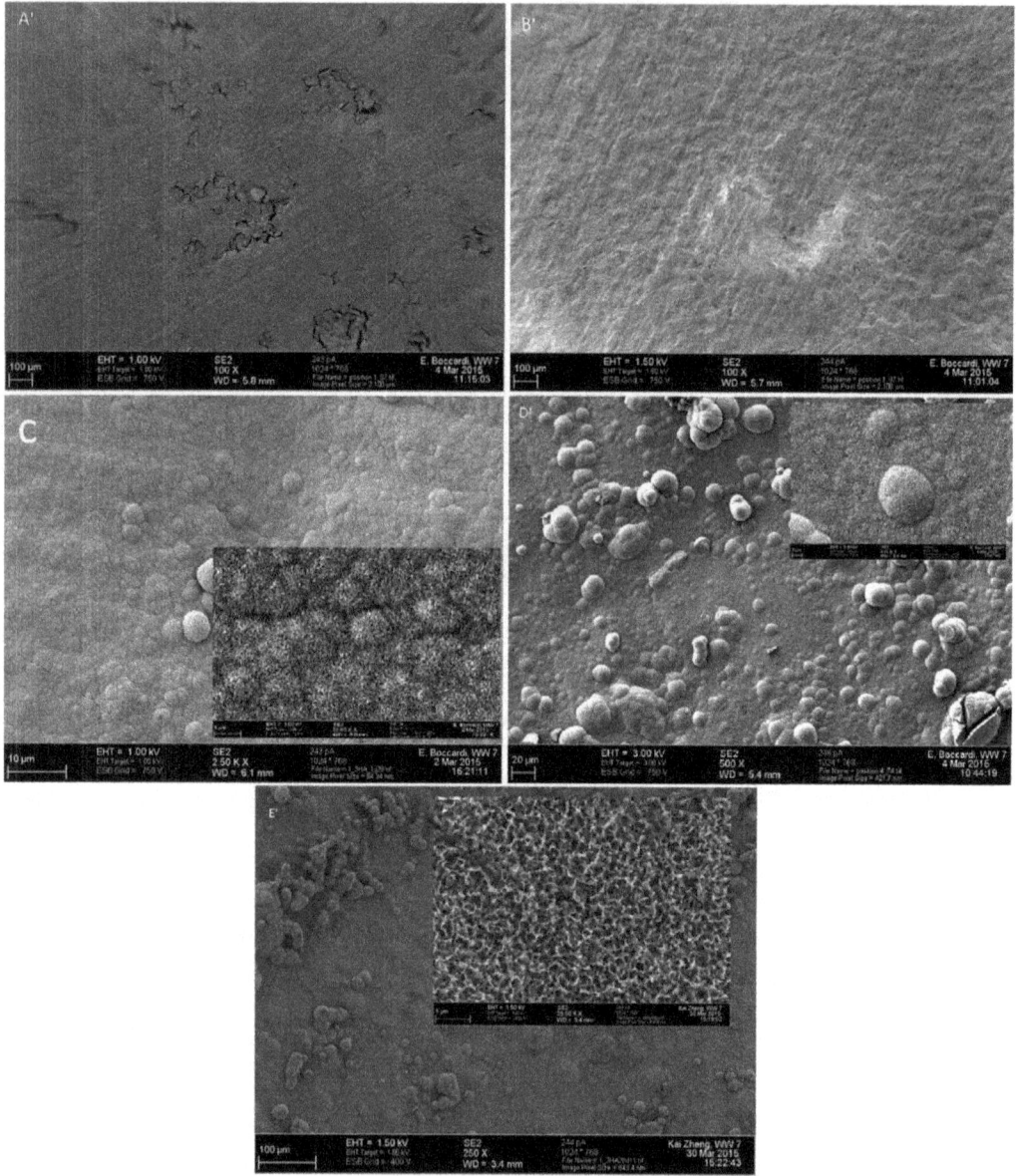

Figure 9   SEM images of 1:3 HA after SBF treatment: (A) untreated; (B) 1 h; (C) 24 h; (D) 7 d; (E) 28 d

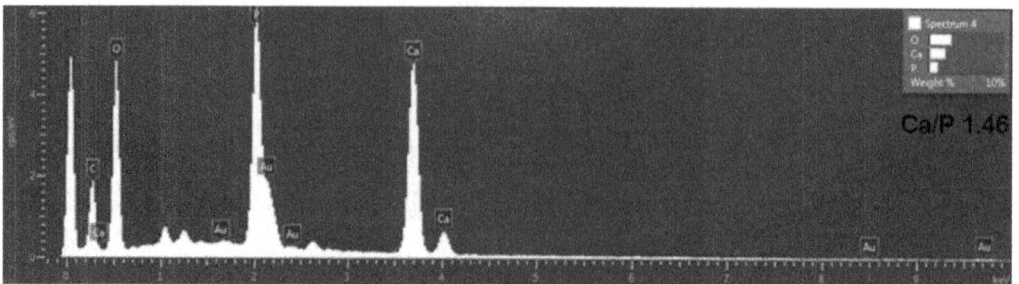

Figure 10   EDS spectrum of 1:2 HA apatite layer, sample (D)

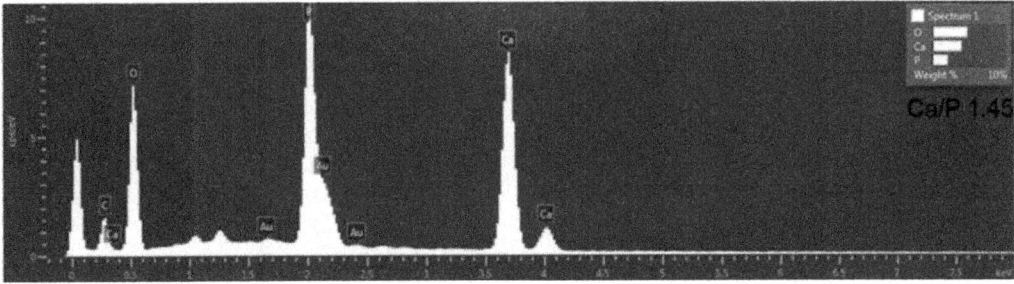

Figure 11   EDS spectrum of 1:3 HA apatite layer, sample (D)

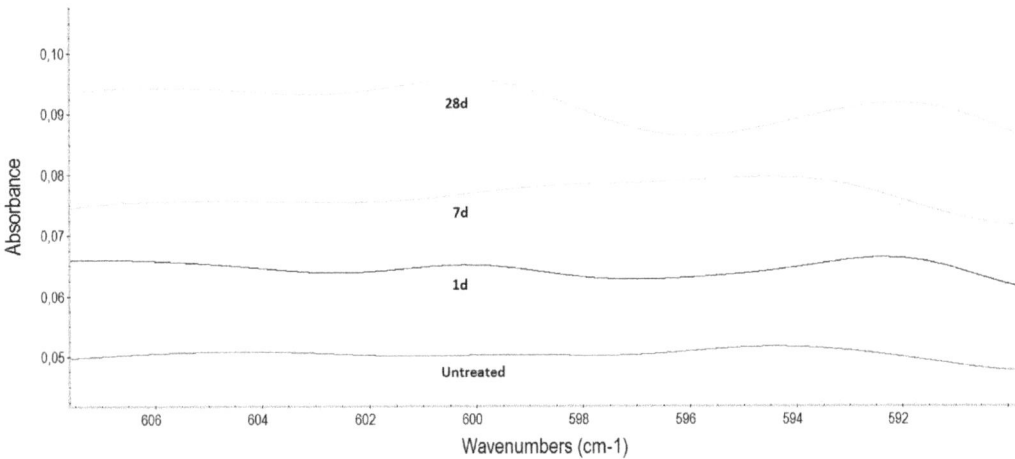

Figure 12   FT-IR spectra of 1:2 HA after SBF treatment

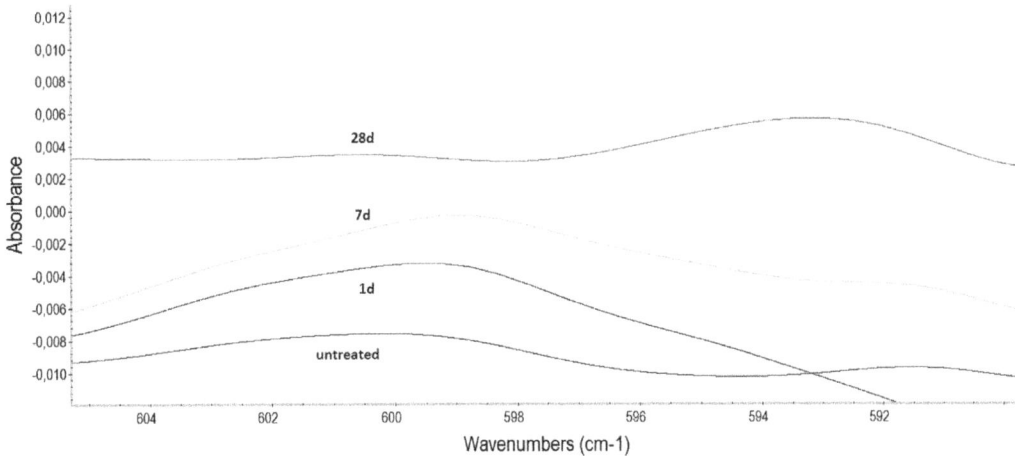

Figure 13   FT-IR spectra of 1:3 HA after SBF treatment

The Nielsen's equation for poorly bonded particles[37] is shown below:

$$\sigma_c = \sigma_m\left(1 - Vp^{\frac{2}{3}}\right)Q \tag{3}$$

where σ_c and σ_m are the strength of the composite and the matrix, Vp is the particles volume fraction and the parameter Q accounts for weaknesses in the structure caused by the discontinuities in stress transfer and generation of stress concentration at the particle/polymer interface. When there is no stress

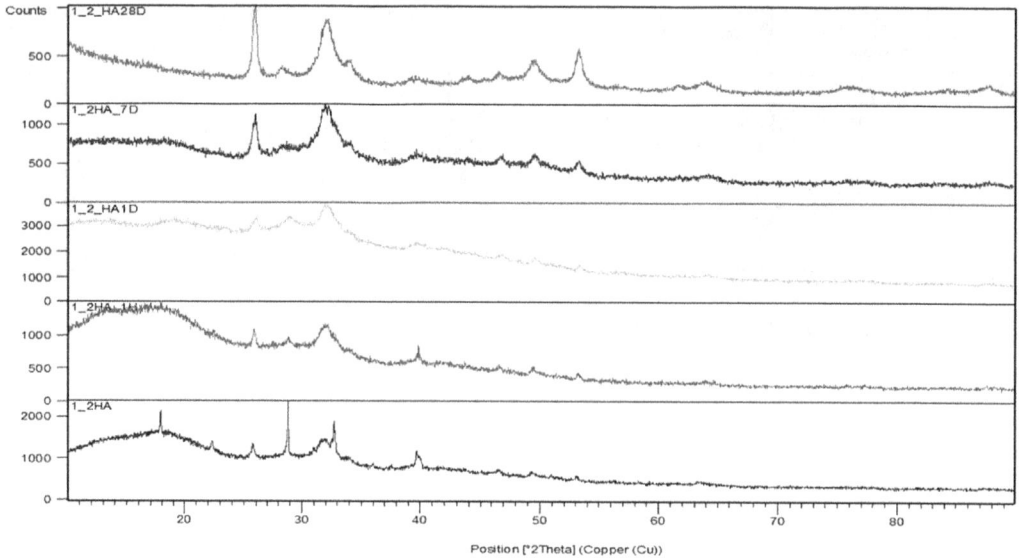

Figure 14   XRD patterns of 1:2 HA after SBF treatment (from the bottom: untreated and after 1 h, 24 h, 7 d, 28 d)

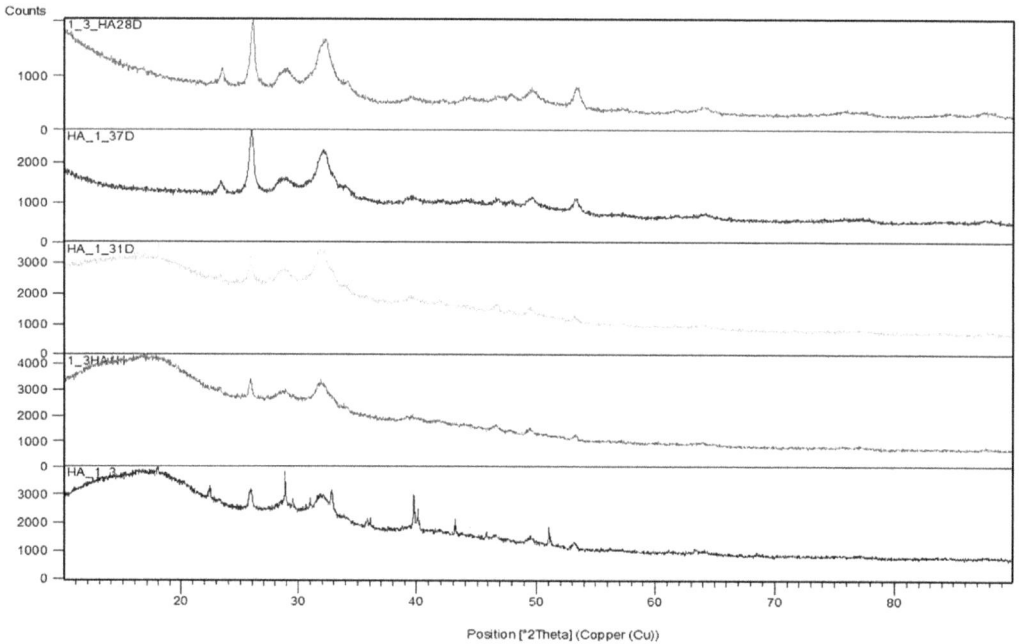

Figure 15   XRD patterns of 1:3 HA after SBF treatment (from the bottom: untreated and after 1 h, 24 h, 7 d, 28 d)

concentration, the value of Q is equal to 1, Q was considered equal to one (case of no stress concentrations) to determine the value of the predicted flexural strength. To determine the stress concentrations in the nanocomposite, the value of Q was determined using the experimental value of σ_c

In case of strong particle-matrix interfacial bonding, Pukanszky's equation[37] can be applied:

$$\sigma_c = \left[\frac{1 - Vp}{1 + 2.5Vp} \sigma_m \right] \exp(BVp) \qquad (4)$$

where B is an empirical constant, which depends on the surface area of particles, particle density, and interfacial bonding energy, B is equal to zero in case of poor interfacial bonding that means that the particles do not carry any load while for good adhesion assumes values around 3.5. The obtained results are listed in Table 5.

The Table clearly shows that using the Nielsen's equation for poorly bonded particles (Q = 1), the strength of the composite decreases by 10% with respect of the unfilled sample;

instead in case of good adhesion, the empirical equation proposed by Pukanszky gives an increment of the flexural strength of 0.7%.

Analyzing the predicted data in comparison with the experimental data, because the strength of the filled polymer decreases by 45%, it can be concluded that the bulk samples are affected by the presence of stress concentration and Q assumes value of 0.61.

Finally, the values of the sample water sorption are listed in Table 6. The data showed that the effect of HA is more remarkable in the case of 1:3 HA as it reduces the water sorption by 21%. This effect could be explained considering the nanoparticles acting as crosslinking points, therefore the cross-linking density of the nanocomposite is higher than the unfilled leading to a smaller water sorption capacity. The effect on the water sorption, however, is mainly determined by the presence of the hydroxyl groups in HEMA, as the samples with PMMA/HEMA ratio 1:3 present a higher value of water sorption.

The SEM micrograph in Fig. 7 shows the presence of cracks on the surface of the nanocomposite samples (1:2 and 1:3 HA) with an average length of about 100 μm. This was probably due to the presence air bubbles that remained trapped in the suspensions during reticulation. These cracks are absent in the unfilled sample and this should be the explanation of the high stress concentrations observed on the nanocomposites in the flexural test.

Concerning bioactivity analysis, SEM micrographs (Figs. 8 and 9) show that after one hour of immersion in SBF, there is no trace of a HA deposit on the surface of both unfilled (here not reported) and composite samples. After one day of immersion, a layer of deposit starts forming exhibiting a typical apatite morphology on the surface of both composite samples 1:2 and 1:3 HA. The short time needed for the apatite layer deposition is related to a high degree of bioactivity, and it could be due to the high specific surface area of the nanorods of hydroxyapatite, as this particular shape presents better adsorption (even HA present in human tooth and bone exhibits the form of nano-polycrystalline hexagonal nanorods[33]) and the presence of the OH groups of HEMA which confers more absorbance to the material.

After seven days of immersion, both the composite samples were almost completely covered by the apatite layer: no trace of a deposit was observed in the unfilled sample (here not reported), while a complete coverage of the surface of the nanocomposites was observed after 28 days of treatment.

EDS analysis (Figs. 10 and 11) of the formed layer allowed its identification as hydroxyapatite, since it shows that it consist of Ca and P with atomic ratio in the range between 1.43 and 1.5 that is close to the Ca/P ratio of natural hydroxyapatite (1.6).

FT-IR spectroscopy (Figs. 12 and 13) shows a small shift of two peaks at around 604 and 592 cm^{-1} in the nanocomposites, which correspond to the P–O vibration of PO_4^{3-} in hydroxyapatite; the shift is more evident in samples treated for 28 days in SBF, while a small signal is also present in the untreated samples due to the presence of HA nanoparticles.

The presence of HA crystallized on the sample surface was also observed in the XRD patterns (Figs. 14 and 15). In the untreated composites, small peaks, related to the HA

nanoparticles present in the polymer matrix, are observed with large percentage of amorphous phase; increasing soaking time in SBF, independently on the PMMA/HEMA ratio, an increase of the amount of crystalline hydroxyapatite phase, that become predominant after 28 days of treatment, can be observed.

Conclusions

PMMA/PHEMA polymer blends with hydroxyapatite as nanofiller were successfully prepared through the 'reactive suspension method'. Nanosized hydroxyapatite was *in situ* synthesized by co-precipitation process in the presence of hydroxyethyl methacrylate as suspending medium reactive toward the subsequent polymerization. FT-IR and XRD analysis confirmed the formation of hydroxyapatite and TEM microscopy revealed that HA nanorods with an aspect ratio value between 5 and 7 were obtained.

DSC analysis showed a significant increase in glass transition temperature in the nanocomposites with respect to the unfilled polymer blends. Dynamic mechanical analysis showed a significant increment in the nanocomposite of the storage modulus measured in the rubbery region (above glass transition temperature) unpredicted by the usual predictive equations which can be attributed to the presence of cross-linking points due to the *in situ* generated particles. An increment of the elastic modulus was also observed at room temperature in compression and three-point bending tests. The presence of hydroxyapatite in the polymer blends resulted in an important decrease in the water sorption values. Excellent results in terms of bioactivity were also obtained in the case of composites containing *in situ* prepared hydroxyapatite.

Acknowledgments

The authors would like to acknowledge Kai Zheng and Dirk Dippold for their contribution in the characterization of the bioactivity of the samples, for the SEM and FT-IR analysis.

Disclosure statement

No potential conflict of interest was reported by the authors.

References

1. L. Cheng, X. Zhou, H. Zhong, X. Deng, Q. Cai and X. Yang: 'NaF-loaded core-shell PAN-PMMA nanofibers as reinforcements for Bis-GMA/TEGDMA restorative resins', *Mater. Sci. Eng. C Mater. Biol. Appl.*, 2014, **34**, 262–269.

2. X. Y. Zhang, X. J. Zhang, Z. L. Huang, B. S. Zhu and R. R. Chen: 'Hybrid effects of zirconia nanoparticles with aluminum borate whiskers on mechanical properties of denture base resin PMMA', *Dent. Mater. J.*, 2014, **33**, 141–146.

3. N. Kojima, M. Yamada, A. Paranjpe, N. Tsukimura, K. Kubo, A. Jewett and T. Ogawa: 'Restored viability and function of dental pulp cells on polymethylmethacrylate (PMMA)-based dental resin supplemented with N-acetyl cysteine (NAC)', *Dent. Mater.*, 2008, **24**, 1686–1693.

4. W. Yu, X. Wang, Q. Tang, M. Guo and J. Zhao: 'Reinforcement of denture base PMMA with ZrO(2) nanotubes', *J. Mech. Behav. Biomed. Mater.*, 2014, **32**, 192–197.

5. I. N. Safi: 'Evaluation the effect of nano-fillers (TiO2, Al2O3, SiO2) addition on glass transition temperature, e-moudulus and coefficient of thermal expansion of acrylic denture base material', *J. Baghdad Coll. Dent.*, 2014, **26**, 37–41.

6. J. C. Zhang, J. Liao, A. C. Mo, Y. B. Li, J. D. Li and X. J. Wang: 'Characterization and human gingival fibroblasts biocompatibility of hydroxyapatite/ PMMA nanocomposites for provisional dental implant restoration', *Appl. Surf. Sci.*, 2008, **255**, 328–330.

7. M. Supova, G. S. Martynkova and K. Barabaszova: 'Effect of Nanofillers Dispersion in Polymer Matrices: A Review', *Sci. Adv. Mater.*, 2011, **3**, 1–25.

8. F. Tan, M. Naciri, D. Dowling and M. Al-Rubeai: '*In vitro* and *in vivo* bioactivity of CoBlast hydroxyapatite coating and the effect of impaction on its osteoconductivity', *Biotechnol. Adv.*, 2012, **30**, 352–362.

9. P. Wutticharoenmongkol, P. Pavasant and P. Supaphol: 'Osteoblastic phenotype expression of MC3T3-E1 cultured on electrospun polycaprolactone fiber mats filled with hydroxyapatite nanoparticles', *Biomacromolecules*, 2007, **8**, 2602–2610.

10. A. P. Marques and R. L. Reis: 'Hydroxyapatite reinforcement of different starch-based polymers affects osteoblast-like cells adhesion/spreading and proliferation', *Mater. Sci. Eng. C*, 2005, **25**, 215–229.

11. L. Shor, S. Guceri, X. Wen, M. Gandhi and W. Sun: 'Fabrication of three-dimensional polycaprolactone/hydroxyapatite tissue scaffolds and osteoblast-scaffold interactions *in vitro*', *Biomaterials*, 2007, **28**, 5291–5297.

12. F. Causa, P. A. Netti, L. Ambrosio, G. Ciapetti, N. Baldini, S. Pagani, et al: 'Poly-epsilon-caprolactone/hydroxyapatite composites for bone regeneration: *in vitro* characterization and human osteoblast response', *J. Biomed. Mater. Res. A.*, 2006, **76**, 151–162.

13. V. Guarino, F. Causa, P. A. Netti, G. Ciapetti, S. Pagani, D. Martini, N. Baldini and L. Ambrosio: 'The role of hydroxyapatite as solid signal on performance of PCL porous scaffolds for bone tissue regeneration', *J. Biomed. Mater. Res. B Appl. Biomater.*, 2008, **86**, 548–557.

14. S. J. Heo, S. E. Kim, Y. T. Hyun, D. H. Kim, H. M. Lee, Y. M. Hwang, S. A. Park and J. W. Shin: '*In vitro* evaluation of poly epsilon-caprolactone/ hydroxyapatite composite as scaffolds for bone tissue engineering with human bone marrow stromal cells', *Key Eng, Mat.*, 2007, **342–343**, 369–372.

15. D. Verma, K. Katti and D. Katti: 'Bioactivity in *in situ* hydroxyapatite-polycaprolactone composites', *J. Biomed. Mater. Res. A.*, 2006, **78**, 772–780.

16. P. Fabbri, F. Bondioli, M. Messori, C. Bartoli, D. Dinucci and F. Chiellini: 'Porous scaffolds of polycaprolactone reinforced with *in situ* generated hydroxyapatite for bone tissue engineering', *J. Mater. Sci. Mater. Med.*, 2010, **21**, 343–351.

17. M. Catauro, M. G. Raucci, F. De Gaetano and A. Marotta: 'Sol-gel synthesis, characterization and bioactivity of polycaprolactone/SiO2 hybrid material', *J. Mater. Sci.*, 2003, **38**, 3097–3102.

18. M. Catauro, M. G. Raucci, F. de Gaetano, A. Buri, A. Marotta and L. Ambrosio: 'Sol-gel synthesis, structure and bioactivity of polycaprolactone/CaO. SiO2 hybrid material', *J. Mater. Sci. Mater. Med.*, 2004, **15**, 991–995.

19. D. Hakimimehr, D. M. Liu and T. Troczynski: 'In-situ preparation of poly(propylene fumarate)–hydroxyapatite composite', *Biomaterials*, 2005, **26**, 7297–7303.

20. S. Z. C. Liou, S. Y. Chen and D. M. Liu: 'Phase development and structural characterization of calcium phosphate ceramics-polyacrylic acid nanocomposites at room temperature in water-methanol mixtures', *J. Mater. Sci.-Mater. M.*, 2004, **15**, 1261–1266.

21. F. Bakan, O. Lacin and H. Sarac: 'A novel low temperature sol-gel synthesis process for thermally stable nano crystalline hydroxyapatite', *Powder Technol.*, 2013, **233**, 295–302.

22. A. Y. Fadeev, R. Helmy and S. Marcinko: 'Self-assembled monolayers of organosilicon hydrides supported on titanium, zirconium, and hafnium dioxides'. *Langmuir*, 2002, **18**, 7521–7529.

23. ISO IS: 'Polymer-based restorative materials', 4th edn, 2009, 4049 Dentistry.

24. ISO IS: 'Determination of compressive properties', 2002, 04 Pastics.

25. T. Kokubo and H. Takadama: 'How useful is SBF in predicting *in vivo* bone bioactivity?' *Biomaterials*, 2006, **27**, 2907–2915.

26. F. W. Liu, X. Z. Jiang, Q. H. Zhang and M. F. Zhu: 'Strong and bioactive dental resin composite containing poly(Bis-GMA) grafted hydroxyapatite whiskers and silica nanoparticles', *Compos. Sci. Technol.*, 2014, **101**, 86–93.

27. R. Kamalian, A. Yazdanpanah, F. Moztarzadeh, R. Ravarian, Z. Moztarzadeh, M. Tahmasbi and M. Mozafari: 'Synthesis and characterization of bioactive glass/forsterite nanocomposites for bone and dental implants', *Ceram.-Silikaty*, 2012, **56**, 331–340.

28. S. M. Abo-Naf, E. S. M. Khalil, E. S. M. El-Sayed, H. A. Zayed and R. A. Youness: 'In vitro bioactivity evaluation, mechanical properties and microstructural characterization of Na2O-CaO-B2O3-P2O5 glasses'. *Spectrochim. Acta A.*, 2015, **144**, 88–98.

29. J. Loof, F. Svahn, T. Jarmar, H. Engqvist and C. H. Pameijer: 'A comparative study of the bioactivity of three materials for dental applications', *Dent. Mater.*, 2008, **24**, 653–659.

30. Q. Chen, I. Thompson and A. R. Boccaccini: 'Bioglass-derived glass-ceramic scaffolds for bone tissue engineering', *Biomaterials*, 2006, **17**, 2414–25.

31. D. N. Ungureanu, N. Angelescu, R.M. Ion, E.V. Stoian and C. Z. Rizescu: 'Synthesis and Characterization of Hydroxyapatite Nanopowders by Chemical Precipitation', in '*Recent Researches in Communications, Automation, Signal Processing, Nanotechnology, Astronomy and Nuclear Physics*', (eds. Z. Bojkovic, J. Kacprzyk, N. Mastorakis, V. Mladenov, R. Revetria, L. A. Zadeh, & A. Zemliak), 296–301; 2011, WSEAS Press, Cambridge, ISBN: 978-960-474-276-9.

32. B. C. E. Idrissia, K. Yamnib, A. Yacoubia and A. Massita: 'A novel method to synthesize nanocrystalline hydroxyapatite: Characterization with X-ray diffraction and infrared spectroscopy', *J. Appl. Chem.*, 2014, **7**, 107–112.

33. S. K. Padmanabhan, A. Balakrishnan, M. C. Chu, Y. J. Lee, T. N. Kim and S. J. Cho: 'Sol-gel synthesis and characterization of hydroxyapatite nanorods', *Particuology*, 2009, **7**, 466–470.

34. N. A. Rangel-Vazquez and T. Leal-Garcia: 'Spectroscopy Analysis of Chemical Modification of Cellulose Fibers', *J. Mex. Chem. Soc.*, 2010, **54**, 192–197.

35. R. Qiao, H. Deng, K. W. Putz and L. C. Brinson: 'Effect of particle agglomeration and interphase on the glass transition temperature of polymer nanocomposites', *J. Polym. Sci. Pol. Phys.*, 2011, **49**, 740–748.

36. D. Morselli, F. Bondioli, M. Sangermano and M. Messori: 'Photo-cured epoxy networks reinforced with TiO2 *in situ* generated by means of non-hydrolytic sol-gel process', *Polymer*, 2012, **53**, 283–290.

37. S. Y. Fu, X. Q. Feng, B. Lauke and Y. W. Mai: 'Effects of particle size, particle/matrix interface adhesion and particle loading on mechanical properties of particulate-polymer composites', *Compos. Part B-Eng.*, 2008, **39**, 933–961.

Permissions

All chapters in this book were first published in Nanocomposites, by Taylor & Francis Online; hereby published with permission under the Creative Commons Attribution License or equivalent. Every chapter published in this book has been scrutinized by our experts. Their significance has been extensively debated. The topics covered herein carry significant findings which will fuel the growth of the discipline. They may even be implemented as practical applications or may be referred to as a beginning point for another development.

The contributors of this book come from diverse backgrounds, making this book a truly international effort. This book will bring forth new frontiers with its revolutionizing research information and detailed analysis of the nascent developments around the world.

We would like to thank all the contributing authors for lending their expertise to make the book truly unique. They have played a crucial role in the development of this book. Without their invaluable contributions this book wouldn't have been possible. They have made vital efforts to compile up to date information on the varied aspects of this subject to make this book a valuable addition to the collection of many professionals and students.

This book was conceptualized with the vision of imparting up-to-date information and advanced data in this field. To ensure the same, a matchless editorial board was set up. Every individual on the board went through rigorous rounds of assessment to prove their worth. After which they invested a large part of their time researching and compiling the most relevant data for our readers.

The editorial board has been involved in producing this book since its inception. They have spent rigorous hours researching and exploring the diverse topics which have resulted in the successful publishing of this book. They have passed on their knowledge of decades through this book. To expedite this challenging task, the publisher supported the team at every step. A small team of assistant editors was also appointed to further simplify the editing procedure and attain best results for the readers.

Apart from the editorial board, the designing team has also invested a significant amount of their time in understanding the subject and creating the most relevant covers. They scrutinized every image to scout for the most suitable representation of the subject and create an appropriate cover for the book.

The publishing team has been an ardent support to the editorial, designing and production team. Their endless efforts to recruit the best for this project, has resulted in the accomplishment of this book. They are a veteran in the field of academics and their pool of knowledge is as vast as their experience in printing. Their expertise and guidance has proved useful at every step. Their uncompromising quality standards have made this book an exceptional effort. Their encouragement from time to time has been an inspiration for everyone.

The publisher and the editorial board hope that this book will prove to be a valuable piece of knowledge for researchers, students, practitioners and scholars across the globe.

List of Contributors

N. Jagannathan, A. R. Anilchandra and C. M. Manjunatha
Fatigue and Structural Integrity Group, Structural Technologies Division, CSIR-National Aerospace Laboratories, Bangalore 560017, India

B. M. Cromer, E. B. Coughlin and A. J. Lesser
Department of Polymer Science and Engineering, University of Massachusetts Amherst, 120 Governors Drive, Amherst, MA 01003, USA

Irmina Samba, Rebeca Hernandez, Nicoletta Rescignano and Carmen Mijangos
Instituto de Ciencia y Tecnología de Polimeros, ICTP-CSIC, Juan de la Cierva 3, 28006 Madrid, Spain

Josè Maria Kenny
Instituto de Ciencia y Tecnología de Polimeros, ICTP-CSIC, Juan de la Cierva 3, 28006 Madrid, Spain
Materials Engineering Center, UdR INSTM, University of Perugia, Strada di Pentima 4 05100, Terni, Italy

Mina Ahani, Marziyeh Khatibzadeh and Mohsen Mohseni
Department of Polymer Engineering and Color Technology, Amirkabir University of Technology, Tehran, Iran

Kai Yang, Maya Endoh, Radha P. Ramasamy, Molly M. Gentleman and Miriam H. Rafailovich
Materials Science and Engineering, Stony Brook University, NY 11794, USA

Thomas A. Butcher and Rebecca Trojanowski
Energy Resources Division, Brookhaven National Laboratory, Upton NY 11973, USA

Golan Abraham Tanami, Liliya Kovalenko and Gad Marom
Casali Center for Applied Chemistry, The Institute of Chemistry and the Center for Nanoscience and Nanotechnology, The Hebrew University of Jerusalem, Jerusalem 91904, Israel

Chung Chueh Chang
Advanced Energy Research and Technology Center (AERTC), State University of New York, Stony Brook, NY 11794, USA

Miriam Rafailovich
Department of Materials Science and Engineering, State University of New York, Stony Brook, NY 11794-2275, USA

Y. Q. Gill, J. Jin and M. Song
Department of Materials, Loughborough University, Loughborough LE11 3TU, UK

Samira Benali, Sabrina Aouadi, Anne-Laure Dechief, Marius Murariu and Philippe Dubois
Center of Innovation and Research in Materials and Polymers (CIRMAP), Laboratory of Polymeric and Composite Materials (LPCM), University of Mons & Materia Nova Research Center, Place du Parc 20, 7000 Mons, Belgium

Nasrin Shadjou
Department of Nanochemistry, Nano Technology Research Center, Urmia University, Urmia, Iran
Faculty of Science, Department of Nano Technology, Urmia University, Urmia, Iran

Mohammad Hasanzadeh
Drug Applied Research Center, Tabriz University of Medical Sciences, Tabriz, Iran
Pharmaceutical Analysis Research Center and Faculty of Pharmacy, Tabriz University of Medical Sciences, Tabriz, Iran

Faeze Talebi
Department of Nanochemistry, Nano Technology Research Center, Urmia University, Urmia, Iran

Ahmad Poursattar Marjani
Faculty of Science, Department of Chemistry, Urmia University, Urmia, Iran

Koon-Yang Lee
The Composites Centre, Department of Aeronautics, Imperial College London, South Kensington Campus, London SW7 2AZ, UK

Alexander Bismarck
Polymer and Composite Engineering (PaCE) Group, Faculty of Chemistry, Institute for Materials Chemistry and Research, University of Vienna, Währingerstraβe 42, Vienna 1090, Austria
Polymer and Composite Engineering (PaCE) Group, Department of Chemical Engineering, Imperial College London, South Kensington Campus, London SW7 2AZ, UK

Hong Jiang, Weixing Yang, Shuiqin Pu, Feng Chen and Qiang Fu
College of Polymer Science and Engineering, State Key Laboratory of Polymer Materials Engineering, Sichuan University, Chengdu, China

Songgang Chai
National Engineering Research Center of Electronic Circuits Base Materials, Guangdong Shengyi Technology Limited Corporation, Dongguan, China

J. F. Feller, M. Castro, H. Bellegou and I. Pillin
Smart Plastics Group, European University of Brittany (UEB), LIMATB-UBS, Lorient, France

K. K. Sadasivuni
Smart Plastics Group, European University of Brittany (UEB), LIMATB-UBS, Lorient, France
Surfaces & Interfaces Group, European University of Brittany (UEB), LIMATB-UBS, Lorient, France

School of Chemical Sciences, Mahatma Gandhi University, Kottayam, India

S. Thomas
Surfaces & Interfaces Group, European University of Brittany (UEB), LIMATB-UBS, Lorient, France

Y. Grohens
School of Chemical Sciences, Mahatma Gandhi University, Kottayam, India

Maria Eriksson and Han Goossens
Laboratory of Polymer Materials, Department of Chemical Engineering and Chemistry, Eindhoven University of Technology, 5600 MB Eindhoven, The Netherlands

Ton Peijs
Laboratory of Polymer Materials, Department of Chemical Engineering and Chemistry, Eindhoven University of Technology, 5600 MB Eindhoven, The Netherlands
Centre for Materials Research & School of Engineering and Materials Science, Queen Mary University of London, Mile End Road, London E1 4NS, UK

T. Lyashenko-Miller and G. Marom
Casali Center of Applied Chemistry, The Institute of Chemistry and the Center for Nanoscience and Nanotechnology, The Hebrew University of Jerusalem, Jerusalem, Israel

J. Fitoussi
Arts et Métiers ParisTech, PIMM – UMR CNRS 8006, 151 Boulevard de l'Hôpital, 75013 Paris, France

Marius Murariu, Anne-Laure Dechief, Rindra Ramy-Ratiarison, Yoann Paint, Jean-Marie Raquez and Philippe Dubois
Center of Innovation and Research in Materials and Polymers (CIRMAP), Laboratory of Polymeric and Composite Materials (LPCM), University of Mons & Materia Nova Research Center, Place du Parc 20, 7000 Mons, Belgium

Jérémy Odent, Jean-Marie Raquez and Philippe Dubois
Laboratory of Polymeric and Composite Materials (LPCM), Center of Innovation and Research in Materials and Polymers (CIRMAP), University of Mons (UMONS), Place du Parc 20, B-7000 Mons, Belgium

Jean-Michel Thomassin and Christine Jérôme
Center for Education and Research on Macromolecules (CERM), University of Liege, Sart-Tilman, Allée de la Chimie 3 B6, B-4000, Liege 1, Belgium

Jean-Michel Gloaguen and Jean-Marc Lefebvre
Unité Maté riaux et Transformations (UMET), UMR CNRS 8207, Université Lille1, Sciences et Technologies/CNRS, Cité Scientifique C6, 59655 Villeneuve d'Ascq, France

Franck Lauro
Industrial and Human Automatic Control and Mechanical Engineering Laboratory (LAMIH), UMR CNRS 8201, University of Valenciennes and Hainaut-Cambresis, Le Mont Houy, BP 311, 59304 Valenciennes Cedex, France

Irene Hassinger
Rensselaer Polytechnic Institute, Troy, New York, USA

Martin Gurka
Institute for Composite Materials, Kaiserslautern, Germany

Han Zhang
School of Engineering and Materials Science, and Materials Research Institute, Queen Mary University of London, Mile End Road, E1 4NS London, UK

Emiliano Bilotti and Ton Peijs
School of Engineering and Materials Science, and Materials Research Institute, Queen Mary University of London, Mile End Road, E1 4NS London, UK

Nanoforce Technology Ltd., Queen Mary University of London, Joseph Priestley Building, Mile End Road, E1 4NS London, UK

Jian Chen
Jiangsu Key Laboratory of Advanced Metallic Materials, School of Materials Science and Engineering, Southeast University, Nanjing 211189, China

Ben D. Beake and Gerard A. Bell
Micro Materials Ltd., Willow House, Ellice Way, Yale Business Village, Wrexham, LL13 7YL, UK

Yalan Tait and Fengge Gao
School of Science and Technology, Nottingham Trent University, Clifton campus, Nottingham, NG11 8NS, UK

Israel Greenfeld and H. Daniel Wagner
Department of Materials and Interfaces, The Weizmann Institute of Science, Rehovot 76100, Israel

Meymanat S. Mohsenzadeh and Mohammad Mazinani
Department of Materials Engineering, Faculty of Engineering, Ferdowsi University of Mashhad, Mashhad, Iran

Seyed Mojtaba Zebarjad
Department of Materials Engineering, Faculty of Engineering, Shiraz University, Shiraz, Iran

Oussama Boumezgane and Massimo Messori
Dipartimento di Ingegneria "Enzo Ferrari", Università di Modena e Reggio Emilia, via P. Vivarelli 10/1, Modena, Italy
Consorzio INSTM, Unità di ricerca di Modena e Reggio Emilia, via Giusti 9, Firenze, Italy

Federica Bondioli
Dipartimento di Ingegneria Industriale, Università di Parma, Parco Area delle Scienze 181/A, Parma, Italy

Sergio Bortolini and Alfredo Natali
Dipartimento Chirurgico, Medico, Odontoiatrico e di Scienze Morfologiche con Interesse Trapiantologico, Oncologico e di Medicina Rigenerativa, Università di Modena e Reggio Emilia, via del Pozzo 71, Modena, Italy

Aldo R. Boccaccini and Elena Boccardi
Department of Materials Science and Engineering, University of Erlangen-Nuremberg,Cauerstraße 6, Erlangen, Germany

Index

www.ingramcontent.com/pod-product-compliance
Lightning Source LLC
Chambersburg PA
CBHW061947190326
41458CB00009B/2808